Sustainable ICTs and Management Systems for Green Computing

Wen-Chen Hu
University of North Dakota, USA

Naima Kaabouch
University of North Dakota, USA

Information Science
REFERENCE

Managing Director: Lindsay Johnston
Senior Editorial Director: Heather A. Probst
Book Production Manager: Sean Woznicki
Development Manager: Joel Gamon
Development Editor: Myla Merkel
Assistant Acquisitions Editor: Kayla Wolfe
Typesetter: Lisandro Gonzalez
Cover Design: Nick Newcomer

Published in the United States of America by
Information Science Reference (an imprint of IGI Global)
701 E. Chocolate Avenue
Hershey PA 17033
Tel: 717-533-8845
Fax: 717-533-8661
E-mail: cust@igi-global.com
Web site: http://www.igi-global.com

Library of Congress Cataloging-in-Publication Data

Sustainable ICTs and management systems for green computing / Wen-Chen Hu and Naima Kaabouch, editors.
p. cm.
Includes bibliographical references and index.
Summary: “This book focuses on information technology using sustainable green computing to reduce energy and resources used”--Provided by publisher.
ISBN 978-1-4666-1839-8 (hardcover) -- ISBN 978-1-4666-1840-4 (ebook) -- ISBN 978-1-4666-1841-1 (print & perpetual access) 1. Telecommunication--Energy conservation. 2. Computer networks--Energy conservation. 3. Electronic data processing--Energy conservation. 4. Green technology. 5. Electronic apparatus and appliances--Environmental aspects. 6. Electronic data processing departments--Energy efficiency. I. Hu, Wen Chen, 1960- II. Kaabouch, Naima, 11959-
TK5102.5.S925 2012
621.381028’6--dc23

2012002991

British Cataloguing in Publication Data
A Cataloguing in Publication record for this book is available from the British Library.

Table of Contents

Section 5
Education and Power Generators

Detailed Table of Contents

Information and Communication Technology (ICT) is an umbrella term that covers the study and business of developing and using technologies to process information and facilitate communications. ICT has the potential to enable a low carbon economy, as pointed out by many studies. One example of the energy (and CO2) saving potential of ICT is illustrated in this chapter. Although there is a net positive effect, the way the actual process is implemented can make a large difference. This leads to the question of how to evaluate different projects or different designs, when it comes to the net sustainability effects. Then a framework is introduced to compare the combined sustainability effects of all the steps in a work process compared to alternative processes. The authors also describe existing indicators and metrics that can be used to fill the framework.

Over the time, the ICT segment has become a major consumer of energy and is, therefore, rapidly becoming a significant source of greenhouse emissions. In addition, global warming is a result of the greenhouse effect, which traps heat, generated by automobiles, that would otherwise escape from Earth. It causes the increase of Earth's average surface temperature. This chapter attempts to account for approximate annual energy consumption and resultant CO2 emissions of this sector. The authors also account for the indirect environmental effects of rapid proliferation of ICT technologies. This work, therefore, highlights potential areas where increased energy efficient measures are urgently required to ensure sustainable development. The positive contribution and potential of ICT in curtailing carbon emissions is also discussed. Towards the end, the authors summarize various initiatives by the scientific community to ensure energy efficiency and sustainable development.

Energy Auto-Tuning (EAT) optimizes energy efficiency of software systems without human intervention. The approach the authors present in this chapter is a self-optimization technique for IT infrastructures, which takes hard- and software components as well as users of software applications into account. It is able to ensure minimal energy consumption for a user request along with a set of non-functional requirements (e.g., the refresh rate of a data extraction tool). To optimize the ratio between utility of end users and the cost in terms of energy consumption, the system needs inherent variability leading to differentiated energy profiles and mechanisms to reconfigure the system at runtime. The proposed approach comprises these mechanisms and introduces an architecture which automatically tunes the energy efficiency of IT systems.

A distributed application has multiple parts that are on different machines, which could be physically close together or be geographically distant. The machines could be all kinds of possibilities such as mainframes, personal computers, workstations, etc. Because the number of distributed applications has been increased greatly, their efficiency becomes critical for energy saving. This chapter classifies applications on distributed systems into computation, communication, and general types of applications. Then, the authors define power consumption models of a server to perform each type of application process. Based on the power consumption models, the authors propose algorithms to select one of the servers for each type of application so that the total power consumption of servers can be reduced, and also evaluate the algorithms in terms of not only performance but also energy consumption.

A cyber-physical energy system is an energy system that is built from and depends on a tight integration of the system's computational and physical components. This chapter focuses on optimal mathematical models for resource allocation. A series of mathematical models is presented in this chapter for solving large-scale energy allocation problems with partially observable states, utility functions, and constrained action is introduced. The proposed techniques use a Linear Programming (LP) approach to determine resource allocations among a set of fuzzy rules that allocates Distributed Energy Resources (DERs) or power sources/sinks and uses to determine improving resource management.

Section 2
Management and Manufacturing

In addition, i-Solutions and Corporate Planning of the University of Southampton, focusing on user confidence level and service improvement, are another two case studies to support. The University of Southampton has achieved cost saving and user confidence with service improvement offered by cloud adoption and services, which have been deployed by several universities in the adoption of CCBF.

Dynamically adapting to manufacturing, environmental, and usage variations is the key to shaving unnecessary design margins, which requires on-chip modules that can sense and configure design parameters both globally and locally to maximize computation efficiency, and maintain this efficiency over the lifetime of the system. This chapter presents an adaptive threshold compensation scheme using a transimpedance amplifier and adaptive body biasing to overcome the effects of temperature variation, reliability degradation, and process variation. The effectiveness and versatility of the scheme are demonstrated with two example applications, one as a temperature aware design to maintain ION to IOFF current ratio, the other as a reliability sensor for Negative Bias Temperature Instability (NBTI).

Multidimensional analysis uses dimensions and measures to analyze data and a supply chain consists of all parties involved, directly or indirectly, in fulfilling a customer request. The most common method of measuring the environmental performance is the international standards of Life Cycle Assessment (LCA). This chapter presents a framework for measuring environmental impact at the item level. By using a Traceability Graph, emissions and resources can be monitored from the data management perspective. The model can be mapped to any precision level of physical tracing. At the most precise level, even a single physical object and its components can be analyzed. From the opposite perspective, the proposed model also supports rough level analysis of products and their histories. In terms of the Traceability Cube, multidimensional analysis can be applied for traceability data.

Billions of cellphones and smartphones are being used. People throw away out-of-date phones all the time, and it causes many environmental problems. This chapter raises the environmental issues of the increasing smartphone market and provides a quantitative analysis on the environmental impact of different life-cycle stages of the smartphones, including the manufacturing stage, using stage, and recycling. To achieve sustainable computing and to best utilize the energy consumed during manufacturing the large number of devices, this chapter demonstrates the methodology and techniques towards reusing smartphones by presenting a case study on reusing smartphones for elementary school education.

A multi-agent system includes a set of software agents that interact to solve problems that are too complicated to be solved by a single software. Power management systems for electronic portable devices are ad-hoc solutions for each device, so solutions are not general enough to be applied to others. A model driven architecture approach is defined in this chapter. It allows defining a general power management systems solution for portable devices. The general solution is based on agent modelling, which is parameterized, so it allows the automatic retrieval of the power management multi-agent system for a particular device giving values to the model's parameters. A laptop equipped with a photovoltaic charger is introduced for simulating power demand and supply in order to discuss power management strategies for reducing electrical consumption.

Map navigation is one of the most popular applications used by mobile users. At the same time, it is also one of the time- and resource-consuming applications. Various methods such as most-recently used and first-in, first-out algorithms are used to reduce the map transmission time and delay. One of the popular methods is online mobile map prefetching and caching. However, the mobility and location features of mobile users are usually left out by these methods. Caching and prefetching maps based on a mobile user's location would greatly reduce the transmission time and hence the battery power consumption. For example, if a user is visiting a town, prefetching the maps of nearby interesting stores and caching the maps of the visited, neighboring landmarks would help the user's visitation experience and save the transmission time. Online mobile map prefetching or caching is useful, but is not widely employed because it involves several different subjects and developers usually are not familiar with all of them. This chapter intends to relieve the problem by introducing essential technologies for online mobile map prefetching and caching so more developers can start working on this kind of project or research. It consists of four themes: (1) green handheld computing, (2) location-based services and programming, (3) map tile system, and (4) location-aware map prefetching and caching methods. A summary is given at the end of this chapter.

Foreword

The impact of global warming and climate change is obvious and can be seen everywhere and anytime. It has changed our daily lives tremendously and costs billions of dollars and many lives every year. For example, natural disasters such as hurricanes and drought happen much more often now. Global warming is caused by many factors from nature and humans. Part of global warming is caused by computers, both hardware and software computing. According to Gartner, a market research company, the global information technology industry generated as much greenhouse gas—about 2% of global CO_2 emissions—as the world's airlines in 2008. As computers are becoming cheaper and more popular, it is foreseen that computers will play a bigger role in the future global warming. Asking people not to use computers or to spend less time on computers is not realistic. One way to mitigate the impact of computers on environment is by applying green computing.

In the past, the amount of energy and resource consumption by computers was significantly ignored. For example, it was common in the past, even today, that office computers were left on overnight. Therefore, simple green-computing methods could be easily deployed and be widely applied to computers and applications such as device/component/computer recycling and power-saving mode. On the other hand, green computing could be complicated and be more effective by using advanced methods such as efficient algorithms and smart scheduling. A book is not able to cover every detail of green computing. This book focuses on two themes: (1) sustainable ICTs (Information and Communication Technologies) and (2) sustainable management systems. ICTs involve the technologies to process information and facilitate communications and management systems including most business systems. These two themes cover a great part of computing and green ICTs, and management systems will substantially reduce the energy and resource consumption.

Today's computer designers and developers must have a green-computing mindset to approach their problems. It is not only for Mother Nature, but also for the trend of computing. Desktop computing no longer dominates the computing methods. Mobile/handheld/smartphone computing is getting extremely popular these days and this kind of computing does not have the luxury of consuming much power. This book, *Sustainable ICTs and Management Systems for Green Computing*, is a timely and urgently needed publication. It provides the most-up-to-date, crucial, and practical information for green computing. Readers learn advanced green-computing knowledge from this book and are able to apply the proposed methods to real-world problems. It is a must-read book for IT students, researchers, scholars, and workers with green computing in mind.

Ruay-Shiung Chang
National Dong Hwa University, Taiwan
February 15, 2012

Ruay-Shiung Chang *received his B.S.E.E. degree from National Taiwan University in 1980 and his Ph.D. degree in Computer Science from National Tsing Hua University in 1988. After graduation, he worked for Chung Shan Institute of Science and Technology and National Taiwan University of Science and Technology. He is now a Vice-President of National Dong Hwa University and a Professor in the Department of Computer Science and Information Engineering. His research interests include Internet, wireless networks, RFID, and cloud computing. He has published more than 70 peer-reviewed journal papers and numerous international conference papers. He is an editor for International Journal of Internet Protocol Technology, Journal of Internet Technology, and Journal of Convergence Information Technology. Dr. Chang is a member of ACM, a senior member of IEEE, and a founding member of Taiwan Institute of Information and Computing Machinery. Dr. Chang also served on the advisory council for the Public Interest Registry (www.pir.org) from 2004 to 2007. In 2009, Dr. Chang received the Outstanding Information Technology Elite Award from the ROC Information Month Committee.*

Preface

This book, *Sustainable ICTs and Management Systems for Green Computing,* collects high-quality research papers and industrial and practice articles in the areas of sustainable ICTs and management systems for green computing from academics and industrialists. It includes research and development results of lasting significance in the theory, design, implementation, analysis, and application of green computing with respect to sustainable ICTs, management systems, and other critical issues. Eighteen excellent articles from forty-eight world-renowned scholars and industry professionals are included in this book, which covers four themes: (1) sustainable ICTs, (2) sustainable management systems, (3) energy-aware mobile handheld devices, and (4) energy-efficient computer vision and image processing. This preface introduces this book, shows the current trends of green computing, gives the organization of this book, and summarizes the discussions.

INTRODUCTION

Global warming and the greenhouse effect are serious problems these days. Many natural disasters like floods and droughts could be triggered or made worse by them. They cost billions of dollars and many lives every year. Computer-related activities contribute to a significant portion of global warming and the greenhouse effect according to the following observations:

- Computers are necessary tools for today's offices. At the same time, computers and monitors have the second highest energy consumption in offices after lighting, based on the report from Dragon Systems Software Limited (2006). They may soon have the highest power consumption in offices because office lighting is changing to energy-saving bulbs quickly.
- Fujitsu Siemens Computers (2009) claimed 85 percent of the electricity that powers PCs is wasted as they sit idly from time to time, especially, during the nights. This problem could be greatly mitigated if power-saving features of computers are used.
- Heat generated by computers and servers raises the temperature in offices. It translates to more use of air conditioning and therefore electricity.

More evidence of the greenhouse effect and global warming caused by computers can be easily found from the Internet. According to the Gartner (2007), a market research company, the global information technology industry generated as much greenhouse gas—about 2% of global CO2 emissions—as the world's airlines in 2008. One way to reduce the greenhouse gas from the industry is green computing, which is the practice of designing, manufacturing, using, and disposing of computers, servers, and associated subsystems for reducing the use of hazardous materials, maximizing energy efficiency, and promoting the recyclability or biodegradability of products. ICTs (Information and Communication Technologies) and management systems account for a significant part of computer systems. Effective green computing could be achieved if sustainable ICTs and management systems are used. This book is the first book to cover sustainable ICTs and management systems for green computing.

Green computing covers a variety of subjects such as materials, environments, electronics, and computing, and all kinds of methods, from simple to complicated, for reducing energy and resource consumption. As a material example, fuel cells are expected to replace lithium-ion batteries in the future for an extended life and energy saving. As an example of simple methods, green computing can be achieved by ensuring computers are turned off at night and enter low power mode when they are idle during the day. On the other hand, an example of complicated green computing methods is prefetching or caching map tiles by using a method of travel-route trajectories. Green-computing workers usually have problems finding the specific books meeting their needs, or such books might not be available. This book is the first book to cover sustainable ICTs and management systems for green computing. It is a collection of green computing articles for reducing energy and resources used. A broader audience could benefit from this book. The intended audience includes industrialists, students, educators, and researchers of information technology, computer science, electrical engineering, material science, and chemistry. It especially benefits the IT personnel of corporations because companies are gradually adapting green computing for their systems. This book will help IT workers smoothly build efficient and effective systems based on their traditional IT knowledge. This book could be used for a textbook of an advanced computer science (or related disciplines) course and could be a reference book for IT professionals and students. Since this book covers the green computing subject systematically, it is also for people desiring to learn the topics on their own.

This book provides rich topics of sustainable green computing with an emphasis on ICTs and management systems from an IT perspective. The rest of this preface is organized as follows:

- *Contemporary green computing methodologies, technologies, and theories*: This section discusses various methodologies, technologies, and theories for green computing. It includes the following subjects: (1) methodologies and paradigms, (2) energy-aware information systems, data centers, and software, (3) energy management and monitors, and (4) applications and others. Each subject and its corresponding topics will be introduced in this section.
- *The organization of this book*: This book consists of 18 chapters divided into five sections: (1) information and communication technologies and systems, (2) management and manufacturing, (3) mobile handheld devices, (4) computer vision and image processing, and (5) education and power generators. Each chapter will be introduced in this section.
- *Summary*: The last section summarizes the methodologies, technologies, and theories of green computing discussed in this preface.

CONTEMPORARY GREEN COMPUTING METHODOLOGIES, TECHNOLOGIES, AND THEORIES

Green computing covers a wide variety of methodologies, technologies, and theories. In a broad sense, all computing methodologies and technologies that help environments can be classified as green computing. This wide coverage not only confuses people, but companies and developers also have problems applying effective green-computing methods to their problems. This preface tries to clarify green computing and suggests major green-computing methodologies and technologies. This section discusses various methodologies, technologies, and theories for green computing. It includes the following topics: (1) methodologies and paradigms, (2) energy-aware information systems, data centers, and software, (3) energy management and monitors, and (4) applications and others. They are introduced next.

Methodologies and Paradigms

This sub-section discusses various kinds of major methodologies and paradigms for green computing including (1) scheduling, resource allocation, and scaling, (2) software and deployment optimization, and (3) virtualization and globalization for energy saving.

- *Scheduling, resource allocation, and scaling*: They are the methods used to maximize the uses of resources such as CPUs, memory, and power. Example methods are round robin scheduling and auction algorithms for resource allocation.
- *Software and deployment optimization*: It is the process to improve a software system or a deployment to make it work more efficiently or use fewer resources such as memory or power. For example, the method of loop unrolling processes "blocks" of several array elements at a time so instruction path lengths are decreased.
- *Virtualization and globalization for energy saving*: Virtualization provides a software layer between the software and hardware and turns the "physical" resources into "logical" ones and globalization is the development of enhancing an integrated global economy. Energy saving can be achieved by using either virtualization or globalization.

Energy-Aware Information Systems, Data Centers, and Software

Running information systems, data centers, and software takes a great chunk of energy consumption by IT. The following topics are related to building energy-aware information systems, data centers, and software:

- *Energy-aware data center design and construction*: According to Koomey (2007), total power used by servers including cooling and auxiliary infrastructure represented about 1.2% of total U.S. electricity consumption in 2005.
- *Power implications for embedded, portable, and mobile computing*: One of the major problems of mobile handheld devices is the short battery life, few hours of operation before needing to be re-charged.
- *Energy-aware bioinformatics and health management*: According to Public Broadcasting Service (PBS, 2008), 16 percent of America's economic activity was devoted to health care in 2005. Bioinformatics and health management are a great part of health care.

Energy Management and Monitors

From a green computing perspective, not only must energy be well managed, but it must also be closely monitored and evaluated. Energy management and monitors include the following topics:

- *Power management of operating system support, power supply, storage, video card, and display*: Operating systems, power supplies, storage, video cards, and displays take a great portion of computer power consumption. Reducing the power consumption by any one of them can greatly save the computer power usage.
- *Monitoring tools for power and performance*: Various energy monitoring systems and tools are available for evaluating power and performance, which could be improved based on the monitoring data.
- *Standards and metrics for green computing*: To evaluate the effectiveness of green computing, standards and metrics must be set so developers and designers can follow them.

Applications and Others

This sub-section includes the topics of green computing applications and the rest subjects:

- *Energy-aware e-learning and lifelong learning*: Distance/lifelong/electronic learning uses many kinds of methods, tools, and software to deliver lectures or facilitate learning. Examples include the Internet, video conferences, and video CDs.
- *Trade-offs between green computing and conventional computing*: Green computing has many benefits, but usually at a price. For example, developing an advanced data prefetching scheme takes more time and human resources compared to developing the one without using data prefetching. Therefore, developers have to find the trade-offs between green computing and conventional computing.
- *Product longevity*: Product longevity may not be equal to product durability. Many times people stop using the products because they do not include the newest features, or companies encourage them to purchase the newer ones. Therefore, properly educating consumers and companies is also an important issue for green computing.
- *Reliability of power-aware computers and networks*: Power-aware computers and networks use advanced methods to reduce power consumption.

ORGANIZATION OF THE BOOK

This book provides timely, critical technologies, methodologies, and theories of green computing with an emphasis on sustainable ICTs and management systems to IT workers and students. It contains 18 chapters divided into five sections: (1) information and communication technologies and systems, (2) management and manufacturing, (3) mobile handheld devices, (4) computer vision and image processing, and (5) education and power generators, which are briefly introduced next.

Information and Communication Technologies and Systems

ICT (Information and Communication Technology) is an umbrella term that covers the study and business of developing and using technologies to process information and facilitate communications. Gartner (2007), a market research company, estimated that the global information technology industry generated as much greenhouse gas—about 2% of global CO2 emissions—as the world's airlines in 2008. ICT has the potential to enable a low carbon economy, as pointed out by many studies. This section includes the following five chapters:

- Chapter 1: "ICT Enabling More Energy Efficient Processes, Including e-Invoicing as a Case": ICT has the potential to enable a low carbon economy, as pointed out by many studies. One example of the energy (and CO_2) saving potential of ICT is illustrated in this chapter. A framework is introduced to compare the combined sustainability effects of all the steps in a work process compared to alternative processes.
- Chapter 2: "Information and Communication Technology Revolution and Global Warming": Over time, the ICT segment has become a major consumer of energy and, therefore, is rapidly becoming a significant source of greenhouse emissions. This chapter attempts to account for approximate annual energy consumption and the resultant CO_2 emissions of this sector. The authors also account for the indirect environmental effects of rapid proliferation of ICT technologies.
- Chapter 3: "Architecture and Mechanisms of Energy Auto-Tuning": Energy Auto-Tuning (EAT) optimizes the energy efficiency of software systems without human intervention. This chapter proposes a self-optimization technique for IT infrastructures, which takes hardware and software components as well as users of software applications into account.
- Chapter 4: "Energy-Efficient Server Selection Algorithms for Distributed Applications": A distributed application has multiple parts that are on different machines, which could be physically close together or be geographically distant. This chapter proposes algorithms to select one server for each type of application so that the total power consumption of servers can be reduced. It also evaluates the algorithms in terms of not only performance but also energy consumption.
- Chapter 5: "Design Models for Resource Allocation in Cyber-Physical Energy Systems": A cyber-physical energy system is an energy system that is built from and depends on a tight integration of the system's computational and physical components. The proposed techniques use a Linear Programming (LP) approach to determine resource allocations among a set of fuzzy rules that allocates Distributed Energy Resources (DERs), or power sources/sinks, and uses it to determine ways of improving resource management.

Management and Manufacturing

Management and manufacturing make up a great portion of industry. This section covers management and manufacturing for green computing and includes the following five chapters:

- Chapter 6: "Green Computing as an Ecological Aid in Industry": Ecology is the study of environmental systems and ecological aid means environmental help. This chapter discusses the use of computing for ecological evaluation in the manufacturing industry.
- Chapter 7: "Green IT Project Management: Optimizing the Value of Green IT Projects within Organizations": Green IT not only processes data but also uses less hazardous materials, maximizes energy efficiency, and promotes the recyclability or biodegradability of products. This chapter aims to define methods to optimize Green IT projects for sustainable value creation within organizations.
- Chapter 8: "Towards a Structured Cloud ROI: The University of Southampton Cost-Saving and User Satisfaction Case Studies": Organizational Sustainability Model (OSM) is used to deliver

specific products by enduring the internal and external pressures of a culture. It combines statistical computation and 3D visualisation to present the Return on Investment (ROI) arising from the adoption of cloud computing by organizations. The University of Southampton has achieved cost-saving and user confidence with service improvement offered by cloud adoption and services.
- Chapter 9: "Reducing Design Margins by Adaptive Compensation for Thermal and Aging Variations": Dynamically adapting to manufacturing, environmental, and usage variations is the key to shaving unnecessary design margins. This chapter presents an adaptive threshold compensation scheme using a transimpedance amplifier and adaptive body biasing to overcome the effects of temperature variation, reliability degradation, and process variation.
- Chapter 10: "Multidimensional Analysis of Supply Chain Environmental Performance": *Multidimensional analysis* uses dimensions and measures to analyze data, and a supply chain consists of all parties involved, directly or indirectly, in fulfilling a customer request. This chapter presents a framework for measuring environmental impact at the item level.

Mobile Handheld Devices

Mobile handheld devices are small, general-purpose, programmable, battery-powered computers that are capable of handling the front end of mobile applications such as mobile Internet browsing and can be operated comfortably while being held in one hand, enabling mobile users to interact directly with mobile applications. This section includes the following three chapters:

- Chapter 11: "Mitigating the Environmental Impact of Smartphones with Device Reuse": People throw away out-of-date phones all the time, and it causes many environmental problems. This chapter demonstrates the methodology and techniques towards reusing smartphones by presenting a case study on reusing smartphones for elementary school education.
- Chapter 12: "Power Management Strategies Based on Multi-Agent Systems for Portable Devices Equipped with Renewable Power Sources: Laptop Case Study": A multi-agent system includes a set of software agents that interact to solve problems that are too complicated to be solved by single software. This chapter introduces a laptop equipped with a photovoltaic charger for simulating power demand and supply in order to discuss power management strategies for reducing electrical consumption.
- Chapter 13: "Essential Technologies for Location-Aware Mobile, Online Map Prefetching, and Caching": Online mobile map prefetching or caching is useful, but is not widely employed because it involves several different subjects, and developers usually are not familiar with all of them. This chapter intends to relieve the problem by introducing essential technologies for mobile map prefetching and caching so more developers can start working on this kind of project or research.

Computer Vision and Image Processing

Computer vision is to process images in order to produce a representation of objects in the world and image processing is the use of computer algorithms to analyze and manipulate a digitized image. This section includes the following three chapters, which discuss computer vision and image processing using green computing:

- Chapter 14: "Green Video Compression for Portable and Low-Power Applications": Video takes up huge space. Therefore, before it is put to use, video normally needs to be compressed. Video compression indicates that the video information is packed into a smaller space. This chapter presents recent advances in the implementation of video compression that allow for energy efficiency and management.
- Chapter 15: "Display Energy Management Based on Eye Tracking": Eye tracking is the measurement of eye activities such as the point of gaze, the motion of eyes, and frequencies of eye blinks. The proposed technology tracks the user's eyes, keeping the display active only if the user looks at its screen. Otherwise, it dims the display or even switches it off to save energy.
- Chapter 16: "Energy-Aware Intelligence in Smart Spaces: A Case Study using Computer Vision and Machine Learning for User-Behavior Analysis": This chapter presents research issues on sensor-based smart environments for energy-aware intelligence. It also gives a study of algorithms for monitoring human activities that provide context awareness to smart environments.

Education and Power Generators

The final two chapters, education and power generators, do not belong to any of the above four sections so they are put into this section. It includes the following two chapters:

- Chapter 17: "The Education Part of Green Computing in Higher Education and Beyond": This chapter addresses current efforts being made to provide this education. Resources available from the federal government, state governments, non-profit groups, trade associations, and colleges and universities are discussed.
- Chapter 18: "Study of Hybrid Sustainable Energy System Based on PEM Fuel Cells and Photovoltaic-Module Power Generator": An energy system comprising two energy sources, photovoltaic and fuel cells, is proposed in this chapter. Photovoltaic cells coupled with electrolytic devices can be used to produce hydrogen and oxygen in a sustainable manner. With the produced hydrogen from the electrolysis process, it is possible to generate electricity through fuel cells.

SUMMARY

Global warming and the greenhouse effect are serious issues these days not only because they cost us billions of dollars in damages, but they also make the Earth hostile toward humans, animals, and plants. The following observations show the global warming and greenhouse effect can no longer be ignored:

- Global surface temperatures have increased about 0.74°C since the late 19th century (National Climatic Data Center, 2012). Global warming causes severe problems such as more hurricanes and droughts and spread of diseases. Another example of the seriousness of global warming is that polar bears now face their greatest threat—extinction—because the melting ice makes their hunting for food much more difficult.
- Global mean sea level was relatively stable over the last several thousand years, but it has changed over the past 100 years—rising at an average rate of 1.7 mm per year (National Climatic Data Center, 2012). Because of that, much land will be covered by water and residents have to be re-lo-

cated. For example, the Maldives president is considering moving the whole country to Australia if the situation gets worse.

- According to Environment Green (2012), global warming and the greenhouse effect cause great damage to the ozone layer. More people will have skin cancer because the ozone layer protects us from the sun's harmful rays.

Greenhouse effects and global warming are caused by many factors, such as harmful emission and global pollution. One of the major factors is the IT industry consuming a high volume of energy for running computers and using a great amount of materials to build computers and IT infrastructure. One way to reduce the greenhouse gasses from the IT industry is green computing. Together, ICTs and management systems account for a large portion of products from the IT industry. This book is, therefore, dedicated to green computing with an emphasis on ICTs and management systems.

Green computing consists of a wide variety of subjects, and it is not possible for a book to cover all those subjects. This book is one of the first few books systematically covering the methodologies, technologies, and theories of green computing. It is also the only green computing book focusing on sustainable ICTs and management systems. There are many benefits of using this book. Some of them are

- Cover the up-to-date information, knowledge, methods, and technologies of green computing with an emphasis on ICTs and management systems.
- Help students, researchers, and professionals master various green-computing technologies.
- Provide a much-needed textbook for a course on green computing.
- Could be used as a reference book for IT, business, computer science, and engineering workers and students, who are interested in green computing.

This book is a collection of high quality articles of green computing from world-renowned scholars and industry professionals. It will help readers better understand green computing and apply it to real world problems. Hope you will enjoy it.

Wen-Chen Hu
University of North Dakota, USA

Naima Kaabouch
University of North Dakota, USA

REFERENCES

Dragon Systems Software Limited. (2006). *Review of computer energy consumption and potential.* Retrieved from January 27, 2012, from http://dssw.co.uk/research/computer_energy_consumption.html.

Environment Green. (**2012**). *The threat from global warming on nature*. Retrieved from February 27, 2012, from http://www.environment-green.com/The_Threat_from_Global_Warming_on_Nature.html.

Fujitsu Siemens Computers. (2009). *Ten surprising facts about green IT*. Retrieved February 14, 2012, from http://www.slideshare.net/nesjo/green-it-10-surprising-facts.

Gartner. (2007). *Gartner estimates ICT industry accounts for 2 percent of global CO2 emissions*. Retrieved January 12, 2012, from http://www.gartner.com/it/page.jsp?id=503867.

Koomey, J. G. (2007). *Estimating total power consumption by servers in the U.S. and the world*. Retrieved March 3, 2012, from http://hightech.lbl.gov/htnews/htn-issue3.html http://hightech.lbl.gov/htnews/htn-issue3.html.

National Climatic Data Center. (2012). *Global warming: Frequently asked questions*. Retrieved from February 24, 2012, from http://www.ncdc.noaa.gov/oa/climate/globalwarming.html.

PBS. (2008, April 18). *Facts & figures: Health care*. Retrieved from March 02, 2012, from http://www.pbs.org/now/shows/416/health-care-facts.html.

Acknowledgment

This book took more than one year to finish, from late 2010 to early 2012. It is a large and difficult—but also enjoyable, memorable, and rewarding—work. The editors spent a great deal of time communicating with (potential) authors via numerous emails, organizing, and managing this book. The successful accomplishment of this book is a credit to many people. The editors thank the authors for their quality works and great efforts of revising their works based on the reviewers' comments. They thank the reviewers who provided such helpful feedback and detailed commentary. Special thanks go to the staff at IGI Global, especially to Myla Merkel, Mehdi Khosrow-Pour, and Jan Travers. Finally, the biggest thanks go to our family members for their love and supports throughout this project.

Wen-Chen Hu
University of North Dakota, USA

Naima Kaabouch
University of North Dakota, USA

February 13, 2012

Section 1
Information and Communication Technologies and Systems

Chapter 1
ICT Enabling More Energy Efficient Processes, Including e-Invoicing as a Case

Freek Bomhof
TNO, The Netherlands

Paula van Hoorik
TNO, The Netherlands

Maarten Hoeve
TNO, The Netherlands

ABSTRACT

ICT has the potential to enable a low carbon economy, as pointed out by many studies. One example of the energy (and CO_2) saving potential of ICT is illustrated in this chapter: how much energy (and emissions) can be saved if the invoicing process is redesigned? Although there is a net positive effect, the way the actual process is implemented can make a large difference. This led to the question of how to evaluate different projects or different designs, when it comes to the net sustainability effects. Then a framework is introduced to compare the combined sustainability effects of all the steps in a work process compared to alternative processes. The authors then describe existing indicators and metrics that can be used to fill the framework.

INTRODUCTION

There are many metrics and frameworks that all address some kind of sustainability in relation to Information and Communication Technologies (ICT). A commonly used definition of 'sustainability' is in the "Brundtland report" (World in Balance, 1997), defining sustainability as "development which meets the needs of the present without compromising the ability of future generations to meet their own needs." Sustainable development is frequently associated with the Triple Bottom Line, economy, society and environment; or more popularly, People, Planet and Profit. We use this Triple Bottom Line as the basis for our assessment of sustainability of ICT.

DOI: 10.4018/978-1-4666-1839-8.ch001

Historically, the application of ICT has been driven mainly by economic factors. Roughly speaking, the *'profit'* part of ICT is an aspect that is already incorporated in the sector. Although there is some debate on this (the Productivity Paradox: productivity growth in companies is not as fast as the growth in ICT performance), we will not elaborate this aspect much deeper.

The *'people'* aspect of ICT comes immediately after the 'profit' aspect. This aspect can be operationalized on various levels:

- Working conditions during production of the software. This includes topics like fair wages, equal opportunities, freedom to join labor unions, etcetera.
- Impact of the ICT on the workplace of people who use the system. The usability of a system is included in this, but also labor conditions, like increased risk of Repetitive Strain Injury (RSI).
- Impact of the ICT on administrative, business or societal processes, analogous to the introduction of the assembly line in manufacturing industry. Will the work become more tedious and lower valued, or will it enable workers to grow and learn?
- The transformative power of ICT on the way people work and behave, like social networks. Examples are the creation or loss of jobs, communication and work patterns that arise.

The *'planet'* aspect of ICT has recently received most attention. This chapter aims to give a view on how to assess the green-ness of processes using ICT. As pointed out in many studies like the Climate Group (2008) and Hilty (2008), ICT has the potential to enable a low carbon economy. In the ICT sector plans and activities to significantly step up the energy efficiency of its products and services are unfolded but ICT's largest influence will be by enabling energy efficiencies in other sectors. That opportunity could deliver carbon savings up to about 20% (Hilty, 2008).

Based on the outcomes of a study on the sustainability impact of e-invoicing (Van Hoorik, et al., 2010) we start with this case to illustrate the effects of different processes on the sustainability outcomes. We show that different process designs have a great impact on sustainability outcomes.

Then ICT sustainability impact framework (Van Hoorik & Bomhof, 2010) is introduced for a structured assessment of sustainability effects. On one hand, it contains the three well known sustainability aspects people, planet and profit. On the other hand, it includes the three types of effects. When assessing the effects of a project or intervention, effects will occur on different levels, according to Hilty (2006) and Berkhout and Hertin (2004).

- Direct effects: the effects that are caused by the existence and use of ICT. For example, the use of energy and materials of producing, using and discarding ICT equipment.
- Indirect effects, also called enabling effects: the effects that are caused by the impact of the usage of ICT on its environment, for example sensors that will switch off the lights when leaving the room.
- System effects: the longer term and behavioral effects, for example by working at home in the morning, people are able to avoid traffic jams and spend less time in the car.

Next to that, rebound effects make it difficult to evaluate the effects of ICT on sustainability. Rebound effects are the effects of ICT that seem to have good advantages in the short term could be undone in the long run (Bomhof, et al., 2009). For example, if people are able to avoid traffic jams in the morning by working at home, they can choose a job that is located further from home. That will increase the number of kilometers spent in the car. This instance of a rebound effect is known as the 'constant travel time hypothesis.'

At the end of the chapter, an overview of indicators needed to assess the "green-ness" of ICT and ICT projects is given. After all, when we want to evaluate which ICT project is the most sustainable, of which design option will have the most sustainable effect, we need indicators that we can measure. We show that indicators for some aspects are easy to measure but give an incomplete picture and other indicators are far more complicated to measure, but essential for a balanced assessment. Finally, we describe a method to estimate values for these not so easy to measure indicators.

E-INVOICING CASE

As an illustration of sustainability effects of ICT and ICT projects we start with a case based on the results of a Dutch study (van Hoorik, 2010) on the sustainability effects of e-invoicing. The topic was the business to government (B2G) invoicing in The Netherlands. The project aimed to shift all these invoices from the current physical (paper) processes to a fully electronic process. Rationale for the project was not only cost reduction, but also the net carbon savings of this transition.

Sustainability Effects of E-Invoicing

A Life Cycle Assessment (LCA) of the two product chains, "paper invoicing" and "digital invoicing" was done and involves the energy usage (in MJ) and the greenhouse gas emissions (measured in CO_2 emissions equivalents) of both chains. The LCA methodology not only includes the direct emissions resulting from energy use, but also the energy required to produce and handle and treat the wastes of the products (paper, envelopes) and equipment (PC, trucks, printers, etc.).

Direct and Indirect Effects

The analyses showed that the paper invoicing has an energy use that is almost twice as great and a CO_2 emission approximately 1.5 times as great as for digital invoicing. As shown in Figures 1 and 2, the savings from switching to digital invoicing

Figure 1. Energy use of paper invoicing versus digital invoicing for B2G

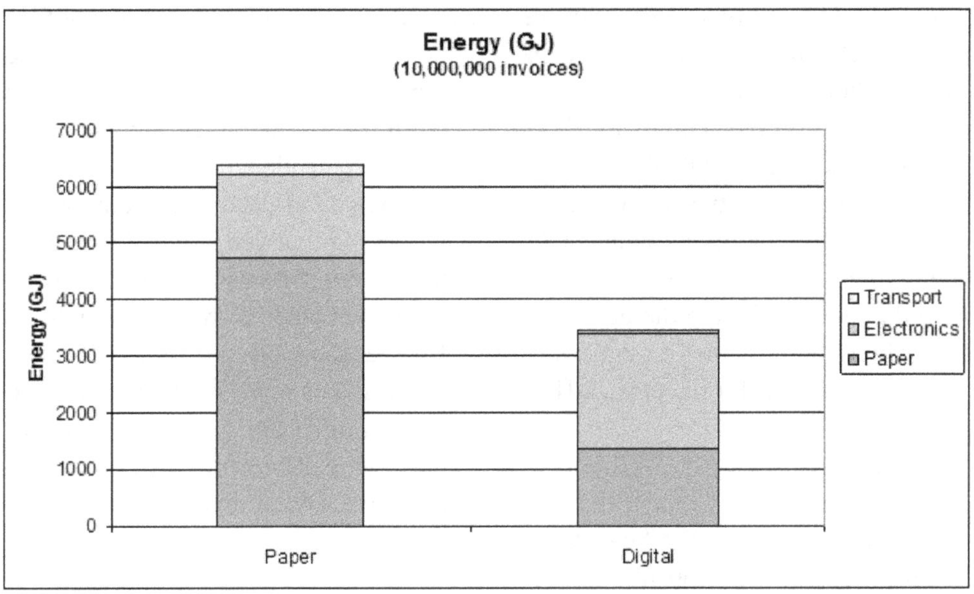

Figure 2. CO_2 emissions of paper invoicing versus digital invoicing for B2G

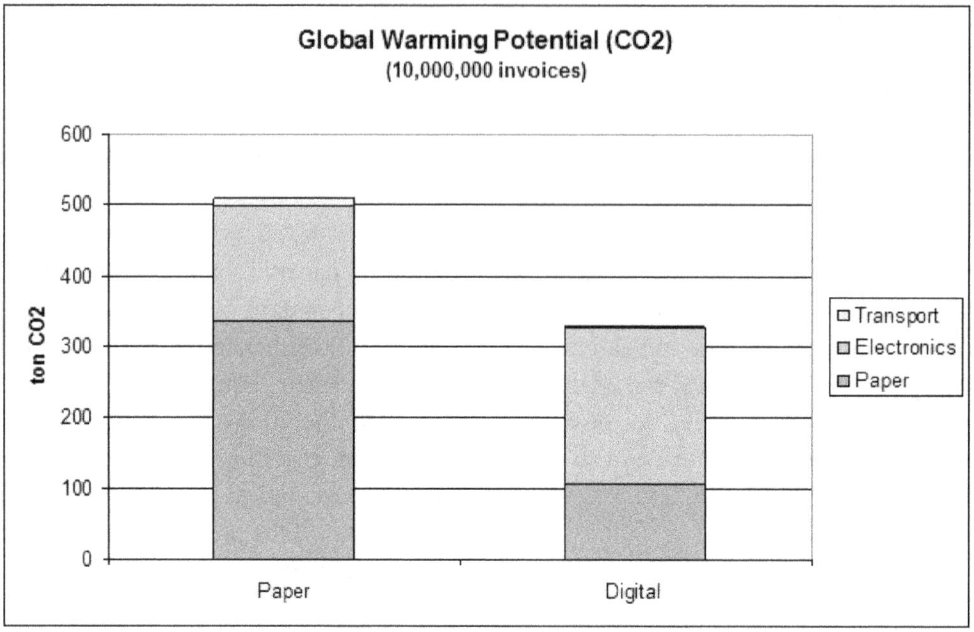

are primarily achieved through savings in paper usage. These savings are barely affected by the increased use of electronics required for digital invoicing. Transport has a negligible impact on the total energy usage and CO_2 emission.

The results of the 10 million invoices sent by businesses to the government (B2G) in The Netherlands amount to energy savings of 3 TJ and almost 200 tonnes of CO_2, which is comparable to the gas use of 55 Dutch households or the CO2 emissions of 20 Dutch residents.

For the total annual volume of invoices, about 3500 million invoices a year, switching to digital invoicing for all invoicing types would have energy savings of approximately 1 PJ and a CO_2 emission savings of almost 70 ktonnes. In relative terms, this would be equivalent to the gas usage of 20,000 Dutch households of the CO_2 emissions of 700 Dutch residents (van Hoorik, et al., 2010).

System Effects

Although system effects turn out to be difficult to divide into people, planet and profit, multiple effects are expected. The Expert Group on e-Invoicing on Electronic invoicing in Europe (2010) stated that e-Invoicing is not dematerialization only—but an enabling platform for automating administrative processes at large—a new paradigm. Interviews with organizations implementing e-invoicing at the moment revealed several system effects.

Internal effects in the company, for instance that because of higher transparency in the processes, a more result driven management style became possible. Another internal effect is that more ICT skills are needed.

External effects, outside the company where the process is changed, are in the case of e-invoicing that less postal services are needed, and the paper and copier business will have smaller turnarounds. Besides these effects leading to smaller demands and less business, in other sectors new business opportunities are created: for instance, cash flow management becomes easier and banks could play an active role in this.

Figure 3. Four different process designs for e-invoicing

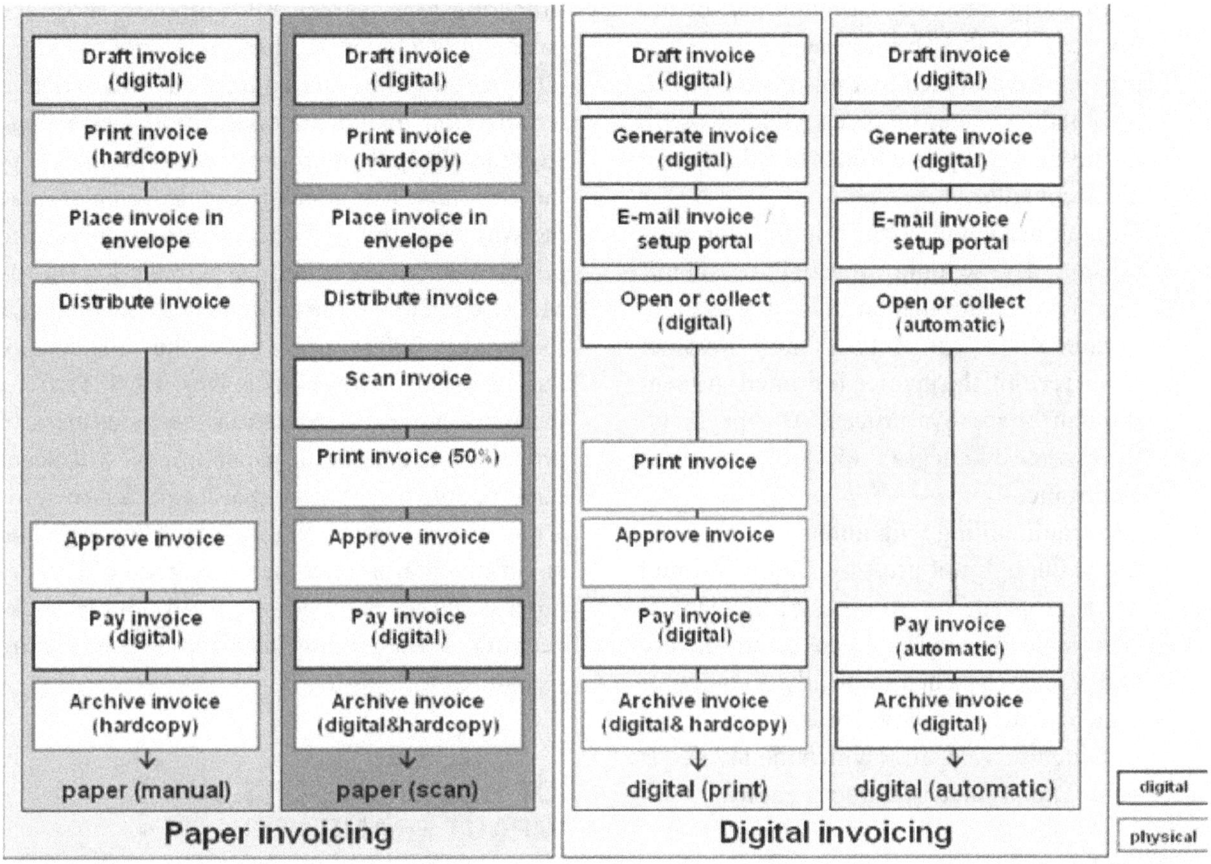

Design Considerations: Different Processes, Different Outcomes

Although it turned out that digital invoicing uses less energy and emits less CO_2, much can be said about the way processes work in practice. Processes can be designed in different ways and it turns out that different designs have considerably different outcomes.

The sustainability effects of two processes of invoicing on paper and two digital processes were analyzed as shown in Figure 3 (van Hoorik, et al., 2010).

Digital and paper invoice processes differ in the way they deliver the invoice and partly in the way invoices are treated and paid. We assume that drafting of the invoice by the company is always done in the same manner, namely on a computer. The settlement of the invoice by the receiving organization can vary. We have analyzed four channels that are used the most in The Netherlands:

1. Invoicing on paper with manual approval. This method of invoicing and handling is by far the most common. After the drafting of the invoice, it is printed, placed in the envelope and sent by mail. In the handling process, the invoice is approved manually with one or more signatures after which the invoice is paid electronically. Finally, the invoice is archived.
2. Invoicing on paper with scanning. The first steps in the process are the same as above but the invoice is now digitalized upon receipt,

using a scanner. Because in most cases the payment process is not digital, part of the invoices are then printed again to be approved, we assumed that this happens with 50% of the scanned invoices. After payment, the invoice is kept in a folder and also stored electronically.

3. Digital invoicing with printing for processing. By sending an email or submit their invoice through an invoicing system, organizations can deliver their invoice. Upon receipt, the invoice is printed and sent through the approval process. After payment, the invoice is kept in a folder and also stored electronically.
4. Electronic billing with automated processing. Although not used frequently yet and not appropriate for all types of invoices, it is possible to receive invoices electronically and process them automatically. A computer system matches the invoice to a contract and provided service, after which the invoice is paid. The invoice is stored digitally.

In order to make the calculations of energy savings, a lot of data is needed. Table 1 presents the data used to analyze the four processes.

Figure 4 shows that the four invoicing processes have very different sustainability outcomes. The energy usage in mega joules is shown for one invoice. The amount of energy used is shown for different items, such as transport, paper, printer, envelop, server and the development of the e-invoicing system. It shows again that envelopes and paper explain a large part of the sustainability outcomes. It also shows the effects of scanning printed invoices when approval processes are not digitalized. In that case, double bookkeeping occurs (digital and hardcopy) and extra prints are created. Finally it also shows that a process with automatic payments and digital approval processes decreases the energy use and CO_2 tremendously by as much as two-thirds.

When designing new processes, it is important to not only look at greener ICT products and green datacenters. Of course, these are important, digital processes like e-invoicing are a form of dematerialization and dematerialization has a huge sustainability potential. The Climate Group (2010) indicates that using technology to dematerialize the way we work and operate across public and private sectors could deliver a reduction of 500 $MtCO_2e$ in 2020—the equivalent of the total ICT footprint in 2002, or just under the emissions of the UK in 2007. However, as shown in this article, there are huge differences in the green-ness of processes. In case of the paper process with scanning, people might think that it is a step towards a greener processes, but in fact it has a greater environmental impact then the 'good old' paper process. Greener processes avoid double bookkeeping, extra printing and copying, and extra signatures for approval.

ICT SUSTAINABILITY IMPACT FRAMEWORK

As we have seen in the e-invoicing case, the amount of metrics that are available to assess the sustainability of ICT is large, and the amount of effects that can be taken into account seems even larger. When we want to improve on ICT's sustainability, this is usually done in a project, and the first thing we have to do is choosing the scope.

A project that involves designing new processes should be assessed on the level of processes. The business requirements to these processes should be taken as the starting point, and within these requirements, the optimal solution has to be designed. The example treated in this section is such a process redesign project, so the level is "IT functions/ processes." Then, the project manager can choose to incorporate other levels as well. The software / application level could be taken into consideration next: is there a choice in the type of software that is used, can the IT architecture be

Table 1. Overview of parameters about paper, electronics, and transport

	Item	Unit	Paper		Digital	
			(man.)	(scan)	(print)	(auto)
Paper	**Invoice**					
	A4	piece	2	2,5	2	0
	Enveloppe					
	C5 enveloppe	piece	1	1	0	0
Electronics	**Draft invoice**					
	PC	min	2,12	2,12	2,12	2,12
	Generate invoice					
	development e-invoicing system	piece	0	0	1	1
	computer server	invoice	0	0	1	1
	PC	min	0	0	0,08	0,08
	printer (70%)	min	0,26	0,26	0	0
	Printing office(30%)	min	0,11	0,11	0	0
	Email invoice					
	PC	min	0	0	0,08	0,08
	computer server	invoice	0	0	1	1
	Network	invoice	0	0	1	1
	Open or collect invoice					
	computer server	invoice	0	0	1	1
	PC	min	0	0	1,24	0
	Pay invoice					
	computer server	invoice	1	1	1	1
	PC	min	0,28	0,28	0,28	0
	Archive invoice					
	computer server	invoice	0	1	1	1
	PC	min	0	0,58	0,33	0
	Printer	invoice	1	1,5	2	0
	Scanner	invoice	0	1	0	0
Transport	**300 km transport of paper(pulp) from storage to store**					
	A4, lorry (>32t)	A4	2	2,5	2	0
	envelope, lorry (>32t)	env	1	1	0	0
	50 km transport paper van store to company					
	A4, lorry (>32t)	A4	2	2,5	2	0
	envelope, lorry (>32t)	env	1	1	0	0
	100 km transport invoices (postal)					
	A4, lorry (>32t)	A4	1	1	0	0
	envelope, lorry (>32t)	env	1	1	0	0
	30 km transport paper to recycling (72%), incinerator (26%) or dump (2%)					
	A4, lorry (>32t)	A4	2	2,5	2	0
	envelope, lorry (>32t)	env	1	1	0	0

Figure 4. Energy usage of all aspects of four processes

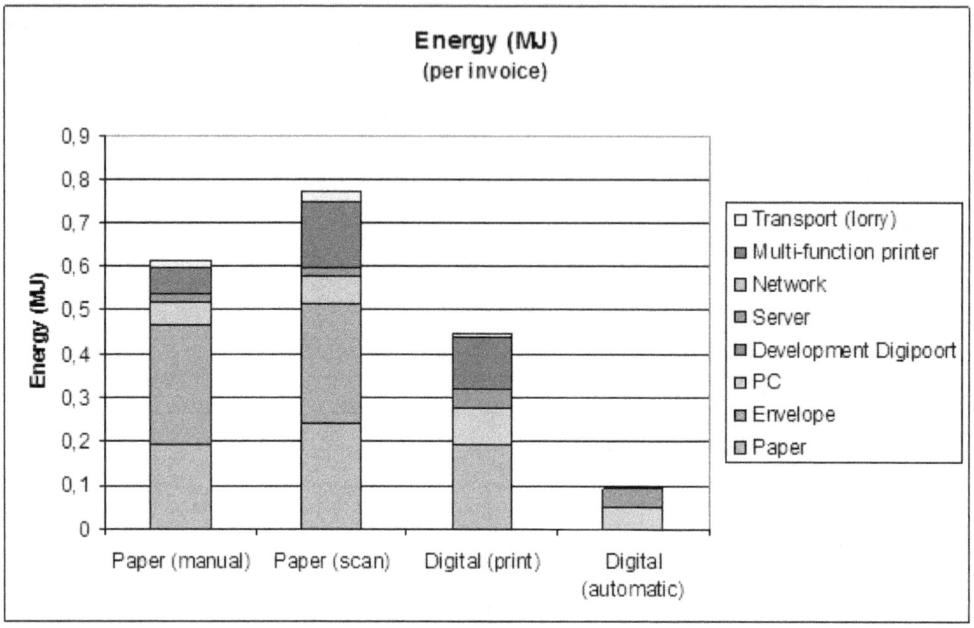

designed in such a way that energy consumption is minimized? And when the IT architecture has been decided upon, the datacenter can be chosen (if there is a choice) so as to have a datacenter with the best PUE performance.

As said, the example case will be on the IT functions / processes level. This means that, for 'planet,' in this case the net total energy and/or embodied carbon will be the main focal point of our analysis. The analysis should contain direct effects, indirect effects, and system effects, all quantified as accurate as possible.

Three Types of Effects: Direct Effects, Indirect Effects, and System Effects

The direct and indirect sustainability effects of ICT usage can be analyzed by using Life Cycle Analysis (LCA) which focuses on production, use and disposal of all materials used in a process. Outcomes are energy usage, greenhouse gasses, and about 15 other environmental indicators. LCA also covers indirect effects.

Direct Effects

The direct sustainability effects of ICT can be subdivided into three main categories:

- Manufacturing of the ICT equipment
- Use of the ICT equipment
- Disposal of the ICT equipment

For the manufacturing and disposal phase, assessing the environmental effects of ICT is usually also done by means of LCA. One of the indicators of LCA is Greenhouse Gas (GHG) emissions. ISO 14064 is an international standard that specifies and guides the quantification and reporting of GHG emissions. ISO 14064 is compatible with the GHG protocol (www.ghgprotocol.org). The LCA also covers the usage phase of a product, but because the usage of ICT has so many different aspects, it is worthwhile to look deeper into that part.

For the use of the ICT Equipment, ultimately it is the energy use, including the energy use of supporting equipment (such as cooling, in data centres) that is of prime importance.

When assessing the direct energy use of new applications in large-scale ICT ecosystems, the main challenge is to calculate what percentage of the energy use of a system can be attributed to this new application.

Indirect Effects

To account for the indirect effects of ICT usage, the challenge is to identify *where* the effects take place. As we learnt from the e-invoicing case below, the easiest is to start with the intended functionality of the ICT, the functionality that is used to justify the business case (profit) of the ICT investments. Indirect effects of ICT usage can be (Berkhout, 2004):

- Information effect: information delivered to users will influence their behaviour. A well-known example is energy feedback displays in households that can have an immediate effect on energy usage up to 8% (Van Dam, 2010) (However, the long-term effect is that this figure will decrease considerably over time.).
- Design and operation of products, services and processes: ICT facilitates the efficient design of products and services.
- Transport and distribution: physical transport can be optimized by applying ICT-supported planning systems.
- Dematerialisation: video conferencing, e-business, summarized by the "move bits, not molecules" slogan.

The measurement of each of these effects can vary considerably. However, ultimately, it is the energy use that counts.

An additional challenge is to identify the rebound effects: effects that have an unanticipated negative influence on the intended positive effect. The occurrence of rebound effects is also called Jevons' paradox: it was identified by the English economist Jevons by the end of the 19th century: increased efficiency in the way coal was used, also increased the demand for coal. This is because the cost of that resource is reduced, so users are tempted to find new opportunities to use this resource. In our time, increased energy efficiency of lighting leads to the same situation.

The rebound effect is apparent both as an indirect and as a system effect (see below). An example is telecommuting: office workers that choose to work at home will indeed reduce energy use from commuting (by car or public transport), but they will use extra energy to light and heat or cool their houses, they will generally use more equipment (extra office furniture and equipment), they can afford to live at larger distances from their work, and they tend to use their cars more often during the day to have short trips, etcetera to go shopping. In Bomhof (2009) it is proposed that a list of these effects can be created by using a checklist containing 'standard' categories:

- Work space and buildings
- User
- Industrial production
- Transport
- Waste
- Energy

When a rebound effect is present, a more detailed quantitative analysis is needed to calculate the net effect of an intervention.

System Effects

System effects are those effects that arise as a result of long term structural changes. Some of these changes are addressed in Berkhout (2004):

- De-materialization: more economic growth is attributed to services, and not in production of goods. In principle, ICT facilitates economic growth without increased usage of materials and energy, because the economic value lies in the information.

Figure 5. ICT sustainability impact framework

ICT sustainability impact framework	Direct	Indirect	System
People	Safety and health Productivity Work quality ...	Human performance Comfort, wellness Job satisfaction ...	Employment Social inclusion Welfare Behavioural change ...
Planet	Energy use, GHG and other LCA indicators for production, se and disposal ...	Information effect Transport/ distribution ← rebound effects ...	Dematerialisation Number of trees ← rebound effects ...
Profit	Costs for design, build test Costs for operating & Maintenance costs Investments in training ...	Shorter process times High efficiency Less errors Higher quality ...	Higher transparency in chains New business opportunities ...

- Rebound effects: also on the system level, rebound effects can be identified. Examples are the world-wide e-waste problem, increased travel because of the 'global village' idea and the biggest one, identified by Hilty (2006): ICT increases national welfare which in turn will increase energy use.

Although these effects are very real, it is very difficult to identify them and to measure them reliably.

When combining the three effects with the triple baseline in one framework, we have an ICT sustainability impact framework as sketched in Figure 5. It can be used to create a balanced assessment of sustainability impact, focusing on more than just energy or CO_2. For each cell, examples of indicators are given.

For a full assessment of the sustainability impact of ICT, one would have to quantify all effects (direct, indirect and system) for the aspects People, Planet, and Profit, including rebound effects. Of course, it is not easy to do this. Usually the 'direct profit' is not difficult. The indirect profit is the realm of more speculation (business cases are based on indirect profits, and are usually estimates), where the system effects on profit are very hard to quantify, although techniques like Real Options Valuation try to do so.

In the next section, we will introduce a number of metrics and indicators that can be used to fill the cells on the 'planet' aspect of the framework. As with 'profit,' the system effects are the hardest to model.

HOW TO ASSESS THE "GREENNESS" OF AN ICT PROJECT?

In most businesses, ICT is introduced and changed as a result of projects. An ICT project and its results can be assessed in many ways. The ICT itself can be assessed by using various indicators, of which we give some examples. Then, the assessment can be done on various levels, and the project *itself* can be assessed as well, and even then, we miss some indicators.

Sustainability of ICT

Below, some metrics are presented that aim to quantify the energy use of various types of ICT. This list is meant as an example and is not exhaustive: for each application, level or sector, a separate metric may be defined.

Energy Efficiency of Network Equipment: ECR

Network equipment consists of all ICT that transmits data from one computer to another. Most network equipment nowadays is dealing with Internet data, based in the Internet Protocol. This is called packet based network equipment. Special (non IP) types of network equipment are found in mobile networks (especially the radio receiver / transmitter) or in analogous television distribution networks.

The Energy Consumption Rating (ECR) (Ceuppens, 2008) initiative is an attempt to provide an efficiency metric that provides information on the energy use of packet based network equipment. The goal of ECR is to define a standard rating that can be used for comparing the energy efficiency of network equipment. The test procedures and methodology to assess the energy efficiency of various network equipment are publicly available.

The ECR gives a number that is expressed in Watts per Gbps. Two ratings are defined:

- The basic ECR rating is defined as the energy efficiency at the maximum achievable network load without packet loss.
- The variable load metric ECR-VL, which is a weighted average of the energy efficiency at network loads between idle and 100%.

The success of the ECR standard is dependent on the adoption by the vendors of network equipment. At this moment the ECR is only supported by Ixia and Juniper Networks.

The ECR is a starting point for deciding what equipment should be used in a network from an energy use point of view. Of course, the network design dictates *how many* equipment is needed. Many efficient routers in a suboptimal designed network can use more energy than fewer inefficient routers in an optimally designed network.

Energy Efficiency of Computer Systems

The EnergyStar standard (EnergyStar, 2009) gives energy efficiency ratings for consumer products. Energy Star includes a specification to rate the efficiency for desktop and notebook computers. This includes an efficiency rating based on separate categories for computer systems (desktops, notebooks, thin-clients, and small servers) as well as a minimal required Power Supply Unit (PSU) efficiency, which is labeled as '80 PLUS': "To qualify for an 80 PLUS sticker, a power supply must achieve at least 80% efficiency at three specified loads (20%, 50%, and 100%) of the maximum rated power of the PSU." The 80PLUS comes in various levels: bronze, silver, gold, and platinum, with increasingly stricter requirements.

EnergyStar / 80Plus addresses one of the largest power drains in computers. However, it does not address the energy efficiency of a computer in terms of "performance per power unit." For microprocessors, the 'Flops per Watt' ratio can used: the number of Floating Point Operations per Watt. This is especially useful for high performance computing, but not for operating server parks.

For computer servers, the SPEC power benchmark evaluates the energy use of a server for a standardized set of operations. It is part of a larger set of benchmarks hosted by the Standard Performance Evaluation Corporation. The test is run by executing Java scripts that will apply various load types ('graduated workload') to the server under test. The results of tests for a lot of vendors are disclosed by SPEC on their website www.spec.org.

A special category of computer systems is formed by embedded computers: hardware and software that is built into systems where some kind of automated operational or real-time task has to be performed, for instance in industrial automation. In many situations, the power supply of embedded systems is critical, because of heat production or because it is battery-based. The Embedded Microprocessor Benchmark Consortium (EEMBC) defines benchmarks for these systems.

Energy Efficiency of Data Centres

The energy efficiency of data centres receives a lot of attention. It is here that the energy use of ICT becomes most apparent. An imperfect, though widely known measure for assessing the energy efficiency of data centres, is the Power Usage Effectiveness (PUE). Another well-known measure is the Datacenter Infrastructure Efficiency (DCiE), which is just the inverse of the PUE.

The PUE is defined by the green grid (GreenGrid, 2007), which is a global consortium of IT companies and professionals with the goal to improve energy efficiency in data centres. The green grid works on the development of standardized and accepted set of metrics, methodologies and processes to achieve this goal.

The Power Usage Effectiveness metric PUE is defined as:

PUE = Total Facility Power / IT Equipment power

The Total Facility Power is the measured power dedicated to the data centre and the IT Equipment power is the power measured of all ICT equipment, mainly computers, network components, and storage. The calculation of PUE is relatively straightforward. Most complexity finds its origin in accurate bookkeeping: the allocation of energy measurements to IT equipment and facility power. The major advantage of PUE is its wide acceptance, and straightforward implementation.

A value of 1.5 in practice is considered as a good value. Facebook announced in April 2011 that their datacentres had a PUE value 1.07.

The PUE methodology has some limitations. The most important limitation is that energy efficiency of hosted ICT infrastructure in data centers is not included: PUE does not provide insight in the effectiveness of individual ICT systems. As an example: Computer systems, which are in a continuous idle state, could possibly be switched off so that energy is saved. From a mere PUE perspective switch of a system could have a negative impact on efficiency because the total IT equipment power consumption is reduced, but probably the total cooling capacity of the data center is not: this means that the ratio "Total Facility Power" versus "IT Equipment Power" becomes less favorable.

The scope of PUE is limited to the hardware, and not the logic behind it. Inefficient software has no impact on the PUE: it assumes that all energy that is going into the IT equipment is 'well spent,' and that only energy going into the supporting infrastructure should be avoided. A bank that runs three different mortgage applications, including test servers and everything else needed to exploit these applications, could have a very favorable PUE. However, integrating all these applications into one and switching on test servers only when needed would decrease the overall energy use much more.

Another example: in an over dimensioned data center, where lots of cooling equipment is running but where only part of the floor is occupied by working IT equipment, it is easy to increase the power efficiency by just adding a bunch of old, unused computers. In that case, hardly extra cooling is needed, but these old computers do use power. This means that the ratio of 'total energy used' versus 'energy used for IT only' becomes more favorable. The main problem is that PUE is a ratio that can be made 'better' by decreasing the nominator, but also by increasing the denominator.

Another limitation is that some sustainability measures, like reusing generated heat though heat pumps, do not have an impact on the efficiency of the data centre itself and therefore have no impact on PUE.

Table 2. OpenDCME model

The data center:	The IT Assets:
• DCiE / PUE • Floor usage • Bypass • Recirculation	• Network architecture efficiency • Storage architecture efficiency • X86 architecture efficiency • RISC architecture efficiency
The tooling: • Network management • Storage management • Compute management • Electrical and mechanical management	The processes: • Change and configuration management • Product lifecycle management • Capacity management • Service Level management

Open Datacenter Measure Efficiency (Open DCME)

OpenDCME (OpenDCME.org) strives to overcome the drawbacks of efficiency ratios, for example the PUE, by focusing on efficiency-inducing policies rather than mere hardware measurements. It strives to be a supplement to the more technical measures like PUE.

The model identifies 16 Key Performance Indicators (KPI) that are grouped into 4 quadrants, as summarized in Table 2.

This approach enables datacenter management to assess more areas that have impact on energy use, to incorporate supporting processes, like product lifecycle management, as well. The scope is limited to datacenters, so this approach is only suitable for 'green it' initiatives. Besides, it cannot be easily used to compare different scenarios.

OpenDCME is started by the Dutch company ManSystems.

ISO 14064

The quantification and reporting of the carbon footprint of an organization is standardized by ISO 14064. It is compatible with the GHG (Green House Gases) protocol (www.ghgprotocol.org).

ISO 14064 provides governments, businesses, regions and other organizations with a set of tools for programs aimed at measuring, quantifying, and reducing greenhouse gas emissions. These standards allow organizations take part in emissions trading schemes using a globally recognized standard. ISO 14064 provides a comparable method of reporting that is internationally considered as "good practice." The standard provides a methodology to quantify GHG emissions at the organization level and the project level. Also includes are guidelines and requirements for validation, verification and certification.

Assessing GHG emission at the organization level is separated into three tiers:

- Direct GHG emissions occur from an organization's own resources, like furnaces, machines or owned vehicles;
- GHG emissions related to purchased electricity;
- Other GHG emissions i.e. related to purchased products and services like transportation.

Outcomes of measured GHG emissions can be shared through platforms like the Carbon Disclosure project (www.cdproject.net, CDP). This includes climate change strategies and reduction targets. The CDP data is available to investors, policy makers, governments, researchers and the public.

Global Reporting Initiative

As much as OpenDCME captures more aspects of energy use in data centers than just kWh's, Global Reporting Initiative (GRI, gri.org) captures more aspects of a company's sustainability than just its carbon footprint. In fact, GRI is an attempt to capture all sustainability aspects of an organization, regardless of size or business.

The guidelines are meant to provide transparent information on the sustainability performance of the organization. Indicators are proposed for economic, environmental, and social performance.

Each indicator is supplied with a protocol that describes how the indicator should be measured. The focus of GRI is to encourage organizations to report on their sustainability indicators, and to make this kind of reporting just as normal as financial reports for the organization's profit are today. Specific protocols for specific sectors are defined.

The indicators in the GRI form a very good basis for judging the net sustainability performance of an organization, yet it does not make an attempt to bring the indicators on a level that enable to compare organizations.

Levels of Sustainability

Looking at the various metrics and frameworks described above, it becomes clear that in an operational situation, the sustainability of ICT can be viewed on various levels, as illustrated in Table 3. The challenge is to find a sustainable optimum on each level, *given the requirements of the next higher level*, and taking the characteristics of the lower level as given.

The challenges on each level are different, and their scope is usually limited to that level. For instance, a data center company may focus on the efficiency of their cooling equipment, but it will not address the way the software is operated. This could be because this software is managed by the datacenter's customers, or because it is another department's responsibility. When looking 'up' from a level, responsibilities are different, so it is not natural to challenge the choices that are made there. For instance, a programmer seldomly ques-

Table 3. Levels of sustainability

Scale		Challenge	Examples	Frameworks
Economies / societies		Combine maximum welfare with maximum sustainability		ISO14064
Cities/ regions		How to balance sustainable working, living and travelling?	Smart cities	ISO14064
Companies		Sustainable operation, footprint reduction		GRI, ISO14064
IT functions / processes		What is the most energy efficient way to reach a business goal?	e-billing	OpenDCME (limited to data centers)
Software / application		Given the functional requirements, what is the most efficient way to accomplish these?	Green coding, application rationalisation	(Joulemeter)
Datacenter		Given the application landscape, what is the minimum of energy needed to support this?	Virtualization	PUE, DCiE
Individual Equipment	Networks: Communication Technology (CT)	Maximum of communication throughput for a minimum of energy	Low Energy Bluetooth	ECR
	Computers: Information Technology (IT)	Maximum of computing power for a minimum of energy	Intel 'Westmere' energy-saving chip	Energy star
	Cooling, lighting, ventilation (facilities)	How to maintain the IT and CT functions with a minimum of energy for supportive technology?	Green Grid recommendations	Energy star

tions the wishes expressed by the user or client. Another mechanism arises when looking 'down.' A company that wants to reduce its footprint, may invest in energy efficient office lighting; but it may overlook the possibilities in their inefficient business processes.

Up to "IT functions / processes" the focus is on ICT. The next level, "companies," and beyond, do not address ICT. On these levels, the scope is extended to for instance changing human behaviour. The example on e-invoicing that was introduced before, is on the 'IT Functions / processes" level.

Missing Measures

Although there is no lack of proposals for indicators, for some ICT related functions, no measures are yet available. We mention two of them below.

Green Coding

One level that has not yet received much attention is the software application level. The topic of "green coding" aims at designing and operating software in such a way that the resulting energy consumption of the hardware that it runs on, is minimized.

The well-known Moore's Law has resulted in ICT hardware that relieves programmers of many restrictions that were previously very important resource restrictions. Bandwidth, processing power, memory and storage are so abundant that for many tasks, there is no real incentive to design software applications in such a way that they make optimal use of these. Requirements like speed of development, portability and maintainability have resulted in software frameworks that have a quite heavy 'footprint.' A typical "hello world" demonstration program in many software application frameworks takes up several megabytes.

Impressive results have been reported, though. The Dutch Social Media website Hyves.nl (comparable to Facebook) reported that making use of a PHP to C++ translator (HipHop, designed by Facebook) alone, resulted in an energy consumption reduction of about 30%.

In very specific application areas, such as mobile computing (where battery power is a major resource limitation) the total energy use of different software designs can be compared. This would encompass standardized test scenarios. For virtualization, where a physical computer server is simulated on a set of powerful central servers, tools like Joulemeter help developers and administrators to monitor the energy use of virtual servers.

The big difficulty with green coding is that it is not easy to define a comparable efficiency measure aimed at energy-aware coding, because all software is different. One could think of "kW per function point," but this does not take into account the usage conditions of the software (heavily used or idle, multi-user or single user, type of hardware, type of operating system, etc.). An indicator based on "useful work" would be beneficial. Currently, the best approach seems to be to stress the need for resource aware design of software.

For software maintenance or redesign, it is however feasible to compare the energy use of the new release to the energy use of the previous release. A standardized set of (preferable automated) test cases on standardized hardware is then needed. In this case, the software can be compared with itself, provided the functionality has not changed so much that it becomes incomparable. An example is the way default settings in Windows XP were changed in Windows Vista, to become much more power efficient.

An interesting experiment of how software tends to become more inefficient in subsequent versions is described in Hilty (2008). The same user task, carried out on computer hardware and operating systems from three generations (1997, 2000, 2003) led to a significant increase in needed CPU time, and in some cases *also* more user interactions (which means that the user efficiency decreased as well).

Figure 6. Sustainability in projects, from Silvius (2010) (Copyright Silvius 2011. Used with permission.)

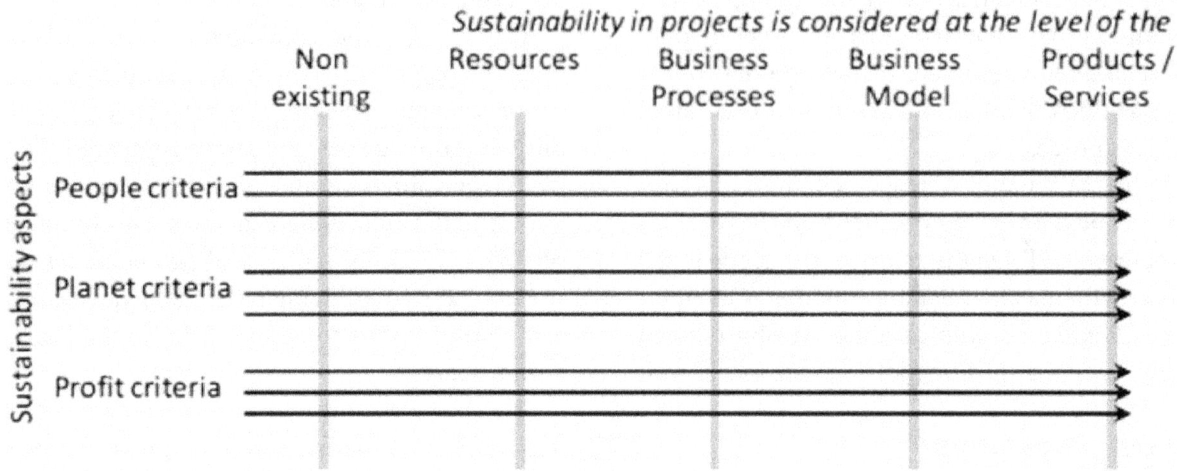

IT Functions and Processes

The same problem (no indicators for green coding) applies to the level of IT functions and processes. When trying to assess the energy use of an IT function or business process, some approach comparable to the Material Input Per Service Unit (MIPS), as described by (Ritthoff, Rohn, & Liedtke, 2003). Unfortunately, the service unit definition is different each time, so many services will not be comparable to each other.

SUSTAINABILITY OF ICT PROJECTS

As argued in Silvius (2010), projects are the vectors of change in an organization, so sustainable organization change has to be founded by sustainable projects.

The Maturity Model for projects assesses a project on criteria that are grouped by the well-known people, planet, and profit pillars. For each criterion, the actual level and the desired level is assessed and plotted on a scale. These scores are based on interviews with project managers and other staff. The conceptual model is depicted in Figure 6.

For all criteria (subdivised into people criteria, planet criteria, and profit criteria), questions are asked by the assessor, and based on the answer, the project's performance for each criterion is ranked as:

- Non existing (no positive impact on sustainability)
- (Positively) affecting the way the company's resources are used
- Affecting the company's business processes
- Changing the company's business model
- Changing the company's products or services (highest sustainability impact)

Based on the differences between actual levels and desired levels, the opportunities for improvement can be identified. Both the project, and the intended project result, are assessed.

More specifically, for ICT projects, this would mean that both the design and construction phase (the project) and the operational phase (exploitation of the project's result) will be assessed. The design and construction phase has not yet received a lot of attention, although there are some examples: some governments that have a 'green procurement' programme, will pose restrictions on the maximum travel distance of subcontractors.

As with physical products, the 'embodied carbon' in a software product may very well comprise a significant percentage of the total carbon footprint; for a typical PC or laptop, this can be as high as 50%. For software, the same comparison may be made. The 'embodied carbon' of a software product would then include:

- The energy needed for operating the development, test and deployment servers, during initial development of the product;
- The energy needed for the same servers, during the maintenance phase of the product.
- Any indirect related energy use, mainly for heating, cooling and lighting the developers' workplace, and, even more indirect, energy use of project-related transport.

Existing metrics on software size (e.g., function points, or the Cocomo method) can be used to calculate the energy needed per function point, or the energy per thousand lines of code (KLOC).

FUTURE RESEARCH DIRECTIONS: TAKING SUSTAINABLE DECISIONS

This chapter concentrated on energy efficiency, but taking energy efficiency into account is only a first step in taking more sustainable investment decisions. All organisations need to decide which projects they will perform with their limited resources. Organisations that aim to be sustainable somehow need to balance People, Planet, and Profit in these decisions. How much profit are they willing to sacrifice for a better planet? Should they invest in people or in planet? To answer such questions further research is required in at least 2 areas:

1. Other aspects of sustainability besides energy efficiency.
2. Weighing and comparing the different aspects of sustainability now and in the future.

Other Aspects of Sustainability

Above we gave many different measures for energy efficiency. But sustainability is broader. For a complete picture, other aspects of People, Planet, and Profit should also be measured. We mention some possible measures below.

Profit

For the 'profit' part, fairly standardized metrics exist, such as Return on Investment (ROI) or Internal Rate of Return (IRR). However, these standardized metrics can still be hard to predict. Even direct financial effects like the costs for design, build, and test are hard to estimate, as evidenced by the many ICT projects going over budget. For indirect effects, the rebound effect can be a complicating factor. System effects are probably hardest to predict, because one would need to predict long-term structural changes in the whole system.

In the e-invoicing case, digital invoices are much more efficient from a financial viewpoint, especially when digitalizing the whole process. E-invoicing creates the opportunity to digitalize more parts of the process, and if automatic payments and digital approval processes are implemented, positive effects are seen not only in energy efficiency but also in profit. For the Dutch building sector, a sector with 450.000 employees in total, sending an invoice on paper, costs about 25 Euros and sending a digital invoice, including digital approval costs 8 Euros. For a sector with 5000 small, medium, and large size organisations, e-invoicing has the potential of saving 300 million Euros a year (www.bouwendNederland.nl). Mai (2010) calculated a theoretical potential saving of 18 Euros per invoice, for organisations that switch completely from stone-age methods to state-of-the- art technology for invoicing. Besides, they expect much more transparency throughout the process that will create higher quality and fewer mistakes in the process. Interviews with organisations implementing e-invoicing also revealed a few rebound effects. One is that respondents expect more invoices to be sent because of mistakes that cannot be solved informally but always need a credit invoice.

Planet

In this chapter, we equated the Planet aspect of sustainability with 'energy use.' Certainly for ICT energy use receives the most attention, but other factors such as toxicity and water use associated with ICT hardware production also are part of the Planet aspect. We even expect that in the future the focus will shift from energy use to sustainable use of raw materials. Though fossil fuel sources will be depleted, alternatives are already available and these will become more attractive, meaning that the energy supply problem will be 'solved' somehow. For the materials use of the ICT industry, much less alternatives are available.

People

The People aspects are diverse and usually hard to quantify. The Global Reporting Initiative (2011), an international network that produces a widely used sustainability reporting framework, has developed a number of standard indicators. For instance, for employment it gives as indicators the total number of new hires, employee turnover and retention rates after paternal leave. For occupational health and safety, they give as indicator rates of injury, occupational diseases, lost days, and absenteeism, and total number of work-related fatalities, but many of their indicators (such as those on human rights) are less relevant to ICT projects in developed countries. Conversely, the Global Reporting Initiative does not provide indicators for more relevant effects like work quality or job and we are not aware of standards that do. More research should be done on how to measure these more qualitative aspects.

The planet aspect also plays a role in the e-invoicing case. In interviews, we found an effect on work quality. One employee said:

'I used to use 50 to 100 sheets of A4 paper each day. With the e-invoicing system I only use one or two sheets.'

Others told us that their job has less administrative routines, which makes their jobs more interesting. A rebound effect was expected on health: by doing much more work on a computer and on screen, there is a risk of Repetitive Stress Injury (RSI). The usability of the system is therefore a crucial aspect to deal with this rebound effect. Organisations that are implementing e-invoicing at the moment, expect that their employees will need more ICT skills. Additionally, a digital process will enable management on results and outcomes. Giving feedback to employees on their results will increase their human performance. In the long run the large scale adoption of a digital process would lead to less paper mail and could lead to a loss of employment in the postal services.

Comparability

Even if one succeeds in quantifying all relevant effects on People, Planet and Profit, one still has to find a way to weigh them in a decision. Should a company choose a solution that is more energy efficient or a solution that uses less rare materials? Is a more energy efficient process also desirable if it lowers profit?

Indicators in the profit category are easy to compare, because they are all expressed in terms of money. For energy there is also a standard unit: tons of CO_2, but the other indicators for People and Planet can be expressed in many different units and cannot easily be compared. The impossibility to compare the different aspects is a known criticism of the triple base line (Robins, 2006).

One possible strategy to overcome this problem is to convert all indicators to a money value. Some governments have taken this quite far. For instance, in its guidelines for cost-benefits analyses of infrastructure project (ECORYS, 2008), the Dutch government has determined that it values a ton of CO_2 at 50 euros and a ton of NO_x at 8 euros. The value CO_2 is based on the costs that would be necessary to save a ton of CO_2 in some other part of the economy. The cost of NO_x is based mainly on the damage it does to people's health. Other such values are based on the citizens' willingness to pay. We do not know of any companies that have expressed how they value the environment in such hard monetary terms. It would be interesting to explore if similar conversion factors can be developed for other organizations besides governments.

Time is also a factor in weighing planet against profit. The damages caused by climate change due to CO_2 are still mostly in the future. Determining how much we should be willing to spend now to avoid them is not easy. For society it may be better to pursue higher economic growth now and invest its proceeds in CO_2 reduction later, than to reduce CO_2 now and immediately face lower economic growth.

To make such a decision we need to weigh the interests of the current generations against future generations. For profit, it is standard practice to discount future costs and revenues, but much of the damage cause by climate change is so far in the future that it would hardly be counted at normal discount rates. In some studies on climate change, such as the Stern Review (Stern, 2007), a very low discount rate is therefore chosen, but this choice is controversial (Nordhaus, 2007). Such controversies show that it is still difficult for organizations to translate Brundlandts definition of sustainability into practical policy and to really take future generations into account in day-to-day decisions.

CONCLUSION

Making an analysis of the sustainability of ICT is a task where it is easy to get lost in an abundance of confusing data. This chapter introduced two notions that help to make the analysis clearer. The first: sustainability can be assessed on various levels. The energy efficiency of the Power Supply Units of individual computers can be an important metric, but only when the 'higher' levels have been addressed adequately: do we need all those computers? The second: distinguishing between People, Planet and Profit, for direct, indirect and system effects. Confusing direct and indirect effects, or overlooking system effects could lead to insufficiently supported sustainability claims: e.g., working from home will decrease energy spent on travelling, but additional energy needed for lighting and heating at home should be taken into the equation as well.

The ICT sustainability impact framework turned out to be a good framework to show all the effects on e-invoicing compared to invoicing on paper. It enables the designer / project manager to keep a 'total overview' vision on the project that is assessed. It showed us that different designs of work processes can have considerably different sustainability outcomes. Just sending digital invoices has some positive effects on sustainability. However, a fully electronic process including digital approval processes and with automatic payments, decreases energy use and CO_2 emissions by another two thirds and creates a much more efficient process with better quality.

REFERENCES

Berkhout, F., & Hertin, J. (2004). De-materialising and re-materialising: digital technologies and the environment. *Futures*, *36*, 903–920. doi:10.1016/j.futures.2004.01.003

Bomhof, F. W., Van Hoorik, P., & Donkers, M. (2009). Systematic analysis of rebound effects for “greening by ict” initiatives. *Communications & Strategies*, *76*, 77–96.

Ceuppens, L., Kharitonov, D., & Sardella, A. (2008). Power saving strategies and technologies in network equipment opportunities and challenges. In *Proceedings of IEEE Symposium on Applications and the Internet*. IEEE Press.

Climate Group. (2008). *Smart 2020: Enabling the low carbon economy in the information age*. London, UK: The Climate Group.

ECORYS. (2008). *Werkwijzer OEI bij MIT-planstudies, bijlage kengetallen*. Rotterdam, The Netherlands: ECORYS.

Energy Star. (2009). *Program requirements for computers version 5.0*. Retrieved from http://www.energystar.gov.

Global Reporting Initiative. (2011). *Sustainability reporting guidelines, version 3.1*. Retrieved from http://www.globalreporting.org.

Green Grid. (2007). *The green grid power efficiency metrics: PUE and DCiE*. Retrieved from http://www.thegreengrid.org.

Harald, B. (2010). *Results of the expert group on e-invoicing*. Paper presented on Electronic Invoicing in Europe. Madrid, Spain.

Herzberg, F. (1971). *Work and the nature of man*. New York, NY: World Publishing.

Hilty, (2006). The relevance of ICTs for environmental sustainability – A prospective simulation study. *Environmental Modelling & Software*, *21*, 1618–1629. doi:10.1016/j.envsoft.2006.05.007

Hilty, L. M. (2008). *Information technology and sustainability - Essays on the relationship between ICT and sustainable development*. Norderstedt, The Netherlands: Books on Demand GmbH.

Kansal, A., Zhao, F., Liu, J., Kothari, N., & Bhattacharya, A. A. (2010). *Virtual machine power metering and provisioning*. Retrieved from http://research.microsoft.com/pubs/120435/JoulemeterVM.pdf.

Mai, H., & Meyer, T. (2010). *E-invoicing: Final step of an efficient invoicing process*. Berlin, Germany: Deutsche Bank Research.

Meulenhoff, P., Bomhof, F. W., & Van Hoorik, P. M. (2010). *Assessing the positive and negative impacts of ICT on people, planet and profit*. Paper presented on ICGreen. Athens, Greece.

Ritthoff, M., Rohn, H., & Liedtke, C. (2003). *Calculating MIPS – Resource productivity of products and services*. Wuppertal, Germany: Wuppertal Institute.

Robins, F. (2006). The challenge of TBL: A responsibility to whom? *Business and Society Review*, *111*(1), 1–14. doi:10.1111/j.1467-8594.2006.00258.x

Silvius, A. J., & Schipper, R. (2010). *A maturity model for integrating sustainability in projects and project management*. Paper presented on 24th IPMA World Congress. Istanbul, Turkey.

Stern, N. (2007). *The economics of climate change: The stern review*. Cambridge, UK: Cambridge University Press.

van Dam, S. S., Bakker, C. A., & Van Hal, J. D. M. (2010). Home energy monitors: Impact over the medium-term. *Building Research and Information*, *38*(5), 458–469. doi:10.1080/09613218.2010.494832

van Hoorik, P. M., & Bomhof, F. W. (2010). *Sustainability impact of ICT – E-invoicing as a case for a generic framework*. Paper presented at the CMI Conference. Copenhagen, Denmark.

van Hoorik, P. M., van Harmelen, T., Head, M., & Meulenhoff, P. (2010). *Energie- en CO_2-besparing door elektronisch factureren.* Paper presented as a TNO report for the Dutch ministry of Economic Affairs. Delft, The Netherlands.

World in Balance. (1987). *Our common future*. Oxford, UK: Oxford University Press. Retrieved from http://worldinbalance.net/intagreements/1987-brundtland.php.

ADDITIONAL READING

ACEEE. (2008). *Information and communication technologies: The power of productivity, how ICT sectors are transforming the economy while driving gains in energy productivity (report number E081).* Washington, DC: American Council for an Energy-Efficient Economy.

Boulanger, P.-M. (2007). Political uses of social indicators: Overview and application to sustainable development indicators. *International Journal of Sustainable Development, 10*(1-2), 14–32. doi:10.1504/IJSD.2007.014411

Brynjolfsson, E. (1993). The productivity paradox of information technology. *Communications of the ACM, 36*(12), 66–77. doi:10.1145/163298.163309

Carbon Disclosure Project. (2011). *Website*. Retrieved from http://www.cdproject.net.

Cocomo. (2011). *Constructive cost model*. Retrieved from http://sunset.usc.edu/csse/research/COCOMOII/cocomo_main.html.

Global Reporting Initiative. (2011). *Website.* Retrieved from http://www.gri.org.

Green Grid. (2011). *On datacenter efficiency (PUE) and design guidelines.* Retrieved from http://www.thegreengrid.org.

Green House Gas Protocol. (2011). *Webpage*. Retrieved from http://www.ghgprotocol.org.

Hilty, L. M. (2008). *Information technology and sustainability - Essays on the relationship between ICT and sustainable development*. Norderstedt, The Netherlands: Books on Demand GmbH.

Jevons, W. S. (1866). *The coal question* (2nd ed.). London, UK: Macmillan and Co.

OpenDCME. (2011). *OpenDCME model for datacenter efficiency*. Retrieved from http://www.opendcme.org.

Schmidt, A., & Hedal Kløverpris, N. (2009). *Environmental impacts from digital solutions as an alternative to conventional paper-based solutions*. Lyngby, Denmark: FORCE Technology Applied Environmental Assessment.

Singh, R. K., Murty, H. R., Gupta, S. K., & Dikshit, A. K. (2009). An overview of sustainability assessment methodologies. *Ecological Indicators, 9*, 189–212. doi:10.1016/j.ecolind.2008.05.011

Software Metrics. (2011). *Function point analysis training course*. Retrieved from http://www.softwaremetrics.com/Function%20Point%20Training%20Booklet%20New.pdf.

SPEC. (2011). *Benchmarking ICT performance*. Retrieved from http://www.spec.org.

Tenner, E. (1997). *Why things bite back*. New York, NY: Vintage Books.

Triplett, J. E. (1999). The solow productivity paradox: What do computers do to productivity. *The Canadian Journal of Economics. Revue Canadienne d'Economique, 32*(2), 309–334. doi:10.2307/136425

KEY TERMS AND DEFINITIONS

ICT: Information and Communication Technologies; all means to electronically process and transfer information.

E-Invoicing: The process where invoices from one organisation to another is processed by means of ICT.

Sustainability: Development which meets the needs of the present without compromising the ability of future generations to meet their own needs.

People, Planet, Profit: The 'triple base line' that is commonly used to operationalize the concept of sustainability.

Indicator: A measurement of a phenomenon that one wants to assess, that is usually numerical but often does not reflect a precise physical measurement process.

Chapter 2
Information and Communication Technology Revolution and Global Warming

Pavel Somavat
CCS Haryana Agricultural University, India

Vinod Namboodiri
Wichita State University, USA

ABSTRACT

Although the Information and Communication Technology (ICT) revolution is hardly 20 years old, it is already directly influencing the lives of more than two thirds of the global population and has played a prominent role in socio-politico-economical development across the societies. However, over the time the ICT segment has become a major consumer of energy, and therefore, it is rapidly becoming a significant source of greenhouse emissions. All the indicators point towards continued growth in ICT. In this chapter, the authors account for the approximate annual energy consumption and resultant CO_2 emissions of this sector. They also account for the indirect environmental effects of rapid proliferation of ICT technologies. This work, therefore, highlights potential areas where increased energy efficient measures are urgently required to ensure sustainable development. The authors also discuss the positive contribution and potential of ICT in curtailing carbon emissions. Towards the end, they summarize various initiatives by the scientific community to ensure energy efficiency and sustainable development. This chapter gives an overview of the extent of the problem, accounts for energy consumption of various ICT sub-segments, and looks into some of the solutions and future directions proposed by researchers to counter this threat.

INTRODUCTION

Rapid advances in electronics, computing and communications engineering has given rise to an Information and Communication Technology (ICT) revolution which has ushered in a paradigm shift in the way we humans live and go about our day to day work. Computers, mobile phones, PDAs, and Internet have greatly transformed our lives and played a significant role in overall development of humanity. Rapid proliferation of Internet connectivity and mobile telecommunications has not only benefitted the developed societies but

DOI: 10.4018/978-1-4666-1839-8.ch002

has greatly assisted in the socio-politico-economic development of underdeveloped societies as well. Hence it comes as no surprise that they are the developing countries of Asia, Africa and Latin America which are the engines driving further growth and expansion in ICT sector.

This revolution however has a down side as well. Over the last 20 years, not only the number of personal computing and communication devices like computers, laptops, mobile phones and smart phones has greatly increased but also the backbone networks and auxiliary infrastructure like data centers have to be expanded to cater to the ever increasing bandwidth demand. All the indicators point out that these numbers will continue to grow. Unknown to many, all these communication and computing devices coupled with the infrastructure needed to support them has become a major consumer of electricity globally. ICT electronics contain rare earths and other hazardous materials like lead, mercury, arsenic, etc., which further harm environment if disposed without proper processing. So it becomes very clear that ICT sector can no more be ignored when it comes to curbing global greenhouse emissions and protecting the environment from irreparable harm.

The following discussion is an attempt to precisely account for the annual energy consumption and resultant greenhouse emissions due to ICT devices, networks and auxiliary infrastructure in order to understand the real extent of the environmental threat. Although the environmental footprints start right from the mining of the materials required to manufacture the components and goes on long even after these devices are disposed, we concentrate only on the *use phase* of the components and devices. ICT is also playing a part in evolution of newer development paradigms ensuring sustainable growth and we also touch a few of these. Finally, we cover some of the latest solutions suggested by the scientific community to improve energy efficiency and pave the wave for sustainable development in ICT.

ACCOUNTING OF OVERALL ICT ENERGY CONSUMPTION

The accounting of ICT sector energy consumption encompasses three major segments that include Personal Devices, Backbone Networks, and Auxiliary Infrastructure. The following are the main constituents of each segment:

1. **Personal Computing Devices:** The devices segment consists of all the personal computing devices around the globe and combined energy consumption associated with their use. These devices include mobiles phones, desktop computers, and laptops.
2. **Backbone Networks:** This segment consists of Internet infrastructure including all the copper/optical fiber networks and devices like routers, switches, firewalls, repeaters, and hubs, etc. Telecommunication networks include all the Mobile Switching Centers (MSCs), Base Transceiving Stations (BTSs), and corresponding support networks.
3. **Auxiliary Infrastructure:** This segment mainly consists of data centers that are centralized repositories containing data storage and telecommunication equipments and is used for storage, management, and dissemination of data associated with a particular organization, business, or agency. As the demand for information over Internet is rapidly increasing, more and more data centers are being established.

Accounting of Personal Computing Devices

As the technologies are becoming more and more affordable, a large number of people across the globe are able to avail the benefits of ICT revolution. Mobile communication is a very good example of wide acceptance and affordability of one such technology. According to (International Telecommunication Union, 2010) there were about 4.6 billion mobile users worldwide with an estimated penetration of 57 per 100 inhabitants towards the end of 2009. According to another study (British Broadcasting Corporation, 2010) the overall number of mobile users crossed the 5 billion mark in mid 2010, so based upon the numbers we can safely assume the current number of mobile phone users to be around 5.3 billion. With the increasing capabilities of these phones and development of newer applications, more and more people are opting for smart phones. According to life time energy consumption study for a mobile phone (Xun, et al., 2010) it has been estimated that during a life time of three years an average mobile phone requires about 75 kWh including raw materials, manufacturing and transportation. Out of these about 22.5 kWh is required during product usage phase. In same work, the authors estimate the lifetime of a mobile phone to be about two years so the annual energy usage for a mobile phone comes out to be 11 kWh per year, which is also, verified by Somavat, Jadhav, and Namboodiri (2010) whereby the average energy consumption of a mobile phone was calculated to be around 1.2W. Hence, the overall global energy consumption due to the usage of 5.3 billion mobile phones per year can be calculated to be 58 million MWh.

Desktop and Laptop computers are other major devices which have witnessed a tremendous surge in their sales and usage over the last decade. According to (Computer Industry Almanac Inc.), despite its sheer size the personal computer industry will witness yearly 40% growth in worldwide computer sales in the period 2006-12. In 2010, mobile computer sales amounted to 169.8 million pieces which far exceeded the desktop sales at 116.6 million. Therefore, the industry is showing increasing trends towards mobility and miniaturization. From the above-mentioned figures, we can assume the total number of computers in use worldwide to be around 1500 million with approximately 950 million desktops and 550 million laptops. The estimated energy consumption of a laptop is around 35W (Somavat, Jadhav, & Namboodiri, 2010) and assuming that a laptop is used at an average of ten hours per day, we can calculate the overall power consumption of all the laptops over a year. Based on above assumption, a single laptop over a year will consume around 127 kWh. For a total of 550 million laptops, this estimated annual energy consumption is around 69 million MWh. Similarly average power consumption of a desktop varies between 100W for a desktop with LCD display to 150 W for a desktop with CRT display (Pickavet, et al., 2008). For our calculations, we can assume the average power consumption for a desktop to be around 130 W. With an average active use of ten hours per day, a single desktop will consume around 474 kW for the whole year. So the overall annual energy consumption by an estimated 950 million desktop computers worldwide can be calculated to be 450 million MWh.

Accounting of Backbone Networks

Internet and mobile communication networks constitute the backbone networks segment for the purpose of accounting of energy consumption. Both Internet and associated telecommunication networks have played a great role in proliferation of ICT revolution across the globe. Internet access in traditionally developed areas have been provided through fiber and copper networks containing devices like switches, routers, firewalls, repeaters and hubs etc. which require energy for their operation. Although optical networks consume least energy when it comes to the comparison with other access technologies (Hinton, Baliga, Feng, Ayre, & Tucker, 2011), they are mainly restricted to backbone networks and the areas with a higher development and larger population densities. Wireless access technologies like WiMAX and UMTS provide a handsome solution to the users away from the urban areas but these are the most energy intensive of the available technologies. Although equally used in developed societies, mobile access technologies are playing an important role in developing countries where wired access technologies have not penetrated in rural and underdeveloped areas due to economical reasons. With the greatly increasing bandwidth requirements, power consumed in the network components like routers and switches is increasing in proportion (Aleksic, 2008). Hence, energy consumption of backbone networks is rapidly increasing and requires particular attention for the development of energy efficient routing and switching techniques.

In a study (Ericcson, 2007) it was estimated that mobile telecom sector contributed to about 0.12% of the global primary energy use. Taking into account rapid growth of mobile infrastructure in developing countries of Asia, Africa, and Latin America, we can safely assume this number to be around 0.2% in 2011. Although mostly saturated, the developed markets are continuously adding the capabilities to the already established networks in form of newest applications and to support higher bandwidths for mobile Internet access. As more and more people in India and China will become subscribers, mobile infrastructure has to be expanded at a rapid pace in these countries that will put a great strain on energy grids. So from the above data and estimates (Somavat, Jadhav, & Namboodiri, 2010) we can calculate the global electricity consumption due to mobile infrastructure in 2011 to be around 40 million MWh. An estimation of overall energy consumption of the Internet is done (Pickavet, et al., 2008) whereby it was calculated that the overall electricity consumed by Internet in 2008 was around 219 million MWh. Keeping in mind the tremendous growth in Internet users and rapidly increasing demand for bandwidth, it is pretty safe to assume a 15% annual growth in energy consumption by Internet infrastructure. So based on the mentioned numbers the overall energy consumption due to Internet infrastructure in 2011 can be estimated to be 333 million MWh.

Accounting of Auxiliary Infrastructure

An estimate (International Telecommunication Union, 2010) puts the total number of Internet users around the world to be 1.7 billion, which is about 26% of total global population. This ever-increasing number of new users coupled with the popularity of Internet based applications, social networking, increasing need for net-based information interchange, audio and video streaming, online gaming, and e-commerce have greatly increased the required bandwidth and demand for storage of huge amounts of data. Tremendous amount of demand means the energy consumption by the data centers doubled between 2000 and 2005 (Koomey, 2008). Microsoft, Facebook, Google, Ebay, Amazon, and Yahoo are some of the companies that have invested heavily in establishment of huge data centers with large amount of rack space for servers. These data centers are big energy guzzlers and can consume up to 100 times the energy consumed by a normal office. Energy is required not only to power the servers and facilitate the data request handling but also to cool the racks as large amount of heat is generated. Therefore, the overall estimate of energy consumption not only includes energy consumed by the devices but also the energy consumed by HVAC system. According to a study (US Department of Energy) keeping in mind the energy efficient practices, the estimated energy consumption by data centers in USA by the end of year 2011 will amount to about 120 million MWh. Detailed studies (Koomey, 2008) suggest that the overall energy consumption by the data centers in 2005 was around 153 million MWh and estimated an annual increase of 16.7% per year. Building on this study, Pickavet et al. (2008) estimated global energy consumption by the data centers in 2010 to be around 350 million MWh. So based on mentioned numbers, we can estimate the data center energy requirements for the year 2011 to be around 406 million MWh.

Figure 1. Overall electric energy consumed by different ICT sub-segments

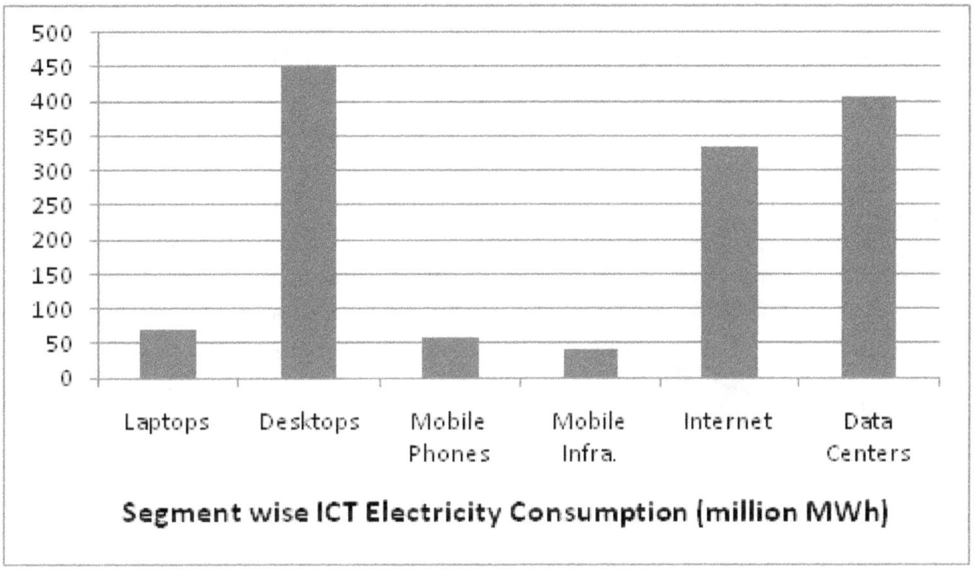

From the above calculations, the overall energy consumption in the personal computing devices segment for the year 2011 comes out to be 577 million MWh, the energy consumption due to backbone infrastructure can be estimated to be 373 million MWh and the energy consumption due to the auxiliary infrastructure comes out to be around 406 million MWh illustrated segment wise in Figure 1. So the total global energy consumption due to combined ICT segment for the year 2011 can be estimated to be 1356 million MWh.

GLOBAL ENERGY CONSUMPTION VS. ICT ENERGY CONSUMPTION

According to a study (International Energy Agency, 2010b), world primary energy demand is set to increase by 36% over a period of 2008 to 2035. This means an average increase of 1.28% per year. In another study (US Census Bureau, 2011) global primary energy consumption in 2010 is estimated to be around 508 quadrillion Btu. Based on these figures the global primary energy consumption for the year 2011 comes out to be 514 quadrillion Btu which is equivalent to 150.63 trillion KWh. According to the same report (US Census Bureau, 2011) the overall global electricity generation in 2008 was about 18.77 trillion kWh. Considering an annual 2% increase every year, this number in 2011 comes out to be around 19.91 trillion kWh. From the above calculations, global electrical energy consumption is about 13.5% of overall global primary energy consumption shown in Figure 2. According to our estimates, overall ICT segment energy consumption in 2011 will be around 1356 million MWh, which is about 1% of the global primary energy consumption and about 7% of the global electricity consumption. Although these numbers appears innocuous in current comparison but keeping in mind the tremendous growth in ICT segment, these numbers are all set to increase in future. Hence, it is high time for the scientific community to focus its attention towards developing energy efficient practices and technologies for sustainable growth in ICT sector.

Figure 2. Electricity as a fraction of global primary energy requirements

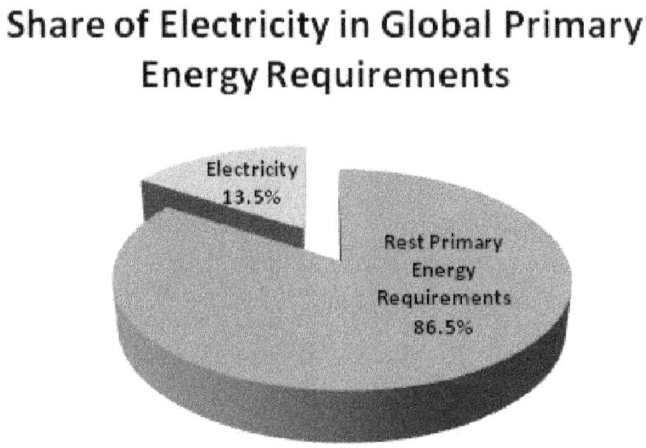

OVERALL ENVIRONMENTAL IMPACT OF ICT SECTOR

ICT industry greatly impacts the environment. The electronic devices and components directly and indirectly affect the environment and these effects can be classified into four different categories:

- Pre-manufacturing energy consumption is the energy consumed during the mining of the metals used to manufacture electronic components. Mining is a very energy intensive process and results in a large amount of CO_2 emissions in addition to generating various other side products harmful to the environment.
- Manufacturing process of these devices is another energy intensive process; sometimes the manufacturing greenhouse emissions exceed the overall greenhouse emissions of the device's active use life. Manufacturing often involves production of harmful wastes.
- Energy consumption by ICT devices during their lifetime is another big issue because of the sheer numbers of the devices and size of support infrastructure. This is the main focus of our current study.
- Energy consumption and hazardous material production during the recycling/disposal phase of the electronic devices.

Pre-Manufacturing Phase

A modern electronic device and its constituent components may contain up to 60 different elements ranging from gold to cobalt, from halogenated compounds to plastics and also heavy metals like lead and selenium. A desktop PC with a CRT display typically contains 20% of its mass in form of iron, 14% aluminum, 7% copper, 6% lead, 2% zinc, and 1% tin besides small quantities of many other metals (Aanstoos, Torres, & Nichols, 1998). Another study (Berkhout & Hertin, 2004) estimates the composition of a typical desktop computer with a CRT display to be metal (43.7%), plastics (23.3%), electronic components (17.3%), and glass (15%). Manufacturing of electronic components starts right from mining, which is another very energy intensive process. According to a report (United Nations Environment Programme, 2009) in the year 2007, almost 1/1000 of the greenhouse emissions were contributed by mining industry. In a study (Mudd, 2007), it was estimated that gold mining typically requires about 143 GJ/kg of gold produced which is equivalent of 39 MWh. According to another study, the mineral industry in Australia utilized approximately 18% of the country's total energy consumption and accounted for greater than 12.3% of the total greenhouse emissions (Department of Climate Change, 2008). Hence, the greenhouse emissions due to ICT sector starts even before the real component manufacturing begins.

Manufacturing Phase

The second phase is the manufacturing of ICT devices and components utilizing products from the mining process. In a study, Williams (2004) found out that manufacturing of a desktop computer with CRT display requires about 6400 MJ, which are equivalent to 1777.7 kWh, which is more than three times the average annual energy consumption during the device use phase. Hence, a computer consumes more energy in the factory than in the use phase assuming that the average lifetime of desktop computer to be around 3 years. Another study (Williams, Ayres, & Heller, 2002) estimated the embodied energy of a typically manufactured small integrated circuit of the size 1.2 cm^2 and having 32 MB memory to be around 9 kWh and requiring 2.3 kg of fossil fuel. So it is clear from the above examples that manufacturing of electronic components and devices is a very energy intensive process therefore we must try to increase the service life of various ICT devices and component rather than disposing them.

Use Phase

Energy consumption by various ICT components during use phase constitutes the most important part of our current discussion. A very large proportion of electricity is generated by burning fossil fuels like coal, petroleum, and gas. Out of all these, coal is the fuel of preference because of its abundance in earth's crust and lower costs although it is the most unclean of the fuels. According to a study (Department of Energy, 2000), the average CO_2 emissions in pounds per kWh for coal is 2.095 (Parliamentary Office of Science and Technology, 2006) for petroleum it is 1.969 and for gas it is 1.321, giving an overall average of 1.297 pounds of CO_2 per kWh of the electricity generated by burning of fossil fuels and other industrial wastes. According to the same study, coal fired power plants generated 51% of overall electric energy in the United States in 1999. According to (International Energy Agency, 2010a) almost 41% of global CO_2 emissions are contributed by electricity and heating sector. So, although electricity is only about 13.5% of global primary energy requirements, its contribution to greenhouse emissions are almost 40%. This disproportionately high number is due to the fact that coal is the main fuel used for generation of electricity. In countries like Australia, China, India, Poland and South Africa, electricity produced by coal fired plants ranges between 69% - 94%. As a matter of fact, these are also the countries which have become development engine of the world and are growing at a rate of more than 8% per annum. In 2008, there was an estimated 29 billion tons of CO_2 emissions globally and since then the emissions have been largely dependent upon the state of global economy. During the financial crisis, greenhouse emissions of the developed economies decreased by 2% while the emissions of developing countries like China and India increased by almost 6% in the same time frame (International Energy Agency, 2010a). Therefore, from the above discussion, we estimate global CO_2 emissions in 2011 to be around 31 billion tons. Hence, 12.4 billion tons of CO_2 emissions in 2011 will be caused solely due to electricity generation as shown in Figure 3, which is barely 13.5% of the global primary energy requirements. This disproportionately large share of electric energy generation towards global CO_2 emissions stresses the need for more efficiency in electricity use. From our calculations 7% of global electricity is currently consumed in ICT sector which means this sector contributes about 868 million tons of annual CO_2 which is almost 3% percent of global CO_2 emissions.

Even though this number is pretty low in comparison but we must keep in mind that these numbers does not include CO_2 emissions caused by the mining industry for the extraction of metals and also the emissions which are caused during the manufacturing of ICT devices and components. One good example where manufacturing process requires more energy than the lifetime use is the case of manufacturing of a desktop computer, which required almost three times of energy used by the device during its life span of three years. Recycling and disposal of electronic equipments also require methods, which need energy consumption. The e-waste is either burned

Figure 3. CO_2 emissions due to electricity consumption vs. all other CO_2 emissions

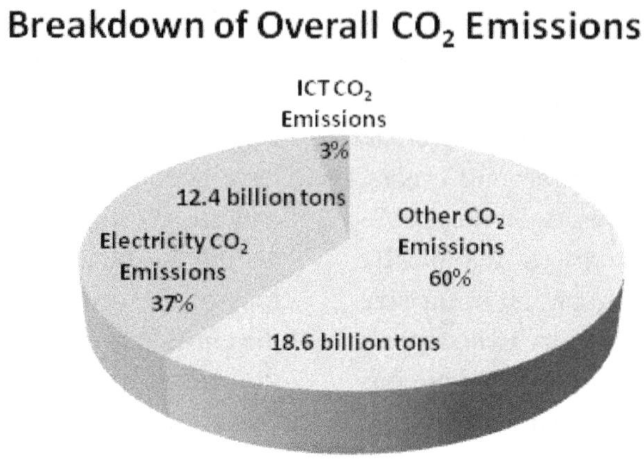

in an open pit or in an incinerator with controlled burning. In both cases, harmful dioxins and furans are generated. With the continued proliferation of ICT in human lives, this contribution towards CO_2 emissions will continue to grow in coming days. So from the above discussion it becomes abundantly clear that any attempts to combat global warming must include special efforts towards decreasing the global electricity consumption as electricity generation contributes to disproportionately higher CO_2 emissions. When it comes to electricity conservation, we must keep a watch on ICT sector, as it is gradually becoming a significant consumer of electricity. So the need of the hour is concerted efforts from the scientific community to work on energy efficiency in ICT sector.

Recycling/Disposal Phase

Disposal of e-waste is another major environmental issue which is a direct side product of ICT revolution. According to a study (Krikke, 2008), every year almost 40 million tons of e-waste is produced and out of which only 20% is properly disposed off and recycled. Computers and mobile phones form bulk of e-waste as their normal life times are shorter (around three years for computers and mobile phones) as compared to TVs, DVDs and other such electronic equipments, which last relatively longer. E-waste has a diverse composition ranging from normal metals like copper, aluminum, and iron to precious metals like gold, silver, and platinum. It also contains heavy metals like mercury, lead, arsenic, cadmium, and selenium in addition to toxic halogenated compounds to act as fire retardants. Developed countries of Europe and North America are the leading producers of e-waste and almost 80% of this waste is exported to developing countries (Schmidt, 2006). China receives almost 70% of all the exported e-waste (Liu, Tanaka, & Matsui, 2006) while other countries importing significant quantities of e-waste are India, Pakistan, Vietnam, Malaysia, and Ghana (Puckett, Westervelt, Gutierrez, & Takamiya, 2005).

E-waste recycling is the process where e-waste components are disassembled and then worked upon to reclaim the potentially useful materials. According to a study (Ladou & Lovegrove, 2008), using recycling we can recover almost 95% of useful materials from a computer and about 45% of materials from CRT monitors. After the removal of all the useful and precious metals, the remaining highly toxic wastes are then mostly dumped into landfills where they keep on polluting the soil and ground water for a long time. Hence, environmentally detrimental effects of e-wastes associated with ICT industry are not just restricted to the manufacturing and the lifetime of the devices but extend even after those devices have long been discarded. One good solution for this problem is that the devices be designed in such a way so that their effective life can be prolonged.

Recycling if done in a proper manner can be a good means of preventing carbon emissions. Metals which can be successfully reclaimed from the e-waste can be utilized once again for manufacturing. Reclaiming metals from the e-waste is less carbon intensive than mining of these metals. According to (Swiss Centre for Life Cycle Inventories, 2006) the CO_2 emissions associated with extraction of precious Platinum Group Metals (PGM) varies from 9380 t CO_2 / t to 13954 t CO_2 / t where as these emissions are 3.4 t CO_2 / t for copper. According to another study (Meskers, Hagelueken, & Van Damme, 2009) at Umicore Precious Metals refining, 300000 tons of e-waste was recycled to reclaim 75000 tons of reusable metal and the CO_2 emissions per ton of the recovered metal was 3.73 t CO_2 / t. This is much less than the average carbon emissions per ton of metal extraction, which is about 17.1 t CO_2 / t for the primary extraction process. Hence recycling the e-waste can be a very attractive option when it comes to preventing CO_2 emissions. Proper recycling procedures can also reduce the dangers of soil and water pollution due to haphazard land filing of these wastes in developing countries.

ENERGY EFFICIENCY IN ICT: GENESIS AND RESEARCH DIRECTIONS

From the above discussion it becomes clear that ICT industry can no more be neglected when it comes to the fight against global warming. The rate, at which the ICT energy requirements are increasing, makes it a potential thrust area for energy efficiency research by the scientific community. ICT encompasses a very diverse field ranging from personal computing devices like computers, mobile phones, smart phones, PDAs to backbone networks like optical and electronic networks containing routers, switches, repeaters etc. This segment also contains the mobile communication networks with all the BTSs, MSCs, etc., and the data centers containing servers to satisfy the data requirements of the Internet generation. Therefore, to be brief, ICT is a very vast field incorporating the latest technological advances in microelectronics, software, manufacturing, signal processing, material sciences, and communication engineering. Because of this diversity of ICT, there is a multitude of opportunities for effecting energy efficiency in various constituent sub-segments.

The potential for energy efficiency in office equipment was identified very early and in 1993 the Energy Star Program which was initiated by United States Environmental Protection Agency (EPA) was extended to include office equipments like computers, printers and monitors. One must note that this program was initiated in 1992 to control the energy consumption and curb the greenhouse emissions from the power plants. Initially the power saving features of the laptops were transferred to desktops and printers with great success and later on governmental regulations made it compulsory for all the federal agencies to buy energy star compliant devices. Therefore, from its humble start as a voluntary program, Energy Star has become a global movement covering almost 40,000 components ranging from commercial to residential markets with thousands of program partners (Brown, Webber, & Koomey, 2000). According to Environmental Protection Agency (2010) in 2009 the program resulted in financial savings over \$17 billion in form of energy bills and simultaneously prevented more than 45 million tons of greenhouse emissions. This huge success of Energy Star program provides an excellent motivation to carry on further work in the direction of energy efficiency in ICT sector.

One important requirement in determining the effectiveness of energy efficient solutions is the measurement of the energy and carbon footprint of the solution. Hence, with improving the energy efficiency of a particular component there is also a need to quantize the effectiveness of that solution. Such metrics will be helpful both for the manufacturers and also for the regulators looking after the energy efficiency issues. The manufacturers can better market their products based on the performance and carbon footprint of the product while the regulators can identify the manufacturers and encourage them to promote efficiency measures (Heddeghem, Vereecken, Pickavet, & Demeester, 2009). Although it is a pretty complex task to precisely account for greenhouse emission footprints, there have been quite a few initiatives towards the standardization. Power Usage Effectiveness (PUE) is one such initiative proposed by a group called "Green Grid" (Belady, Rawson, Pfleuger, & Cader, 2008). PUE is basically a measure of how efficiently a data center uses the power or how much power is actually used by the data servers in the data center. It is the ratio of total power consumed in the data center to the power delivered to the servers and other computing equipments. Energy Consumption Rating (ECR) is another such initiative, which provides a framework for measuring the energy efficiency of network and telecommunication devices (Energy Consumption Rating Initiative, 2009). Yet another metrics is Performance per Watt, which is used in computing to measure the energy efficiency of particular computer architecture or a hardware component. It measures the computation rate of computer for every watt of power consumed by the system. The performance metrics vary from Floating Point Operations Per Second (FLOPS), Million Instructions Per Second (MIPS), or any other performance benchmark score depending upon the domain (Segars, 2001).

Energy Efficiency Research at Hardware/Component Level

The quest for energy efficiency starts right from the component level. It has become pretty clear that developing energy efficient solution is critical for the survival of semiconductor industry. A good example of this type of development is multicore technology by Intel. The latest Blade and Rack servers incorporate this multicore technology and deliver much higher performances and have lesser footprint. Most simple and attractive idea for device power management includes temporarily shutting down the equipment when there is no utilization and to slowdown processing when less performance is enough. Some other innovations proposed by Intel include strained silicon technology, which increases the transistor performance by 30% for same level of energy consumption. Mobile voltage positioning and Speed Step technologies also minimizes the processor power consumption. Speed Step dynamically scales frequency and voltage depending upon demand while in second case processor voltage is adjusted depending upon the demand (Higgs, 2007). In another study (Dreslinski, Wieckowski, Blaauw, Sylvester, & Mudge, 2010), Near Threshold Technology (NTS) has been proposed where the CMOS based digital processors are operated at way below typical supply voltage of 200 mV (Soeleman & Roy, 1999), NTS offers up to 10 times or higher energy efficiency.

Other solutions for effecting energy efficiency at chip level are Dynamic Voltage Scaling (DVM) and Dynamic Frequency Scaling (DFS). Combined together these are called Dynamic Voltage and Frequency Management (DVFM). In a CMOS based design, power dissipated is proportional to the square of voltage multiplied by frequency therefore dynamically decreasing these parameters depending upon application can save power in energy restrained environments like laptops and mobile phones. In DVM, the voltage of the component is scaled up or down depending upon the immediate requirements. Lower voltage decreases the power consumption but affects the performance. DFS, also called *CPU throttling,* is a technique whereby the microprocessor frequency can be adjusted depending upon the processing requirements. Decreasing the frequency results in lesser number of instructions carried per unit time. The frequency of a processor can be decreased to conserve energy or to decrease the amount of heat generated when the processing requirements are lower. Generally, both DVM and DFS are used in a combination to save power when the processing requirements on a component are less. Power supplies are the devices, which convert ac power from the grid into dc power with different voltages required to run computers and servers. Highly efficient power supplies have the potential to save considerable amount of energy. 80 plus certified power supplies ensures 80% or greater conversion efficiency at various loads with a power factor of 0.9 and above (Plug Load Solutions, 2011). Another interesting concept is asymmetrical multi-core design, which allows the processor to operate under heavy loads at a performance level of nearly two equal cores while consuming less energy and with lower heat dissipation (Ghiasi & Grunwald). Low power logic circuit design (Starzyk & He, 2007) is another promising approach, which intends to do away with the switching losses associated with CMOS logic circuits.

Energy Efficiency Research at Backbone Networks Level

Because of the speed with which the bandwidth requirements are increasing, backbone networks are under consistent strain not only to satisfy the user demands but also to do this in an energy efficient way. The need for energy efficiency in the Internet was stressed way back in 2003 (Gupta & Singh, 2003). There is a twofold motivation to incorporate energy efficiency in backbone networks, firstly due to increasing power bills the telecom companies are forced to seek energy efficient solutions for cutting expenditure and secondly to reduce the carbon footprint of the networks. In addition to these, there are legitimate concerns about the future networks being able to handle the increasing bandwidth requirements in an energy efficient manner. In the network core, one solution to achieve energy efficiency is to selectively switch off the idle network elements when the network utilization is low (during night-time) while at the same time retaining the ability to traffic the remaining data (Zhang, Chowdhury, Tornatore, & Mukerjee, 2010). In a comparative study (Aleksic, 2009) analyzing the architectures of high capacity future networks, it was found out that future routers and switches employing optical technologies will be more energy efficient than their electronic counterparts. Another study (Baliga, Ayre, Hinton, & Tucker, 2007) recommended the use of pure optical switching networks to satisfy the future bandwidth requirements in an energy efficient manner. It was realized that circuit switching has a higher efficiency in optical domain and in order to visualize a highly energy efficient architecture, a hybrid system involving packet, circuit and burst switching must be realized in core network. According to International Telecommunication Union (2008) the telecommunications industry is migrating to a single unified IP-based next generation network (NGN) which will significantly decrease the number of switching centers. The switching devices will be more tolerant to climatic conditions thereby removing the necessity of air conditioning. In addition, the NGNs will use latest standards like VDSL2 (ITU-T G.993.2) which has provisions for three different power modes (full, low power, and sleep) while the older VDSL had only full power mode. By introducing these innovations, ITU indents to decrease the carbon footprint of the telecom sector in future networks.

There have been attempts towards more scalable networks with simple device architecture where the traffic is forwarded at packet level. A proposal for traffic lookup has been suggested (Roberts, 2009) whereby next generation routers will forward the traffic at flow levels. On the similar lines (Baldi & Ofek, 2009) suggested a synchronous time based IP switching which offers advance synchronization of the routers and traffic scheduling. This approach is based on pipeline forwarding and will reduce header processing and decrease the complexities in buffer size, speed up the switching fabric and memory access bandwidth is also improved. Sleep/Standby approaches are also proposed for the networks on the lines of power management whereby the network components or parts are allowed to go to low energy states when there is no utilization. However, a node in sleep mode is against the concept of 24-hour availability and has performance constraints. Solutions have been suggested (Christensen, Gunaratne, Nordman, & George, 2005) in form of a proxy named Network Connectivity Proxy (NCP) to keep the track of nodes in sleep mode and waking them up at the time of need. Study regarding the possibility of implementing low power idle mode in Optical Network Units (ONUs) has been done (Shing-Wa, Valcarenghi, She-Hwa, Campelo, Yamashita, & Kazovsky, 2009) for energy efficiency in Passive Optical Networks. Results from the analysis show that there are significant energy savings while using the proposed architectures, which reduce clock recovery overheads. Another novel scheme has been proposed (Park, Choi, Lee, & Park, 2008) where the power consumed by Home Gateway architecture is decreased by using sleep and wake-up triggers activated by the presence of real user service traffic on the interface.

Mobile Communications sector provides a different sort of challenge when it comes to energy efficiency. With more than half of the global population already, the users of mobile phones and additional millions becoming subscribers every year in countries like China and India, this sector is becoming a major consumer of electrical energy. One of the main factors for the popularity of mobile phones is development of highly compact and energy efficient end terminals. Traditionally energy efficiency issues were neglected in access networks and in typical mobile environment, BTS or Base Stations are major consumers of energy (Gruber, Blume, Ferling, Zeller, Imran, & Strinati, 2009). However, with the penetration of mobile communications into hitherto underdeveloped areas, the providers have to set up BTSs away from the power grids where diesel generators and alternate sources are required to power the system. Hence, additional costs have lately provided the motivation to the service providers to seek energy efficiency in access networks. For the operation of many off-grid sites, energy provisioning may cost up to 50% of total operational costs (Correia, et al., 2010). In case of mobile phones, major CO_2 emissions are caused during the manufacturing phase while usage phase energy consumption is less because of design optimization and relatively lesser life time. Access networks are the biggest culprits in the mobile communication as they have minimal provisioning for energy efficiency. A glaring example is DC power consumption of a typical three-sector site, which consumes about 50% of the peak power even under zero load conditions (Ferling, et al., 2010). Mobile networks normally have slowly changing load pattern as well as highly dynamic fluctuations whereas the systems are configured rather statically (Auer, et al., 2010). This means there is a good scope of network energy efficiency with energy efficient Radio Resource Management (RRM).

Good alternate energy solutions in form of solar or eolic generators have been proposed for BTSs situated in off-grid areas to cater the need of rural populations (Galluccio, Leonardi, Morabito, & Palazzo, 2008). Power consumption analysis of typical radio access equipment shows that significant amount of power (50-80%) is consumed by the power amplifier including the feeder (Correia, et al., 2010). However, power consumption in cellular networks is independent of the traffic load. Therefore, the need of the hour is development of advanced power amplifiers with power management concepts incorporating adaptability to changing traffic. In GSM/WCDMA networks one good option is to selectively shut down and turn on some of the transceivers, cells and backhaul links during night time when the traffic is low (Ericcson, 2007). In a study (Bambos, 1998) the potential of adaptive power control in wireless networking has been assessed for the design of energy efficient network. Another approach using cooperative scheduling has been proposed (Han, Himsoon, Siriwongpairat, & Liu, 2005) where energy efficiency analysis in case of simple two user scenario has been done. The effects of smaller cell size on the energy efficiency of mobile communication has been studied (Leem, Baek, & Sung, 2010) and the results show that as the cell size decreases the energy efficiency and the system capacity increases even though the numbers of BTSs need to be increased. A joint European initiative with 15 partners from 10 different countries called "Energy Aware Radio and Network Technologies" (EARTH) is a leading force in the direction of mobile communication energy efficiency research (EARTH, 2010).

Energy Efficiency Research in Auxiliary Infrastructure

Popularity of multimedia rich Internet applications, online multimedia sharing and demand for online video is forcing the companies and service providers to invest very heavily in data centers. There are indications that by 2014, 91% of all the Internet traffic will consist of video data in form of Internet TV, video on demand, Internet and P2P (Cisco, 2010). The concept of cloud computing is also based up on massive computing and storage capacity provided by these data centers. Because of this tremendous increase in data demand, data centers have become a major contributor to greenhouse emissions. The number of servers in a data center has reached the order of 10,000 and much larger data centers with up to 150,000 servers are in the line (Church, Greenberg, & Hamilton, 2008). Not only the numbers of the data centers and the servers have gone up, the power density per rack has also exploded over last few years. In 2004, the power density per rack was about 10 kW/rack (Church, Greenberg, & Hamilton, 2008) which increased to 55 kW/rack with in a short span of 2-3 years (Ranganathan, Leech, Irwin, & Chase, 2006). However, with various efficiency measures in place, the current consumption is between 25-30 KW/rack. So it becomes very clear from the above numbers that when it comes to tackling greenhouse emissions, data centers present quite a challenge.

Almost 50% of energy in data centers is required for cooling (US Environmental Protection Agency, 2007) and another major fraction is wasted on idle or under-utilized resources. Currently the average server utilization in traditional data centers lie between 20-30% as the hardware devices draw about 50% of the peak power in their idle state (Barroso & Holzle, 2007). In order to overcome this shortcoming, adjustable power states namely active and inactive have been widely implemented in newer devices as specified in Advanced Configuration and Power Interface (ACPI) (Benini, Bogliolo, & De Micheli, 2000). Dynamic Voltage and Frequency Scaling (DVFS) is used to reduce the device voltage during the times of lower utilization. A similar concept has been used in the disks to achieve low power active state (Carrera, Pinheiro, & Bianchini, 2003). Energy efficiency can be attained by keeping more servers or devices in inactive stage for longer periods of time. Load concentration can be used in disk arrays where cache discs can be used to keep frequently requested data leaving the other discs to be inactive (Zhu, Shankar, & Zhou, 2004). Server consolidation at the data center level can be achieved by transferring the workload to small number of servers, which are active while others can be made inactive. This concept manifests itself in virtualization, which separates applications from the underlying hardware and supports resource sharing. According to a study investigating the energy efficiency of virtualization in data centers, (Pretorius, Ghassemian, & Ierotheou, 2010) it was found out that virtualization prevents almost 30% of greenhouse emissions.

As higher density server designs like in blade architecture are coming to the fore, the power density associated with the servers has also increased and therefore there are ever-present chances of thermal failures. Hence, there is a need for cooling the server racks which results in disproportionately higher energy consumption in data center cooling systems. Variable speed fans have been proposed to ensure that the fans are started and their speed can be controlled according to the requirement of dissipation of heat from the server racks. In blade architecture, a number of fans are employed in a single rack and each fan is controlled according to the area specific thermal condition (Tolia, Wang, Marwah, Bash, Ranganathan, & Zhu, 2008). Water based cooling method is much better than the air based cooling but it has not gained much popularity due to complexity. A detailed study (Hamann, 2008) presented a sensing and measurement technique to ascertain data center heat distribution in 3-D. Results can be used to employ efficient cooling. There are losses due to AC to DC power conversion in the data centers, which can be reduced by improving the efficiency of power provisioning. Since the implementation of energy efficient states is less effective in components other than CPU, experts are calling for radical redesign in memory and disk subsystems (Barroso & Holzle, 2007). Another interesting proposal involves the study of energy efficiency of various content dissemination strategies and recommended content-centric networking model over the host-oriented model (Lee, Rimac, Kilper, & Hilt, 2011).

Therefore, with this we conclude our study of latest research efforts in the direction of energy efficiency at component/device, backbone networks and auxiliary infrastructure levels. The need of the hour is concerted efforts from the researchers to work on and develop newer methodologies to combat greenhouse emissions in the ICT sector and to allow for continued sustainable development.

CONTRIBUTION OF ICT IN CURTAILING CARBON EMISSIONS

Although on one hand the expansion in ICT sector presents a formidable challenge when it comes to the carbon emissions, on the other hand ICT also provides a great opportunity to curb carbon emissions and ensure sustainable development. ICT has ensured newer growth paradigms which promise greater energy efficiency. True potential of the savings due to ICT can be gauged from the report (Global e-Sustainability Initiative, 2008) which states "ICT could cut global business as usual greenhouse emissions by 15% and save up to € 600 billion by 2020." Some of the products and services available in markets that employ the power of ICT to curb carbon emissions are home energy monitoring kits, HVAC automation systems, e-health delivery system, eco driving software solutions, telecommuting and telepresence. Various different segments where the potential and outreach of the ICT can be utilized to ensure sustainable development are as under:

1. **Spreading Environmental Awareness**: Huge penetration of Internet and mobile communication can be used as a tool to spread environmental awareness among all the sections of the society. Using these technologies, environment related information can be disseminated among the urban/rural areas and people can be informed and motivated to make a contribution towards cutting carbon emissions.
2. **Smart Logistics:** Logistics is another area where the outreach of the ICT can be utilized to ensure financial and carbon savings. Better route and load planning can be done using custom made programs and hence

considerable fuel savings can be realized. Management software can also assist in increasing the overall inventory efficiency and decreasing the number of vehicles travelling with less than optimum cargo. Other benefits include minimum packaging requirements and reduction in inventory space.

3. **Smart Buildings:** ICT can be employed in constructing smarter structures for residential and commercial use. These buildings incorporate solutions that maximize the energy efficiency and building management solutions for optimal HVAC control. Structure design can be improved to optimize space utilization and elements like smart lighting control and ventilation on demand can be incorporated. Hence smart buildings can ensure considerable carbon savings in the future.
4. **Smart Grids:** Observation and intelligent management of power distribution system is another area where ICT can be greatly utilized for improving efficiency. Smart Grids consisting of various sensors and controls can ensure efficient transmission and distribution of energy and can precisely detect the position of any faults or outages. In the developing countries like India, transmission and distribution losses contribute to almost 30-40% of the total energy used (Hanna, 2010). These numbers also include electricity theft by the users during distribution. Hence ICT enabled Smart grids can be used not only to ensure efficient distribution but also to keep a tab on unscrupulous users.
5. **Dematerialization:** Rapid proliferation of high-speed broadband has ensured the so-called "Dematerialization" whereby the traditional way of doing the things have been made redundant. Some of the examples of dematerialization include replacement of paper based billing system with e-billing, video and audio conferencing which curtails the need for air travel; telecommuting involves working from house rather than travelling to the office, e-commerce, e-taxation and e-papers etc. According to a study (Tenhunen & Penttinen, 2010) use of e-invoicing instead of traditional paper based invoices saves almost 63% of carbon emissions per invoice. Hence we must ensure that the traditional paper based invoicing mechanisms be replaced with more efficient e-invoices when it comes to utility bills, banking records and for other commercial transactions. Similarly, a large amount of carbon emissions can be preventing by using video and audio conferencing rather than traveling to distant places. Telecommuting is another promising way to work from home and thus making every day travel to the office redundant. Similar other initiatives like e-commerce, e-banking and e-health delivery systems are reducing the need for travel and the paper work involved thereby ensuring energy efficiency.

Hence, from the above discussion it is abundantly clear that ICT itself is making significant contributions to the sustainable development. However, the true potential in this regard is still not realized and further work is required to explore and implement newer paradigms for containing carbon emissions.

CONCLUSION

ICT sector is expanding rapidly, becoming a cause of real environmental concern due to disproportionately high greenhouse emissions associated with use of electric energy. Although electricity is only 13.5% of global primary energy requirements, its contribution to greenhouse emissions is almost 40%. This is mainly because of predominant use of coal as a fuel. According to our calculations for 2011, almost 7% of the global electricity will be consumed by ICT sector and this will result in 3% of global greenhouse emissions. These figures only account for the *use phase* of the ICT devices and corresponding infrastructure and do not give a complete *lifecycle view*. If the energy consumed during the pre-manufacturing, manufacturing and recycling is considered, these numbers will be much higher. Therefore, if the rate at which ICT sector is currently expanding goes on without the introduction of energy efficiency measures, this will be a real environmental concern and existential threat will loom over many ICT related industries.

Therefore, the need of the hour is better understanding of the gravity of the problem and concerted efforts from the scientific community to introduce newer energy efficient paradigms and practices to ensure sustainable growth. Since ICT is a very vast field and energy consumption is also distributed across a number of components and processes, efforts are required from people with different specializations to achieve the overall energy efficiency. In some cases manufacturing of the devices result in more CO_2 emissions than the total lifetime use, so the average lifetime of the devices must be increased. Recycling and reuse of the constituent material results in considerable greenhouse savings, this practice should be encouraged. ICT itself has a great potential to contribute towards sustainable development. E-billing, e-commerce, telecommuting, and telepresence must be promoted. At the end, some of the latest energy efficient solutions proposed across the ICT sub-segments have been covered to acquaint the reader with the latest trends and future directions in ICT energy efficiency research.

REFERENCES

Aanstoos, T., Torres, V., & Nichols, S. (1998). Energy model for end-of-life computer disposition. *IEEE Transactions on Components, Packaging and Manufacturing Technology*, *21*(4), 295–301. doi:10.1109/3476.739174

Aleksic, S. (2008). Power consumption issues in future high-performance. In *Proceedings of the 10th Anniversary International Conference on Transparent Optical Networks,* (pp. 194-198). IEEE Press.

Aleksic, S. (2009). Analysis of power consumption in future high-capacity network nodes. *Journal of Optical Communications and Networking*, *1*(3), 245–258. doi:10.1364/JOCN.1.000245

Auer, G., Gódor, I., Hévizi, L., Imran, M., Malmodin, J., Fazekas, P., et al. (2010). Enablers for energy efficient wireless networks. In *Proceedings of the 72nd Vehicular Traffic Conference,* (pp. 1-5). IEEE Press.

Baldi, M., & Ofek, Y. (2009). Time for a greener internet. In *Proceedings of Green Communications (IEEE ICC 2009)*. Dresden, Germany: IEEE Press.

Baliga, J., Ayre, R., Hinton, K., & Tucker, R. (2007). Photonic switching and the energy bottleneck. In *Proceedings of International Conference on Photonics in Switching.* San Francisco, CA: IEEE.

Bambos, N. (1998). Toward power-sensitive network architectures in wireless communications: Concepts, issues, and design aspects. *IEEE Personal Communications*, *5*(3), 50–59. doi:10.1109/98.683739

Barroso, L., & Holzle, U. (2007). The case for energy-proportional computing. *IEEE Computer*, *40*(12), 33–37. doi:10.1109/MC.2007.443

Belady, C., Rawson, A., Pfleuger, J., & Cader, T. (2008). *Green grid data center power efficiency metrics: PUE and DCIE.* Retrieved May 10, 2011, from http://www.thegreengrid.org/~/media/WhitePapers/White_Paper_6__PUE_and_DCiE_Eff_Metrics_30_December_2008.ashx?lang=en.

Benini, L., Bogliolo, A., & De Micheli, G. (2000). A survey of design techniques for system-level dynamic power management. *IEEE Transactions on Very Large Scale Integration (VLSI) Systems*, *8*(3), 299–316.

Berkhout, F., & Hertin, J. (2004). De-materialising and re-materializing: Digital technologies and the environment. *Futures*, *36*(8), 903–920. doi:10.1016/j.futures.2004.01.003

British Broadcasting Corporation. (2010, July 9). *News technology*. Retrieved May 4, 2011, from http://www.bbc.co.uk/news/10569081.

Brown, R., Webber, C., & Koomey, J. (2000). *Status and future directions of the ENERGY STAR program.* Washington, DC: ACEEE Summer Study.

Carrera, E., Pinheiro, E., & Bianchini, R. (2003). Conserving disk energy in network servers. In *Proceedings of 17th International Conference on Supercomputing.* San Francisco, CA: ACM.

Christensen, K., Gunaratne, C., Nordman, B., & George, A. (2005). Managing energy consumption costs in desktop PCs and LAN switches with proxying split TCP connections, and scaling of link speed. *International Journal of Network Management*, *15*(5), 297–310. doi:10.1002/nem.565

Church, K., Greenberg, A., & Hamilton, J. (2008). On delivering embarrasingly distributed cloud services. In *Proceedings of ACM HotNets.* Calgary, Canada: ACM.

Cisco. (2010). *Cisco visual networking index.* Retrieved May 10, 2011, from http://www.cisco.com/en/US/solutions/collateral/ns341/ns525/ns537/ns705/ns827/white_paper_c11-481360_ns827_Networking_Solutions_White_Paper.html.

Computer Industry Almanac Inc. (2011). *Computer industry almanac*. Retrieved May 4, 2011, from http://www.c-i-a.com/worldwideuseexec.htm.

Correia, L., Zeller, D., Blume, O., Ferling, D., Jading, Y., & Godor, I. (2010). Challenges and enabling technologies for energy aware mobile radio networks. *IEEE Communications Magazine*, *48*(11), 66–72. doi:10.1109/MCOM.2010.5621969

Department of Climate Change. (2008). *Australia's national greenhouse accounts: National inventory by economic sector 2006.* Canberra, Australia: Commonwealth of Australia, Department of Climate Change.

Department of Energy. (2000). *Carbon dioxide emissions from the generation of electric power in United States*. Washington, DC: Environmental Protection Agency, Department of Energy.

Dreslinski, R., Wieckowski, M., Blaauw, D., Sylvester, D., & Mudge, T. (2010). Near-threshold computing: Reclaiming Moore's law through energy efficient integrated circuits. *Proceedings of the IEEE*, *98*(2), 253–266. doi:10.1109/JPROC.2009.2034764

EARTH. (2010). *Energy Aware Radio and Network Technologies*. Retrieved May 10, 2011, from https://www.ict-earth.eu/

Energy Consumption Rating Initiative. (2009). *About the ECR.* Retrieved May 10, 2011, from http://www.ecrinitiative.org/about/.

Environmental Protection Agency. (2010). *ENERGY STAR overview of 2009 achievements.* Retrieved May 10, 2011, from http://www.energystar.gov/ia/partners/annualreports/2009_achievements.pdf.

Ericcson. (2007). *Sutainable energy use in mobile communications.* Retrieved May 8, 2011, from http://www.ericsson.com/campaign/sustainable_mobile_communications/downloads/sustainable_energy.pdf.

Ferling, D., Bohn, T., Zeller, D., Frenger, P., Gódor, I., & Jading, Y. (2010). Energy efficiency approaches for radio nodes. In *Proceedings of the Future Network & Mobile Summit, 2010*, 1–9.

Galluccio, L., Leonardi, A., Morabito, G., & Palazzo, S. (2008). Timely and energy-efficient communications in rural infostation systems. *IEEE Wireless Communications*, *15*(3), 48–53. doi:10.1109/MWC.2008.4547522

Ghiasi, S., & Grunwald, D. (2011). *Aide de camp: Asymmetric dual core design for power and energy reduction.* Retrieved May 10, 2011, from http://citeseerx.ist.psu.edu/viewdoc/summary?doi=10.1.1.86.6806.

Global e-Sustainability Initiative. (2008). *SMART 2020: Enabling the low carbon economy in the information age.* Retrieved August 22, 2011, from http://www.smart2020.org/_assets/files/Smart2020UnitedStatesReportAddendum.pdf.

Gruber, M., Blume, O., Ferling, D., Zeller, D., Imran, M., & Strinati, E. (2009). EARTH — Energy aware radio and network technologies. In *Proceedings of the 20th International Symposium on Personal, Indoor and Mobile Radio Communications,* (pp. 1-5). Cannes, France: IEEE Press.

Gupta, M., & Singh, S. (2003). Greening of the internet. In *Proceedings of ACM SIGCOMM,* (pp. 361-371). Karlsruhe, Germany: ACM Press.

Hamann, H. (2008). A measurement-based method for improving data center energy efficiency. In *Proceedings of the International Conference on Sensor Networks, Ubiquitous and Trustworthy Computing,* (pp. 312-313). IEEE Press.

Han, Z., Himsoon, T., Siriwongpairat, W., & Liu, K. (2005). Energy-efficient cooperative transmission over multiuser OFDM networks: Who helps whom and how to cooperate. In *Proceedings of the Wireless Communications and Networking Conference,* (pp. 1030-1035). IEEE Press.

Hanna, N. K. (2010). *E-transformation: Enabling new development strategies*. Berlin, Germany: Springer Science. doi:10.1007/978-1-4419-1185-8

Heddeghem, W., Vereecken, W., Pickavet, M., & Demeester, P. (2009). Energy in ICT - Trends and research directions. In *Proceedings of the 3nd International Symposium on Advanced Networks and Telecommunication Systems.* New Delhi, India: IEEE Press.

Higgs, T. (2007). Energy efficient computing. In *Proceedings of the International Symposium on Electronics & the Environment,* (pp. 210-215). Orlando, FL: IEEE Press.

Hinton, K., Baliga, J., Feng, M., Ayre, R., & Tucker, R. (2011). Power consumption and energy efficiency in the internet. *IEEE Network*, *25*(2), 6–12. doi:10.1109/MNET.2011.5730522

International Energy Agency. (2010a). *CO2 emssions from fuel combustion highlights*. Paris, France: International Energy Agency.

International Energy Agency. (2010b). *World energy outlook 2010 factsheet.* Paris, France: OECD/IEA.

International Telecommunication Union. (2008). *ICTs and climate change*. Geneva, Switzerland: ITU.

International Telecommunication Union. (2010). *ITU.* Retrieved April 5, 2011, from http://www.itu.int/newsroom/press_releases/2010/pdf/PR08_ExecSum.pdf.

Koomey, J. G. (2008). Worldwide electricity used in data centers. *Environmental Research Letters*. Retrieved from http://files.campus.edublogs.org/uowblogs.com/dist/d/75/files/2010/04/Koomey.pdf.

Krikke, J. (2008). Recycling e-waste: The sky is the limit. *IT Professional*, *10*(1), 51–55. doi:10.1109/MITP.2008.16

Ladou, J., & Lovegrove, S. (2008). Export of electronics equipment waste. *International Journal of Hygiene and Environmental Health, 14*(1), 1–10.

Lee, U., Rimac, I., Kilper, D., & Hilt, V. (2011). Toward energy-efficient content dissemination. *IEEE Network, 25*(2), 14–19. doi:10.1109/MNET.2011.5730523

Leem, H., Baek, S. Y., & Sung, D. K. (2010). The effects of cell size on energy saving, system capacity, and per-energy capacity. In *Proceedings of the Wireless Communications and Networking Conference,* (pp. 1-6). IEEE Press.

Liu, X., Tanaka, M., & Matsui, Y. (2006). Generation amount prediction and material flow analysis of electronic waste: A case study in Beijing, China. *Waste Management & Research, 24*(5), 434–445. doi:10.1177/0734242X06067449

Meskers, C., Hagelueken, C., & Van Damme, G. (2009). Green recycling of EEE: Special and precious metal recovery from EEE. In *Proceedings of Extraction & Processing Division of The Minerals, Metals & Materials Society* (pp. 1131–1136). San Francisco, CA: TMS.

Mudd, G. (2007). Global trends in gold mining: Towards quantifying environmental and resource sustainability? *Resources Policy, 32*(1-2), 42–56. doi:10.1016/j.resourpol.2007.05.002

Park, W., Choi, C., Lee, H., & Park, K. (2008). Energy efficient home gateway based on user service traffic in always-on home network environment. In *Proceedings of International Conference Advances in Electronics and Micro-Electronics,* (pp. 121-125). Valencia, Spain: IEEE.

Parliamentary Office of Science and Technology. (2006). *Carbon footprint of electricity generation*. London, UK: UK Parliament.

Pickavet, M. Vereecken., W., Demeyer, S., Audenart, P., Vermeulen, B., Develder, C., et al. (2008). Worldwide energy needs for ICT: The rise of power-aware networking. In *Proceedings of the 2nd International Symposium on Advanced Networks and Telecommunication Systems,* (pp. 1-3). Mumbai, India: IEEE Press.

Plug Load Solutions. (2011). *80 plus certified power supplies & manufacturers*. Retrieved August 21, 2011, from http://www.plugloadsolutions.com/80PlusPowerSupplies.aspx.

Pretorius, M., Ghassemian, M., & Ierotheou, C. (2010). An investigation into energy efficiency of data centre virtualisation. In *Proceedings of the International Conference on P2P, Parallel, Grid, Cloud and Internet Computing,* (pp. 157-163). IEEE Press.

Puckett, J., Westervelt, S., Gutierrez, R., & Takamiya, Y. (2005). *The digital dump: Exporting re-use and abuse to Africa.* Seattle, WA: Basel Action Network.

Ranganathan, P., Leech, P., Irwin, D., & Chase, J. (2006). Ensemble-level power management for dense blade servers. In *Proceedings of the 33rd International Symposium on Computer Architecture,* (pp. 66-77). IEEE Press.

Roberts, L. (2009). A radical new router. *IEEE Spectrum, 46*(7), 34–39. doi:10.1109/MSPEC.2009.5109450

Schmidt, W. (2006). Unfair trade - E-waste in Africa. *Environmental Health Perspectives, 114*(4), 232–235. doi:10.1289/ehp.114-a232

Segars, S. (2001). *Low power design techniques for microprocessors.* Retrieved May 10, 2011, from http://www.asicfpga.com/site_upgrade/asicfpga/pds/ip_pds_files/SS_ISSCC2001.pdf.

Shing-Wa, W., Valcarenghi, L., She-Hwa, Y., Campelo, D., Yamashita, S., & Kazovsky, L. (2009). Sleep mode for energy saving PONs: Advantages and drawbacks. In *Proceeding of 2nd Green Communications Workshop*. Honolulu, HI: IEEE Press.

Soeleman, H., & Roy, K. (1999). Ultra-low power digital subthreshold logic circuits. In *Proceedings of the International Symposium on Low Power Electronics Design,* (pp. 94-96). IEEE/ACM.

Somavat, P., Jadhav, S., & Namboodiri, V. (2010). Energy consumption of personal computing including portable communication devices. In *Proceedings of 1st International Conference on Energy-Efficient Computing and Networking,* (pp. 141-149). New York, NY: ACM Press.

Starzyk, J., & He, H. (2007). A novel low-power logic circuit design scheme. *IEEE Transactions on Circuits and Wystems. II, Express Briefs*, *54*(2), 176–180. doi:10.1109/TCSII.2006.883093

Swiss Centre for Life Cycle Inventories. (2006). *Ecoinvest centre.* Retrieved May 10, 2011, from http://www.ecoinvent.ch/.

Tenhunen, M., & Penttinen, E. (2010). Assessing the carbon footprint of paper vs. electronic invoicing. In *Proceedings of the 21st Australasian Conference on Information Systems.* Brisbane, Australia: IEEE.

Tolia, N., Wang, Z., Marwah, M., Bash, C., Ranganathan, P., & Zhu, X. (2008). Delivering energy proportionality with non energy-proportional systems: Optimizing the ensemble. In *Proceedings of Conference on Power Aware Computing and Systems.* ACM Press.

United Nations Environment Programme. (2009). *Recycling - From e-waste to resources*. Berlin, Germany: Oktoberdruck AG.

US Census Bureau. (2011). *Statistical abstract of the United States: 2011*. Washington, DC: US Census Bureau.

US Department of Energy. (2011). *Federal energy management program.* Retrieved May 6, 2011, from http://www1.eere.energy.gov/femp/program/dc_energy_consumption.html.

US Environmental Protection Agency. (2007). *Report to congress on server and data center energy efficiency*. Washington, DC: EPA.

Williams, E. (2004). Energy intensity of computer manufacturing: Hybrid assessment combining process and economic input-output methods. *Environmental Science & Technology, 38*(22), 6166–6174. doi:10.1021/es035152j

Williams, E., Ayres, R., & Heller, M. (2002). The 1.7 kilogram microchip: Energy and material use in the production of semiconductor devices. *Environmental Science & Technology, 36*(2), 5504–5510. doi:10.1021/es025643o

Xun, L., Ortiz, P., Browne, J., Franklin, D., Oliver, J., Geyer, R., et al. (2010). Smartphone evolution and reuse: Establishing a more sustainable model. In *Proceedings of the 39th International Conference on Parallel Processing,* (pp. 476-484). San Diego, CA: IEEE Press.

Zhang, Y., Chowdhury, P., Tornatore, M., & Mukerjee, B. (2010). Energy efficiency in telecom optical networks. *IEEE Communications Surveys & Tutorials, 12*(4), 441–458. doi:10.1109/SURV.2011.062410.00034

Zhu, Q., Shankar, A., & Zhou, Y. (2004). PB-LRU: A self-tuning power aware storage cache replacement algorithm for conserving disc energy. In *Proceedings of the 18th International Conference on Supercomputing.* Malo, CA: ACM.

ADDITIONAL READING

Antonopoulus, N., & Gillam, L. (Eds.). (2010). *Cloud computing: Principles, systems and applications*. London, UK: Springer.

Burd, T. D., & Brodersen, R. W. (2002). *Energy efficient microprocessor design*. Dordrecht, The Netherlands: Kluwer Academic Publishers. doi:10.1007/978-1-4615-0875-5

Cafaro, M., & Aloisio, G. (2010). *Grids, clouds and virtualization*. London, UK: Springer.

Falsafi, B., & Vijaykumar, T. N. (2000). Power-aware computer systems. In *Proceedings of the Second International Workshop, PACS 2002*. Berlin, Germany: Springer.

Gloekler, T., & Meyr, H. (2004). *Design of energy-efficient application-specific instruction set processors (ASIPs)*. Dordrecht, The Netherlands: Kluwer Academic Publishers.

Goldworm, B., & Skamarock, A. (2007). *Blade servers and virtualization: Transforming enterprise computing while cutting costs*. New York, NY: John Wiley & Sons.

Hatziargyriou, N., Dimeas, A., Tomsti, T., & Weidlich, A. (2011). *Energy-efficient computing and networking*. Berlin, Germany: Springer.

IEA. (2007). *Tracking industrial energy efficiency and CO2 emissions*. Paris, France: OECD/IEA.

Kaxiras, S., & Martonosi, M. (2008). *Computer architecture techniques for power-efficiency*. New York, NY: Morgan & Claypool Publishers.

Kranzlmueller, D., & Toja Min, A. (Eds.). (2011). *Information and communication on technology for the fight against global warming*. Berlin, Germany: Springer.

Kreith, F., & West, R. E. (1997). *CRC handbook of energy efficiency*. Boca Raton, FL: CRC Press.

Mahmood, Z., & Hill, R. (Eds.). (2011). *Cloud computing for enterprise architectures*. London, UK: Springer.

Oklobdzija, V. G., & Krishnamurthy, R. K. (Eds.). (2006). *High-performance energy-efficient microprocessor design*. Dordrecht, The Netherlands: Springer. doi:10.1007/978-0-387-34047-0

Ou, J., & Prasanna, V. K. (2010). *Energy efficient hardware-software co-synthesis using reconfigurable hardware*. Boca Raton, FL: Chapman & Hall/CRC Press.

Panda, P. R., Shrivastava, A., Silpa, B. V., & Gummidipudi, K. (2010). *Power-efficient system design*. Berlin, Germany: Springer. doi:10.1007/978-1-4419-6388-8

Roebuck, K. (2011). *Energy efficient ethernet (802.3az): High-impact technology - What you need to know: Definitions, adoptions, impact, benefits, maturity, vendors*. Brisbane, Australia: Emereo Pty Ltd.

Schmitz, M. T., Al-Hashimi, B. M., & Eles, P. (2005). *System-level design techniques for energy-efficient embedded systems*. Dordrecht, The Netherlands: Kluwer Academic Publishers.

Schulz, G. (2009). *The green and virtual data center*. Boca Raton, FL: CRC Press. doi:10.1201/9781420086676

Talvi, B., & Heinzelman, W. (2006). *Energy-efficient real-time data communications*. Berlin, Germany: Springer.

Zang, L. (Ed.). (2011). *Energy efficiency and renewable energy through nanotechnology*. Berlin, Germany: Springer.

KEY TERMS AND DEFINITIONS

Backbone Networks: These consist of the underlying communication infrastructure which supports both wired and wireless communications.

Data Centers: These are electronic repositories using to store digital information to be accessed by internet users.

E-Waste: Electronic waste usually consists of discarded electronic items which may pose a considerable threat to the environment if disposed improperly.

Global Warming: Average rise in near earth atmospheric and sea temperatures mainly due to human activities threatening the very existence of the planet.

Greenhouse Emissions: A vast array of gases including carbon dioxide and methane which trap solar radiation and raises global temperatures.

Information and Communication Technology: It is an extended synonym used for implying a combination of information technology and communications engineering encompassing both devices and software.

Virtualization: the process of creating a virtual copy of any hardware, operating system or a storage device for ensuring energy efficiency.

Chapter 3
Architecture and Mechanisms of Energy Auto-Tuning

Sebastian Götz
Technische Universität Dresden, Germany

Claas Wilke
Technische Universität Dresden, Germany

Sebastian Cech
Technische Universität Dresden, Germany

Uwe Aßmann
Technische Universität Dresden, Germany

ABSTRACT

Energy efficiency of IT infrastructures has been a well-discussed research topic for several decades. The resulting approaches include hardware optimizations, resource management in operating systems, network protocols, and many more. The approach the authors present in this chapter is a self-optimization technique for IT infrastructures, which takes hard- and software components as well as users of software applications into account. It is able to ensure minimal energy consumption for a user request along with a set of non-functional requirements (e.g., the refresh rate of a data extraction tool). To optimize the ratio between utility of end users and the cost in terms of energy consumption, the system needs inherent variability leading to differentiated energy profiles and mechanisms to reconfigure the system at runtime. The authors present their approach called Energy Auto-Tuning (EAT) comprised of these mechanisms and an architecture which automatically tunes the energy efficiency of IT systems.

1. INTRODUCTION

Today, developing energy efficient software has become one of the major challenges for software engineering. According to several studies, the carbon dioxide emissions caused by Information and Communication Technology (ICT) were estimated as two percent of world-wide CO2 emissions in 2007 and further increases are expected (Gartner, Inc., 2007; Climate Group, 2008). For ICT hardware, like CPUs, hard drives and network devices, several energy optimization techniques have been explored (Hewlett-Packard;

DOI: 10.4018/978-1-4666-1839-8.ch003

Intel; Microsoft; Phoenix Technologies; Toshiba, 2010). However, approaches taking software into account have been investigated in a less intensive way. Representative approaches are Fei, Zhong, and Jha (2008), Kansal and Zhao (2008), and Seo, Malek, and Medvidovic (2008).

In this chapter, we propose Energy Auto-Tuning (EAT) as a solution for energy-optimizing software systems. EAT bases on Component-Based Software Development (CBSD) and takes both software and hardware components into account. Software components can provide different qualities (e.g., different refresh rates or resolutions) and can be deployed on different hardware devices (e.g., different servers). At runtime, the EAT system decides where to deploy software components once a user requests a provided service. Furthermore, the EAT system is capable to select the best of multiple existing implementations (variants of components) providing different quality of services and consuming different amounts of energy. The optimal variant is selected and deployed at the optimal server, its energy consumption.

The basic principle of Auto-Tuning (AT) can be described as a control loop. An AT system, running on a hardware infrastructure, monitors and controls itself. For each user request, the system reflects if it is able to provide the highest possible user utility for the least possible cost in terms of efforts like energy consumption in its current configuration. A system configuration in our terms is a set of software component instances deployed on component-containers, which run on servers (or, more general, computing entities). To improve efficiency, the AT system does not only need to know its current configuration, but needs to be able to compute other (probably more optimal) configurations and to reconfigure itself at runtime. Notably, reconfiguration is an activity of the system, which implies costs by itself. Hence, the effort imposed by reconfiguration needs to be

Figure 1. Control loop of auto-tuning systems. Adapted from Dobson et al. (2006).

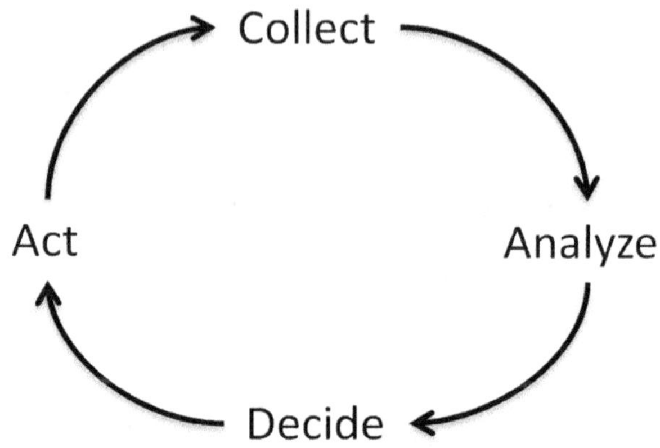

taken into account additionally. The basic steps of an AT system are depicted in Figure 1.

They are: *collect* information about the system in its current configuration, *analyze*, if the system is able to optimally serve the user request, *decide* for a reconfiguration if another, more efficient configuration exists, and *act*, i.e. perform the actual reconfiguration. The general objective function of the overall system is described by Formula 1:

$$efficiency = \frac{\sum_{i=1}^{n}(utility_i \cdot w_i^u)}{\sum_{j=1}^{m}(effort_j \cdot w_j^e)}$$

Efficiency is defined as the ratio of the sum of weighted utilities and the sum of weighted efforts to achieve these utilities. The number of utilities is denoted by *n*, whereas the number of efforts is denoted by *m*. The variables w_i^u and w_j^e (i.e., weights) are used to prioritize certain utilities and efforts respectively. Energy efficiency is a special case of this formula, where the only effort of interest is the energy consumption (cf. Formula 2).

$$efficiency_{energy}(F,W,C,C') = \frac{\sum_{i=1}^{n}(utility_i(F,C') \cdot w_i^u)}{energy(F,C') \cdot w^e + energy(RC(C,C'),C)} \Big| w_1^u \ldots w_n^u, w^e \in W$$

This formula reflecting our approach considers the user demands (i.e., a set of weights w_i^u) provided by the user requesting a system feature *F*, the current system configuration *C* and the target (i.e., possibly new) configuration *C'*. *RC* is a special system feature, which reconfigures the system. It takes the current and target configuration as input. Both utility and energy are functions deriving the achievable utility and energy respectably for a feature on a system configuration. Both take the requested feature and the current system configuration as input. Systems following this objective function are energy auto-tuning systems.

Energy efficiency in our approach is achieved by constantly switching to the system configuration, which has the highest value (Formula 2). That is for a system configuration *C*, a specified user request *F* and a set of user demands *W*, another system configuration *C'* is chosen, if the ratio of utility and energy results in the highest value in comparison to all other system configurations. In other words, the achievable utility for the user request in the new system configuration is compared with the energy consumption implied by it as well as by switching from the current to the new configuration. The user demands can be expressed as weights, because they are a set of non-functional requirements, which have to be fulfilled for a user request (functional requirement). To effectively use the formula above, the system needs to derive or know all possible system configurations to compute their efficiency a fixed feature, user demand, and current configuration.

To save energy our approach utilizes two techniques: selection and consolidation. Selection denotes the deployment of those software implementations, which use fewer resources compared to other implementations. Because we consider user utility in addition, our approach will not necessarily decide for the least energy consuming implementation, but negotiate the tradeoff between utility and energy consumption as expressed by the formula above. The second technique to save energy is consolidation of workload. As has been shown previously, for example in Tsirogiannis, Harizopoulos, and Shah (2010), servers are most energy efficient when they are highly utilized. Hence, our approach derives mappings of software to hardware in such a way that only few resources are used, but these resources are highly utilized. In other words, the work to be performed by the system is consolidated on as few resources as possible. Notably, user utility again poses a tradeoff, which is negotiated by our approach. Using fewer resources typically leads to reductions of Non-Functional Properties (NFPs) (e.g., response time).

To realize EAT we developed the Three-layer Energy Auto-tuning Runtime Environment (THEATRE) originally presented in Götz, Wilke, Schmidt, Cech, and Assmann (2010). The runtime process behind THEATRE is depicted in Figure 2. The THEATRE manages a landscape of present hardware components (e.g., servers) on which a set of software components can be deployed and instantiated to provide services that can be requested and invoked by users of the system. The THEATRE consists of three central managers—each representing one of its three layers—maintaining the system and a model that represents THEATRE's knowledge of the system at runtime. The THEATRE receives and executes user requests (user layer, Global User Manager [GUM]), reconfigures the software (software layer, Global Energy Manger [GEM]) and its deployment on the hardware (resource layer, Global Resource Manager [GRM]) energy efficiency. The five steps of the THEATRE process are shortly

Figure 2. The specialized control loop for energy auto-tuning (EAT)

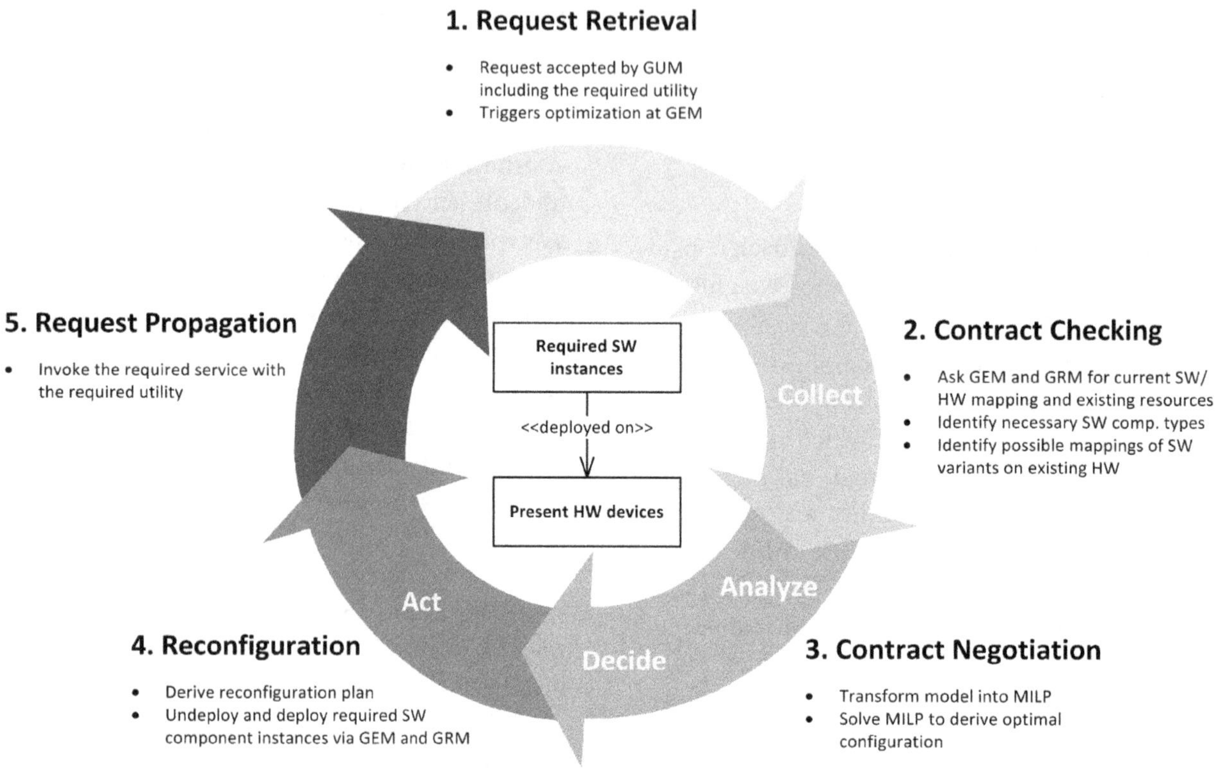

outlined below and described in more detail in the remainder of this chapter.

1. *Request Retrieval:* If a user requests a service, the service request is retrieved by the first central manager of the THEATRE, the GUM. Besides the service the user intents to invoke, the user can specify NFPs he requires for the service (e.g., minimum refresh rates).
2. *Contract Checking:* Subsequently, the GUM delegates the request to the GEM to check if the system has to be reconfigured to efficiently serve the user request (e.g., if new software component instances have to be deployed or parts of the system have to be reconfigured to provide the requested NFPs). During contract checking, the GEM searches for the required software components (which can have multiple implementation variants providing different NFPs) and requests the GRM for the currently present hardware devices and their available resources (e.g., free disk space, memory, or CPU cycles). The GEM computes all possible mappings of existing software variants onto the existing hardware landscape. How we model the available software and hardware variants and their dependencies using our Cool Component Model (CCM) and our Energy Contract Language (ECL) is outlined in Section 2. The contract checking phase itself is described in Section 3.1.
3. *Contract Negotiation:* The requirements of the user request together with the possible software-to-hardware mappings are transformed into a Mixed Integer Linear Program (MILP) that is passed to a MILP solver to compute the optimal system configuration, the user's demands, and the optimal energy

consumption rate. Contract negotiation is explained in more detail in Section 3.2.

4. *Reconfiguration:* The result from the contract negotiation phase is used to compute a reconfiguration plan denoting which software component variants have to be undeployed, moved, and deployed to provide the requested service and utility. Afterwards, the reconfiguration plan is executed by the GEM. Reconfiguration is explained in more detail in Section 3.3.
5. *Request propagation:* The original user request is forwarded to the now energy-optimal deployed software landscape providing the required service and utility.

The attentive reader might wonder how the THEATRE and its managers know about the available software component implementations, hardware devices and how the knowledge about their energy consumption is monitored and maintained to allow the necessary energy consumption prediction for each possible configuration. This process called energy assessment runs in parallel to the EAT control loop and allows to always register new software implementations and to plug or unplug hardware devices. How energy assessment works is explained in detail in Section 3.4.

Besides the presentation of our modeling techniques in Section 2 and the THEATRE process in Section 3, the remainder of this chapter discusses related work in Section 4. Future work and challenges regarding our EAT approach are pointed out in Section 5 and finally, in Section 6, we conclude the chapter.

2. ARCHITECTURE OF ENERGY AUTO-TUNING SYSTEMS

In this section we will elaborate on the architecture of THEATRE by discussing the CCM and the ECL. A general overview of the concepts of both the CCM and the ECL is given in Figure 3. The CCM is used to describe the structure of both Software (SW) and Hardware (HW) components. Furthermore, it allows the definition of NFPs provided by different SW and HW component types. In addition, the CCM allows for the definition of behavior for SW and HW components, their utilization, and energy consumption. Finally, the CCM is used to model individual implementations of SW and

Figure 3. The concepts of CCM, ECL and their interconnections

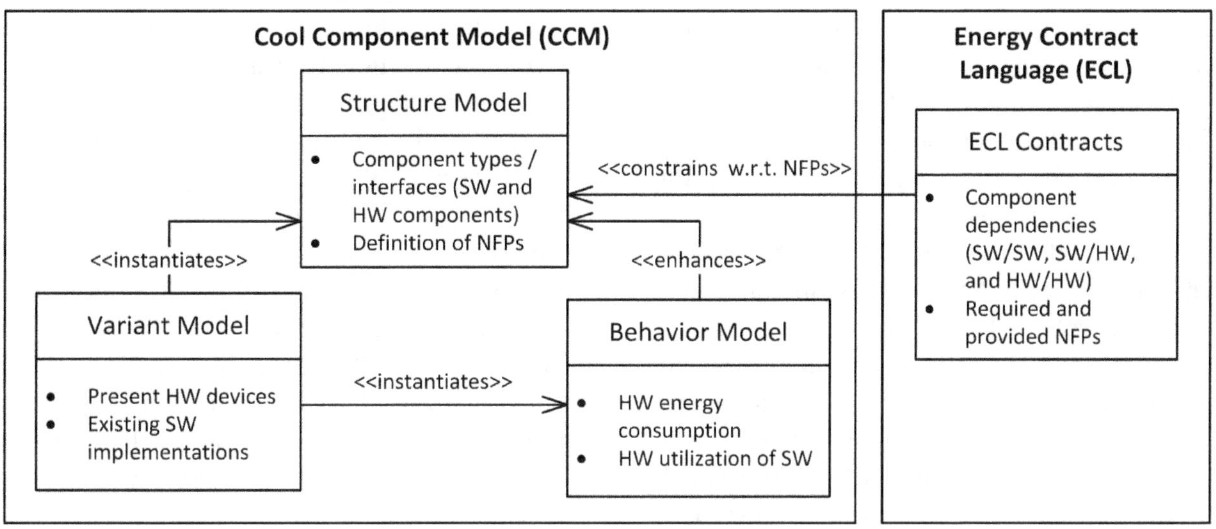

HW component types (i.e., variants). The ECL is used to describe provided and required NFPs of individual variants as well as their dependencies to other SW and HW component types. A detailed discussion of the CCM and ECL concepts is given in the following subsections.

2.1. The Cool Component Model

Component-Based Software Development (CBSD) provides a structured way to construct modular software systems. Szyperski et al. define a software component as "… a unit of composition with contractually specified interfaces and explicit context dependencies only" (Szyperski, Gruntz, & Murer, 2002). In consequence, components explicitly define what they provide to their environment (i.e., interfaces) and in turn what they require (i.e., context dependencies). Component models extend this basic definition by further constraints on what constituents a component needs to comprise and how components can be composed. Our CCM is one such extension, which focuses on variability in terms of multiple implementations (i.e., variants) of component types and the modeling of NFPs, especially energy consumption. Moreover, the CCM does not restrict components to be software artifacts, but allows for the modeling of hardware resources as components, too. Another key characteristic of the CCM is its explicit meta-level. That is the ability to define component types describing components. For software components this meta-level is well known as it allows for the declaration of multiple implementations per component. For hardware resources, the meta-level allows to define types of resources that may exist in an IT infrastructure, like servers, hard disks, but also mobile phones or service robots. The CCM provides three kinds of models to capture an EAT system: structure, variant, and behavior models. Structure models describe the aforementioned meta-level. Instances of this meta-level are described by variant models, which allow for the declaration of component implementations and concrete hardware landscapes. Finally, behavior models are responsible to capture the behavior of components' energy consumption. Their behavior is expressed in terms of templates, which are based on energy state charts. In the following subsections we will first outline the structure and variant models, followed by a discussion about the modeling of NFPs and the concept of energy behavior templates.

Structure and Variant Models

The CCM distinguishes between component types and variants in terms of structure and variant models respectively. To provide an overview of these models, we introduce a simple stock tracking application as a running example in this chapter. This example scenario will be used in Section 3 to explain how certain EAT mechanisms work.

The stock tracking application's task is to extract and analyze stock quotations from several stock exchanges, to analyze the quotations and to present the results. Figure 4 shows the structure model capturing the software structure of this application consisting of software component types, port types and connector types between port types of different software component types. In the following, elements of the meta-level are formatted in typewriter, whereas variants and NFPs are formatted in italic font. The Extractor type collects a history of stock quotations from several companies. It is used by the Analyzer type, which may provide several algorithms to analyze arbitrary data. In turn the Analyzer type is used by the Presenter type to generate a graphical representation of the analyzed data. For each software component type several NFPs are specified that are required for EAT at runtime. The implementations of one software component type have the same NFPs, which differ in their values. For example, each Extractor implementation has a *refresh rate*, but its actual value differs for the different variants. Details regarding these NFPs are explained in the next subsection.

Figure 4. Structure and variants of the stock tracking application scenario

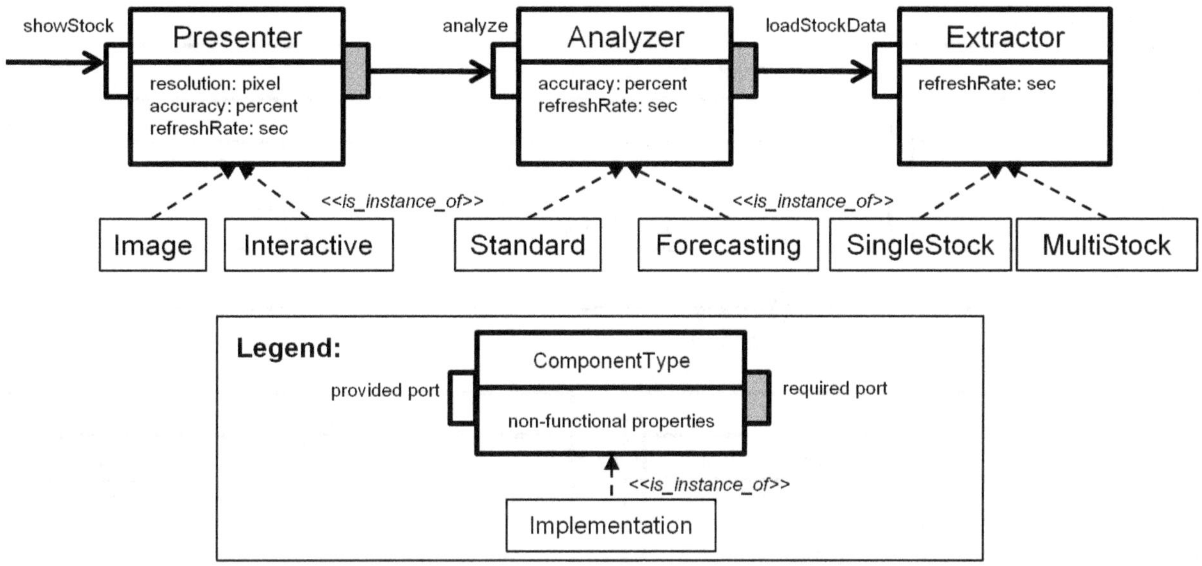

For each software component type, several implementations (i.e., variants) exist, which are shown in Figure 4, too. Concrete variants refer to their type and provide values for their functional and non-functional properties. In our scenario, there are two variants of the Presenter type, an *Image* presenter and an *Interactive* presenter. The first provides a simple image covering the analyzed data, whereas the latter provides additional zooming functionality to the user. For the Analyzer type there is a *Standard* variant, which is able to analyze historical data and a *Forecasting* variant, which provides additional forecasting abilities for stock quotations. Finally, a *SingleStock* and a *MultiStock* extractor exist that differ in the number of stock exchanges they can utilize to retrieve stock quotation data. The *SingleStock* extractor is able to collect data from one stock exchange center only (e.g., Frankfurt). In contrast, the *MultiStock* extractor takes several stock exchange centers into account (e.g., Frankfurt, Tokyo, and New York).

For energy-aware optimization, the resources of an IT landscape need to be captured, too. This is because they are utilized by the execution of software components. In other words, software consumes energy by utilizing specific hardware resources during its execution. The IT infrastructure is modeled similar to software component types. There is a structure model describing types of resources that may be part of an infrastructure and a variant model describing concrete resources (variants of a resource type).

The structure and variant model for our example are depicted in the upper and lower part of Figure 5 respectively. The structure model defines that an IT infrastructure consists of one or more servers. Each Server may have one or more CPUs, network cards (Net), RAM chips and hard disks (HDD). For Net and Server two port types are specified and connected via port connector types. This indicates that each Server is able to connect to other Servers to exchange data. Similar to the modeling of software component types several hardware-specific NFPs are defined in the structure model (e.g., the Net device provides a *bandwidth*). The concrete infrastructure consists of two server resources. *Server 1* and *2* are variants of the type Server defined in the structure model (for clarity the *is_instance_of* relationship is not

Figure 5. Structure model of an IT infrastructure

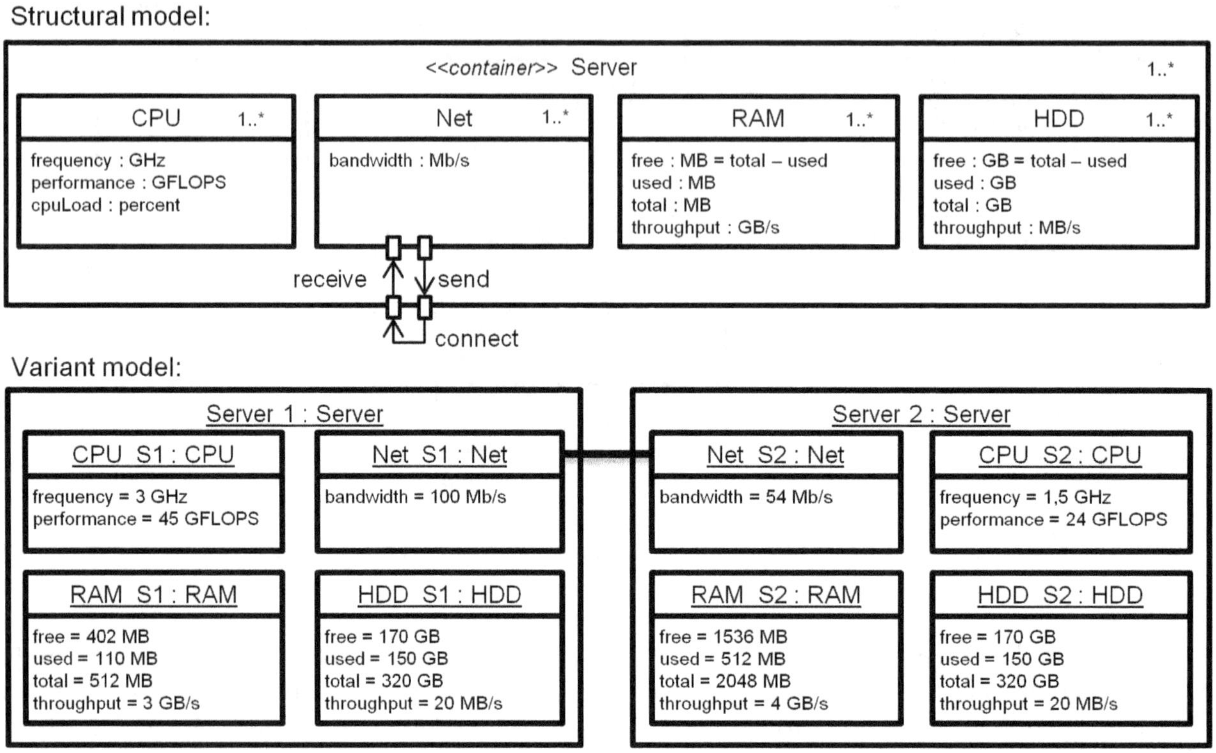

shown in Figure 5). According to the structure model, Servers have to have at least one concrete CPU, Net, RAM and HDD resource each. However, they can differ in the concrete values of their NFPs.

Notably, any kind of IT infrastructure can be captured by a structure model. A mobile phone would be modeled quite similar to servers. The HDD might be replaced by a memory stick and further resource types, like a Bluetooth interface and a GPS device, might be added. A central concept in our approach is the mapping of software component implementations to resources. Therefore, entities of structure models can be marked as containers. In Figure 5, the server is marked as such. In a structure model, which contains mobile phones in addition to servers, both servers and mobile phones would be marked as containers. As structure models can be defined by the developers of EAT systems, our approach, hence, offers a high degree of flexibility and extensibility, the kind of system entities.

Non-Functional Properties

The modeling of NFPs is a key requirement to optimize energy efficiency of complex HW/SW systems. We refer the interested reader to Chung, Nixon, Yu, and Myopoulos (2000) and Chung and do Prado Leite (2009) for a discussion about functional and non-functional properties in software engineering. Energy consumption is a specific NFP of hardware components, being aggregated for software components as an NFP, too. Considering our stock tracking application introduced above, further examples of NFPs are the *refresh rate* provided by the Extractor, the *accuracy* of the Analyzer and the *resolution* provided by the Presenter component. In general, each NFP refers or belongs to a component. The *refresh rate* is a property of an Extractor component and the *resolution* a property of a Presenter component. Hence, in the CCM, NFPs are defined for component types in structure models. They usually have a unit and an increasing or decreasing ordering relation (not shown in our example for clarity). For example, the *refresh rate* for stock quotations is measured in seconds and has a decreasing order (i.e., the shorter the *refresh rate*, the better the utility). The *accuracy* of the Analyzer is measured in percent and has an ascending order (i.e., the higher the *accuracy*, the better the utility). As presented in Figure 4 and Figure 5, such properties can be defined for soft- and hardware component types. Typical properties of hardware resources are the total *size* of a hard disk drive, the maximum *bandwidth* of a network device and the *performance* of a CPU (e.g., measured in Floating Point Operations Per Second, FLOPS).

Notably, NFPs are defined on the meta-level (i.e., for hard- or software component types). Their actual values can be determined at runtime for a specific instance (resource or implementation) only. For example, the maximum *throughput* of a network device is specific for each concrete device and the size of an HDD is specific for each concrete HDD. We distinguish three types of NFPs in the CCM: static instance, monitored, and calculated properties. If a value of an NFP is immutable for a given hard- or software instance, we call it a static instance property. The *size* of an HDD is an example for this kind of property. If a value of an NFP can only be derived at runtime by profiling the whole (or a part of the) system, we call it a monitored property. In our example, the remaining *free* disk space of an HDD is an example for this kind of property. Finally, if a property's value can be derived from other properties, we call it a calculated property. For example, the *used* disk space of an HDD can be derived by subtracting the *free* disk space from the (*total*) size of the HDD. All three kinds of NFPs can be realized using the same technique to retrieve their value: calling the runtime environment, which is responsible to return the static instance and monitored values as well as to perform required calculations.

Energy Behavior Templates (Parametric Energy State Charts)

To model the behavior of hardware resources we use the concept of Energy State Charts (ESCs), as introduced in Benini, Hodgson, and Siegel (1998). ESCs base on the existence of power saving or performance modes of hardware resources and extend classical state charts (Harel, 1987) in two ways. Each state and transition is qualified by an energy consumption rate. Transitions are additionally qualified by a delay. Take the power saving modes of an HDD as an example. A simplified energy state chart for an HDD, as depicted in Figure 6, describes an *idle* mode, a *busy* mode and a *sleep* mode. The HDD will consume more energy if it is *busy*, than when it is *idle* or even *asleep*. Notably, if a read or write request (*rw*) needs to be handled and the HDD is *asleep* it takes time and energy to spin the disk up (i.e., to switch from *sleep* to *busy* mode). A switch from *idle* to *busy* mode will take fewer time and energy.

The energy consumption rate and the delays are specific to each concrete resource, but the general behavior defined by an energy state chart can be used for multiple resources. In consequence, the CCM contains energy behavior templates. These templates are energy state charts, which have placeholders for energy consumption rate, delays, and custom defined placeholders. In Figure 6, the energy consumption rate (EC) is depicted by the parameters EC_{busy}, EC_{idle} and EC_{sleep} respectively. The delay of a transition as well as the corresponding energy consumption rate is depicted in Figure 6 for clarity only for the transition from sleep to busy by the parameters $EC_{sleep2busy}$ and $Delay_{sleep2busy}$.

Declaring an instance of a resource type includes the selection of an energy behavior template and to provide values for the placeholders (cost parameters). These values are not necessarily just

Figure 6. Energy behavior template for an HDD

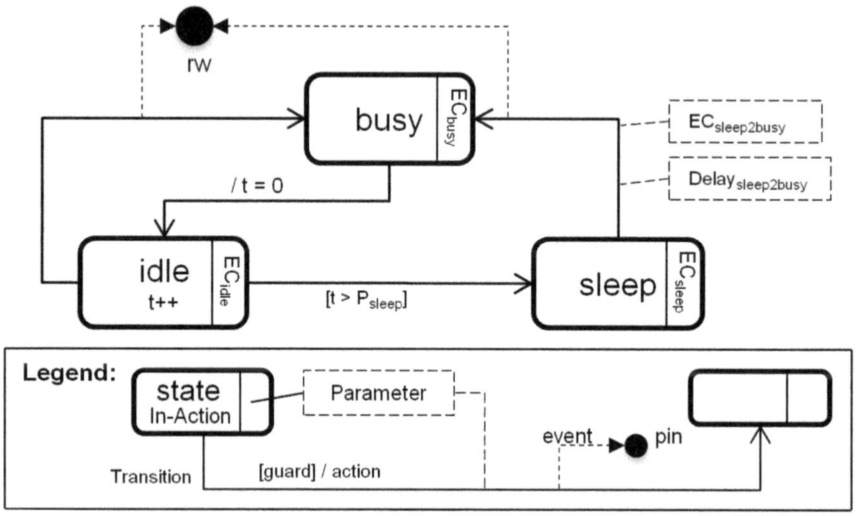

numbers, but can be mathematical expressions using the variables defined in the energy state chart as well as the declared cost parameters. A behavior template for an HDD could include a variable counting how often the disk has been spun up or down. This variable could be used in a mathematical expression to calculate the power consumption in a certain state of the state chart to consider the mechanical wear.

The purpose of energy behavior templates is to derive the energy consumption implied by a specified workload. For that purpose, we built a simulation tool, which will be presented in Section 3.4.

2.2. The Energy Contract Language

As mentioned above, component types (software as well as hardware) can define NFPs declaring the qualities they provide at runtime. For each implementation of a component, these values are either bound to concrete values or these values are computed at runtime. However, to configure and auto-tune an application at runtime the dependencies between component types have to be declared in addition. For example, for a Presenter implementation it has to be declared which NFPs of the Analyzer component are required to provide certain *accuracy*. To express such dependencies we developed the Energy Contract Language (ECL). The ECL allows for the definition of contracts for components specifying (probably) multiple quality modes providing and requiring different NFPs (i.e., qualities).

We refer the interested reader to Beugnard, Jézéquel, Plouzeau, and Watkins (1999) for a fundamental discussion on contracts in the context of component-based systems.

Box 1 shows an example ECL contract for the *Image* variant of the Presenter component. The contract consists of two different quality modes (formatted in typewrite in the following), a low and a highQuality mode (the latter not shown completely). Within each mode, three different kinds of clauses can be specified:

1. Requirements on other software components (e.g., lines 4 – 7 express that the lowQuality mode requires an Analyzer implementation with an *accuracy* of at least 50% and a *refreshRate* of at least 300 seconds).
2. Requirements on other hardware components (e.g., lines 9 – 11 express that the lowQuality

Box 1. Example ECL contract for the image variant of the presenter component

```
 1 contract Image implements software Presenter {
 2
 3  mode lowQuality {
 4   requires component Analyzer {
 5    accuracy min: 0.5
 6    refreshRate min: 300
 7   }
 8
 9   requires resource CPU {
10    frequency min: 400
11   }
12
13    provides accuracy min: 0.5
14    provides refreshRate min: 300
15    provides imageWidth min: 800
16    provides imageHeight min: 600
17   }
18
19  mode highQuality {
20   requires component Analyzer {
21    accuracy min: 0.9
22    refreshRate min: 60
23   }
24
25    /* More requirements and provisions here ... */
26  }
27 }
```

mode requires a CPU with a *frequency* of at least 400 MHz).

3. Provided NFPs (e.g., lines 15 – 16 express that the lowQuality mode provides a resolution of 800 * 600 pixels).

Although not visible in the example contract of Box 1, all requirements and provisions are typed values as the NFPs are specified using units like seconds or MHz as specified in the structure model. Besides minimum amounts required of specific NFPs, maxima and intervals can be specified as well. Similar contracts as shown for the *Image* variant have to be provided for all software component types' implementations. Thus, similar contracts exist for the *Interactive* Presenter variant, as well as all Analyzer and Extractor variants. Contracts defined for hardware resources are imaginable as well (although not contained in the provided example). However, hardware components can only declare dependencies to other hardware components, as they are located "below" the software within the execution platform. Once ECL contracts exist for all components, users can request a service including an expected quality as shown in Box 2.

Box 2. Example user request including demands

```
call Presenter.showStock expecting {
    accuracy min: 0.5
    refreshRate min: 200
}
```

Depending on the user requirements and the existing component variants, the EAT system will analyze and auto-tune the application maximum utility for minimal energy consumption. After this section discussed the architecture and modeling of EAT systems, the mechanisms behind EAT as introduced by Figure 2 (cf. page 3) are elaborated in the next section.

3. MECHANISMS OF ENERGY AUTO-TUNING

In this section, we will elaborate on various mechanisms required by an energy auto-tuning system. As a starting point, in Section 3.1, we will sketch a mechanism to determine and assess all possible system configurations in terms of user demand fulfillment. In Section 3.2, we will discuss an iterative approach combining energy and utility assessment to select the optimal system variant taking reconfiguration efforts into account. Afterwards, in Section 3.3, we will discuss the process of determining and performing actions required to reconfigure the system. Finally, in Section 3.4, we will discuss a mechanism to derive the energy consumption imposed by invoking a system feature (i.e., energy assessment).

3.1. Utility Assessment and Contract Checking

In this subsection, we focus on the question how to derive configurations, which are able to serve the user's requests and demands (i.e., utility assessment by contract checking).

To determine valid system configurations in terms of software component implementations mapped to hardware resources, we developed a mechanism we call contract checking following the definition of Quality-of-Service-level contracts by Beugnard, Jézéquel, and Plouzeau (2010) along with Meyer's design by contract principle (Meyer, 1992). The mechanism can be divided into five steps:

1. Quality mode selection,
2. Resolving software dependencies,
3. Collection of resource requirements,
4. Determination of all SW/HW mappings, and
5. Identification of valid configurations, which are those able to serve the user's request and demands (i.e., can be mapped onto available hardware).

The process is initiated by the user, which sends a request as shown in Box 2 (cf. page 10) to the system. The demands of this request are compared with the provided NFPs of each contract, which has been defined for the component type the request is referring to. In the example of Section 2.2 step (1) will identify the highQuality mode of the *Image* implementation (cf. Box 1 on page 9) as being able to serve the user demands. The lowQuality mode does not fit, because a *refreshRate* of at least 200 seconds has been requested by the user, but the lowQuality mode only ensures a minimum *refreshRate* of 300 seconds.

The second step recursively resolves the dependencies to software components as declared in the modes, which have been identified before. In the example, the dependency to the Analyzer component will be resolved. This is done by handling the dependency like a user request. That is, software component dependencies are interpreted as calls for which valid modes (step [1]) have to be identified. This is possible, because dependencies between software components specify that the dependent component type will call another component type. The result of resolving all dependencies are quality paths. These are sets of quality modes of each implementation required to serve the user's demands. In the example, each quality path consists of three implementation-quality mode pairs: one mode per Presenter, Analyzer and Extractor implementation. The reason is, that one implementation of each component is required to serve the user's request.

The resource requirements can be directly extracted from the respective contracts of each identified quality path. The determination of all SW/HW mappings is realized by building the cross-product of implementations and resources marked as containers. Which resources actually exist is either modeled statically or can be retrieved from the GRM, which is responsible to manage all resources of the infrastructure. Thus, steps (3) and (4) are straight forward. Finally, step (5) investigates each configuration determined in step (4) the possibility to map the implementations to resources.

Multiple optimizations to this general mechanism are possible. For example, we combined the steps (4) and (5) to skip mappings of which we know that they will not support the user's request, which is possible due to the order in which the mappings are checked. Nevertheless, contract checking has exponential complexity, because each dependency between software components duplicates the number of quality paths by the number of existing component implementations and quality modes of the required software component. Thus, the approach is suboptimal for systems with many dependencies between software components and long dependency chains. Nevertheless, contract checking does not need to be repeatedly performed unless the hard- or software landscape changes. Notably, the approach is iterative, meaning that the addition of new resources or software components does not invalidate the results of existing mappings. The removal of a resource or software component only leads to the removal of all mappings, which include the respective entity, but does not affect any other mapping.

In summary, contract checking is a mechanism to assess system configurations, user utility, which filters all configurations, which cannot be deployed on a provided infrastructure. In principle, multiple working configurations can be identified. Contract negotiation, which will be explained in the next subsection, provides means to order these configurations and thus, to identify the best of them.

3.2. Contract Negotiation

Negotiation of contracts is the central mechanism of EAT systems. Its main task is to identify the optimal system configuration for a given user request and a set of user demands. More generally, contract negotiation needs to consider multiple concurrent user requests and demands as well as all predictable future user requests. We call this mechanism contract negotiation, because it negotiates the tradeoff between high user utility (serving user demands) and low energy consumption by utilizing the specified contracts of the EAT system. Contracts implicitly define this tradeoff by specifying provided and required NFPs. They explicitly state resource requirements and thus, implicitly energy consumption.

To identify the optimal system configuration we use the energy assessment mechanism presented in Section 3.4, which includes the assessment of system reconfiguration efforts, and the utility assessment mechanism presented in the previous subsection. Energy assessment leads to a numerical result, namely the amount of energy, which will be consumed by a specified user request. In contrast, utility assessment results in a set of valid mappings of implementations to container resources. To quantify these mappings in terms of utility, the provided NFPs of each mapping are used, which can be extracted from the contracts identified by the respective quality paths.

The set of all possible system configurations can be described as a constraint system, which has three kinds of variables:

1. Boolean variables indicating whether or not a software component implementation shall be mapped to a resource,
2. Resource usage variables and
3. Software-related NFP variables.

The determination of the optimal system configuration is, hence, a constraint solving optimization problem. The translation between resource usage and implied energy consumption is realized by computed factors of the objective function. The objective function is to minimize resource usage weighted by these factors. The minimum utility requested by the user is expressed by constraints. Fortunately, our optimization problem can be formulated using linear constraints only. Hence, we are able to specify our optimization problem as a linear program. The requirement of Boolean variables in addition to floating point variables classifies our problem as Mixed Integer Linear Program (MILP). An introduction to linear programming and related constraint solving techniques can be found in Nemhauser and Wolsey (1988). For a more general discussion on algorithms including linear and dynamic programming we refer the interested reader to Dasgupta, Papadimitriou, and Vazirani (2007). A detailed elaboration on the implementation of contract negotiation, the concrete objective function, and all constraint types can be found in Götz, Wilke, Cech, and Assmann (2011). A successful application of linear programming in the context of smart energy grids has been shown in Ranganathan and Nygard (2010).

3.3. System Reconfiguration

A further mechanism required by EAT systems is the ability to determine and execute all actions required to reconfigure the system from a source configuration to a target configuration. The source configuration is a set of software component implementations as well as their mapping to the currently underlying IT infrastructure. The target configuration is the optimal system configuration determined by contract negotiation. The source and the target configuration have to be compared in order to derive a reconfiguration plan consisting of atomic reconfiguration steps to be executed. Deriving a reconfiguration plan includes the following steps:

1. Identification of changes at the level of software components,
2. Identification of changes regarding the SW/HW mapping, and
3. Derivation and execution of a reconfiguration plan.

In the following paragraphs, we explain these three steps based on the example introduced in Section 2.1. Figure 7 depicts the available implementations of the three software component types and both servers in the infrastructure. Assume a source configuration consisting of the software component implementations *Image* (Presenter type), *Standard* (Analyzer type) and *SingleStock* (Extractor type). The *Image* and *SingleStock* implementations are deployed on *Server 2*. The *Standard* implementation (Analyzer type) is deployed on *Server 1*. In Figure 7 this is indicated by the gray dash-dotted line between implementations and both servers of the IT infrastructure. By using contract negotiation the following software component implementations are detected as target configuration: *Interactive* (Presenter type), *Forecasting* (Analyzer type) and *SingleStock* (Extractor type). The *Interactive* and *Forecasting* implementations have to be deployed

Figure 7. Example of the source and target configuration of the stock tracking application

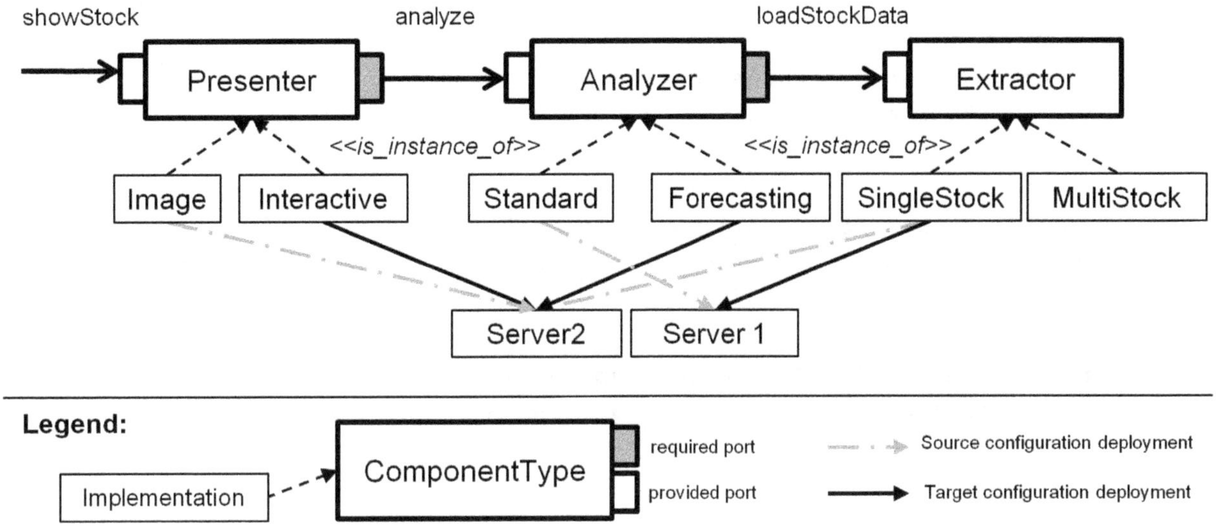

on *Server 2* and the *SingleStock* implementation has to be deployed on *Server 1* (indicated by the black bold line between implementations and servers in Figure 7). In terms of the CCM, each configuration can be described as a variant model of software components. System reconfiguration compares both variant models and derives a reconfiguration plan.

In the first step the source and the target variant model are compared their software component types. In cases where the source and the target variant model differ in the selected implementations for the same component type, a reconfiguration step for this component type is necessary. Consider the source and target configuration mentioned above. The first step evaluates that for the Presenter type the *Image* variant (provided by the source configuration) has to be replaced by the *Interactive* implementation (provided by the target configuration) and for the Analyzer type the *Standard* implementation has to be replaced by the *Forecasting* implementation.

The second step is responsible to discover component types in both configurations for which component migrations are necessary (i.e., which components have to be redeployed on another server). For this purpose, each software component implementation in the source and target variant model provides specific deployment information, which is compared to derive the required component migrations. In our example the result is that the implementation behind the Analyzer type has to be migrated from *Server 1* to *Server 2* and the implementation of the Extractor type has to be migrated from *Server 2* to *Server 1*.

In step (3), the knowledge gathered in step (1) and step (2) is combined in order to extract a reconfiguration plan that describes a set of reconfigurations for each software component type. Such a reconfiguration plan has to be executed by THEATRE to reconfigure the system into an optimal configuration. There are three cases for reconfiguration resulting in three different kinds of reconfiguration steps. These cases are explained based on the reconfiguration plan derived from the source and the target configuration explained above, depicted in Box 3.

The reconfiguration for the Extractor type requires that the *SingleStock* implementation has to be migrated from *Server 2* to *Server 1*, because there are only differences, the deployment information in the source and target variant model.

Box 3. Example of a reconfiguration plan

```
reconfigurationPlan {
    reconfiguration for SWComponentType Extractor {
    migrate SWComponent SingleStock from Server2 to Server1
    }

    reconfiguration for SWComponentType Analyzer {
    replace SWComponent Standard on Server1 with Forecasting on Server2
    }

    reconfiguration for SWComponentType Presenter {
    replace SWComponent Image on Server2 with Interactive
    }
}
```

This information is gathered in step (2). Component migration means that the component state of the *SingleStock* implementation has to be stored on *Server 1,* thereafter *SingleStock* has to be deployed to *Server 2,* and finally the component state from *Server 1* has to be restored on *Server 2*. The Presenter type requires a local replacement of the *Image* implementation by the *Interactive* implementation. Hence, the component state of *Image* on *Server 2* has to be saved and restored in the *Interactive* implementation after its deployment has been finalized. A local replacement is necessary in cases where differences between the source and the target variant model only exist at software component level. That is, when the deployment information of both variants of a type is equal, whereas different variants of a software component type are detected in the first step. The last kind of reconfiguration is a remote replacement. It is necessary in cases where a different software component implementation (result of step [1]) and different deployment information (result of step [2]) are recognized. In our example this case occurs for the Analyzer type because of the result of step (1) that the *Standard* analyzer has to be replaced by the *Forecasting* analyzer and both implementations have different deployment information, which is determined in step (2).

The employed component containers do not necessarily have to provide the functionality to save and restore the state of software components subject to migration (state migration). The only requirement for containers is their ability to deploy and undeploy software components. State migration and the execution of reconfiguration scripts are realized by the GEM of the THEATRE.

3.4. Energy Assessment

An essential part of energy optimization is to determine how much energy is or will be consumed by the execution of a feature in a certain system configuration or, in short, the process of energy assessment. To determine system configurations, which are better than other configurations in terms of energy efficiency for a given user request and set of user demands, a mechanism to predict the required energy to perform the request on a given system configuration is needed. For that purpose, we developed a simulation approach. Furthermore, we developed an approach to derive the resource usage implied by the execution of a feature, including the reconfiguration as a system feature. To bridge the gap between resource usage and energy consumption simulation, we propose a fine-grained resource usage analysis resulting in workloads for our simulator. In the following, we will discuss these steps in detail.

Simulation of Energy State Charts

To predict future energy consumption we developed a simulator for ESCs. A simulation requires a workload (stimulus) for which the consumed energy (response) shall be predicted. For that reason we introduced two immutable variables to energy state charts per default: total energy consumption (*energy*) and a global timer (*time*). The energy variable is used to accumulate the consumed *energy* over (simulation) time. The *time* represents the current simulation time. ESCs have in- and out-pins. An in-pin denotes an event the ESC is waiting for. Out-pins denote events the ESC will fire. This way multiple ESCs can be connected using corresponding in- and out-pins (e.g., the ESC depicted in Figure 6 on page 8 contains an in-pin for read/write [*rw*] events). The simulation itself is a loop, which stops after the last event of the workload has been processed or at a user-defined point in time. The loop counts the simulation *time* up and checks whether there are enabled transitions starting at the current state. A transition is enabled, if the condition of it evaluates to true, the event it is waiting for is defined in the workload at the current simulation time, and/or neither condition nor (in-)events are defined for that transition. If no transition is enabled, the simulator interprets the energy rate expression defined in the current state and adds the resulting value to the *energy* variable. Else, the transition is fired, the current state is changed to the target state of the transition and the energy consumed by following the transition (delay * energy) is added to the *energy* variable. Notably, the delay of transitions leads to jumps in simulation time. This includes the danger to skip events defined in the workload by accident. Nevertheless, the workloads on hardware resources are not defined by the user, but are derived using our resource usage analysis mechanism, which will be presented in the next subsection. Thus, events, which occur while a resource is busy by transitioning to another state, are not skipped by accident. Indeed, such events are ignored, because the system is incapable of processing them at the specified point in time.

A drawback of using ESC simulation is their complexity in terms of parallel execution. The main energy consumers in a typical server are CPU, HDD, network devices and, if present, Graphical Processing Units (GPU). These resources are used in parallel and, hence, consume energy in parallel. As indicated before, parallel execution of ESCs can be emulated by connecting their in- and out-pins. That is an ESC fires an event using an out-pin, where another ESC is waiting for that event by an in-pin. The fired events are logged in a global workload. Nevertheless, each ESC has to be simulated in sequence. In the easiest case, an order of interconnected ESCs can be derived. This is done by analyzing the connected in- and out-pins. An ESC which has no out-pins cannot introduce a dependency itself, but can depend itself on another ESC. In the worst case, the ESCs are simulated in time-slices. That is an ESC α depends on an ESC β, which depends on α, too, where a close analysis of the dependency reveals that β is waiting for an event of α at time t_1 and α is waiting for an event of β at time t_2 after t_1. To simulate both ESCs multiple possibilities exist. First, α can be simulated until t_2 followed by a complete simulation of β and a simulation of α from t_2 to the end. Second, β is simulated until t_1 followed by α, which is simulated till the end and the simulation of β from t_1 till the end. Both approaches imply similar efforts and enable parallel simulation of ESCs. The determination of the workload required as an input for the simulation will be discussed in the following subsection.

Resource Usage Analysis of Software Components

The energy consumed by software depends on how the software utilizes which hardware resources. Hence, to optimize the energy efficiency of heterogeneous IT landscapes, our EAT approach requires knowledge about the resource usage of software components during their execution. This knowledge is used to derive a workload for ESCs. In combination with the simulation approach explained above, the energy consumption implied by executing a feature of a software component can be estimated and used as a factor in the objective function of the MILP during contract negotiation as explained in Section 3.2. To gather utilization data of certain hardware resources we propose the Java Resource Usage Profiling Infrastructure (JRUPI) (Süttner, 2011) as an approach to extract such data at runtime (i.e., during the execution of software). JRUPI is a flexible and portable framework for profiling and analyzing the resource usage data of arbitrary software component features. Based on collected resource usage data, a mathematical model is derived that describes the resource usage behavior of a certain feature of a software component variant. Assuming Java methods as such a feature, the model is defined by considering the metadata of method parameters as variables. Imagine a list of String objects as a parameter of a method. In this case, variables for the mathematical model might be the mean length of Strings and the length of the list.

JRUPI uses dynamic tracing and profiling techniques (Sun, 2008) to capture resource usage data at runtime. The mathematical model of captured data is derived in a separate offline analysis phase. Figure 8 shows how JRUPI collects resource usage data. At the level of the operating system so called *profilers* are responsible to monitor the resource usage of the Java Virtual Machine (JVM) process. So far, profilers for CPUs, network devices and hard disk drives were implemented by using the dynamic tracing framework SystemTap (Jacob, Larson, Leitao, & da Silva, 2010). Considering the fact that profilers collect resource usage data of the JVM process, all running Java threads are monitored by profilers. However, in our case we are interested in collecting resource usage data, a certain software component. This problem can be solved by using dynamic bytecode manipulations techniques. At runtime, a software component is represented as bytecode and is executed in a JVM. Like the JVM process is monitored by profilers, the considered software component is monitored, too. The bytecode of methods is instrumented in a manner that at each entry and exit point of a method an event is triggered at the operating system level. This event is captured and evaluated by the system profilers and indicates that subsequent resource usage is correlated to the Java method, which triggered the event. Nevertheless, by using dynamic bytecode instrumentation, profilers are able to selectively monitor events from the considered software component and not from the whole JVM process. From a technical point of view, dynamic bytecode instrumentation is realized using the bytecode manipulation framework BTrace (Oracle, 2010). As mentioned above, an additional offline analysis phase is required to derive a mathematical resource usage model. During offline analysis profiling data is further processed by filtering, aggregating, and calculating statistical parameters. The resource usage model is derived automatically by using the analysis tool Eureqa (Schmidt & Lipson, 2009), which uses symbolic regression to detect formulas in a given set of data. Finally, the resulting (mathematical) formulas are used to derive the workload for ESCs.

To summarize this section, energy auto-tuning systems require several mechanisms to optimize the energy efficiency of IT systems. This includes the ability to estimate the energy consumption consumed by software components for processing a request. Based on this information possible software variants as well as the optimal mapping to hardware resources can be estimated by contract checking and contract negotiation. In order to

Figure 8. Resource usage profiling with JRUPI. Adapted from Süttner (2011).

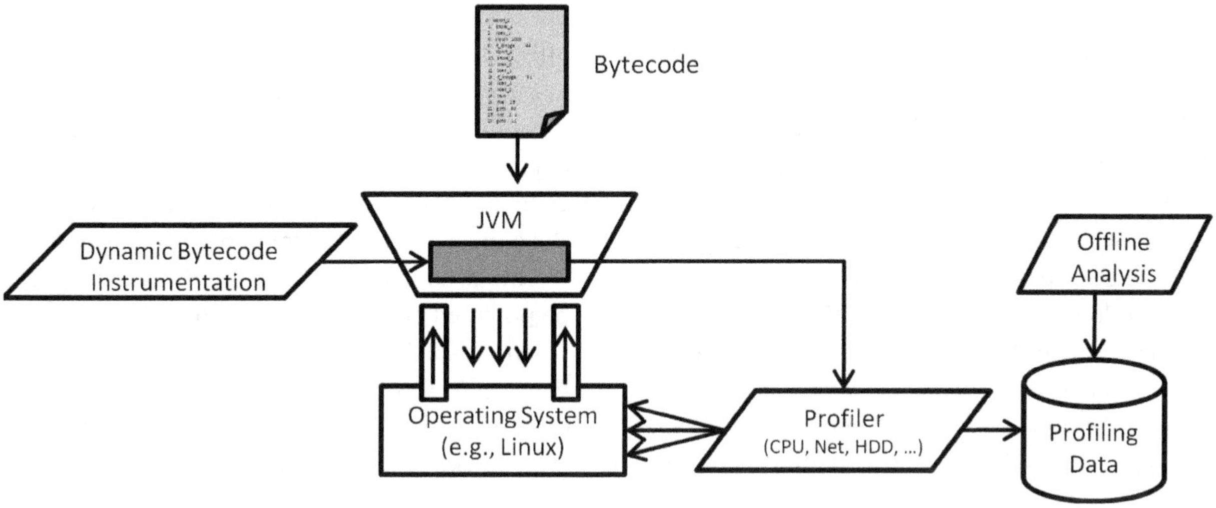

realize the optimal configuration at runtime a reconfiguration plan has to be derived and executed.

4. RELATED WORK

This section presents related work from the domains of energy auto-tuning, multi-quality component models as well as resource utilization, and energy consumption analysis of software applications. Our general idea of the THEATRE has been elaborated in Götz, Wilke, Schmidt, Cech, and Assmann (2010). The resource managers of the THEATRE are specified in Götz, Wilke, Schmidt, Cech, Waltsgott, and Fritzsche (2010). A detailed introduction into profiling concepts of JRUPI can be found in Süttner (2011). A first fully operational proof-of-concept implementation of EAT has been realized by Püschel (2011).

4.1. Auto-Tuning

The term auto-tuning emerged from the field of high performance computing where it typically represents the automatic optimization of an algorithm (e.g., matrix multiplication or Fourier transform): its performance, its input data (e.g., matrix size), and/or its underlying hardware. The major idea behind auto-tuning can be summarized by the auto-tuning control loop as depicted in Figure 1 on page 2. Algorithm implementations with auto-tuning facilities can be distinguished into approaches supporting either optimization at installation time, optimization at runtime, or both. ATLAS (Whaley & Dongarra, 1997) optimizes at installation time by generating optimized code for the underlying hardware considering cache sizes and the amount of available CPU registers. FFTW (Frigo & Johnson, 1998)—a library for Fourier transform—optimizes at runtime by composing several code snippets providing similar functionality with different performance properties depending on the underlying hardware. At runtime, the snippet's performance is analyzed and according to it, they are selected and composed to improve execution performance. Whereas ATLAS and FFTW focus on the optimization of specific algorithms, their performance, our EAT approach focuses on optimizing complete software applications—probably distributed onto several servers—the tradeoff between energy consumption and user utility (i.e., energy efficiency).

Flinn et al. developed an approach for energy efficient execution of multiple applications on mobile devices. They measured the application's resource usage and derived mathematical models allowing for resource and thus, energy consumption approximation (Flinn & Satyanarayanan, 2004). At runtime, these formulas are used to decide whether to execute an application on the mobile device or on an external server. The approach does not always select the best solution as the decision for the optimal configuration is based on heuristic solvers. Nevertheless, Flinn's approach can be considered as an EAT approach, the application's deployment at runtime. However, it is tailored to the local versus remote execution scenario. Our approach is of much broader scope as it considers workload consolidation and migration of implementations between computing entities in general.

Another auto-tuning approach was developed by Lachenmann, Marrón, Minder, and Rothermel (2007). They developed the programming language Levels that can be used for the development of sensor networks. Levels allow defining different levels of Quality of Service (QoS) for sensors (e.g., data collection and transmission, or only message forwarding from other sensors). The different levels are annotated in the source code and for each such code block the energy consumption rate is annotated as well. At runtime, the system decides which QoS level each sensor should provide its current battery status. This way, Levels can help to improve the lifetime and operability of sensor networks. The Levels approach can be considered as a tailored EAT approach as it focuses on the specific domain of sensor networks.

A further approach for QoS optimization of software applications on mobile devices has been developed by Fei, Zhong, and Jha (2008). The development of software applications includes the modeling of different power modes and their provided QoS, respectively. The user can prioritize the QoS he is expecting and the operating system schedules the applications, the required utility and energy efficiency. Thus, Fei et al. negotiate the tradeoff between energy consumption and user utility similar to our EAT approach. In contrast to our approach, Fei et al. focus on multiple applications on a single mobile device whereas our approach focuses on multiple applications executed on multiple mobile and stationary devices. We expect to gain larger energy savings (e.g., by workload consolidation) as unutilized devices can be powered down during runtime.

Microsoft's research project Cuanta focuses on developing a mechanism to operate clouds of servers in an energy-optimal manner. An investigation how workloads can be consolidated in a server cloud scenario to optimally configure resource utilization, i.e. energy consumption showed that energy-optimal operation is situated between low and high utilization of server resources as low utilization leads to avoidable power consumption during idle phases whereas high utilization leads to energy consumption due to execution delays (Srikantaiah, Kansal, & Zhao, 2008). Furthermore, the work proposes an algorithm to optimally pack workloads on servers performance and energy consumption. After estimating the optimal configuration, utilization, workloads are deployed to servers in an energy-optimal manner. Thus, Cuanta can be considered as an energy auto-tuning approach. However, in contrast to our approach software component interaction is not considered. Furthermore, migration and reconfiguration which are essential for auto-tuning are not part of the Cuanta approach.

4.2. Modeling Power Consumption and Multi-Quality Component Models

Besides our CCM, several other research projects have focused on developing component models that integrate NFPs into modeling as well as modeling the energy consumption of (at least hardware) components. A detailed discussion on power modeling in the context of servers is presented in Rivoire (2008). In addition, we refer the interested reader to Aagedal (2001), which contains fundamental insights into the concepts required in multi-quality component models.

The COMQUAD research project developed a component model including NFP descriptions. Components were separated into their specifications and implementations, to allow modeling of multiple implementation variants for the same component specification (Göbel, Pohl, Röttger, & Zschaler, 2004). Besides, CQML+ contracts (Röttger & Zschaler, 2003) were used to specify NFPs and coarse-grain resource dependencies for NFP-aware scheduling of software components on single-server applications. Both the COMQUAD component model and CQML+ majorly inspired our work on CCM and ECL as well as the design of our resource and energy managers.

The consecutive research projects MADAM and MUSIC developed a component model for self-adaptive applications on mobile devices (Geihs, Khan, Reichle, Solberg, Hallenstein, & Merral, 2006). As this component model supports modeling of multiple implementation variants as well as NFPs, it can be considered as another inspiration of our CCM. In contrast to our approach, MADAM/MUSIC's component model does not consider the distribution of software applications on multiple servers (or mobile devices) and does not support energy optimization of complete hardware landscapes.

A third major influence for the development of CCM was the HRC component model emerging from the SPEEDS project (The SPEEDS Consortium, 2009). HRC is currently revised and improved within the CESAR research project (Baumgart, 2010; Baumgart, Reinkemeier, Rettberg, Stierand, Thaden, & Weber, 2010; Armengaud, et al., 2011). It focuses on the component-based development for embedded systems including the capability for both software and hardware modeling and their behavior. HRC uses contracts to specify NFPs. Each contract consists of requirements and provisions and can be aligned to a certain viewpoint (e.g., real-time or safety). Whereas CCM focuses on EAT at runtime, HRC focuses on verification and testing of embedded systems at design time. However, both approaches use contracts to specify NFPs of software and hardware components.

A further, closely related research project is DIVA. The key focus of this project was on the management of dynamic adaptive systems, where especially the problem of exponential growth of potential system configurations has been investigated. DIVA provides a solution by combining methods from aspect-oriented programming/modeling (Kiczales, Lamping, Mendhekar, Maeda, Lopes, & Loingtier, 1997) and MDSD (Morin, Barais, Nain, & Jézéquel, 2009). The DIVA approach allows for automatic adaptation of a system at runtime supporting goal-based optimization of NFPs as well as rule-based reconfiguration of the system (Fleurey & Solberg, 2009). The key difference to our approach is the granularity of the optimization problem. DIVA symbolizes the impact of implementations on NFPs (i.e., the free size of memory is represented by symbols like *LOW*, *MEDIUM*, and *HIGH* and the impact of an implementation can only be expressed as being low, medium, and so on, too). Our approach supports sub-symbolic information in addition (i.e., the actual value of free size of memory in megabyte). We encapsulate symbolization in our contracts, where quality modes denote the symbols and the expressions on the required NFPs describe the domain of the symbol. However, reasoning on sub-symbolic information is less efficient, due to the raised complexity, it allows to derive finer-grained configurations (e.g., a configuration could include not just the information which CPU to use, but the [optimal] frequency this CPU should have). Such fine-grained information allows reducing energy consumption in addition to the course-grain decision of which resources to use. Current hardware is usually far from being energy-proportional (Barroso & Hölzle, 2007), which is reflected by a very high baseload electricity and a narrow working area.

Imagine, for example, a server consuming 100W being idle and 120W at full load. In consequence, energy savings can mostly be achieved by selecting or turning off the right resources. In such a scenario symbolic reasoning, as in DIVA, is feasible, but especially for the next generation of hardware, which is supposed to be more and more energy-proportional (Borkar & Chien, 2011) there is a need for additional, finer-grained energy optimizations. Therefore, in contrast to DIVA, our approach allows for sub-symbolic as well as symbolic reasoning.

Another modeling approach focusing on NFPs is the Modeling and Analysis of Real-Time Embedded Systems (MARTE) profile (Object Management Group, 2011) for UML that extends UML with capabilities to describe NFPs for both, software and hardware components. Typically, NFPs defined in MARTE are real-time and performance properties. However, the definition of power consumption rates is possible as well. Arpinen et al. presented an extension for the MARTE profile (Arpinen, Salminen, Hämäläinen, & Hännikäinen, 2010) that allows defining power state machines for hardware devices that are similar to the energy state charts used in CCM for energy assessment of hardware devices. Nevertheless, they cover the behavior of SW components only by mapping use cases to system states and have no template mechanism for reuse. A further approach using UML and MARTE for the modeling of systems that can be reconfigured NFPs at runtime has been presented by Quadri, Gamatié, Boulet, and Dekeyser (2010). Although they mention power consumption as an NFP, the details of their power consumption model are not further elaborated. Furthermore, their approach focuses on hardware-near embedded systems whereas our approach focuses on component-based software applications deployed on sever landscapes.

4.3. Resource Utilization and Energy Assessment of Software Applications

The research of software's resource utilization or even its energy consumption is a challenging research topic and an important prerequisite for EAT.

A general introduction into energy-aware computing and different aspects that influence the energy consumption of software and hardware applications can be found in Ellis (2007). An introduction into performance evaluation and profiling can be found in Fortier and Michel (2003).

Lafond et al. investigated how to predict the average energy consumption of Java-based applications. They profiled Java bytecode instructions by executing them on a specific JVM and hardware landscape (Lafond & Lilius, 2006). They were able to measure and predict the energy consumption rate for a large subset of the Java bytecode instruction set and evaluated their results on several benchmarks.

Similar work has been done by Seo, Malek, and Medvidovic (2007, 2008). Again, the energy consumption of Java bytecode instructions was profiled resulting in a framework for the prediction of Java applications' energy consumption. Besides profiling and prediction, the predicted values were compared with actual consumption rates by measuring the consumption through the application's execution resulting in a variance of less than five percent between predicted and measured energy consumption rates.

Another framework for the resource utilization analysis of Java applications has been built by (Navas, Méndez-Lojo, and Hermenegildo (2006, 2008). They developed a framework that employs formal methods and control-flow graph analysis to derive formulas for a Java application's resource utilization depending on its input parameters. The user has to specify expected formulas for the NFPs he is interested in (e.g., an expected correlation of a method's input parameters and its energy consumption) and the framework will optimize this formula resulting in a more precise prediction for the NFPs. The framework's energy prediction is based on the results from Lafond et al. for energy prediction (Lafond & Lilius, 2006). The approach of Navas et al. can be compared to our approach for a Java program's resource utilization analysis. However, our approach uses profiling able to derive formulas for resource utilization without hints specified by the user.

A further approach profiling and prediction of applications' energy consumption has been presented by Kansal and Zhao (2008). They developed the Windows tool JouleMeter that allows predicting the energy consumption of processes running on a Windows desktop PC. The tool is calibrated using either the battery sensor of the PC running in battery power mode or an external power measuring tool when running on a continuous power supply. Afterwards, JouleMeter is able to estimate the power consumption of processes by profiling their resource utilization (e.g., CPU usage and memory allocation).

5. FUTURE WORK

Regarding future work and remaining challenges for EAT we discuss three important parts in the following. These comprise extensions at the user layer of THEATRE, improved simulation and profiling mechanisms and an evaluation of EAT.

Currently, our starting point for energy optimization is a single feature request to an EAT-enabled application. Such requests are combined with quality requirements of a single user in a textual language. However, in realistic distributed applications there is a multitude of users having different quality requirements. Considering each single user for EAT in such a case is not manageable at runtime. Hence, we plan to investigate multi-user models in the future.

A key challenge for energy auto-tuning is to estimate the energy consumption of an application for a given user request. For this purpose, we use ESCs in combination with the JRUPI framework. An important drawback of ESCs is that they do not support parallelism. Instead, the simulation of multiple ESCs in parallel has to be transformed into sequential execution traces. Although for small amounts of only few ESCs this transformation does not pose a problem, it does not scale. As an alternative, we plan to investigate Petri nets (Petri, 1962), because they provide means for parallel execution and a variety of tools to simulate Petri nets already exists.

An important limitation of JRUPI in its current state of development is that the resource usage model cannot be used to derive fine-grained workloads (e.g., for the ESC simulator or Petri nets). This is because JRUPI considers resource usage in general and not over time. This may lead to an inaccurate workload and in consequence to variances of error in the simulation result. Therefore, we plan to extend JRUPI in a sense that extracted resource usage models are based on time, so that the workload for the simulator can be derived more accurately. That is the offline analysis of JRUPI will be extended to answer the question, which resources have been utilized over which periods of time. Another important drawback of this approach is that resource usage models depend on concrete resources. Imagine two servers and a software component variant that writes a certain amount of data to a HDD. The first server has just a single HDD, whereas the second server has multiple HDDs organized in a RAID system. In order to write the same amount of data by the software component, the two servers will have different resource usage statistics, because different resources are utilized. Hence, for each server a separate resource usage model needs to be determined (as long as they do not consist of completely identical hardware). Consider a software component having N implementations and M servers (with different configurations). To fully cover such a HW/SW system, N * M resource usage models need to be determined. JRUPI uses non-functional benchmarks written by the component developers to utilize the resources for profiling in a meaningful way. We plan to automate the process of benchmark execution, profiling and resource usage model determination, so all needed models can be computed on the target (i.e., productive) infrastructure once per server before the EAT system is started the first time for productive use.

A prototype of THEATRE (based on OSGi) and a modeling tool suite for CCM and ECL (based on Eclipse) have already been developed. Nevertheless, to evaluate our EAT approach we need to measure the energy consumption of the whole IT infrastructure as well as of single hardware resources. Measured results have to be compared with the predicted energy consumption based on CCM models, ECL contracts and user requests. We started the implementation of an energy-consumption profiling and testing framework for Java applications (Wilke, Götz, Reimann, & Aßmann, 2011) that will be further improved and extended in future work. Another important point is to consider the energy consumption caused by energy auto-tuning itself. This comprises energy effects due to the optimization process itself, energy effects due to system reconfiguration as well as energy effects due to collecting data about the EAT system at runtime. Results of this evaluation should be considered in further refinements of EAT mechanisms.

As a case study, we are implementing the stock tracking application described in Section 2. Further, we plan to implement a second case study based on a video application. Such energy-aware applications are executed in the THEATRE, whose basic infrastructure (i.e., corresponding managers) is implemented as OSGi components (OSGi Alliance, 2011). So far, resource managers are able to collect resource specific data at server level and to deliver an infrastructure model to the global energy manager. Energy managers are able to monitor deployed software component types and software components which are OSGi bundles containing specific metadata. User managers are not considered as far as we are currently focusing on a single user model. For this purpose, the already implemented textual request language is sufficient. The energy auto-tuning mechanisms explained in this chapter are also implemented prototypically. Hence, the current implementation of THEATRE allows to use data from other layers and to execute the EAT control loop except for the acting phase. The latter requires an interpreter for the reconfiguration plan that is part of current work. In fact, it is possible to estimate the best software/hardware mapping of a given application based on simplified assumptions regarding the energy consumption of software components.

6. CONCLUSION

This chapter provided an overview of our approach for Energy Auto-Tuning (EAT), which allows for self-optimization the tradeoff between the contradicting goals of reducing energy consumption and improving user utility. It enables the modeling and development of complex hardware/software systems with special focus on NFPs. We elaborated on the architectural elements of EAT systems in Section 2. These include the CCM, our component model to capture software components and resources of the underlying IT infrastructure as well as their energy-related behavior and the ECL, our contract language to express dependencies between software components and between software components and resources in terms of NFPs. In Section 3 we discussed mechanisms required by EAT systems. These include

1. Utility assessment by contract checking: determining system configurations, which are able to fulfill the utility requirements of the user,
2. Contract negotiation: which aims to derive the optimal of the previously identified system configurations,

3. Reconfiguration: which derives a plan of required actions to switch from one configuration to another, and
4. Energy assessment: answering the question how much energy will be consumed by invoking a feature in a certain system configuration.

The EAT approach covers a wide range of research questions from different areas: Profiling energy usage of feature invocations, simulation and self-optimization as well as -adaptation to name but a few. Not all research questions EAT system have been discussed conclusively as we have shown in the previous section. We plan to do further research in these areas in the research projects CoolSoftware, ZESSY/QualiTune and the Collaborative Research Center HAEC, which integrates research groups from electrical engineering and computer science.

ACKNOWLEDGMENT

This research is part of the projects ZESSY #080951806, CoolSoftware #FKZ13N10782 and the DFG-funded Collaborative Research Center 912 (HAEC). The ZESSY project is funded by the European Social Fund and Federal State of Saxony. The project CoolSoftware is part of the Leading-Edge Cluster "Cool Silicon," which is sponsored by the Federal Ministry of Education and Research (BMBF) within the scope of its Leading-Edge Cluster Competition.

REFERENCES

Armengaud, E., Zoier, M., Baumgart, A., Biehl, M., Chen, D., Griessnig, G., et al. (2011). Model-based toolchain for the efficient development of safety-relevant automotive embedded systems. In *Proceedings of the SAE 2011 World Congress & Exhibition*. SAE.

Arpinen, T., Salminen, E., Hämäläinen, T., & Hännikäinen, M. (2010). Extension to MARTE profile for modeling dynamic power management of embedded systems. In *Proceedings of the 1st Workshop on Model Based Engineering for Embedded Systems Design (M-BED), co-located with DATE 2010*. Dresden, Germany: IEEE.

Baumgart, A. (2010). A common meta-model for the interoperation of tools with heterogeneous data models. In *Proceedings of the ECMFA 2010 - 3rd Workshop on Model-Driven Tool & Process Integration*. ECMFA.

Baumgart, A., Reinkemeier, P., Rettberg, A., Stierand, I., Thaden, E., & Weber, R. (2010). A model-based design methodology with contracts to enhance the development process of safety-critical systems. In *Proceedings of the Software Technologies for Embedded and Ubiquitous Systems,* (pp. 59-70). Berlin, Germany: Springer.

Climate Group. (2008). *SMART 2020: Enabling the low carbon economy in the information age. Paper presented as Report on Behalf of the Global eSustainability Initiative (GeSI)*. London, UK: Climate Group.

Fei, Y., Zhong, L., & Jha, N. K. (2008). An energy-aware framework for dynamic software management in mobile computing systems. *ACM Transactions on Embedded Computer Systems*, *7*, 1–31. doi:10.1145/1347375.1347380

Fleurey, F., & Solberg, A. (2009). A domain specific modeling language supporting specification, simulation and execution of dynamic adaptive systems. In *Proceedings of Model Driven Engineering and Systems* (pp. 606–621). Berlin, Germany: Springer. doi:10.1007/978-3-642-04425-0_47

Flinn, J., & Satyanarayanan, M. (2004). Managing battery lifetime with energy-aware adaptation. *ACM Transactions on Computer Systems*, *22*, 137–179. doi:10.1145/986533.986534

Fortier, P. J., & Michel, H. E. (2003). *Computer systems performance evaluation and prediction*. Burlington, MA: Digital Press.

Frigo, M., & Johnson, S. G. (1998). FFTW: An adaptive software architecture for the FFT. In *Proceedings of the 1998 IEEE International Conference on Acoustics Speech and Signal Processing,* (pp. 1381-1384). IEEE Press.

Gartner, Inc. (2007). *Gartner estimates ICT industry accounts for 2 percent of global CO2 emmisions*. Press Release. New York, NY: Gartner.

Geihs, K., Khan, M. U., Reichle, R., Solberg, A., Hallenstein, S., & Merral, S. (2006). Modeling of component-based adaptive distributed applications. In *Proceedings of the 2006 ACM Symposium on Applied Computing,* (pp. 718-722). ACM Press.

Harel, D. (1987). Statecharts: A visual formalism for complex systems. *Science of Computer Programming, 8*, 231–274. doi:10.1016/0167-6423(87)90035-9

Hewlett-Packard, I. Microsoft, Phoenix Technologies, & Toshiba. (2010). *Advanced configuration and power interface specification, revision 4.0a*. Washington, DC: HP, Intel, Microsoft, Phoenix, Toshiba.

Kansal, A., & Zhao, F. (2008). Fine-grained energy profiling for power-aware application design. In *Proceedings of the First Workshop on Hot Topics in Measurement and Modeling of Computer Systems (HotMetrics08) at ACM Sigmetrics.* ACM Press.

Kiczales, G., Lamping, J., Mendhekar, A., Maeda, C., Lopes, C., & Loingtier, J.-M. (1997). Aspect-oriented programming. In *Proceedings of the Object-Oriented Programming,* (pp. 220-242). Berlin, Germany: Springer.

Lachenmann, A., Marrón, P. J., Minder, D., & Rothermel, K. (2007). Meeting lifetime goals with energy levels. In *Proceedings of the 5th International Conference on Embedded Networked Sensor Systems,* (pp. 131-144). ACM Press.

Lafond, S., & Lilius, J. (2006). An energy consumption model for an embedded Java virtual machine. In *Proceedings of Architecture of Computing Systems - ARCS 2006* (pp. 311–325). Berlin, Germany: Springer. doi:10.1007/11682127_22

Morin, B., Barais, O., Nain, G., & Jézéquel, J.-M. (2009). Taming dynamically adaptive systems using models and aspects. In *Proceedings of the 31st International Conference on Software Engineering (ICSE),* (pp. 122-132). Washington, DC: IEEE Computer Society.

Navas, J., Méndez-Lojo, M., & Hermenegildo, M. (2006). *Customizable resource usage analysis for Java bytecode - Deliverable 2.6 - Preliminary report on advanced resource policies*. Java.

Navas, J., Mendez-Lojo, M., & Hermenegildo, M. (2008). Safe upper-bounds inference of energy consumption for Java bytecode applications. In *Proceedings of The Sixth NASA Langley Formal Methods Workshop*, (pp. 29-32). NASA.

Object Management Group. (2011). *UML profile for MARTE: Modeling and analysis of real-time embedded systems, version 1.1.* Retrieved September 9, 2011, from http://www.omg.org/spec/MARTE/.

Oracle. (2010). *BTrace user's guide*. Retrieved July 21, 2011, from http://kenai.com/projects/btrace/pages/UserGuide.

OSGi Alliance. (2011). *OSGi service platform release 4*. Retrieved July 29, 2011, from http://www.osgi.org/Release4/.

Petri, C. A. (1962). *Kommunikation mit Automaten.* Bonn, Germany: Institut für instrumentelle Mathematik der Universität Bonn.

Quadri, I., Gamatié, A., Boulet, P., & Dekeyser, J. (2010). *Modeling of configurations for embedded system implementations in MARTE*. Paper presented at the 1st Workshop on Model Based Engineering for Embedded Systems Design (M-BED), co-located with DATE 2010. Dresden, Germany.

Ranganathan, P., & Nygard, K. (2010). An optimal resource assignment problem in smart grid. In *Proceedings of the Second International Conference on Future Computational Technologies and Applications (FUTURE COMPUTING)*, (pp. 28-34). Lisbon, Portugal: IARIA.

Schmidt, M., & Lipson, H. (2009). Distilling free-form natural laws from experimental data. *Science*, *324*(5923), 81–85. doi:10.1126/science.1165893

Seo, C., Malek, S., & Medvidovic, N. (2007). An energy consumption framework for distributed Java-based systems. In *Proceedings of the 22nd IEEE/ACM International Conference on Automated Software Engineering*. Atlanta, GA: IEEE Press.

Seo, C., Malek, S., & Medvidovic, N. (2008). Estimating the energy consumption in pervasive Java-based systems. In *Proceedings of the 2008 Sixth Annual IEEE International Conference on Pervasive Computing and Communications*. IEEE Press.

SPEEDS Consortium. (2009). *D.2.1.5 SPEEDS L-1 meta-model*. Retrieved from http://speeds.eu.com/downloads/SPEEDS Meta-Model.pdf.

Srikantaiah, S., Kansal, A., & Zhao, F. (2008). Energy aware consolidation for cloud computing. In *Proceedings of the 2008 Conference on Power Aware Computing and Systems*. IEEE.

Sun. (2008). *Solaris dynamic tracing guide*. Retrieved May 10, 2011, from http://download.oracle.com/docs/cd/E19253-01/817-6223/.

Tsirogiannis, D., Harizopoulos, S., & Shah, M. A. (2010). Analyzing the energy efficiency of a database server. In *Proceedings of the 2010 International Conference on Management of Data (SIGMOD 2010)*, (pp. 231–242). New York, NY: ACM Press.

Whaley, R. C., & Dongarra, J. (1997). *Automatically tuned linear algebra software*. Knoxville, TN: University of Tennessee.

ADDITIONAL READING

Aagedal, J. Ø. (2001). *Quality of service support in development of distributed systems*. PhD Thesis. Oslo, Norway: University of Oslo.

Barroso, L. A., & Hölzle, U. (2007). The case for energy-proportional computing. *IEEE Computer*, *40*(12), 33–37. doi:10.1109/MC.2007.443

Benini, L., Hodgson, R., & Siegel, P. (1998). System-level power estimation and optimization. In *Proceedings of the International Symposium on Low Power Electronics and Design (ISLPED-98)*, (pp. 173–178). New York, NY: ACM Press.

Beugnard, A., Jézéquel, J.-M., & Plouzeau, N. (2010). Contract aware components, 10 years after. *Electronic Proceedings in Theoretical Computer Science*, *37*, 86–100. doi:10.4204/EPTCS.37.1

Beugnard, A., Jézéquel, J.-M., Plouzeau, N., & Watkins, D. (1999). Making components contract aware. *IEEE Computer*, *32*, 38–45. doi:10.1109/2.774917

Borkar, S., & Chien, A. A. (2011). The future of microprocessors. *Communications of the ACM*, *54*, 67–77. doi:10.1145/1941487.1941507

Chung, L., & do Prado Leite, J. C. (2009). On non-functional requirements in software engineering. In *Proceedings of Conceptual Modeling: Foundations and Applications* (pp. 363–379). Berlin, Germany: Springer. doi:10.1007/978-3-642-02463-4_19

Chung, L., Nixon, B. A., Yu, E., & Mylopoulos, J. (2000). *Non-functional requirements in software engineering*. Norwell, MA: Kluwer Academic Publishers.

Dasgupta, S., Papadimitriou, C., & Vazirani, U. (2007). *Algorithms*. Columbus, OH: McGraw-Hill Higher Education.

Dobson, S., Denazis, S., Fernández, A., Gärti, D., Gelenbe, E., & Massacci, F. (2006). A survey of autonomic communications. *ACM Transactions on Autonomous and Adaptive Systems, 1*(2), 223–259. doi:10.1145/1186778.1186782

Ellis, C. (2007). *Controlling energy demands in mobile computing systems*. New York, NY: Morgan and Claypool Publishers.

Göbel, S., Pohl, C., Röttger, S., & Zschaler, S. (2004). The COMQUAD component model: Enabling dynamic selection of implementations by weaving non-functional aspects. In *Proceedings of the 3rd International Conference on Aspect-Oriented Software Development,* (pp. 74-82). Lancaster, UK: IEEE.

Götz, S., Wilke, C., Cech, S., & Assmann, U. (2011). To appear: Runtime variability management for energy-efficient software by contract negotiation. In *Proceedings of 6th International Workshop Models@run.time*. Models@run.time.

Götz, S., Wilke, C., Schmidt, M., Cech, S., & Assmann, U. (2010). Towards energy auto tuning. In *Proceedings of First Annual International Conference on Green Information Technology (GREEN IT),* (pp. 122-129). Singapore: GSTF.

Götz, S., Wilke, C., Schmidt, M., Cech, S., Waltsgott, J., & Fritzsche, R. (2010). THEATRE resource manager interface specification: *Vol. 1. 0, technical report TUD-FI10-08*. Dresden, Germany: Technische Universität Dresden.

Jacob, B., Larson, P., Leitao, B. H., & da Silva, S. A. (2010). *SystemTap: Instrumenting the Linux kernel for analyzing performance and functional problems*. Seattle, WA: IBM Redbooks.

Meyer, B. (1992). Applying "design by contract". *Computer, 25*(10), 40–51. doi:10.1109/2.161279

Meyer, B. (1997). *Object-oriented software construction* (2nd ed.). New York, NY: Prentice Hall.

Nemhauser, G. L., & Wolsey, L. A. (1988). *Integer and combinatorial optimization*. New York, NY: Wiley-Interscience.

Püschel, G. (2011). *Energieeffizienz in workflow-systemen*. Diploma Thesis. Dresden, Germany: Technische Universität Dresden.

Rivoire, S. M. (2008). *Models and metrics for energy-ef□cient computer systems*. PhD Thesis. Palo Alto, CA: Stanford University.

Röttger, S., & Zschaler, S. (2003). Enhancements to CQML. In *Proceedings of the 1st International Workshop on Quality of Service in Component-Based Software Engineering,* (pp. 43-56). IEEE.

Süttner, P. (2011). *Abstrakte verhaltensbeschreibung von CCM softwarekomponenten*. Diploma Thesis. Dresden, Germany: Technische Universität Dresden.

Szyperski, C., Gruntz, D., & Murer, S. (2002). *Component software - Beyond object-oriented programming*. Boston, MA: Addison-Wesley / ACM Press.

Wilke, C., Götz, S., Reimann, J., & Aßmann, U. (2011). Vision paper: Towards model-based energy testing. In *Proceedings of the ACM/IEEE 14th International Conference on Model Driven Languages and Systems (MODELS2011),* (pp. 480-489). Wellington, New Zealand: Springer.

KEY TERMS AND DEFINITIONS

Non-Functional Property: A characteristic, which reflects *how* a component operates. For example, the response time, expressing how fast an operation of a component is.

Quality Contract: Defines dependencies between software and hardware components in terms of provided and required non-functional properties.

Contract Checking: The process of determining all system configurations, which fulfill the constraints defined in quality contracts.

Contract Negotiation: The process of computing the optimal system configuration, which serves the user's request, ensures the user's minimum requirements and requires the least energy.

User Utility: A function mapping non-functional properties—which are perceptible by the user (e.g., frame rate)—to a finite set of levels of satisfaction (e.g., video playback is perceived by the user as fluent, weak hesitant or strong hesitant).

System Configuration: A selection of (software) component implementations and their mapping onto system resources.

Energy Assessment: The process of determining how much energy is required by a certain feature of the system in a specified system configuration.

Self-Reconfiguration: The ability of a (component-based) system, to adjust itself at runtime by changing its configuration at runtime.

Chapter 4
Energy-Efficient Server Selection Algorithms for Distributed Applications

Tomoya Enokido
Rissho University, Japan

Ailixier Aikebaier
Seikei University, Japan

Makoto Takizawa
Seikei University, Japan

ABSTRACT

A wide range of distributed applications are developed on various types of scalable information systems like Peer-to-Peer (P2P) and cloud computing systems. Various types of distributed algorithms are so far developed to minimize the computation time and response time and to maximize the throughput. On the other hand, it is getting more significant to realize energy-aware information systems for reducing the consumption of natural resources. In order to realize the energy-aware information systems, it is critical to reduce the total electrical power consumption of computers and networks. In this chapter, the authors classify applications on distributed systems into computation, communication, and general types of applications. Then, they define power consumption models of a server to perform each type of application process. Based on the power consumption models, the authors propose algorithms to select one of the servers for each type of application so that the total power consumption of servers can be reduced and also evaluate the algorithms in terms of not only performance but also energy consumption.

1. INTRODUCTION

Information systems are now becoming scalable like Peer-to-Peer (P2P) (Buford, Yu, & Lua, 2009; Perera, 2008) and cloud systems (Grossman, 2009; Zhang & Zhou, 2009). Types of computational devices like server computers and sensor nodes are interconnected in networks like wireless and wired networks. There are various kinds of distributed applications like Web applications

DOI: 10.4018/978-1-4666-1839-8.ch004

(Bianchini & Carrera, 2000), database applications (Ferraiolo, Kuhn, & Chandramouli, 2007), and Google file systems (Ghemawat, Gobioff, & Leung, 2003) on distributed systems. In order to realize distributed applications, various types of distributed algorithms are so far developed. For example, algorithms for allocating computation resources like CPU and memory to processes and synchronizing multiple conflicting processes are discussed to minimize the computation time and response time, maximize the throughput, and minimize the memory space (Rahimi & Haug, 2010; Enokido & Takizawa, 2010). On the other hand, it is getting more significant to realize energy-aware information systems for reducing the consumption of natural resources and resolve air pollution on the Earth as discussed in Green IT (Hemmert, 2010; Murphy, Sterling, & Dekate, 2010). It is critical to reduce the total electrical power consumption of information systems in order to realize energy-aware information systems. Various kinds of hardware technologies like low-power consumption CPUs are now being developed. We do not discuss the hardware specification of each server like CPU and memory at the micro level. We rather consider how much electronic power a whole server consumes at a macro level. We measure how much electric power a server consumes to perform each type of application by using the power meter Watts up?.Net (Watts up? Products, 2011) where the power consumption of each server can be measured every one second.

In this chapter, we consider a cluster of server computers and discuss how to reduce the total power consumption of the server computers. In the cluster system, a client issues a request to a load balancer. Each time the load balancer receives a request from a client, the load balancer selects a server in the cluster and forwards the request to the server. On receipt of the request, the server performs the request as an application process and sends a response to the client. We classify applications, which are performed on the cluster system into three types: computation, communication, and general types of applications. In the computation type applications, a client issues a request to a server and the server mainly consumes computation resources to process the request. Database applications are typical examples of the computation type applications. In the communication type applications, a server mainly consumes network resources to transmit a large volume of data to a client like File Transfer Protocol (FTP) applications. In the general type of applications, applications are composed of both the computation and communication modules. Apache Web server (Apache HTTP Server Version 2.0 Documentation, 2011) using the Deflate function (Apache Module mod_deflate, 2011) is a typical example of the general type of applications. Suppose a client issues a request to a Web server, which is implemented by Apache 2.0 to obtain a Web page. On receipt of the request, the Web server obtains data in requested Web pages and compresses the data into a file by using the Deflate module. Then, the Web server transmits the compressed file to the requesting client.

First, we discuss a simple computation model and a power consumption model of a server for a computation type application. In the simple computation model, any application process is performed on a server with maximum clock frequency and the more number of application processes are concurrently performed, the larger computation resources is consumed. According to the simple computation model, we discuss how to estimate when current application processes terminate on a server. In the simple power consumption model, if at least one application process is performed on a server, the electric power is maximally consumed on the server. Based on the simple computation model and the power consumption model, we propose two types of server selection algorithms to reduce the total power consumption of servers for computation type applications: *Computation Laxity Based* (CLB) and *Power Consumption Laxity Based* (PCLB) algorithms. In the CLB algorithm, a server which can most early finish processing a request is selected for a requesting client. In the PCLB algorithm, a server which consumes the minimum power consumption is selected for a requesting client. Then, we evaluate the CLB and PCLB algorithms in terms of total power consumption and elapse time compared with the basic Round-Robin (RR) algorithm (Job Scheduling Algorithms in Linux Virtual Server, 2011).

Next, we discuss a file transmission model and a power consumption model for a communication type application. We define two types of file transmission models: server bound and client bound models. In the server-bound model, the total amount of maximum receipt rates of all the current clients on a server is larger than the effective transmission rate of the server. Hence, the more number of clients transmit requests to the server, the longer it takes to transmit files to the clients. On the other hand, the total amount of maximum receipt rates of all the current clients on a server is smaller than the effective transmission rate of a server in the client bound model. Hence, every client can receive a file at the maximum receipt rate. Then, we measure how much electric power a server spends to transmit files to clients. According to the experimentation, we define the power consumption model for communication type applications. In the power consumption model, the power consumption rate of a server to transmit files to clients is lineally increased for the total transmission rate. Based on the file transmission model and the power consumption model, we propose three types of server selection algorithms to reduce the total power consumption of servers for communication type applications: Power Consumption-Based (PCB), Transmission Rate-Based (TRB), and Extended Power Consumption-Based (EPCB) algorithms. In the TRB algorithm, a server is selected for each requesting client where the transmission rate of the server to transmit a file to the client is the largest. In the PCB algorithm, a server is selected for a client where the power consumption to transmit a file to the client is smallest. In the PCB algorithm, we consider only the power consumption of a server transmitting a file to a client at time when the server starts transmitting the file. However, the server might be transmitting files to other clients at the same time. Hence, the power consumption where the server transmits files to other clients has to be considered. In the EPCB algorithm, the total amount of power consumption of a server to transmit files to every requesting client is considered to make more accurate the estimation of the power consumption. Then, we evaluate the TRB, PCB, and EPCB algorithms in terms of the total power consumption and elapse time compared with the RR algorithm.

Finally, we discuss three models, the computation, transmission, and power consumption models for general type of applications. A general type of application is composed of computation and transmission processes. Here, computation and transmission processes might be concurrently performed with each other. Hence, we have to take into account the concurrent execution of the computation and transmission processes to define the computation and transmission models. We measure how much electric power a server consumes to perform general type application processes. According to the experimentation, we extend the simple computation model to genera type applications. We find that the transmission model defined for communication type application can be adopted for general type application according to the experimentation. In addition, we define the simple power consumption model for general type of applications. In the power consumption model, if at least one computation process is performed on a server, the server maximally consumes the electric power. If only transmission processes are performed on a server, the electric power of the server depends on the total amount of the transmission rate of the server. According to the computation model, transmission model, and power consumption model, we propose the Extended Power Consumption Laxity-Based

(EPCLB) algorithm to select a server for general type of applications. In the EPCLB algorithm, a server which consumes the minimum power consumption is selected for a requesting client. We evaluate the EPCB algorithm in terms of the total power consumption and elapse time compared with the RR algorithm.

In section 2, we show the related works. In section 3, we define a cluster of servers, which we consider in this chapter. In section 4, we define simple computation and power consumption models for computation type applications. Then, we propose the CLB and PCLB algorithms to select a server in a set of servers so as to not only satisfy some constraint like deadline one but also reduce the total power consumption of servers. We evaluate the CLB and PCLB algorithms in terms of total power consumption compared with the traditional RR algorithm. In section 5, we define the file transmission model and the power consumption model for communication type applications. Then, we propose the PCB, TRB, and EPCB algorithms to select a server for reducing the total power consumption of servers. We evaluate the PCB, TRB, and EPCB algorithms in terms of total power consumption compared with the traditional RR algorithm. In section 6, we discuss the computation, transmission, and power consumption models for the general type application. Then, we propose the EPCLB algorithm to reduce the total power consumption of servers. We evaluate the EPCLB algorithm compared with the traditional RR algorithm.

2. RELATED WORKS

In the green IT technologies (Hemmert, 2010; Murphy, Sterling, & Dekate, 2010), the total electric power consumption of computers and networks has to be reduced. In wireless sensor networks (Yang, Mino, Spaho, Barolli, Durresi, & Xhafa, 2011; Liu, Zhao, & Li, 2011) and ad hoc networks (Ikeda, Kulla, Hiyama, Barolli, & Takizawa, 2011; Durresi, Durresi, Paruchuri, & Barolli, 2011), routing algorithms to reduce the power consumption of battery in each node are discussed. A cloud computing (Grossman, 2009; Zhang & Zhou, 2009) is now getting popular in various types of distributed applications. A cloud computing system is composed of a huge number of server computers like Google file systems (Ghemawat, Gobioff, & Leung, 2003). Various types of load-balancing algorithms are discussed for distributed systems. Bevilacqua discussed a method to obtain efficient load balancing for data parallel applications through dynamic data assignment and a simple priority mechanism on a heterogeneous cluster system (Bevilacqua, 1999). Colajanni *et al.* discussed a load balancing algorithm which adjusts the time-to-live values of the DNS based scheduler for geographically distributed heterogeneous Web servers (Colajanni, Cardellini, & Yu, 1998). The Linux Virtual Server (LVS) is a software to build a cluster system and some types of load balance algorithms are implemented like round-robin and least-connection algorithms (Job Scheduling Algorithms in Linux Virtual Server, 2011). By these algorithms, the load of each server is balanced to maximize the throughput but do not take into account the reducing of the power consumption. On the other hand, Rajamani *et al.* identified the key factors in the system-workload context that impact energy saving policies in a cluster system and discussed how to reduce the total power consumption of servers by turning off servers (Rajamani & Lefurgy, 2003). Yang *et al.* proposed an optimized method to find the required number of Web servers in a cluster system (Yang, Xiong, Aikebaier, Enokido, & Takizawa, 2009). Biancini *et al.* discussed how to reduce the power consumption of a data center with a cluster of homogeneous (Bianchini & Rajamony, 2004) and heterogeneous (Heath, Diniz, Carrera, Meira, & Bianchini, 2005) server computers by turning off servers which are not required for handling a collection of requests. However, it is difficult to dynamically turn on and off servers in some types of systems like mission critical systems. In addition, computers cannot be turned off by other persons different from the owners in some types of computation models like Peer-to-Peer (P2P) and grid models. In this chapter, we do not consider the approach to turning off servers to reduce the total power consumption of servers. The approach can be adapted for not only cluster systems but also other scalable systems like P2P and grid systems.

3. A CLUSTER OF SERVERS

Let S be a set of multiple servers $s_1, ..., s_n$ ($n \geq 1$) each of which holds a full replica of data d as shown in Figure 1. Let C be a set of clients $c_1, ..., c_m$ ($m \geq 1$) which issue requests to servers in the server set S to obtain the data. We assume the networks are reliable and synchronous (Fischer, Lynch, & Paterson, 1985). A client c_s first issues a request to a load balancer K (Colajanni, Cardellini, & Yu, 1998; Bevilacqua, 1999; Job Scheduling Algorithms in Linux Virtual Server, 2011). The load balancer K selects one server s_t in the set S and forwards the the request to the server s_t. Here, the server s_t spends computation and communication resources to perform the request. For example, the CPU resources are consumed to encode multimedia data (Steinmetz & Nahrstedt, 2004) in Web applications. On receipt of a request, the server s_t performs a process p_i of the request. A term *process* means an *application process* on a server to handle a request issued by a client in this chapter. In the general type of applications, a process p_i manipulates the data d and then transmits a reply file f to the client c_s. A process p_i is thus composed of a pair of a *computation subprocess* cp_i and a *transmission subprocess* tp_i, i.e. $p_i = \langle cp_i, tp_i \rangle$. In the computation type applications, a process p_i is composed of only computation subprocess cp_i, i.e. $p_i = cp_i$. In the communication type applications, a process p_i is composed of only transmission subprocess, i.e. $p_i = tp_i$. A process being performed is *current* at time τ. A process, which already terminates before time τ is *previous* at time τ. Thus, there are a set P of application processes $p_1, ..., p_l$ ($l \geq 1$) to be performed on servers in the set S.

4. SERVER SELECTION ALGORITHMS FOR COMPUTATION TYPE APPLICATIONS

In this section, we define a pair of models, the computation and power consumption models of a server to perform computation type application process. Based on the power consumption model, we propose two types of server selection algorithms to select one of servers for computation

Figure 1. A cluster of servers

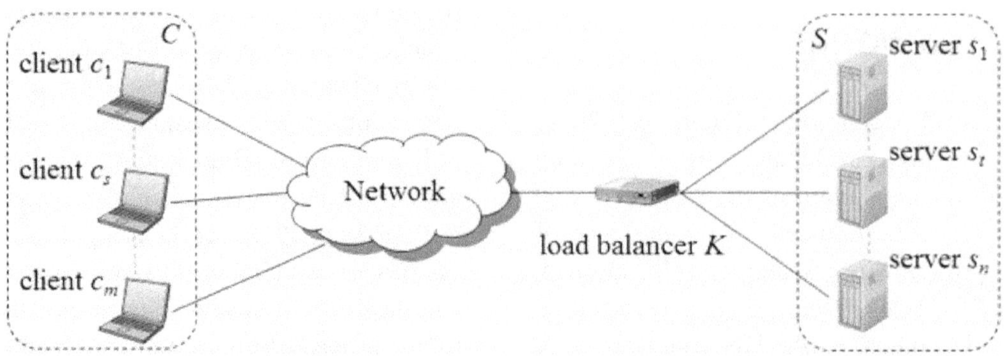

type applications so that the total power consumption of servers can be reduced.

4.1. Computation Model for Computation Type Applications

4.1.1. Computation Rate of a Server

Let $CPt(\tau)$ ($\subseteq P$) be a set of current processes on a server st at time τ. $N_t(\tau)$ shows the number of current processes, $N_t(\tau) = |CP_t(\tau)|$. If $CP_t(\tau) = \{p_i\}$ and $N_t(\tau) = 1$, a process p_i is *exclusively* performed on a server s_t at time τ. If $p_i \subset CP_t(\tau)$ and $N_t(\tau) > 1$, a process p_i is *interleaved* with another process $p_j \subset CP_t(\tau)$ $(i \neq j)$. Let $CP(\tau)$ show a set of current processes on every server at time τ, $CP(\tau) = \bigcup_{t=1,\ldots,n} CP_t(\tau)$. Suppose a process p_i is performed on a server s_t. Here, we consider the following parameters:

- T_{ti} = total computation time of a process p_i on a server s_t.
- $minT_{ti}$ = minimum computation time of a process p_i which is exclusively performed on a server s_t $(1 \leq minT_{ti} \leq T_{ti})$.
- $maxT_i = max(minT_{1i}, \ldots, minT_{ni})$.
- $minT_i = min(minT_{1i}, \ldots, minT_{ni})$.
- $F_{ti} = 1 / T_{ti}$ = *average computation rate* (*ACR*) of a process p_i on a server s_t $(0 < F_{ti} \leq 1 / minT_{ti} \leq 1)$ [1/tu].
- $maxF_{ti} = 1 / minT_{ti}$ = maximum *ACR* of a process p_i on a server s_t.
- $maxF_i = max\ (maxF_{1i}, \ldots, maxF_{ni})$.
- $minF_i = min\ (maxF_{1i}, \ldots, maxF_{ni})$.

If a process p_i is exclusively performed on the fastest server s_t and the slowest server s_u, $minT_i = minT_{ti}$ and $maxT_i = minT_{ui}$, respectively. We assume it takes at least one time unit [tu] to perform a process on any server. F_{ti} shows how many percentages of the total amount of computation of a process p_i are performed on a server s_t for one time unit.

The more number of processes are performed, the longer it takes to perform each of the processes on a server s_t. Let $\alpha_t(\tau)$ indicate the *computation degradation rate* of a server s_t at time τ $(0 \leq \alpha_t(\tau) \leq 1)$ [1/tu]. Here, $\alpha_t(\tau_1) \leq \alpha_t(\tau_2) \leq 1$ if $N_t(\tau_1) \leq N_t(\tau_2)$ for every pair of different time τ_1 and τ_2. $\alpha_t(\tau) = 1$ if $N_t(\tau) \leq 1$. In this chapter, $\alpha_t(\tau)$ is assumed to be $\varepsilon_t^{N_t(\tau)-1}$ where $0 \leq \varepsilon_t \leq 1$. Suppose it takes 50 [sec] to exclusively perform a process p_i on a server s_t. Here, $T_{ti} = minT_{ti} = 50$ [sec]. Here, $F_{ti} = maxF_{ti} = 1/50 = 0.02$ [1/sec]. Suppose that another process p_u where $minT_{tu} = 50$ starts at the same time $s\tau$ as the process p_i. If $\varepsilon_t = 0.98$, $\alpha_t(\tau) = \varepsilon_t = 0.98$ since $N_t(\tau) = 2$ for $\tau \geq s\tau$. Hence, $T_{tu} = 1.02 \cdot minT_{tu} = 51$ [sec] and $F_{tu} = 1/51$. We define the *normalized computation rate* (*NCR*) $f_{ti}(\tau)$ of a process p_i on a server s_t at time τ as follows:

Normalized Computation Rate (NCR)

$$f_{ti}(\tau) = \begin{cases} \alpha_t(\tau) \cdot maxF_{ti}\ /\ maxF_i\ [1/\text{tu}]. \\ \alpha_t(\tau) \cdot minT_i\ /\ minT_{ti}\ [1/\text{tu}]. \end{cases} \quad (1)$$

The maximum *NRC* $maxf_{ti}$ shows $maxF_{ti} / maxF_i$. $0 \leq f_{ti}(\tau) \leq maxf_{ti} \leq 1$. The *NCR* $f_{ti}(\tau)$ shows how many number of steps of a process p_i are performed on a server s_t at time τ. $maxf_t$ is defined to be the maximum computation rate of a server s_t. Here, $maxf_{ti}(\tau) \leq maxf_t$ for every process p_i and time τ.

Next, suppose that a process p_i starts and terminates on a server s_t at time $s\tau_{ti}$ and $e\tau_{ti}$, respectively. Here, $T_{ti} = e\tau_{ti} - s\tau_{ti}$. The following formula holds:

$$\int_{s\tau_{ti}}^{e\tau_{ti}} f_{ti}(\tau)\, d\tau \;=\; minT_i \;\cdot\; \int_{s\tau_{ti}}^{e\tau_{ti}} \frac{\alpha_t(\tau)}{minT_{ti}} d\tau = minT_i. \quad (2)$$

If a process p_i is exclusively performed on the server s_t, $T_{ti} = e\tau_{ti} - s\tau_{ti} = minT_{ti}$. If other processes are performed, i.e. $\alpha_t(\tau) < 1$, $T_{ti} = e\tau_{ti} - s\tau_{ti} = minT_{ti}\ /\ \alpha > minT_{ti}$ where $\alpha = \int_{s\tau_{ti}}^{e\tau_{ti}} \alpha_t(\tau) d\tau\ /\ (e\tau_{ti} - s\tau_{ti})$ (< 1). Here, $minT_i$

shows the total amount of computation to be performed by a process p_i. If a process p_i includes more number of execution steps than another process p_u, $minT_i > minT_u$. The more number of processes are concurrently performed with the process p_i on a server s_k at time τ, the smaller the NCR $f_{ki}(\tau)$ is. Suppose a process p_i starts at time $s\tau_{ki}$ and terminates at time $e\tau'_{ki}$. Here, $T_{ki} = e\tau'_{ki} - s\tau_{ki} > minT_i$ and $\int_{s\tau_{ki}}^{e\tau'_{ki}} f_{ki}(\tau)d\tau = minT_i$. The computation laxity $cl_{ti}(\tau) = minT_i - \int_{s\tau_{ti}}^{\tau} f_{ti}(x)dx$ shows how much computation a server s_t has to spend to perform up a process p_i at time τ: $cl_{ti}(s\tau_{ti}) = minT_i$ and $cl_{ti}(e\tau_{ti}) = 0$. If the process p_i would be exclusively performed on the server s_t, the process p_i is expected to terminate at time $e\tau_{ti} = \tau + cl_{ti}(\tau)$.

4.1.2. Knots of Processes

Each process p_i starts on a server s_t at time $s\tau_{ti}$ and terminates at time $e\tau_{ti}$. A process p_i is referred to as *precedes* another process p_j on a server s_t ($p_i \rightarrow_t p_j$) if $e\tau_{ti} < s\tau_{tj}$. A process p_i is *interleaved* with another process p_j on a server s_t ($p_i |_t p_j$) iff $e\tau_{tj} \geq e\tau_{ti} \geq s\tau_{ti}$ or $e\tau_{tj} \geq s\tau_{ti} \geq s\tau_{tj}$. A process p_i is *connected* with another process p_j on a server s_t ($p_i \leftrightarrow_t p_j$) iff (1) p_i is interleaved with p_j ($p_i |_t p_j$) or (2) p_i is interleaved with some process p_k which is connected with p_j ($p_i |_t p_k$ and $p_k \leftrightarrow_t p_j$).

A *schedule* sch_t is a partially ordered set of processes performed on a server s_t in the precedent relation $\rightarrow_t$ and in the connected relation $\leftrightarrow_t$. Here, a *knot* $K_t(p_i)$ is a closure subset of the processes in a schedule sch_t which are connected with a process p_i, i.e. $K_t(p_i) = \{p_j \mid p_j \leftrightarrow_t p_i\}$. The schedule sch_t is divided into pairwise disjointing knots $K_{t1}, \ldots, K_{tl_t}$. Let p_j and p_k are a pair of processes in a knot $K_t(p_i)$ where p_j first starts at time $s\tau_{tj}$ and p_k lastly terminates at time $e\tau_{tk}$. The execution time TK_t of the knot $K_t(p_i)$ is $e\tau_{tk} - s\tau_{tj}$. Let $KP_t(\tau)$ be a *current knot* which is a set of current or previous processes which are connected with at least one current process in the set $CP_t(\tau)$ of current processes at time τ.

4.1.3. Simple Computation Model

We consider a computation model which each server s_t satisfies the following properties (Enokido, Aikebaier, & Takizawa, 2010):

1. $maxf_{ti} = maxf_{tu} = maxf_t$ for every pair of different processes p_i and p_j.
2. $\sum_{p_i \in P_t(\tau)} f_{ti}(\tau) = \alpha_t(\tau) \cdot maxf_t$.

The first condition shows that any process is performed on a server with the maximum clock frequency. The second condition means that the more number of processes are concurrently performed, the larger computation resource is consumed due to the increased overhead like process switch. Let K_t be a knot in a schedule sch_t of a server s_t, where $s\tau$ and $e\tau$ are the starting time and termination time, respectively. The execution time TK_t of the knot K_t is $\sum_{p_i \in K_t} minT_{ti} / \alpha$ where $\alpha = \int_{s\tau}^{e\tau} \alpha_t(\tau)d\tau \, / \, (e\tau - s\tau)$.

Let us consider three processes p_1, p_2, and p_3 on a server s_t as shown in Figure 2 (1). First, suppose that the processes are serially performed, i.e. $e\tau_{t1} = s\tau_{t2}$ and $e\tau_{t2} = s\tau_{t3}$. A knot K_t is $\{p_1\}$ at time τ ($s\tau_{t1} \leq \tau < e\tau_{t1}$), $K_t = \{p_1, p_2\}$ for time τ ($s\tau_{t2} \leq \tau < e\tau_{t2}$), and $K_t = \{p_1, p_2, p_3\}$ for time τ ($s\tau_{t3} \leq \tau < e\tau_{t3}$). Here, the execution time TK_t of the knot K_t is $e\tau_{t3} - s\tau_{t1} = minT_{t1} + minT_{t2} + minT_{t3}$ since $\alpha_t(\tau) = 1$ for $s\tau_{t1} \leq \tau \leq e\tau_{t3}$.

Next, suppose three processes p_1, p_2, and p_3 start at same time $s\tau$ and terminate at same time $e\tau$ as shown in Figure 2 (2). Here, $\alpha_t(\tau) = \varepsilon_t^{3-1} = \varepsilon_t^2$ for $s\tau \leq \tau \leq e\tau$. $TK_t = (minT_{t1} + minT_{t2} + minT_{t3}) / \varepsilon_t^2$. For $\varepsilon_t = 0.99$, $TK_t = 1.02 \cdot (minT_{t1} + minT_{t2} + minT_{t3})$. Let us consider a knot K_t where the processes p_1, p_2, and p_3 are concurrently performed. The process p_1 first starts at time $s\tau_{t1}$ and the process p_3 lastly terminates at time $e\tau_{t3}$ as shown

Figure 2. Execution time of current knot

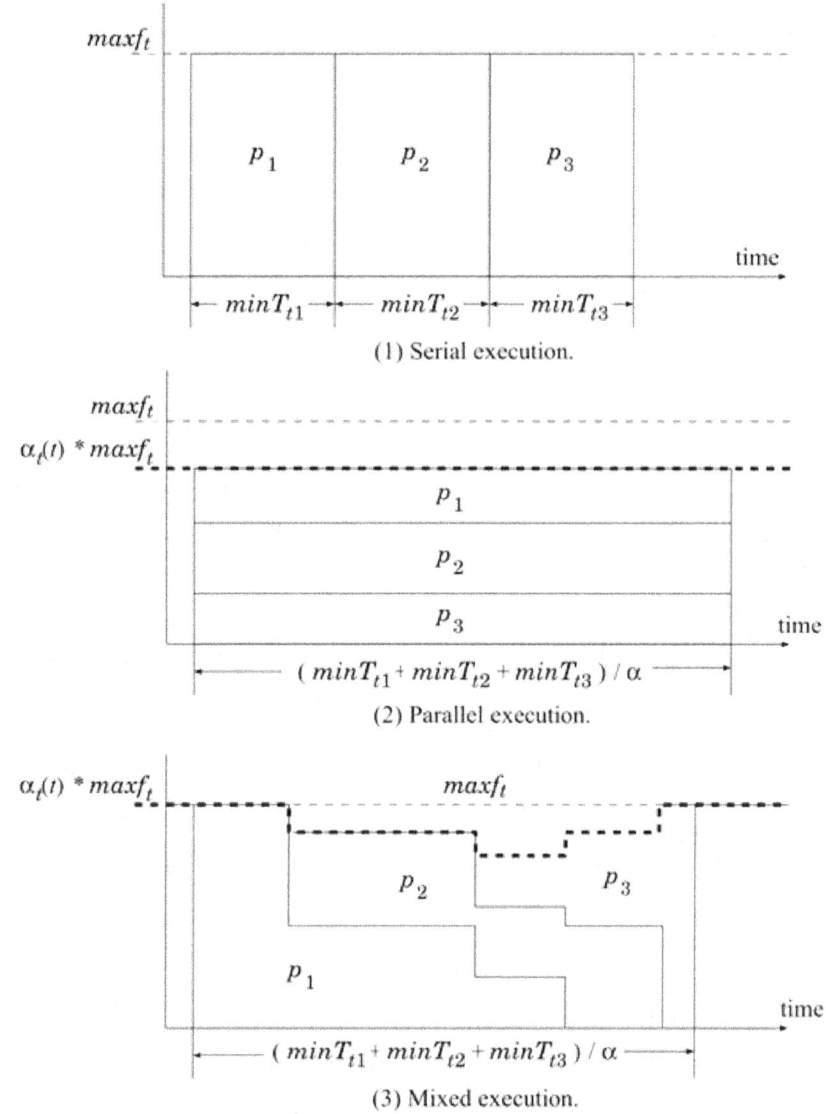

in Figure 2 (3). A knot K_t is $\{p_1, p_2, p_3\}$. Here, $TK_t = e\tau_{t3} - s\tau_{t1} = (minT_{t1} + minT_{t2} + minT_{t3}) / \alpha$. $\alpha = [(s\tau_{t2} - s\tau_{t1}) + (s\tau_{t3} - s\tau_{t2})\varepsilon_t + (e\tau_{t1} - s\tau_{t3})\varepsilon_t^2 + (e\tau_{t2} - e\tau_{t1})\varepsilon_t + (e\tau_{t3} - e\tau_{t2})] / (e\tau_{t3} - s\tau_{t1})$.

It depends on the scheduling algorithm how much the NCR $f_{ti}(\tau)$ of each current process p_i is at time τ. In this chapter, we assume CPU resource is equally allocated to every current process as follows:

$$f_{ti}(\tau) = \alpha_t(\tau) \cdot maxf_t \,/\, |\, CP_t(\tau) \,|\,. \qquad (3)$$

4.1.4. Estimated Termination Time

We discuss how to estimate a termination time of a current knot $KP_t(\tau) = \{p_{t1}, ..., p_{tl_t}\}$ of processes on a server s_t, where the starting time is $s\tau$. Let $CP_t(\tau)$ be a set of current processes being performed on a server s_t at time τ. Suppose a process p_i starts on a server s_t at time τ, i.e. $s\tau_{ti} = \tau$. Let

Figure 3. Events in current knot

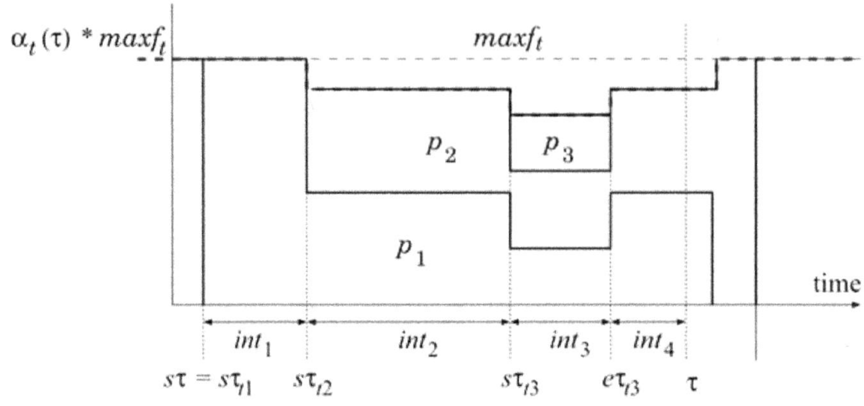

$EV_t(\tau)$ be a set of starting time and termination time of processes which start or terminate in the current knot $KP_t(\tau)$ between time $s\tau$ and current time τ on the server s_t. Here, each time element e_t in the set $EV_t(\tau)$ is referred to as an *event* on the knot $KP_t(\tau)$. Events in $EV_t(\tau)$ are sorted in the ascending order. $EV_t(\tau) = \langle e_1, ..., e_l \rangle$ where $e_1 \leq ... \leq e_l$. For example, $EV_t(\tau) = \langle s\tau_{t1}, s\tau_{t2}, s\tau_{t3}, e\tau_{t3}, \tau \rangle$ in Figure 3. If there is no current knot $KP_t(\tau)$ on a server s_t at time τ, a process p_i is the first process on the new current knot $KP_t(\tau)$ and $EV_t(\tau) = \langle s\tau_{ti} \rangle$ where $s\tau_{ti} = \tau$. Let int_k be time interval between an event e_k and a succeeding event e_{k+1} in $EV_t(\tau)$, i.e. $int_k = e_{k+1} - e_k$ ($1 \leq k \leq l - 1$ and $l \geq 2$). $int_1 = 0$ if $l = 1$. Let AC_{tik} be the amount of computation of a process p_i which is performed on a server s_t during the interval int_k. $N_t(int_k)$ indicates the number of processes which are being performed in an interval int_k. AC_{tik} is given as follows:

$$AC_{tik} = \frac{\varepsilon_t^{N_t(int_k)-1} \cdot maxf_t}{N_t(int_k)} \cdot int_k. \tag{4}$$

For example, the number $N_t(int_2)$ of concurrent processes during the time interval int_2 is 2 in Figure 3. Then, AC_{t12} of a process p_1 in the time interval int_2 on the server s_t is $(\varepsilon_t \cdot maxf_t) / 2 \cdot int_2$.

Let $INT_t(\tau)$ be a set of time intervals $\{int_1, ..., int_{l-1}\}$ between time $s\tau$ and time τ on the current knot $KP_t(\tau)$. The total amount $TAC_{ti}(\tau)$ of computation which is performed on a server s_t between time $s\tau$ and τ for each process p_i in $KP_t(\tau)$ is $TAC_{ti}(\tau) = \sum_{k=1}^{l-1} AC_{tik}$. The total amount of computation to be performed to terminate a process p_i is $maxf_t \cdot minT_{ti}$ on a server s_t. The residual computation quantity RCQ_{ti} shows how much computation has to be spent to terminate a process p_i on the server s_t after time τ. RCQ_{ti} is given as follows:

$$RCQ_{ti} = (maxf_t \cdot minT_{ti}) - TAC_{ti}. \tag{5}$$

A tuple $\langle i, RCQ_{ti} \rangle$ is a pair of an identifier i of a process p_i and a residual computation quantity RCQ_{ti} of the process p_i which has to be performed after time τ. If $RCQ_{ti} \geq 0$, a tuple $\langle i, RCQ_{ti} \rangle$ of a process p_i is stored in a set RCQ_t. If $RCQ_{ti} \leq 0$, a tuple $\langle i, RCQ_{ti} \rangle$ of a process p_i is not stored in RCQ_t since the process p_i has terminated before time τ.

$EstT_{ti}(\tau)$ shows an estimated termination time of a process p_i where the process p_i is performed at $NCRf_{ti}(\tau)$ on the server s_t at time τ. Here, $EstT_{ti}(\tau) = (\varepsilon_t^{N_t(\tau)-1} \cdot maxf_t \,/\, N_t(\tau))^{-1} \cdot RCQ_{ti}$. $minEstT_{ti}(\tau)$ is the minimum value in $EstT_{ti}(\tau)$ where $p_i \subset RCQ_t$ at time τ. Let $ET_t(\tau)$ be an estimated termination time of a current knot $KP_t(\tau)$ where a server s_t is selected for a process p_j at time τ. $ET_{p_j}(\tau)$ shows an *estimated termination time* of the process p_j

Figure 4. Calculation of the estimated termination time

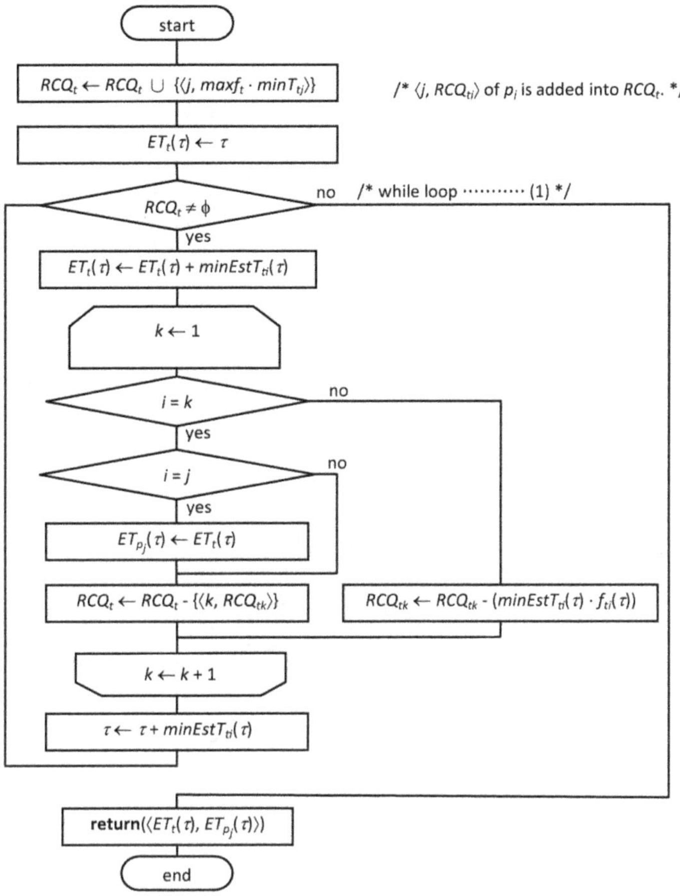

on a server s_t at time τ. A pair of $ET_t(\tau)$ and $ET_{p_j}(\tau)$ are calculated as shown in Figure 4.

Suppose a server s_t is selected for a process p_4 at time τ as shown in Figure 5. Here, we assume that $RCQ_t=\{\langle p_1, 4\rangle, \langle p_4, 8\rangle, \langle p_2, 16\rangle\}$. In addition, $maxf_t = 1$ and $\varepsilon_t = 0.99$. Here, the *while* loop is performed until RCQ_t gets empty, i.e. three times (see Figure 4 line (1)). In the first round, $maxf_t(\tau) = \varepsilon_t^{3\text{-}1} \cdot 1 = 0.99^2 \cdot 1 = 0.98$, $EstT_{t1} = 12.5$, $EstT_{t4} = 25$, $EstT_{t2} = 50$, and $ET_t(\tau) = \tau + 12.5$, respectively, as shown in Figure 5 (1). Then, RCQ_t is changed to $\{\langle p_4, 4\rangle, \langle p_2, 12\rangle\}$ since $\langle p_1, 4\rangle$ is removed and RCQ_{t4} and RCQ_{t2} are changed with $8 - (0.32 \cdot 12.5)$ and $16 - (0.32 \cdot 12.5)$, respectively. In the second round, $maxf_t(\tau) = \varepsilon_t^{2\text{-}1} \cdot 1 = 0.99 \cdot 1 = 0.99$, $EstT_{t4} = 8.1$, $EstT_{t2} = 24.3$, and $ET_t(\tau) = \tau + 12.5 + 8.1 = \tau + 20.6$, respectively, as shown in Figure 5 (2). Here, the process p_4 is the new process for which the server s_t is selected at time τ. Therefore, $ET_{p_4}(\tau) = \tau + 20.6$. Then, RCQ_t is changed to $\{\langle p_2, 8\rangle\}$. In the third round, $maxf_t(\tau) = \varepsilon_t^{0} \cdot 1 = 1$, $EstT_{t2} = 8$, and $ET_t(\tau) = \tau + 12.5 + 8.1 + 8 = \tau + 28.6$, respectively, as shown in Figure 5 (3). Hence, the estimated termination time $ET_t(\tau)$ of a current knot $KP_t(\tau)$ is $\tau + 28.6$ and the estimated termination time $ET_{p_4}(\tau)$ of a process p_4 is $\tau + 20.6$.

Figure 5. Estimated termination time

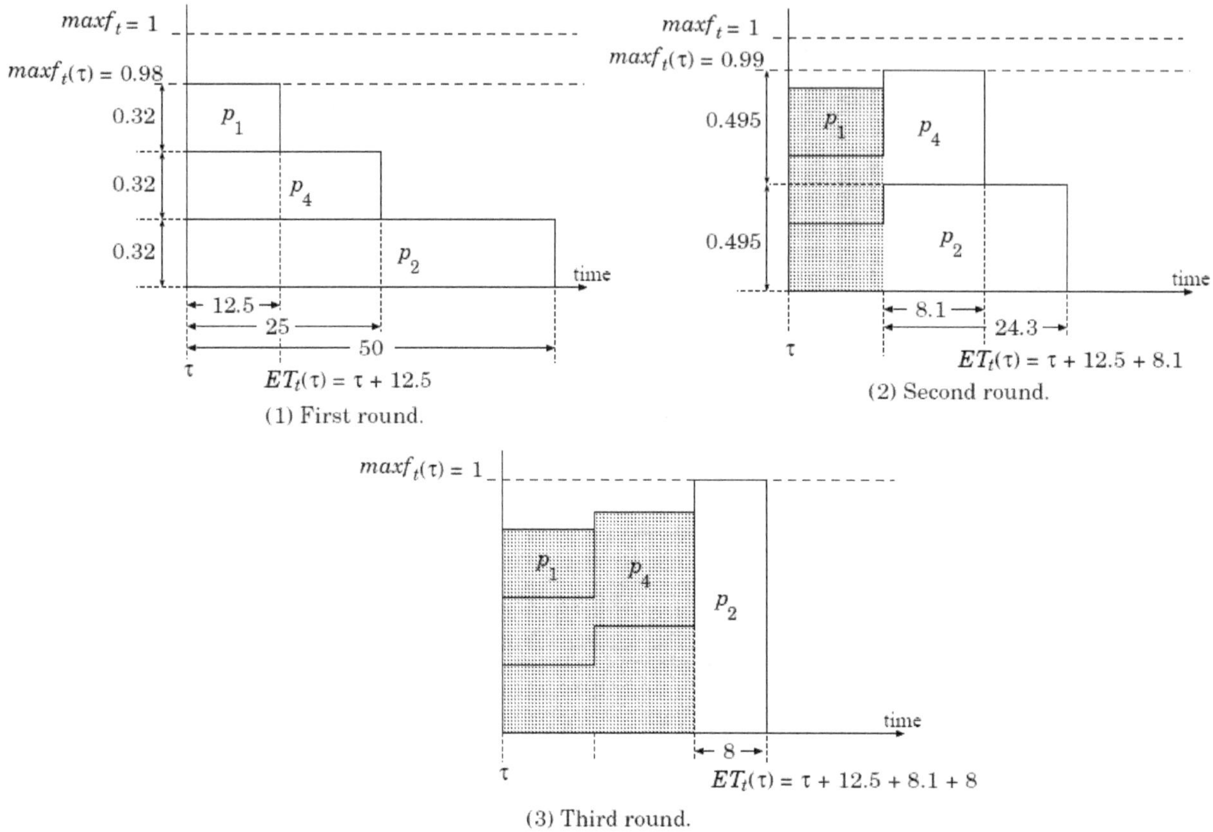

4.2. Power Consumption Model for Computation Type Applications

We discuss the power consumption model for computation type applications. Here, we consider the following parameters:

- $E_t(\tau)$ = electric power consumption rate of a server s_t at time τ [W] ($t = 1, \ldots, n$).
- $maxE_t$ = maximum electric power consumption rate of a server s_t.
- $minE_t$ = minimum electric power consumption rate of a server s_t, i.e. s_t is in idle state.
- $maxE = max(maxE_1, \ldots, maxE_n)$.
- $minE = min(minE_1, \ldots, minE_n)$.

We define the *normalized power consumption rate* (*NPCR*) $e_t(\tau)$ of a server s_t at time τ as follows:

$$e_t(\tau) = E_t(\tau) \,/\, maxE \quad (\leq 1\ ). \tag{6}$$

Let $mine_t$ and $maxe_t$ show the minimum power consumption rate $minE_t \,/\, maxE$ and the maximum power consumption rate $maxE_t \,/\, maxE$ of a server s_t, respectively. The *NPCR* $e_t(\tau)$ shows a ratio of how much a server s_t consumes electric power to a server which mostly consumes the electric power at time τ. If the fastest server s_t maximumly spends the electric power with the maximum clock frequency, $e_t(\tau) = maxe_t = 1$. We define a *simple power consumption model* for a server s_t as follows:

$$e_t(\tau) = \begin{cases} maxe_t, & \text{if } N_t(\tau) \geq 1. \\ mine_t, & \text{otherwise.} \end{cases} \tag{7}$$

Figure 6. Power consumption laxity (PCL)

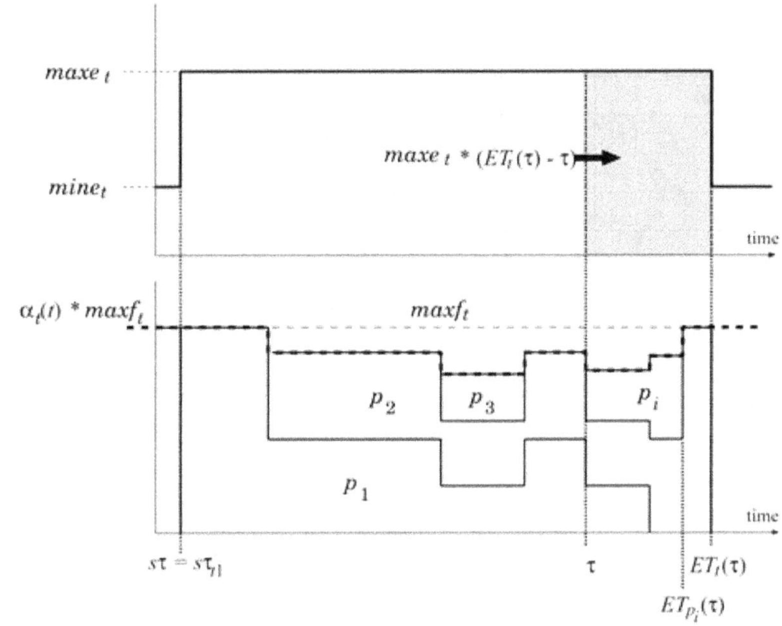

This means, if at least one process is performed on a server s_t, the electric power is maximally consumed on the server s_t. If no process is performed on the server s_t, $e_t(\tau) = mine_t$. The *normalized power consumption* $NPC_t(\tau_1, \tau_2)$ [Ws] of a server s_t from time τ_1 to time τ_2 is given as follows:

$$NPC_t(\tau_1,\ \tau_2) = \int_{\tau_1}^{\tau_2} e_t(\tau)\ d\tau. \tag{8}$$

4.3. Power Consumption Laxity Based (PCLB) Algorithm

Suppose a process p_i is issued to a load balancer K at time τ. First, the *Power Consumption Laxity (PCL)* $le_t(\tau)$ of each server s_t is estimated at time τ (Figure 6). The *PCL* $le_t(\tau)$ is given as follows:

$$le_t(\tau) = maxe_t \cdot (ET_t(\tau) - \tau). \tag{9}$$

In the *power consumption laxity-based* (PCLB) algorithm, a server s_t which consumes the minimum power consumption is selected for a process p_i, i.e. a server s_t whose *PCL* $le_t(\boldsymbol{\tau})$ is minimum is selected in the server set S. Here, a pair of $ET_t(\boldsymbol{\tau})$ and $ET_{pj}(\boldsymbol{\tau})$ are calculated as shown in Figure 4. At time $\boldsymbol{\tau}$, a server s_t is selected for a process p_i by the following procedure PCLB:

```
PCLB(τ, S, p_i) {
le_t(τ) = maxe_t · (ET_t(τ) - τ) for each
server s_t in S;
server = s_t where le_t(τ) is minimum;
return(server);
}
```

4.4. Computation Laxity Based (CLB) Algorithm

The *Computation Laxity* (*CL*) $cl_{ti}(\tau)$ of a process p_i shows how long it takes to perform up the process p_i at time τ on a server s_t. Suppose a process p_i is issued to the load balancer K at time τ. The *CL* $cl_{ti}(\tau)$ is given as follows:

$$cl_{ti}(\tau) = ET_{p_i}(\tau) - \tau. \tag{10}$$

Table 1. Servers

Server	S_1	S_2	S_3	S_4	S_5
$maxf_t$	1.000	0.998	0.859	0.840	0.839
$minE_t$ ($mine_t$)	149 (0.76)	141 (0.72)	133 (0.68)	129 (0.65)	108 (0.55)
$maxE_t$ ($maxe_t$)	197 (1)	186 (0.94)	160 (0.81)	156 (0.79)	136 (0.69)
Server	S_6	S_7	S_8	S_9	S_{10}
$maxf_t$	0.833	0.729	0.716	0.713	0.700
$minE_t$ ($mine_t$)	98 (0.5)	139 (0.71)	147 (0.75)	108 (0.55)	96 (0.49)
$maxE_t$ ($maxe_t$)	125 (0.63)	182 (0.92)	196 (0.99)	136 (0.69)	123 (0.62)

In the *Computation Laxity-Based* (CLB) algorithm, a server s_t which can most early terminate a process p_i is selected for the process p_i at time τ by the following procedure CLB:

```
CLB(τ, S, p_i) {
cl_ti(τ) = ET_pi(τ) - τ  for each server
s_t in S;
server = s_t where cl_ti(τ) is minimum;
return(server);
}
```

4.5. Evaluation of the PCLB and CLB Algorithms

We evaluate the PCLB and CLB algorithms in terms of total power consumption compared with the RR algorithm through the simulation. There are ten servers $s_1, ..., s_{10}$ as shown in Table 1, $S = \{s_1, ..., s_{10}\}$. The servers in the server set S are ordered with respect to the processing speed, i.e. maximum computation rate. Here, the server s_1 is the fastest server and the server s_{10} is the slowest server. The *NCR* $maxf_t$ of each server s_t is randomly selected between 1 and 0.7 [1/sec]. The maximum computation rate $maxf_1 = 1$ and $maxf_{10} = 0.7$. The computation degradation rate ε_t of each server s_t is randomly selected between 0.99 and 0.95. The minimum power consumption rate $minE_t$ of each server s_t is randomly selected between 85 and 150 [W]. The maximum power consumption rate $minE_t$ of each server s_t is randomly selected between 1.2 times and 1.4 times of minimum power consumption rate $minE_t$. Here, the maximum NPCR $maxe_1 = 1$ and minimum NPCR $mine_1 = 0.76$. In this section, we take one [sec] as one time unit.

Totally 600 number of processes, $P = \{p_1, ..., p_{600}\}$ are issued to the servers in the set S during 1,000 seconds. The minimum computation time $minT_i$ of each process p_i on the fastest server is randomly selected between 2 and 10 as a real number. For example, if $minT_i$ of a process p_i is 5.24 [sec], it takes six seconds for the fastest server s_1 to perform the process p_i. Here, $minT_{ti} = minT_i / maxf_t$ for each server s_t. The starting time of each process is randomly selected between 0 and 1,000 seconds in the simulation time. Figure 7 shows a traffic pattern, i.e. how many number of processes start at time τ. In simulations, the PCLB, CLB, and RR algorithms are performed on the same traffic pattern. The maximum number of processes starting at the same time is 5 and the average number is 0.6 at each second.

In the RR algorithm, the servers are ordered as $s_1, ..., s_{10}$. A request is first issued to the fastest server s_1. Until the server s_1 is overloaded, requests are issued to the server s_1. Then, requests are issued to the server s_2. Thus, requests are issued to the servers in the set S.

Figure 8 shows the total power consumption [Ws] of the servers in the set S at each time τ. Table 2 shows the total power consumption, which is obtained by subtracting the power consumption of servers in the idle state from the total power

Figure 7. Total number of processes starting at time τ

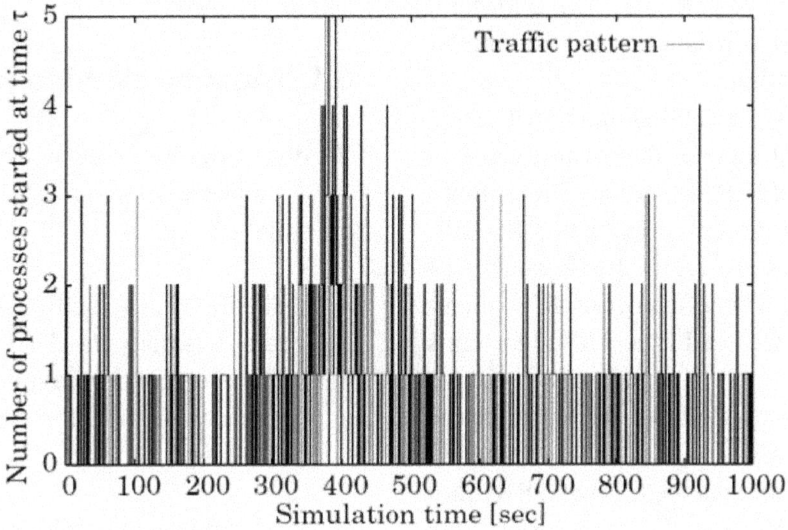

Figure 8. Total power consumption [Ws]

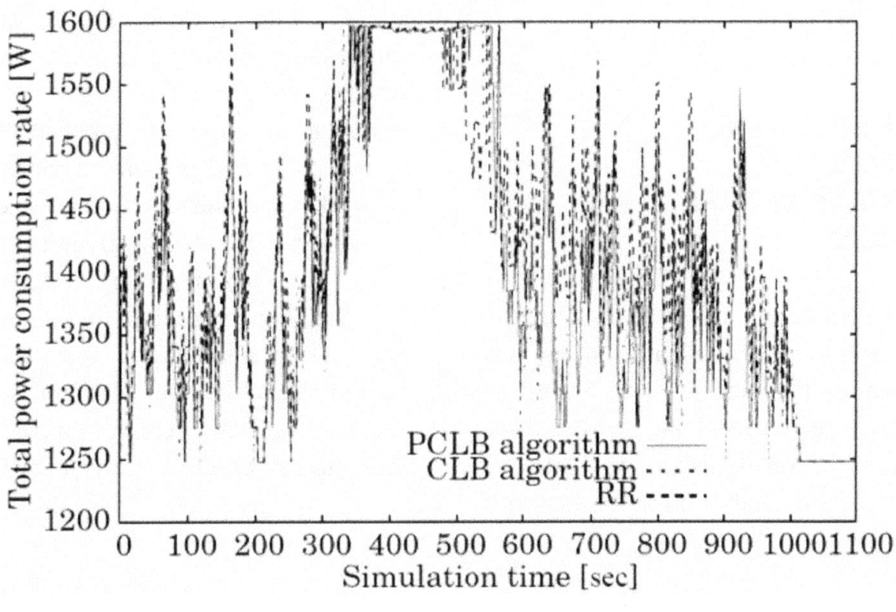

consumption during the simulation, i.e. the power consumption for only performing totally 600 processes. In the PCLB algorithm, a server s_t which minimally consumes the power consumption for each process p_i is selected. In the CLB algorithm, a server s_t where a process p_i can most early terminate is selected. The total power consumption of the PCLB algorithm is 167,608 [Ws]. The total power consumption of the servers in the CLB algorithm is 177,762 [Ws]. The total power consumption in the RR algorithm is 190,938 [Ws]. In the PCLB algorithm, the power consumption can be more reduced than the CLB and RR algorithms. In the PCLB algorithm, 5.7% and 12.2% of the power consumption can be smaller than the CLB and RR algorithms, respectively. In the CLB algorithm, 6.9% of the power consumption can be smaller than the RR algorithm.

Table 3 shows the total simulation time to terminate the total number 600 of processes in the PCLB, CLB, and RR algorithms. The termination times of all the processes in the PCLB, CLB, and RR algorithms are 1,014, 1,010, and 1,014 [sec], respectively. The difference of the termination time among PCLB, CLB, and RR algorithms is neglectable.

From the evaluation results, we conclude the total power consumption can be more reduced in the PCLB algorithm than the CLB and RR algorithms and the difference of termination time among the PCLB, CLB, and RR algorithms is almost neglectable. Therefore, the PCLB algorithm is more useful than the CLB and RR algorithms.

5. SERVER SELECTION ALGORITHMS FOR COMMUNICATION TYPE APPLICATIONS

In this section, we define a pair of models, the transmission and power consumption models for communication type applications. Based on the power consumption model, we propose three types of server selection algorithms to select one of servers for communication type applications so that the total power consumption of servers can be reduced.

5.1. Transmission Model

We discuss how to select a server in the server set S for a client c_s so that the following constraints are satisfied:

- The file f has to be transmitted to the client so as to satisfy the deadline constraint.
- The power consumption of a selected server s_t to transfer the file f has to be minimized.

$tr_{ts}(\tau)$ shows a transmission rate of a server s_t to transmit a file f_s to a client c_s at time τ. Suppose a server s_t concurrently sends files $f_1, ..., f_m$ to a set $CT_t(\tau)$ of clients $c_1, ..., c_m$ at transmission rates $tr_{t1}(\tau), ..., tr_{tm}(\tau)$ $(m \geq 1)$, respectively, at time τ. b_{ts} shows the maximum bandwidth [bps] between a server s_t and a client c_s. Let $Maxtr_t$ be the maximum transmission rate [bps] of the server s_t $(\leq b_{ts})$ which is smaller than the bandwidth b_{ts} of the network. Here, the total transmission rate $tr_t(\tau)$ of the server s_t at time τ is given as $tr_t(\tau) = tr_{t1}(\tau) + ... + tr_{tm}(\tau)$. Here, $0 \leq tr_t(\tau) \leq Maxtr_t$.

Each client c_s receives messages at receipt rate $rr_s(\tau)$ at time τ. Let $Maxrr_s$ indicate the maximum receipt rate of the client c_s. Here, $tr_{ts}(\tau) \leq Maxrr_s$. We assume each client c_s receives a file from at most one server at rate $Maxrr_s (= rr_s(\tau))$. The server

Table 2. Total power consumption [Ws]

	PCLB	CLB	RR
Total power consumption [Ws]	167,608	177,762	190,938

Table 3. Termination time [sec]

	PCLB	CLB	RR
Termination time [sec]	1,014	1,010	1,014

s_t allocates each client c_s with transmission rate $tr_{ts}(\tau)$ so that $tr_{ts}(\tau) \leq Maxrr_s$ at time τ.

Let TR_{ts} be the total transmission time of a file f_s from a server s_t to a client c_s. If the server s_t sends files to other clients concurrently with the client c_s, the transmission time TR_{ts} is increased. Let $minTR_{ts}$ show the minimum transmission time $|f_s| / min(Maxrr_s, Maxtr_t)$ [sec] of a file f_s from a server s_t to a client c_s where $|f_s|$ indicates the size [bit] of the file f_s. $TR_{ts} \geq minTR_{ts}$. Suppose the server s_t starts and ends transmitting a file f_s to the client c_s at time $s\tau$ and $e\tau$, respectively. Here, $\int_{s\tau}^{e\tau} tr_{ts}(\tau)d\tau = |f_s|$ and the transmission time TR_{ts} is $(e\tau - s\tau)$. If the server s_t sends only the file f_s to the client c_s at time τ, $tr_{ts}(\tau) = min(Maxrr_s, Maxtr_t)$ [bps]. The transmission laxity $lt_{ts}(\tau)$ is $|f_s| - \int_{\tau}^{e\tau} tr_{ts}(\tau)d\tau$ [bit] at time τ, i.e. how many bits of a file f_s the server s_t still has to transmit to the client c_s at time τ. First, we consider a model where a server s_t satisfies the following properties:

Server-Bound Model

If $Maxrr_1 + ... + Maxrr_m \geq \sigma_t(\tau) \cdot Maxtr_t$, $\sum_{c_s \in CT_t(\tau)} tr_{ts}(\tau) = \sigma_t(\tau) \cdot Maxtr_t$ for every time τ.

Here, $\sigma_t(\tau)$ (≤ 1) is the *transmission degradation* factor of a server s_t. Here, we assume $\sigma_t(\tau) = \gamma_t^{1-|CTt(\tau)|}$ $(0 < \gamma_t \leq 1)$ at time τ. The more number of clients a server s_t transmits, the longer it takes. The *effective transmission rate* of the server s_t is $\sigma_t(\tau) \cdot Maxtr_t$.

Let us consider three files f_1, f_2, and f_3 which a server s_t sends to clients c_1, c_2, and c_3, respectively. First,

suppose that the server s_t serially sends the files f_1, f_2, and f_3 to the clients c_1, c_2, and c_3, i.e. $e\tau_{t1} = s\tau_{t2}$ and $e\tau_{t2} = s\tau_{t3}$ as shown in Figure 9 (1). Here, the transmission time TR_t is $e\tau_{t3} - s\tau_{t1} = minTR_{t1} + minTR_{t2} + minTR_{t3}$. Next, suppose the server s_t starts transmitting three files f_1, f_2, and f_3 at time $s\tau$ and terminates at time $e\tau$ as shown in Figure 9 (2). Here, since three files are concurrently transmitted, $CT_t(\tau) = 3$ and $\gamma_t^{-2} \cdot TR_t = \gamma_t^{-2} \cdot (minTR_{t1} + minTR_{t2} + minTR_{t3})$. For $\gamma_t = 0.98$, it takes about 1.4% longer time than the serial transmission.

On the other hand, we consider another environment where some client c_s cannot receive a file from a server s_t at maximum transmission rate $Maxtr_t$, i.e. $Maxrr_s < Maxtr_t$. Hence, the transmission rate $tr_{ts}(\tau)$ is the maximum receipt rate $Maxrr_s$ of the client c_s.

Client-Bound Model

If $Maxrr_1 + ... + Maxrr_m \leq \sigma_t(\tau) \cdot Maxtr_t$, $\sum_{c_s \in CT_t(\tau)} tr_{ts}(\tau) = Maxrr_1 + ... + Maxrr_m$ for every time τ.

Even if every client c_{ts} receives a file at maximum rate $Maxrr_s$, the effective transmission rate is not degraded.

5.2. Experimentations

5.2.1. Environment

We measure how much electric power a computer spends to transmit files to other computers. As shown in Figure 10, a pair of server computers s_1 and s_2 are interconnected with a pair of client computers c_1 and c_2 in 1Gbps networks. The server s_1 is equipped with a one-core CPU. The server s_2 is composed of a pair of two-core CPUs. That is, the bandwidth b_{ts} from a server s_t to a client c_s is 1Gbps ($t = 1, 2$). Each client c_s downloads a file f from one of the servers. The size of the file f is 43,051,806 bytes long. Here, we measure the total power consumption of the servers s_1 and s_2.

For each server s_t, we consider two types of experimentations, one-client ($1C_t$) and two-client ($2C_t$) environments ($t = 1, 2$). In the $1C_t$ environment, one client, say c_1 downloads the file f from the server s_t. In the $2C_t$ environment, a pair of the clients c_1 and c_2 concurrently download the file f from the server s_t.

Figure 9. Transmission time

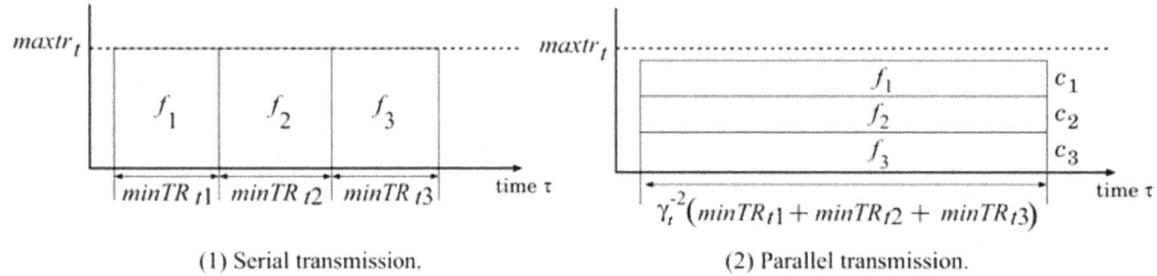

5.2.2. Experimental Results

A server s_t consumes the electric power to transmit files to clients while clients consume less amount of electric power. In the $1C_1$ environment, the server s_1 transmits a file f to one client, say c_1 at rate tr_{11}. Here, the server s_1 is composed of one one-core CPU. The maximum transmission rate $Maxtr_1$ is 160 [Mbps] in the network of bandwidth b_{11} = 1G [bps]. In the $2C_1$ environment, the server s_1 concurrently transmits the file f to a pair of clients c_1 and c_2. Here, $tr_1 = tr_{11} + tr_{12}$. Figure 11 (1) shows the power consumption rate of the server s_1 for the total transmission rate tr_1. At the higher rate tr_1 the server s_1 transmits the file f, the larger amount of power consumption the server s_1 consumes. We obtain the approximated formula $PC_1(tr)$ to show the power consumption rate of a server s_1 for total transmission rate tr [Mbps] by using the least-squares method to the experimental results. In Figure 11 (1), the bold dotted line shows the approximated power consumption of the server s_1 where one client downloads the file f from the server s_1. The dotted line shows the approximated power consumption of the server s_1 where a pair of clients c_1 and c_2 concurrently

Figure 10. Environment

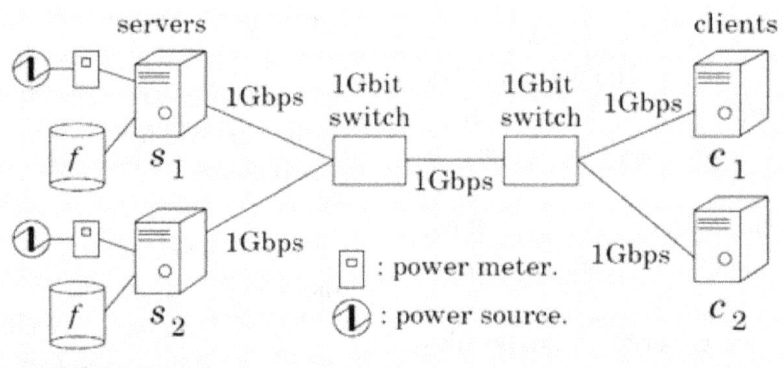

Server	s_1	s_2
Number of CPUs	1	2
Number of cores / CPU	1	2
CPU	AMD Athlon 1648B (2.7GHz)	AMD Opteron 270 (2GHz)
Memory	4,096MB	4096MB
DISK	150GB 7200rpm	74GB 10000rpm x 2 RAID1
NIC	Broadcom Gbit Ether (1Gbps)	Nvidia Ether Controler (1Gbps)

download the file f from the server s_1. Let $PC_1^1(tr)$ and $PC_1^2(tr)$ be the power consumption rates in the $1C_1$ and $2C_1$ environments, respectively, at total transmission rate tr. In a single-CPU server s_t, the power consumption rate $PC_t(tr)$ is proportional to the total transmission rate tr.

$$1C_1: \quad PC_1^1(tr) = 0.11tr + 4.15 \quad [\mathrm{W}] .$$
$$2C_1: \quad PC_1^2(tr) = 0.12tr + 4.43 \quad [\mathrm{W}].$$

Next, we consider another server s_2 which is composed of a pair of two-core CPUs. Here, the maximum transmission rate $Maxtr_2$ of the server s_2 is 450 [Mbps]. We measure the power consumption rate for the total transmission rate tr_2 for $1C_2$ and $2C_2$. Figure 11 (2) shows the power consumption rate [W] of the server s_2 for the total transmission rate tr. Following Figure 11 (2), the power consumption rate of the server s_2 also depends on the total transmission rate tr_2 like $1C_1$. At the higher rate the server s_2 transmits, the larger power consumption s_2 consumes. The approximated formulas $PC_2^1(tr)$ and $PC_2^2(tr)$ of the power consumption rate of the server s_2 for total transmission rate tr [Mbps] are given in the $1C_2$ and $2C_2$ environments as follows:

$$1C_2: \quad PC_2^1(tr) = 0.02tr + 3.02 \quad [\mathrm{W}] .$$
$$2C_2: \quad PC_2^2(tr) = 0.03tr + 3.34 \quad [\mathrm{W}].$$

The increase rate of the power consumption of the server s_2 in $2C_2$ is about 1.5 times larger than $1C_2$. Compared with the one-CPU case $1C_t$, the power consumption rate is not so much increased for the increase of transmission rate in the two-CPU case $2C_t$.

Following the experiments, the power consumption rate $PC_t(tr)$ of a server s_t is lineally increased for transmission rate $tr_t(\tau)$ ($0 \leq tr_t(\tau) \leq Maxtr_t$) as follows:

$$PC_t(tr) = \beta_t(m) \cdot \delta_t \cdot tr_t(\tau) + minE_t. \qquad (11)$$

Here, δ_t is the power consumption to transmit one Mbits [W/Mb] for the $1C_t$ environment. δ_t depends on a server type s_t. As shown in Figure 11, the more number of clients, the larger amount of electric power is consumed. $\beta(m)$ shows how much power consumption is increased for the number m of clients, $\beta(m) \geq 1$ and $\beta_t(m) \geq \beta_t(m-1)$. There is a fixed point $maxm_t$ such that $\beta_t(maxm_t - 1) \leq \beta_t(maxm_t) = \beta_t(maxm_t + h)$ for $h > 0$. $minE_t$ gives the minimum power consumption rate of the server s_t where no file is transmitted. $\beta_t(maxm_t) \cdot \delta_t \cdot Maxtr_t + minE_t$ gives the maximum power consumption rate $maxE_t$ of the server s_t.

Figure 11. Power consumption rate [W]

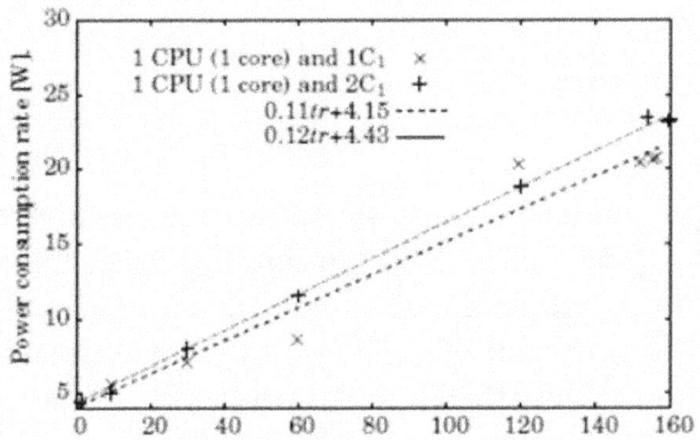

(1) One-CPU : Power consumption rate [W].

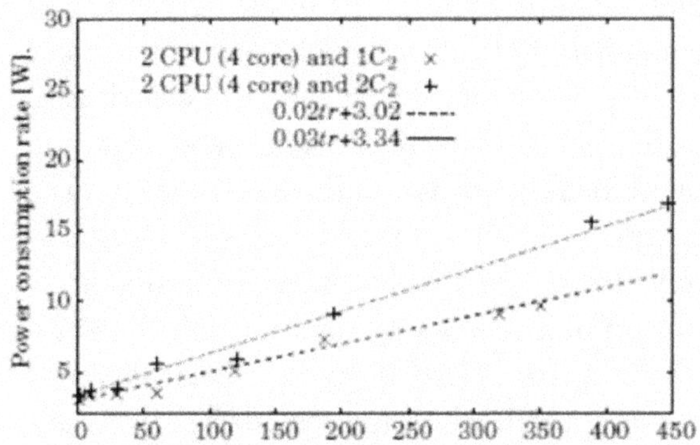

(2) Two-CPU : Power consumption rate [W].

5.3. Power Consumption Model for Communication Type Applications

We would like to discuss how much electrical power a server s_t consumes to transfer a file to a client c_s. Let $PT_t(\tau)$ show the electric power consumption rate [W] of a server s_t for transmitting files to clients at time τ ($t = 1, ..., n$). $maxE_t$ and $minE_t$ indicate the maximum and minimum electric power consumption of a server s_t, respectively. Here, $minE_t \leq PT_t(\tau) \leq maxE_t$. $maxE$ and $minE$ show $max(maxE_1, ..., maxE_n)$ and $min(minE_1, ..., minE_n)$, respectively. Here, we assume that only file transfer applications are performed on each server s_t. The electric power consumption rate $PT_t(\tau)$ of a server s_t at time τ is given as follows:

$$PT_t(\tau) = PC_t(tr_t(\tau)) = \beta_t(|CT_t(\tau)|) \cdot \delta_t \cdot tr_t(\tau) + minE_t. \quad (12)$$

The power consumption $TPC_t(\tau_1, \tau_2)$ [Ws] of a server s_t from time τ_1 to time τ_2 is given as follows:

$$TPC_t(\tau_1, \tau_2) = \int_{\tau_1}^{\tau_2} PT_t(\tau) d\tau. \quad (13)$$

5.4. Server Selection Algorithms

5.4.1. Transmission Rates

At time τ, the maximum transmission rate $maxtr_t(\tau)$ of a server s_t depends on the transmission degradation factor $\sigma_t(\tau) = \gamma_t^{1-|CTt(\tau)|}$ of the server s_t, i.e. the number of clients to which the server s_t concurrently transmits files at time τ. Each time a new request is issued by a client c_s and a current request for a client c_s is terminated at time τ, $CT_t(\tau) = CT_t(\tau) \cup \{c_s\}$ and $CT_t(\tau) = CT_t(\tau) - \{c_s\}$, respectively. Here, the maximum transmission rate $maxtr_t(\tau)$ of a server s_t at time τ is calculated as $\gamma_t^{1-|CTt(\tau)|} \cdot Maxtr_t$. Here, $0 < \gamma_t \leq 1$. Each server s_t can transmit a file with rate at least $tr_{ts}(\tau) = maxtr_t(\tau) / |CT_t(\tau)|$ to a client c_s in the set $CT_t(\tau)$. Here, if the maximum receipt rate $Maxrr_s$ of a client c_s is larger than the transmission rate $tr_{ts}(\tau)$, the server s_t transmits a file to a client c_s at rate $tr_{ts}(\tau)$ at time τ. Otherwise, the server s_t transmits at rate $Maxrr_s$. Here, the unused part of the maximum transmission rate $R_{ts}(\tau)$ of the server s_t for the client c_s at time τ (= $tr_{ts}(\tau)$ - $Maxrr_s$) can be used for other clients. $V_{ts}(\tau)$ shows an insufficient transmission rate of a server s_t for a client c_s at time τ (= $Maxrr_s - tr_{ts}(\tau)$). $V_t(\tau)$ and $R_t(\tau)$ show the total amount of insufficient and unused transmission rate of a server s_t at time τ, respectively. $V_t(\tau) = \sum_{c_s \in CT_t(\tau)} V_{ts}(\tau)$ and $R_t(\tau) = \sum_{c_s \in CT_t(\tau)} R_{ts}(\tau)$. In order to more efficiently utilize the total transmission rate $maxtr_t(\tau)$, the transmission rate $tr_{ts}(\tau)$ of a server s_t for a client c_s at time τ is calculated as shown in Figure 12.

5.4.2. Transmission Rate-Based (TRB) Algorithm

We discuss how a load balancer K selects a server s_t for a client c_s in the server set S. In the Transmission Rate-Based (TRB) algorithm, a server s_t is selected for a client c_s where the transmission rate $tr_{ts}(\tau)$ of the server s_t to transmit a file f to a client c_s is the largest. Here, a server s_t is selected for a client c_s in the TRB algorithm at time τ as follows:

```
TRB(c_s, τ) {
tr_ts(τ) is calculated for each server
s_t in S;    /* tr_ts(τ) is calculated as
shown in Figure 12.*/
server = s_t where tr_ts(τ) is maximum;
return(server);
}
```

In reality, the transmission rate between a server s_t and a client c_s is dynamically changed in the network since the transmission rate of the server s_t is dynamically changed based on the number of clients. Therefore, it is not easy to estimate the

Figure 12. Calculation of transmission rate

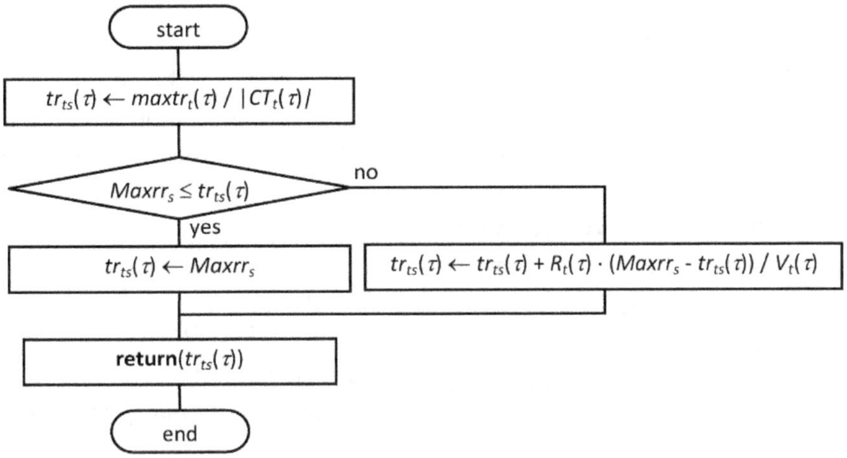

transmission rate of the server s_t to a client c_s from the practical point of view.

5.4.3. Power Consumption-Based (PCB) Algorithm

In the *Power Consumption-Based* (PCB) algorithm, a server s_t is selected for the client c_s where the power consumption to transmit a file f to a client c_s is the smallest. Here, $|f| / tr_{ts}(\tau)$ is an estimated transmission time at time t when a server s_t starts transmitting a file f to a client c_s with a transmission rate $tr_{ts}(\tau)$. The power consumption rate $PT_t(\tau)$ of each server s_t at time τ is $\beta_t(|CT_t(\tau)|) \cdot \delta_t \cdot tr_{ts}(\tau) + minE_t$ as discussed in the preceding section. It is not easy to estimate how much electric power the server s_t consumes to transmit a file f to the client c_s since there might be other clients that receive files. Here, the estimated change of power consumption $EPT_{ts}(\tau)$ [Ws] of a server s_t for transmitting a file f to a client c_s at time τ when s_t starts transmitting f is defined as follows:

$$EPT_{ts}(\tau) = (|f| / tr_{ts}(\tau)) \cdot \beta_t(|CT_t(\tau)|) \cdot \delta_t \cdot tr_{ts}(\tau) = |f| \cdot \beta_t(|CT_t(\tau)|) \cdot \delta_t. \tag{14}$$

Here, a server s_t is selected for a client c_s in the PCB algorithm by using $EPT_{ts}(\tau)$ at time τ as follows:

```
PCB (c_s, τ) {
EPT_ts(τ) = | f | · β_t(| CT_t(τ) |) · δ_t
for each server s_t in S;
server = s_t where EPT_ts(τ) is maximum;
return(server);
}
```

In the PCB algorithm, a server s_t is selected for a client c_s without considering the transmission rate between the server s_t and the client c_s.

5.4.4. Extended Power Consumption-Based (EPCB) Algorithm

In the PCB algorithm, only the power consumption where a server s_t transmits a file f to a client c_s at time τ when the server s_t starts transmitting the file f is considered. i.e. the estimated change of power consumption $EPT_{ts}(\tau) = |f| \cdot \beta_t(|CT_t(\tau)|) \cdot \delta_t$. The server s_t might be transmitting the file f to other clients at time τ. The power consumption where the server s_t transmits the file f to other clients has to be considered for make more accurate the estimation of the power consumption.

For this, the total amount of power consumption of a server to transmit files to every requesting client is considered in the extended power consumption-based (EPCB) algorithm. $lt_{ts}(\tau)$ shows the transmission laxity of a file f_s where a server s_t has to transmit the file f to a client c_s at time τ as discussed in the preceding subsection. Here, the total transmission laxity of a server s_t at time τ is $\sum_{c_s \in CT_t(\tau)} lt_{ts}(\tau)$. Here, the estimated change of power consumption $EPT_{ts}(\tau)$ [Ws] of a server s_t for transmitting a file f to a client c_s at time τ when s_t starts transmitting the file f is defined as follows:

$$EPT_{ts}(\tau) = \sum_{c_s \in CT_t(\tau)} lt_{ts}(\tau) \cdot \beta_t(|CT_t(\tau)|) \cdot \delta_t. \quad (15)$$

Here, a server s_t is selected for a client c_s in the EPCB algorithm by using $EPT_{ts}(\tau)$ at time τ as follows:

```
EPCB (c_s, τ) {
EPT_ts(τ) = Σ_{c_s∈CT_t(τ)} lt_ts(τ) · β_t(| CT_t(τ)
|) · δ_t  for each server s_t in S;
server = s_t where EPT_ts(τ) is maximum;
return(server);
}
```

For example, there are a pair of servers s_1 and s_2. The power consumption coefficients δ_1 and δ_2 to transmit one [Mbit] for one client of servers s_1 and s_2 are 0.09 and 0.07, respectively. The server s_1 is selected by a client c_1 ($CT_1(\tau) = \{c_1\}$) and the server s_2 is selected by a client c_2 ($CT_2(\tau) = \{c_2\}$) at time τ, respectively. Suppose a client c_3 issues a new request to transmit a file f whose size is one Gbytes to a load balancer K at time τ. The transmission laxities $lt_{11}(\tau)$ and $lt_{22}(\tau)$ are 0.1 and 0.9 [GByte], respectively, at time τ. In the PCB algorithm, a server s_t which has the minimum value of the formula $|f| \cdot \beta_t(|CT_t(\tau)|) \cdot \delta_t$ is selected. Suppose the increasing rates $\beta_1(2)$ and $\beta_2(2)$ of the power consumption of the servers s_1 and s_2 are 1.2 and 1.1, respectively. Here, $|f| \cdot \beta_1(|CT_1(\tau)|) \cdot \delta_1 = 1 \cdot 1.2 \cdot 0.09 = 0.108$. $|f| \cdot \beta_2(|CT_2(\tau)|) \cdot \delta_2 = 1 \cdot 1.1 \cdot 0.07 = 0.077$. Therefore, the server s_2 is selected for a client c_3 in the PCB algorithm. On the other hand, a server s_t which has the minimum value of the formula $\sum_{c_s \in CT_t(\tau)} lt_{ts}(\tau) \cdot \beta_t(|CT_t(\tau)|) \cdot \delta_t$ is selected in the EPCB algorithm. Here, $\sum_{c_s \in CT_1(\tau)} lt_{1s}(\tau) \cdot \beta_1(|CT_1(\tau)|) \cdot \delta_1 = 1.1 \cdot 1.2 \cdot 0.09 = 0.119$. $\sum_{c_s \in CT_2(\tau)} lt_{2s}(\tau) \cdot \beta_2(|CT_2(\tau)|) \cdot \delta_2 = 1.9 \cdot 1.1 \cdot 0.07 = 0.146$. Therefore, the server s_1 is selected for a client c_3 in the EPCB algorithm.

5.5. Evaluation of the TRB, PCB, and EPCB Algorithms

We evaluate the TRB, PCB, and EPCB algorithms in terms of the total amount of power consumption and elapse time compared with the RR algorithms through the simulation. In the evaluation, there are five servers s_1, s_2, s_3, s_4, and s_5 as shown in Table 4, $S = \{s_1, s_2, s_3, s_4, s_5\}$. The power consumption coefficient δ_t to transmit one Mbits for one client of each server s_t is randomly selected between 0.02 and 0.11 [W/Mb] based on the experimental results. The increasing rate of the power consumption $\beta_t(m)$ for the number m of clients is randomly selected between 1.09 and 1.5. The minimum power consumption rate $minE_t$ of each server s_t is randomly selected between 3 and 4 [W]. The maximum transmission rate $Maxtr_t$ of each server s_t is randomly selected between 150 and 450 [Mbps]. Each server s_t has a replica of a file f. The size of the file f is one giga-byte.

Totally 100 clients download the file f from one server s_t in the server set S. The maximum receipt rate $Maxrr_s$ of each client c_s is randomly selected between 1 and 100 [Mbps]. Each client c_s issues a transfer request of the file f to a load balancer K at time st_s. Here, the starting time st_s of each client c_s is randomly selected between 1 and 3,600 [sec] at the simulation time. Each client

Table 4. Servers

Servers	δ_t	β_t(m)	minE$_t$ [W]	Maxtr$_t$ [Mbps]
S_1	0.03	1.259	3.39	406
S_2	0.05	1.195	3.17	401
S_3	0.03	1.285	3.12	249
S_4	0.09	1.117	3.90	231
S_5	0.02	1.162	3.02	171

c_s issues one request at time $s\tau_s$ in the simulation. In the simulations of the EPCB, TRB, PCB, and RR algorithms, the starting time $s\tau_s$ of the file transmission to each client c_s is the same.

Figure 13 shows the total power consumption rate [W] of the servers s_1, ..., s_5 at each time. Table 5 shows the total power consumptions of the servers in the EPCB, PCB, TRB, and RR algorithms. The total power consumptions of the EPCB, PCB, TRB, and RR algorithms are 204,973, 209,831, 250,176, and 1,058,238 [Ws], respectively. The total amount of power consumption in the RR algorithm is larger than the EPCB, PCB, and TRB algorithms. In the TBR algorithm, a server s_t is selected for a client c_s, whose transmission rate for the client c_s is the largest. Then, the total amount of power consumption of the TRB algorithm is larger than the EPCB and PCB algorithms. In the EPCB and PCB algorithms, a server s_t is selected for a client c_s, whose power consumption is the smallest to transmit a file f to the client c_s. The total amount of power consumption of the EPCB algorithm is smaller than the PCB algorithm because the estimation of the power consumption is enhanced. The total power consumption of the EPCB algorithm is reduced about 2% than the PCB algorithm.

Table 6 shows the elapse time in the EPCB, PCB, TRB, and RR algorithms. The elapse time are 9,134 [sec], 9,160 [sec], 9,087 [sec], and 44,178 [sec] in the EPCB, PCB, TRB, and RR algorithms, respectively. The elapse time of the RR algorithm is larger than the EPCB, PCB, and TRB algorithm. In the TRB algorithm, a server s_t which can supply the maximum transmission rate is selected for a client c_s. Hence, the elapse time of the TRB algorithm is smaller than the EPCB and PCB algorithms. However, the difference of the elapse time among EPCB, PCB, and TRB algorithms is negligible.

In the EPCB and PCB algorithms, a server s_t is selected for a client c_s without considering the transmission rate between the server s_t and the client c_s. On the other hand, a server s_t is selected for a client c_s based on the estimated transmission rate in the TRB algorithm. From the evaluation results, we consider the total power consumption can be more reduced in the EPCB algorithm than the PCB and TRB algorithms and the difference of the elapse time among the EPCB, PCB, and TRB algorithms is neglectable. In reality, the transmission rate between a server s_t and a client c_s is dynamically changed in the network since the transmission rate of a server s_t is dynamically changed based on the number of clients. It is not easy to estimate the transmission rate of the server s_t to a client c_s from the practical point of view. A server s_t for a client c_s can be selected without considering the transmission rate between the server s_t and the client c_s in the EPCB and PCB algorithms. In addition, the EPCB algorithm can be reduced the power consumption than the PCB algorithm. Therefore, the EPCB algorithm is simpler and more useful than the TRB and PCB algorithms.

Figure 13. Total power consumption [Ws]

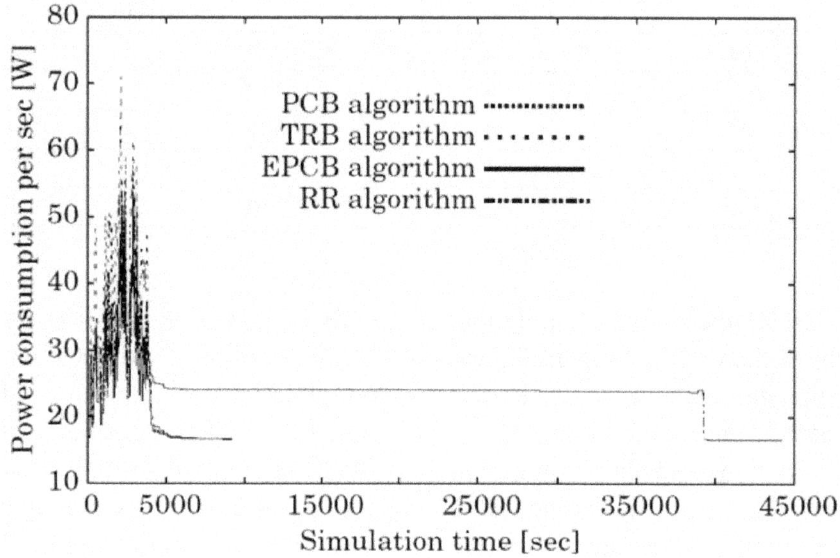

6. SERVER SELECTION ALGORITHMS FOR GENERAL TYPE OF APPLICATIONS

In this section, we define three models, the computation, transmission, and power consumption models for general type of applications. Based on the power consumption model, we propose an *Extended Power Consumption Laxity-Based* (EPCLB) algorithm to select one of servers to perform general type of application processes so that the total power consumption of the servers can be reduced.

6.1. Experimentations

6.1.1. Environment

We consider a general type of application process, which is composed of both the computation and communication modules. As shown in Figure 14, a server s_1 with data d of 544 [MB] is interconnected with three clients c_1, c_2, and c_3 in one-Gbps networks. On receipt of the request from a client c_s, the server s_1 compresses the data d to a file f of 530 [MB] by using the Deflate function (Apache Module mod_deflate, 2011) and then transmits the file f to c_s.

Table 5. Total power consumption [Ws]

EPCB	PCB	TRB	RR
204,973 [Ws]	209,831 [Ws]	250,176 [Ws]	1,058,238 [Ws]

Table 6. Elapse time

EPCB	PCB	TRB	RR
9,134 [sec]	9,160 [sec]	9,087 [sec]	44,178 [sec]

Figure 14. Environment

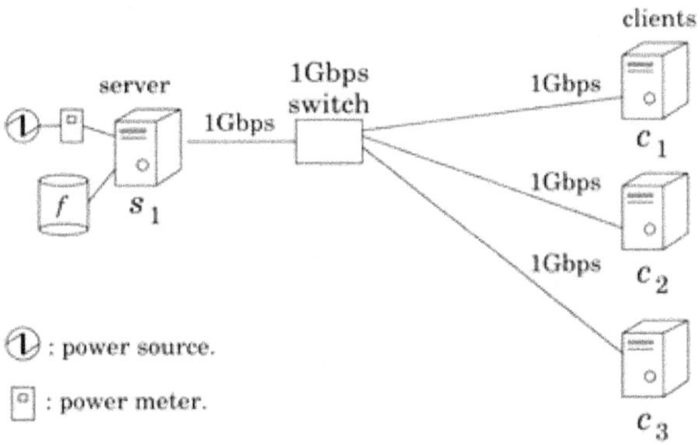

Server s_1

Number of CPU and cores	1 CPU and 1 core / CPU
CPU	AMD Athlon 1648B (2.7GHz)
Memory	4,096 MB
NIC	Broadcom Gbit Ether (1Gbps)

6.1.2. Concurrent Execution of Requests

First, we measure how much electric power the server s_t consumes to perform requests in the following five types of experimentations every one second as shown in Figure 14.

E1. A client c_1 issues a request to the server s_1.
E2. A client c_2 issues a request to the server s_1.
E3. A client c_1 issues a request to the server s_1.
E4. A pair of clients c_1 and c_2 concurrently issue requests to the server s_1.
E5. Three clients c_1, c_2, and c_3 concurrently issue requests to the server s_1.

Table 7 summarizes the computation time for compressing data, transmission rate for each client, and Power Consumption Rate (PCR) of the server s_1 in each experimentation. It takes 46, 44, and 45 [sec] to compress the data d in the experimentations E1, E2, and E3, respectively. A pair of the compression time in the experimentations E4 and E5 are 93 and 146 [sec], respectively. It takes 2.1 and 3.2 times longer time to compress the data d in the experimentations E4 and E5 where multiple clients issue requests, respectively, than E1, E2, and E3 where a single client issues a request. Figure 15 shows the power consumption rate [W] for elapse time [sec]. The server s_1 consumes the electric power to compress the data d at rates 120, 119, 119, 120, and 121 [W] in the experimentations E1, E2, E3, E4, and E5, respectively. The experimentations imply the simple computation model holds for compressing the data as discussed in the section 4. That is, any process is performed on a server with the maximum clock frequency.

In the experimentations E1, E2, and E3, the average transmission rates with the clients c_1, c_2, and c_3 are 241, 140, and 124 [Mbps], respectively. The average power consumption rates during transmission in the experimentations E1, E2, and E3 are 105, 98, and 97 [W], respectively. In the experimentation E4, the average transmission rates for the clients c_1 and c_2 are 234 and 131 [Mbps], respectively. The total transmission rate at which the server s_1 concurrently transmits the file f to the clients c_1 and c_2 is 365 (= 234 + 131) [Mbps]. Here, the average power consumption rate is 112 [W]. The average power consumption rate is 99 [W] where the server s_1 transmits to the client c_2. In the experimentation E5, the average transmission rates for the clients c_1, c_2, and c_3 are 219, 137, and 61 [Mbps], respectively. Here, the total transmission rate is 417 (= 219 + 137 + 61) [Mbps]. This means that the maximum transmission rate of the server s_1 is 417[Mbps]. The transmission rate for the client c_3 (61 [Mbps]) is 49% smaller than the transmission rate (124 [Mbps]) in the experimentation E3. Since the total rate of

Table 7. Concurrent execution of requests

Exp	Compression time [sec]	PCR of compression [W]	Transmission rate [Mbps]	PCR of transmission [W]
E1	46	120	241	105
E2	44	119	140	98
E3	45	119	124	97
E4	93	120	234 (c_1) 131 (c_2)	112 (c_1 and c_2) 99 (c_2)
E5	146	121	291 (c_1) 137 (c_2) 61 (c_3)	116 (c_1, c_2, and c_3) 106 (c_2 and c_3) 95 (c_3)

the maximum receipt rates of three clients is larger than the maximum transmission rate (417 [Mbps]) of the server s_1, the server s_1 cannot transmit the files to the client c_3 at maximum receipt rate.

The average power consumption rates of the server s_1 to transmit to three clients c_1, c_2, and c_3 at 417 [Mbps] and to a pair of clients c_2 and c_3 at 198 [Mbps] are 116 and 106 [W], respectively. The average power consumption rate to the client c_3 at 61 [Mbps] is 95 [W]. The power consumption rate of the server s_1 depends on the transmission rate as discussed in the section 5. In addition, the maximum power consumption rate for transmitting data is smaller than compressing the data.

6.1.3. Concurrent Execution of Compression and Transmission

We consider how much amount of power a server s_1 consumes to concurrently perform computation and transmission processes. The following three types of experimentations are performed concurrently with a computation process to compress data:

CT1. The server s_1 transmits a file f to only a client c_1.

CT2. The server s_1 concurrently transmits a file f to a pair of clients c_1 and c_2.

Figure 15. Concurrent execution of requests

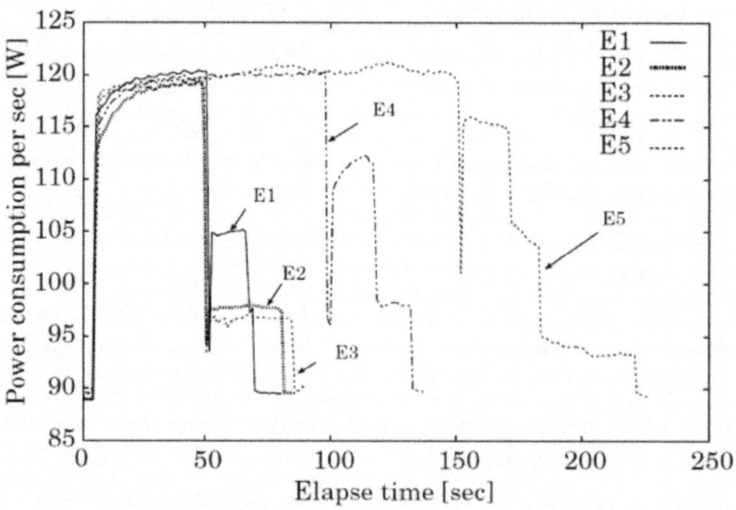

CT3. The server s_1 concurrently transmits a file f to three clients c_1, c_2, and c_3.

Figure 16 shows the power consumption rate of a server s_1 for elapse time. Table 8 summarizes time for computation and transmission and power consumption. In the experimentations CT1, CT2, and CT3, the transmission time is shorter than the compression time. The power consumption rates of the server s_1 in the experimentations CT1, CT2, and CT3 are 120, 121, and 122 [W], respectively. This means that even if the server s_1 transmits the file f concurrently to multiple clients, the server s_1 consumes the maximum power consumption as long as at least one compression process is performed. On the other hand, the more number of concurrent processes, the longer compression time. For example, it takes 46, 55, and 62 [sec] to compress the data d in the experimentations CT1, CT2, and CT3, respectively. This means it takes 1.2 times longer time to compress the data d if one more transmission process is additionally executed. It takes similar time to transmit the file f to each of clients c_1, c_2, and c_3 in the experimentations CT1, CT2, and CT3 as presented in the preceding subsection. Hence, even if compression processes are performed concurrently with a transmission process, it takes almost the same time to transmit data.

6.2. System Model of General Type of Applications

6.2.1. Computation Model for General Type of Application

$P_t(\tau)$ shows a set of current processes on a server s_t at time τ. $P(\tau)$ is $\bigcup_{t=1,\ldots,n} P_t(\tau)$. $NP_t(\tau) = | P_t(\tau) |$. Let $CP_t(\tau)$ and $CT_t(\tau)$ be sets of current computation and transmission processes on a server s_t at time τ, respectively. $NC_t(\tau) = | CP_t(\tau) |$ and $NT_t(\tau) = | CT_t(\tau) |$. $| P_t(\tau) | = | CP_t(\tau) | + | CT_t(\tau) |$. For $CP_t(\tau) = \{ cp_i \}$ and $P_t(\tau) = \{ p_i \}$, $NC_t(\tau) = 1$ and $NT_t(\tau) = 0$. Here, a computation process cp_i is exclusively performed on a server s_t at time τ. Suppose a computation process cp_i is performed on a server s_t. Here, we consider the following parameters for the computation process cp_i:

- T_{ti} = total computation time of the computation process cp_i on a server s_t.
- $minT_{ti}$ = minimum computation time of a computation process cp_i where cp_i is exclusively performed on a server s_t ($1 \leq minT_{ti} \leq T_{ti}$).
- $maxT_i = max(minT_{1i}, \ldots, minT_{ni})$.
- $minT_i = min(minT_{1i}, \ldots, minT_{ni})$.

Figure 16. Concurrent executions of compressions and transmissions

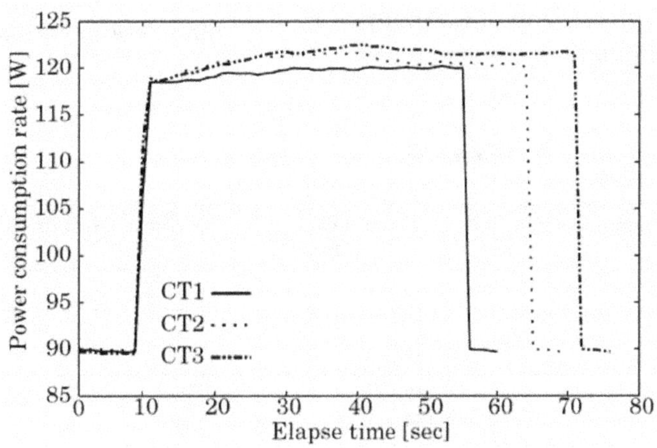

- $F_{ti} = 1 / T_{ti}$ = *average computation rate* (*ACR*) of a computation process cp_i on a server s_t $(0 < F_{ti} \leq 1 / minT_{ti} \leq 1)$ [1/tu].
- $maxF_{ti} = 1 / minT_{ti}$ = maximum *ACR* of a computation process cp_i on a server s_t.
- $maxF_i = max\ (maxF_{1i}, ..., maxF_{ni})$.
- $minF_i = min\ (maxF_{1i}, ..., maxF_{ni})$.

Let $\alpha_t(\tau)$ be *the computation degradation rate* of a server s_t at time τ $(0 \leq \alpha_t(\tau) \leq 1)$ [1/tu] as discussed in the preceding section 4. $\alpha_t(\tau_1) \leq \alpha_t(\tau_2) \leq 1$ if $NC_t(\tau_1) \leq NC_t(\tau_2)$. $\alpha_t(\tau) = 1$ if $NC_t(\tau) \leq 1$. $\alpha_t(\tau)$ is assumed to be $\varepsilon_t^{NC_t(\tau)-1}$ where $0 \leq \varepsilon_t \leq 1$. Let $\theta_t(\tau)$ indicate the *transmission-computation degradation rate* of the server s_t when multiple transmission processes are concurrently performed $(0 \leq \theta_t(\tau) \leq 1)$ [1/tu]. $\theta_t(\tau_1) \leq \theta_t(\tau_2) \leq 1$ if $NT_t(\tau_1) \leq NT_t(\tau_2)$. $\theta_t(\tau) = 1$ if $NT_t(\tau) \leq 1$. In this chapter, $\theta_t(\tau)$ is assumed to be $\vartheta_t^{NT_t(\tau)-1}$ where $0 \leq \vartheta_t \leq 1$. We define the *normalized computation rate* (*NCR*) $f_{ti}(\tau)$ of a computation process cp_i on a server s_t at time τ as follows:

Normalized Computation Rate (NCR) for General Type of Applications

$$f_{ti}(\tau) = \begin{cases} \alpha_t(\tau) \cdot \theta_t(\tau) \cdot maxF_{ti} \ / \ maxF_i \ [1/\text{tu}]. \\ \alpha_t(\tau) \cdot \theta_t(\tau) \cdot minT_i \ / \ minT_{ti} \ [1/\text{tu}]. \end{cases} \quad (16)$$

The maximum *NRC* $maxf_{ti}$ is $maxF_{ti} / maxF_i$. $0 \leq f_{ti}(t) \leq maxf_{ti} \leq 1$. The *NCR* $f_{ti}(\tau)$ shows how many number of steps of a computation process cp_i are performed on a server s_t at time τ. Here, we assume computation resource is uniformly allocated to every current computation process as follows:

$$f_{ti}(\tau) = \alpha_t(\tau) \cdot \theta_t(\tau) \cdot maxf_i / \mid NC_t(\tau) \mid . \quad (17)$$

Suppose a computation process cp_i is exclusively performed on every server at time τ. Here, $f_{ti}(\tau) = 1$ for the fastest server s_t. If a server s_t is faster than another server s_u and the computation process cp_i is exclusively performed on the servers s_t and s_u, $f_{ui}(\tau) < f_{ti}(\tau) \leq 1$. The normalized computation rate $f_t(\tau)$ of the server s_t is $\sum_{cp_i \in CP_t(\tau)} f_{ti}(\tau).f_{ti}(\tau) = maxf_i$ if cp_i is exclusively performed on the server s_t. Next, suppose that a computation process cp_i starts and terminates on a server s_t at time $s\tau_{ti}$ and $e\tau_{ti}$, respectively. Here, $T_{ti} = e\tau_{ti} - s\tau_{ti}$. Here, $\int_{s\tau_{ti}}^{e\tau_{ti}} f_{ti}(\tau)d\tau = minT_i$. The computation laxity $lc_{ti}(\tau) = minT_i - \int_{s\tau_{ti}}^{\tau} f_{ti}(\tau)d\tau$ shows how much computation the server s_t has to spend to perform the computation process cp_i at time τ.

We define a *simple computation model* which each server s_t satisfies the following properties for general type of applications:

Simple Computation Model for General Type of Applications

1. $maxf_{ti} = maxf_{tj} = maxf_t$ for every pair of different computation processes cp_i and cp_j on a server s_t.

Table 8. Concurrent executions of compressions and transmissions

Exp.	Compression time [sec]	Transmission rate [Mbps]	Transmission time [sec]	Power consumption rate [W]
CT1	46	240	18	120
CT2	55	217 (c_1) 134 (c_2)	20 32	121
CT3	63	212 (c_1) 134 (c_2) 71 (c_3)	20 32 60	122

2. $f_t(\tau) = \alpha_t(\tau) \cdot \theta_t(\tau) \cdot maxf_t$.

The first condition shows that any process is performed on a server with the maximum clock frequency. The second condition means that the more number of computation and transmission processes are concurrently performed, the larger computation resource is consumed for overhead.

6.2.2. Transmission Model for General Type of Applications

According to the experimental results discussed in the subsection 6.1, the transmission model discussed in the subsection 5.1 can be adopted as the transmission model for general type of applications. We consider two models where a server satisfies the following properties for general type of applications:

Server-Bound Model

If $Maxrr_1 + ... + Maxrr_m \geq \sigma_t(\tau) \cdot Maxtr_t$, $\sum_{c_{ts} \in CT_t(\tau)} tr_{ts}(\tau) = \sigma_t(\tau) \cdot Maxtr_t$ for every time τ.

Client-Bound Model

If $Maxrr_1 + ... + Maxrr_m \leq \sigma_t(\tau) \cdot Maxtr_t$, $\sum_{c_{ts} \in CT_t(\tau)} tr_{ts}(\tau) = Maxrr_1 + ... + Maxrr_m$ for every time τ.

6.3. Power Consumption Model for General Type of Applications

Let $maxE_t$ and $minE_t$ be the maximum and minimum electric power consumption of a server s_t, respectively. $E_t(\tau)$ shows the electric power consumption rate [W] of a server s_t at time τ (t = 1, …, n), $minE_t \leq E_t(\tau) \leq maxE_t$. The electric power consumption rate $PT_t(\tau)$ of a server s_t at time τ when only transmission processes are being performed is given as follows:

$$PT_t(\tau) = \beta(|NT_t(\tau)|) \cdot \delta_t \cdot tr_t(\tau) + minE_t. \quad (18)$$

Here, δ_t is the power consumption rate [W/Mb] for a server s_t to transmit one Mbits. For $m = |NT_t(\tau)|$, $\beta_t(m)$ shows how much power consumption is increased for the number m of transmission processes, $\beta(m) \geq 1$ and $\beta_t(m) \geq \beta_t(m - 1)$. There is a fixed point $maxm_t$ such that $\beta_t(maxm_t - 1) \leq \beta_t(maxm_t) = \beta_t(maxm_t + h)$ for $h > 0$. $maxPT_t = \beta_t(maxm_t) \cdot \delta_t \cdot Maxtr_t + minE_t$.

We define a *simple power consumption model* for genera applications as follows:

Simple Power Consumption Model for General Type of Applications

1. $maxE_t \geq maxPT_t$.

2.

$$Et(\tau) = \begin{cases} maxE_t & \text{if } NC_t(\tau) \geq 1. \\ PT_t(tr_t(\tau)) & \text{if } NC_t(\tau) = 0 \text{ and } NT_t(\tau) \geq 1. \\ minE_t & \text{otherwise.} \end{cases} \quad (19)$$

If at least one computation process cp_i is performed, the electric power is maximally consumed on a server s_t. If only transmission processes are performed, the electric power consumption depends on the transmission rate of a server s_t.

6.4. Extended Power Consumption Laxity Based (EPCLB) Algorithm

6.4.1. Estimation of Power Consumption Laxity

We discuss how to estimate the total power consumption laxity $lpc_t(\tau)$ of a server s_t at time τ. cp_{ts} and tp_{ts} show computation and transmission processes to perform a request from a client c_s on a server s_t, respectively. Here, $lc_{ts}(\tau)$ shows the computation laxity where a server s_t has to spend to perform up a computation process cp_{ts} at time τ. $lt_{ts}(\tau)$ shows the transmission laxity of a file f_s where a server s_t has to transmit the file f to a client c_s at time τ. The total computation laxity at time τ of a server s_t is $\sum_{cp_{ts} \in CP_t(\tau)} lc_{ts}(\tau)$. The total transmission laxity of a server s_t at time τ is $\sum_{tp_{ts} \in CT_t(\tau)} lt_{ts}(\tau)$. Let $LC_t(\tau)$ be a collection of computation laxity of every computation process of a server s_t at time τ. Let $LT_t(\tau)$ be a collection of transmission laxity of transmission processes of a server s_t at time τ.

Suppose a server s_t receives a request from a client c_u at time τ. Then, $CP_t(\tau) = CP_t(\tau) \cup \{cp_{tu}\}$. $EstC_{ts}(\tau)$ is an estimated termination time of a computation process cp_{ts} on a server s_t at time τ. Here, $EstC_{ts}(\tau) = (maxf_t(\tau) / NC_t(\tau))^{-1} \cdot lc_{ts}(\tau)$. $minEstC_t(\tau)$ is an estimated minimum termination time of a computation process cp_{ts} which is terminated on a server s_t at first after time τ. $EstT_{ts}(\tau)$ is the estimated termination time of a transmission process tp_{ts} on a server s_t at time τ. $EstT_{ts}(\tau) = lt_{ts}(\tau) / tr_{ts}(\tau)$. The transmission rate $tr_{ts}(\tau)$ is calculated as shown in Figure 12. $minEstT_t(\tau)$ is the estimated minimum termination time of a transmission process tp_{ts} which is at first terminated on a server s_t after time τ.

λ shows the estimated termination time to perform up all concurrent computation processes on a server s_t at time τ. According to the formula (19), the server s_t consumes the maximum power consumption rate $maxE_t$ between λ and τ since at least one computation process is performed on the server s_t between λ and τ. After time λ, only transmission processes are performed on the server s_t. The total transmission laxity of the server s_t at time λ is needed to obtain the estimated change of the total power consumption $EPT_t(\lambda)$ of the server s_t for transmitting files at time λ. Here, the estimated termination time λ and a set of transmission laxity $LT_t(\lambda)$ of transmission processes of the server s_t at time λ are estimated as shown in Figure 17.

Suppose a pair of computation processes cp_{t1} and cp_{t2}, and a pair of transmission processes tp_{t3} and tp_{t4} are performed on a server s_t at time τ in Figure 18 (1). Here, $\lambda = \tau$. $CP_t(\lambda) = \{cp_{t1}, cp_{t2}\}$ and $CT_t(\lambda) = \{tp_{t3}, tp_{t4}\}$. Here, we assume that $LC_t(\lambda) = \{lc_{t1}(\lambda) = 5, lc_{t2}(\lambda) = 8\}$ and $LT_t(\lambda) = \{lt_{t3}(\lambda) = 600, lt_{t4}(\lambda) = 900\}$ [Mbits] at time τ. In addition, $|f| = 100$ [Mbits], $maxf_t = 1.0$, $Maxtr_t = 100$ [Mbps], $\varepsilon = 0.9$, $\vartheta = 0.9$, $\gamma = 0.9$, $Maxrr_1 = 60$ [Mbps], $Maxrr_2 = 50$ [Mbps], $Maxrr_3 = 30$ [Mbps], and $Maxrr_4 = 60$ [Mbps]. Here, *while* loop is performed two times. In the first round, $maxf_t(\lambda) = \varepsilon^{2-1} \cdot \vartheta^{2-1} \cdot maxf_t = 0.9 \cdot 0.9 \cdot 1.0 = 0.81$ and $maxtr_t(\lambda) = \gamma^{2-1} \cdot Maxtr_t = 0.9 \cdot 100 = 90$ [Mbps]. Here, $maxf_{t1}(\lambda) = maxf_{t2}(\lambda) = 15 / 2 = 0.405$. $maxtr_{t3}(\lambda) = 30$ [Mbps] and $maxtr_{t4}(\lambda) = 60$ [Mbps]. Then, $minEstC_t(\lambda) = 12.3$ and $minEstT_t(\lambda) = 15$. Here, $min(minEstC_t(\lambda), minEstT_t(\tau)) = minEstC_t(\lambda) = 12.3$. Then, $CP_t(\lambda)$ and $CT_t(\lambda)$ are changed to $\{cp_{t2}\}$ and $\{tp_{t1}, tp_{t3}, tp_{t4}\}$, respectively. In addition, $LC_t(\lambda)$ and $LT_t(\lambda)$ are also changed to $\{lc_{t2}(\lambda) = 3\}$ and $\{lt_{t1}(\lambda) = 1{,}000, lt_{t3}(\lambda) = 231, lt_{t4}(\lambda) = 162\}$ [Mbits]. Hence, $\lambda = \lambda + 12.3$. In the second round, $maxf_t(\lambda) = \varepsilon^{1-1} \cdot \vartheta^{3-1} \cdot maxf_t = 1 \cdot 0.81 \cdot 1.0 = 0.81$ and $maxtr_t(\lambda) = \gamma^{3-1} \cdot Maxtr_t = 0.81 \cdot 100 = 81$ [Mbps] [see Figure 18 (2)]. Here, $maxf_{t2}(\lambda) = 0.81$ and $maxtr_{t1}(\lambda) = maxtr_{t3}(\lambda) = maxtr_{t4}(\lambda) = 27$ [Mbps]. Then, $minEstC_t(\lambda) = 3.7$ and $minEstT_t(\lambda) = 6.0$. $min(minEstC_t(\lambda), minEstT_t(\tau)) = minEstC_t(\lambda) = 3.7$. Hence, $\lambda = \lambda + 12.3 + 3.7 = \lambda + 16$ and $LT_t(\lambda) = \{lt_{t1}(\lambda) = 900.1, lt_{t3}(\lambda) = 131.1, lt_{t4}(\lambda) = 62.1\}$ [Mbits], respectively.

6.4.2. Extended Power Consumption Laxity-Based (EPCLB) Algorithm

Suppose a client c_s issues a request to a load balancer K at time τ. If the load balancer K selects a server s_t and the request is performed on the server s_t, the power consumption of the server s_t may be changed. The estimated change $EPT_{ts}(\tau)$ [Ws] of power consumption of a server s_t for transmitting a file f_t to a client c_s at time τ is $\sum_{tp_{ts} \in CT_t(\tau)} lt_{ts}(\tau) \cdot \beta(|CT_t(\tau)|) \cdot \delta_t$. The Computation Power Consumption Laxity (CPCL) $cle_t(\tau)$

Figure 17. The estimated termination time λ and a set of transmission laxity $LT_t(\lambda)$

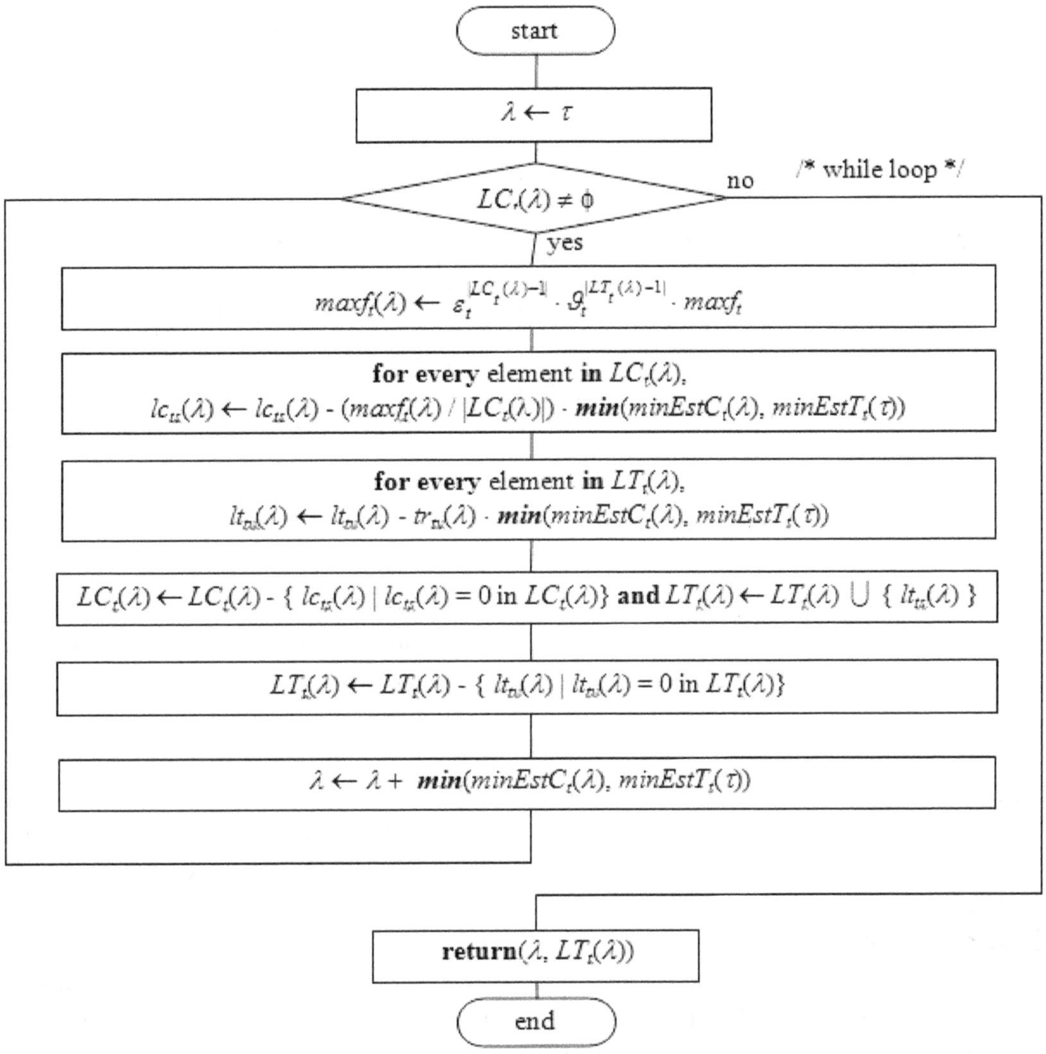

$= maxE_t \cdot (\lambda - \tau)$ shows how much power a server s_t has to consume to perform up all the computation processes at time τ. The total power consumption laxity (TPCL) $tle_t(\tau) = maxE_t \cdot (\lambda - \tau) + EPT_{ts}(\tau)$ shows how much power a server s_t has to consume to perform up all the computation and transmission processes at time τ.

In the Extended Power Consumption Laxity-Based (EPCLB) algorithm, a server s_t which consumes the minimum power consumption is selected for a client c_s, i.e. a server s_t whose $tle_t(\tau)$ is minimum is selected in the server set S. At time τ, a server s_t is selected for a client c_s by the following procedure EPCLB():

```
EPCLB() {
    for every server s_t in S {
⟨λ, LT_t(λ)⟩ is calculated;   /* ⟨λ,
LT_t(λ)⟩ is calculated as shown in
Figure 17.*/
        cle_t(τ) = maxE_t · (λ - τ);
EPT_ts(τ) = Σ_{tp_ts∈CT_t(λ)} lt_ts(λ) · β_t(| LT_t(λ)
```

Figure 18. Estimated termination time

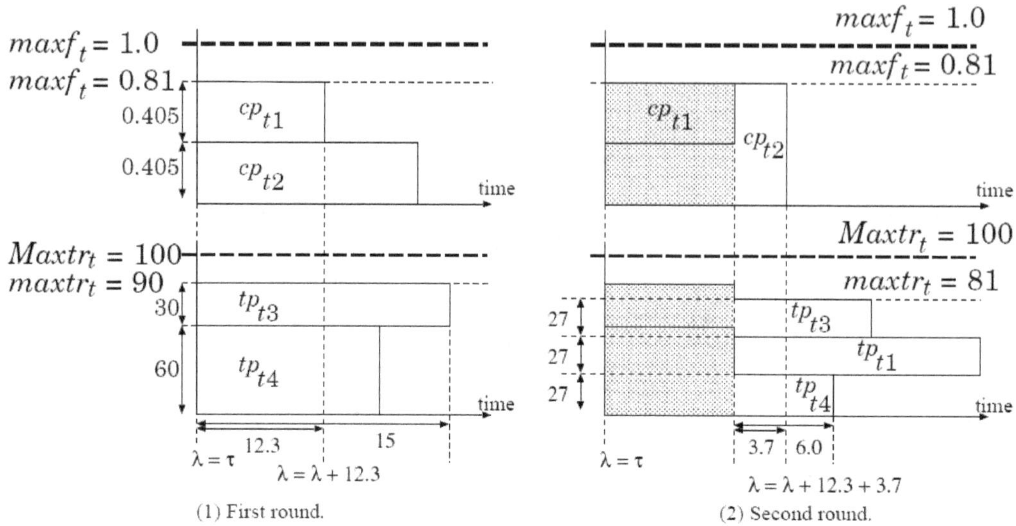


```
|) · δ_t ;
        tle_t(τ) = cle_t(τ) + EPT_ts(τ);
    }
    server = s_t where tle_t(τ) is mini-
mum;
    return(server);
}
```

6.5. Evaluation of the EPCLB Algorithm

We evaluate the EPCLB algorithm in terms of total power consumption compared with the basic Round-Robin (RR) algorithm (Job Scheduling Algorithms in Linux Virtual Server, 2011) through the simulation. In this section, we take one [sec] as one time unit. There are ten servers s_1, ... s_{10} as shown in Table 9, $S = \{s_1, ... s_{10}\}$. The servers in the server set S are ordered with respect to the processing speed, i.e. maximum computation rate. Here, the server s_1 is the fastest server and the server s_{10} is the slowest server. Each server s_t holds full replicas of three pieces of data d_1, d_2, and d_3. The sizes of data d_1, d_2, and d_3 are 556, 278, and 112 [MByte], respectively. Each time a server s_t receives a request to obtain one d_h of the three pieces data, the server s_t compresses the requested data d_h to a file f_h and then transmits the compressed file f_h to a requesting client c_s. Here, the minimum computation time $minT_{1di}\}$ to compress data d_1, d_2, and d_3 on the fastest server s_1 are 42, 20, and 4 [sec], respectively. The sizes of compressed file of the three data d_1, d_2, and d_3 are 500, 250, and 100 [MB], respectively. The minimum power consumption rate $minE_t$ of each server s_t is randomly selected between 87 and 150 [W]. The maximum power consumption rate $maxE_t$ of each server s_t is randomly selected between 1.2 times and 1.4 times higher than the minimum power consumption rate $minE_t$. The maximum Normalized Computation Rate (NCR) $maxf_t$ of each server s_t is randomly selected between 1 and 0.7 [1/sec] (t = 1, ..., 10). The maximum computation rate $maxf_1 = 1$ and $maxf_{10} = 0.73$. The computation degradation rate ε_t is randomly selected between 0.99 and 0.95. The transmission-computation degradation rate ϑ_t is randomly selected between 0.83 and 0.89. The transmission degradation factor is randomly selected between 0.98 and 0.97. The power consumption coefficient δ_t for each server s_t to transmit one Mbits is randomly selected between 0.02 and 0.11 [W/Mb]. The increasing rate of the power consumption $\beta_t(m)$ for the number m of clients to which s_t concurrently transmits files is randomly selected between 1.09 and 1.41. The maximum transmission rate $Maxtr_t$ of each server s_t is randomly selected between 175 and 450 [Mbps]. The maximum power consumption rate $maxPT_t(tr)$ for transmission process of each server s_t is 95% of the $maxE_t$.

Totally one thousands clients issue requests of data d_h to a load balancer K at time $s\tau_s$. Each client c_s issues one request at time $s\tau_s$ in the simulation. The maximum receipt rate $Maxrr_s$ of each client c_s is randomly selected between 1 and 100 [Mbps]. Each client c_s randomly selects data d_h from three types of data. The starting time $s\tau_s$ of each client c_s is randomly selected between 0 and 9,000 [sec] at simulation time. Figure 19 shows a traffic pattern, i.e. how many number of clients start at time. In the simulation, the EPCLB and RR algorithms are performed for the same traffic pattern.

Figure 20 shows the total power consumption [Ws] of the servers in the set S at time τ. Table 10 shows the total power consumption, which is obtained by subtracting the minimum power consumption of server in the idle state from the total power consumption during the simulation, i.e. the power consumption for only performing totally 1,000 requests. The total power consumption of the EPCLB algorithm is 769,937 [Ws]. The total power consumption in the RR algorithm is 1,092,420 [Ws]. In the EPCLB algorithm, 29% of the power consumption can be reduced compared with the RR algorithm.

Table 11 shows the elapse time to terminate the total number 1,000 of requests in the EPCLB and RR algorithms. The termination times of all the processes in the EPCLB and RR algorithms are 8,634 [sec] and 8,635 [sec], respectively. The difference of the termination time between EPCLB and RR algorithms is neglectable.

From the evaluation results, we conclude the total power consumption can be more reduced in the EPCLB algorithm than the RR algorithm and the difference of termination time among the EPCLB and RR algorithms is almost neglectable. Therefore, the EPCLB algorithm is more useful than the RR algorithm.

7. FUTURE RESEARCH DIRECTIONS

In this chapter, we proposed the PCLB, EPCB, and EPCLB algorithms for computation, communication, and general type applications, respectively, to reduce the total power consumption of servers. According to the simulation results, the total power consumption of servers can be reduced in the

Table 9. Servers

Server	Max f_t	E_t	ϑ_t	$maxE_t$	$minE_t$	δ_t	$\beta_t(m)$	$Maxtr_t$	γ_t	$maxPC_t(tr)$
s_1	1	0.95	0.83	210	150	0.11	1.48	390	0.98	200
s_2	0.99	0.98	0.89	104	87	0.02	1.1	450	0.98	99
s_3	0.94	0.97	0.87	131	101	0.04	1.19	390	0.98	125
s_4	0.91	0.98	0.88	113	94	0.02	1.21	418	0.97	107
s_5	0.88	0.96	0.85	176	135	0.08	1.32	280	0.98	167
s_6	0.85	0.98	0.87	150	115	0.05	1.41	356	0.97	142
s_7	0.82	0.96	0.84	190	136	0.07	1.25	356	0.98	181
s_8	0.79	0.99	0.88	107	89	0.02	1.09	440	0.98	101
s_9	0.76	0.96	0.83	159	122	0.03	1.1	304	0.97	151
s_{10}	0.73	0.95	0.83	200	143	0.10	1.39	175	0.97	190

PCLB, EPCB, and EPCLB algorithms than the RR algorithm. In the future, we have to evaluate the PCLB, EPCB, and EPCLB algorithms by adopting real applications. In this chapter, we assume the networks are reliable and synchronous. However, the networks are unreliable or asynchronous in real environment. In addition, some faults like application process and network faults might occur. In order to adopt the PCLB, EPCB, and EPCLB algorithms to real environment, it is necessary to improve the PCLB, EPCB, and EPCLB algorithm by making the algorithms fault-tolerant.

8. CONCLUSION

In this chapter, we classified applications on a cluster system into three types: computation, communication, and general type of applications. We defined computation, communication, and power consumption models for each type of application based on the measurement of a power consumption of servers. According to the computation, communication, and power consumption models, we proposed algorithms to select one of servers in a server cluster for each type of application so that

Figure 19. Traffic pattern

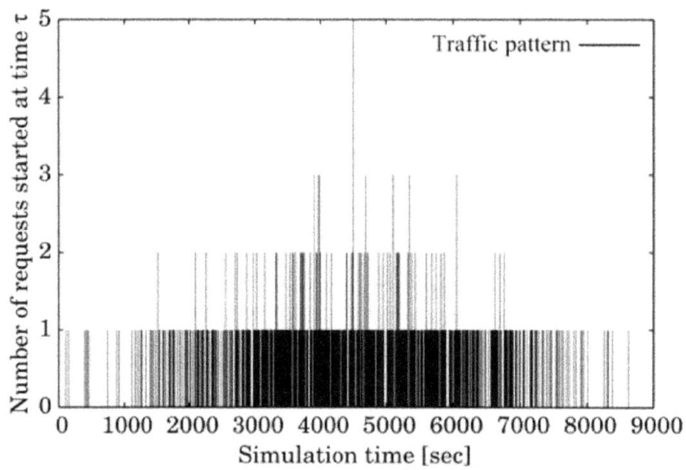

Figure 20. Total power consumption [Ws]

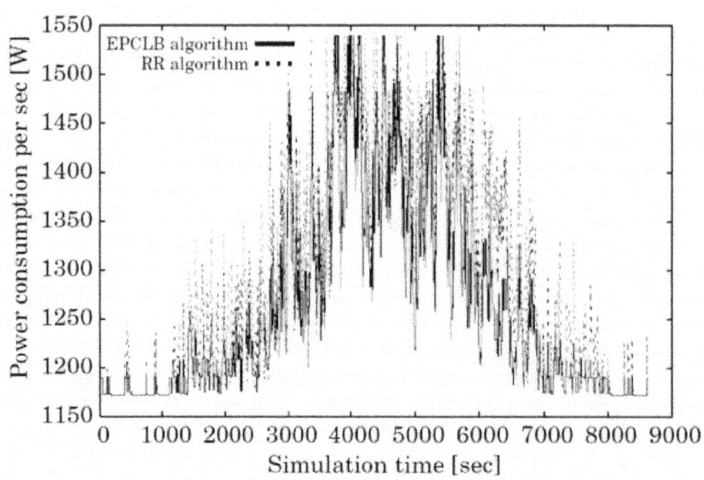

Table 10. Total power consumption [Ws]

EPCLB	RR
769,937 [Ws]	1,092,420 [Ws]

Table 11. Elapse time

EPCLB	RR
8,634 [sec]	8,635 [sec]

the total power consumption of servers can be reduced. We evaluated the proposed algorithms in terms of the total power consumption and elapse time compared with the RR algorithm. The evaluation results showed the proposed algorithms can reduce the total power consumption of servers on each type of application.

REFERENCES

Apache. (2011a). *Apache http server version 2.0 documentation*. Retrieved April 29, 20011, from http://httpd.apache.org/docs/2.0/en/.

Apache. (2011b). *Apache module mod_deflate*. Retrieved April 29, 2011, from http://httpd.apache.org/docs/2.0/en/mod/mod_deflate.html.

Bevilacqua, A. (1999). A dynamic load balancing metthod on a heterogeneous cluster of workstations. *Informatica*, *23*(1), 49–56.

Bianchini, R., & Carrera, E. V. (2000). Analytical and experimental evaluation of cluster-based network servers. *World Wide Web (Bussum)*, *3*(4), 215–229. doi:10.1023/A:1018734706376

Bianchini, R., & Rajamony, R. (2004). Power and energy management for server systems. *Computer*, *37*(11), 68–76. doi:10.1109/MC.2004.217

Buford, F. J., Yu, H., & Lua, K. E. (2009). *P2P networking and applications*. New York, NY: Morgan Kaufmann.

Colajanni, M., Cardellini, V., & Yu, P. S. (1998). Dynamic load balancing in geographically distributed heterogeneous web servers. In *Proceedings of the 18th IEEE International Conference on Distributed Computing Systems (ICDCS 1998)*, (pp. 295-303). Washington, DC: IEEE Computer Society.

Durresi, A., Durresi, M., Paruchuri, V., & Barolli, L. (2011). Ad hoc communications for emergency conditions. In *Proceedings of the 25th IEEE International Conference on Advanced Information Networking and Applications (AINA-2011)*, (pp. 787-794). Washington, DC: IEEE Computer Society.

Enokido, T., Aikebaier, A., & Takizawa, M. (2010). A model for reducing power consumption in peer-to-peer systems. *IEEE Systems Journal*, *4*(2), 221–229. doi:10.1109/JSYST.2010.2047296

Enokido, T., & Takizawa, M. (2010). A purpose-based synchronization protocol for secure information flow control. *Journal of Computer Systems Science and Engineering*, *25*(2), 25–32.

Ferraiolo, D. F., Kuhn, D., & Chandramouli, R. (2007). *Role-based access control*. Norwood, MA: Artech House.

Fischer, M. J., Lynch, N. A., & Paterson, M. S. (1985). Impossibility of distributed consensus with one faulty process. *Journal of the ACM*, *32*(2), 374–382. doi:10.1145/3149.214121

Ghemawat, S., Gobioff, H., & Leung, S. (2003). The Google file system. In *Proceedings of the 19th ACM Symposium on Operating Systems Principles (SOSP 2003),* (pp. 29-43). New York, NY: ACM Press.

Grossman, R. L. (2009). The case for cloud computing. *IT Professional, 11*(2), 23–27. doi:10.1109/MITP.2009.40

Heath, T., Diniz, B., Carrera, E. V., Meira, W. J., & Bianchini, R. (2005). Energy conservation in heterogeneous server clusters. In *Proceedings of the 10th ACM SIGPLAN Symposium on Principles and Practice of Parallel Programming (PPoPP 2005),* (pp. 186-195). New York, NY: ACM Press.

Hemmert, S. (2010). Green HPC: From nice to necessity. *Computing in Science & Engineering, 12*(6), 8–10. doi:10.1109/MCSE.2010.134

Ikeda, M., Kulla, E., Hiyama, M., Barolli, L., & Takizawa, M. (2011). Experimental results of a MANET testbet in indoor stairs environment. In *Proceedings of the 25th IEEE International Conference on Advanced Information Networking and Applications (AINA-2011),* (pp. 779-786). Washington, DC: IEEE Computer Society.

Liu, X., Zhao, H., & Li, X. (2011). EPC: Energy-aware probability-based clustering algorithm for correlated data gathering in wireless sensor networks. In *Proceedings of the 25th IEEE International Conference on Advanced Information Networking and Applications (AINA-2011),* (pp. 419-426). Washington, DC: IEEE Computer Society.

Lunix Virtual Server. (2011). *Job scheduling algorithms in linux virtual server*. Retrieved April 29, 2011, from http://www.linuxvirtualserver.org/docs/scheduling.html.

Murphy, R., Sterling, T., & Dekate, C. (2010). Advanced architectures and execution models to support green computing. *Computing in Science & Engineering, 12*(6), 38–47. doi:10.1109/MCSE.2010.124

Perera, G. (2008). *New search paradigms and power management for peer-to-peer file sharing.* Saarbrücken, Germany: VDM Verlag.

Rahimi, S. K., & Haug, F. S. (2010). *Distributed database management systems*. Hoboken, NJ: John Wiley & Sons. doi:10.1002/9780470602379

Rajamani, K., & Lefurgy, C. (2003). On evaluating request-distribution schemes for saving energy in server clusters. In *Proceedings of the 2003 IEEE International Symposium on Performance Analysis of Systems and Software (ISPASS 2003),* (pp. 111-122). Washington, DC: IEEE Computer Society.

Steinmetz, R., & Nahrstedt, K. (2004). *Multimedia systems*. New York, NY: Springer-Verlag.

Watts Up Meters. (2011). *Watts up? Products*. Retrieved April 29, 2011, from https://www.wattsupmeters.com/secure/products.php?pn=0&wai=0&spec=2.

Yang, T., Mino, G., Spaho, E., Barolli, L., Durresi, A., & Xhafa, F. (2011). A simulation system for multi mobile events in wireless sensor networks. In *Proceedings of the 25th IEEE International Conference on Advanced Information Networking and Applications (AINA-2011),* (pp. 411-418). Washington, DC: IEEE Computer Society.

Yang, Y., Xiong, N., Aikebaier, A., Enokido, T., & Takizawa, M. (2009). Minimizing power consumption with performance efficiency constraint in web server clusters. In *Proceedings of the 12th International Conference on Network-Based Information Systems (NBiS2009),* (pp. 45-51). Washington, DC: IEEE Computer Society.

Zhang, L., & Zhou, Q. (2009). CCOA: Cloud computing open architecture. In *Proceedings of the IEEE International Conference on Web Services,* (pp. 607-616). Los Angeles, CA: IEEE Computer Society.

ADDITIONAL READING

Aikebaier, A., Barolli, V., Enokido, T., & Takizawa, M. (2008). A protocol for making an agreement in a peer-to-peer (P2P) society. *Journal of Interconnection Networks, 9*(4), 317–335. doi:10.1142/S0219265908002308

Barolli, L., Ikeda, M., Xhafa, F., & Durresi, A. (2010). A testbed for MANETs: Implementation, experiences and learned lessons. *IEEE Systems Journal, 4*(2), 243–252. doi:10.1109/JSYST.2010.2047174

Coulouris, G., Dollimore, J., & Kindberg, T. (2005). *Distributed systems concepts and design.* Harlow, MA: Addison Wesley.

Durresi, A., Zhang, P., Durresi, M., & Barolli, L. (2010). Architecture for mobile heterogeneous multi domain networks. *Mobile Information Systems, 6*(1), 49–63.

Enokido, T. (2007). Concurrency control based on role concept. *International Journal of High Performance Computing and Networking, 5*(1/2), 75–83. doi:10.1504/IJHPCN.2007.015766

Enokido, T., Aikebaier, A., & Takizawa, M. (2010). Energy-efficient server selection algorithms for network applications. In *Proceedings of the International Conference on Broadband, Wireless Computing, Communication and Applications (BWCCA2010),* (pp. 159-166). Washington, DC: IEEE Computer Society.

Enokido, T., Aikebaier, A., & Takizawa, M. (2011). An integrated power consumption model for communication and transaction based applications. In *Proceedings of the 25th IEEE International Conference on Advanced Information Networking and Applications (AINA-2011),* (pp. 695-702). Washington, DC: IEEE Computer Society.

Enokido, T., Suzuki, K., Aikebaier, A., & Takizawa, M. (2010). Algorithms for reducing the total power consumption in data communication-based applications. In *Proceedings of the 24th IEEE International Conference on Advanced Information Networking and Applications (AINA2010),* (pp. 142-149). Washington, DC: IEEE Computer Society.

Ghosh, S. (2006). *Distributed systems: An algorithmic approach.* New York, NY: Chapman & Hall/CRC.

Guting, R. H., & Schneider, M. (2005). *Moving objects databases.* New York, NY: Morgan Kaufmann Publishers.

Hassan, M., & Jain, R. (2004). *High performance TCP/IP networking concepts, issues, and solutions.* Upper Saddle River, NJ: Pearson Education International.

Jia, W., & Zhou, W. (2005). *Distributed network systems from concepts to implementations.* New York, NY: Springer.

Kawanami, S., Enokido, T., & Takizawa, M. (2009). Heterogeneous clock group protocol for causally ordered delivery of messages. *International Journal of Wireless and Mobile Computing, 3*(4), 247–254. doi:10.1504/IJWMC.2009.029344

Kshemkalyani, A. D., & Singhal, M. (2008). *Distributed computing principales, algorithms, and systems.* Cambridge, UK: Cambridge University Press.

Kumar, V. (2006). *Mobile database systems*. New York, NY: Wiley-Interscience. doi:10.1002/0470048298

Lin, W., & Veeravalli, B. (2003). *Object management in distributed database systems for stationary and mobile computing environments*. Boston, MA: Kluwer Academic Publishers.

Marchese, M. (2007). *QoS over heterogeneous networks*. West Sussex, UK: John Wiley & Sons, Ltd. doi:10.1002/9780470058763

Puder, A., Romer, K., & Pilhofer, F. (2006). *Distributed systems architecture: A middleware approach*. Amsterdam, The Netherlands: Morgan Kaufmann Pubishers.

Rao, K. R., Bojkovic, Z. S., & Milovanovic, D. A. (2009). *Wireless multimedia communications convergence, DSP, QoS, and security*. New York, NY: CRC Press.

Rittinghouse, H. W., & Ransome, J. F. (2010). *Cloud computing implementation, management, and security*. New York, NY: CRC Press.

Tanaka, Y., Enokido, T., & Takizawa, M. (2009). Transactional agents on distributed object systems. *International Journal of High Performance Computing and Networking, 6*(2), 148–159. doi:10.1504/IJHPCN.2009.027465

Tsuneizumi, I., Aikebaier, A., Ikeda, M., Enokido, T., & Takizawa, M. (2010). A scalable group communication protocol with hybrid clocks. *Concurrency and Computation, 23*(5), 477–490. doi:10.1002/cpe.1650

Vidyarthi, D. P., Sarker, B. K., Tripathi, A. K., & Yang, L. T. (2009). *Scheduling in distributed computing systems analysis, design, and models*. New York, NY: Springer. doi:10.1007/978-0-387-74483-4

Watanabe, K., Nakajima, Y., Enokido, T., & Takizawa, M. (2007). Ranking factors in peer-to-peer overlay networks. *ACM Transaction on Autonomous and Adaptive Systems, 2*(3), 11:1-11:26.

Zou, X., Ramamurthy, B., & Magliveras, S. S. (2005). *Secure group communications over data networks*. New York, NY: Springer.

KEY TERMS AND DEFINITIONS

Computation Laxity: The computation laxity shows how much computation a server s_t has to spend to perform up a process p_i at time τ.

Transmission Laxity: The transmission laxity shows how many bits of a file f_s the server s_t still has to transmit to the client c_s at time τ.

Total Power Consumption Laxity: The total power consumption laxity shows how much power a server s_t has to consume to perform up all the computation and transmission processes at time τ.

Computation Laxity-based (CLB) Algorithm: An algorithm to select a server s_t whose computation laxity is minimum in a server set S for computation type applications.

Power Consumption Laxity-based (PCLB) Algorithm: An algorithm to select a server s_t whose power consumption laxity is minimum in a server set S for computation type applications.

Power Consumption-based (PCB) Algorithm: An algorithm to select a server s_t whose power consumption to transmit a file f to a requesting client c_s is minimum for communication type applications.

Transmission Rate-based (TRB) Algorithm: an algorithm to select a server s_t whose transmission rate for a requesting client c_s is maximum for communication type applications.

Extended Power Consumption-based (EPCB) Algorithm: An algorithm to enhance the estimation of the power consumption in the PCB algorithm for communication type applications.

Extended Power Consumption Laxity-based (EPCLB) Algorithm: An algorithm to select a server s_t for general type of applications.

Chapter 5
Design Models for Resource Allocation in Cyber-Physical Energy Systems

Prakash Ranganathan
University of North Dakota, USA

Kendall Nygard
North Dakota State University, USA

ABSTRACT

Today's and tomorrow's smart grid systems are made more efficient, cleaner, and reliable by "smart" control mechanisms and decision models that deliver information to consumers so they can better manage energy resources. The rapidly changing needs and opportunities of today's electric grid market require unprecedented levels of interoperability to integrate diverse information systems to share knowledge and collaborate among sub-devices or sub-systems in the grid. This book chapter focuses on optimal mathematical models for resource allocation. A series of mathematical models is presented in this book chapter for solving large-scale energy allocation problems with partially observable states, utility functions, and constrained action is introduced. The authors' techniques use a Linear Programming (LP) approach to determine resource allocations among a set of fuzzy rules that allocates Distributed Energy Resources (DER's) or power sources/sinks and uses to determine improving resource management.

INTRODUCTION

Dynamic real-time power systems often operate in continuously changing environments such as adverse weather conditions, sudden transformer failures, malfunctioning of a sub-system of a transmission or distribution network. These disruptions along with the complexity of our power network systems, cause the energy demand and loads of a power system to fluctuate, potentially resulting in widespread outages and huge price spikes. Electrical energy is traditionally generated at large power plants and transferred in bulk to substations that in turn distribute power to residential, commercial, and industrial customers. A wide variety of devices are involved and utilized,

DOI: 10.4018/978-1-4666-1839-8.ch005

including transformers, sensors, controllers, and, more recently, smart meters at points of consumption. A successful Smart Grid must be highly instrumented, data rich, networked, and integrated and managed as a complete "end to end" system. However, centralized control is clearly impossible, leading to the challenge of distributing control and still providing high performance and efficiency.

Data from the North American Electric Reliability Council (NERC) and analyses from the Electric Power Research Institute (EPRI) indicate that average outages from 1984 to the present time have affected nearly 700,000 customers per event annually as noted by Masood Amin (2004). Smaller outages occur much more frequently and affect tens to hundreds of thousands of customers every few weeks or months, while larger outages occur every two to nine years and affect millions. Although preventing these outages remain challenging, such changes (increases or decreases) in demand by consumers can often be offset by Distributed Energy Resources (DERs) which are renewable resources like solar and wind based power to satisfy the shortages or reduce the outage levels. In our work, we consider the use of such DER based standby mechanisms and formulate to support their issue.

The book chapter is divided into two sections. Section 1 focuses on resource optimization methods by formulating generalized Linear Programming models and section 2 details the applicability of branch and bound method in allocating DER resources. In section 2, we specifically apply an Iterative Binary Integer Linear Programming (IBILP) technique similar to Wolsey (1998) to optimally assign DERs to a region based on criteria such as power levels, demands and preferences.

BACKGROUND

Green Energy solutions that are described as "smart"—from smart meters, smart buildings to smart appliances to the Smart Grid—have intelligent sensors to measure temperature or other variables to receive and transmit data, memory chips to store the information and process it, and manage the power flow to adjust energy loads. In building a Smart Grid self-healing model, there are multiple considerations that are important to include as explained by Amin (2008). Some considerations pertain to the physical infrastructure, such as the generators, busses, relays, and transmission lines should be considered. Other considerations pertain to the cyber infrastructure, such as the communication networks, storage, protocols, and procedures for management of the grid. There was been several studies that pushes for advanced secure infrastructure of the grid as described by Amin and Stringer (2008).

The current electric grid in the U.S. is about 100 years old and while incremental upgrades to it have been made over the last few decades, the advancements in fields such as Software and Internet Technologies, wireless technologies, have been slow to transition into the grid. It is the convergence of a number of factors that provides for a perfect storm today that is expected to make a revolutionary change in the nature of the electrical grid.

The various factors that led to futuristic self-healing grid are as follows:

- Rapid reduction in limited natural resources
- Growth in population
- Rising cost of energy
- Current technology advances make it possible
- Increase in Global warming and climatic conditions
- Changes in the utility business operations worldwide
- Deregulation in various parts of the world
- Better awareness and education among consumers
- Increase in renewable, distributed, and smaller power generation
- Increase in power storage capability

- Radio Frequencies ID (RFIDs) and sensors having narrowed the virtual distance between physical and cyber world

As a result, there is tremendous pressure at the government, industry and consumer level to innovate and various entities are working and have articulated their visions into what the futuristic grid should look like.

Today's grid features a typical centralized approach where few powerful central stations broadcast energy to the different consumers. However, as renewable energy resources pave their way, we expect that future users will not be simply consumers of energy but also producers and they reflect the ability to interchangeably slip to these roles. The smart grid defined in Amin (2008) is an emerging concept targeting to provide the next-generation electricity network that will boast advanced configurability, reactive-ness, and self manageability. It is a complex infrastructure depicting system of system characteristics such as interdisciplinary nature, operational and managerial independence of its elements, geographical distribution, high heterogeneity of the networked systems as well as emergent behavior and evolutionary development. It is expected to be the key part in a global ecosystem of interacting entities, whose cooperation will give birth to innovative cross-industry services. One of the key driving forces behind these efforts is energy efficiency and better management of the available resources (locally and globally). In order to achieve this, fine-grained monitoring and management is needed.

Primary considerations for self healing cyber grid architecture are given below:

Identification of malfunctions. When an electricity supply/demand problem occurs, it must be accurately and rapidly identified in terms of its location, type, and severity. Smart meters that are programmed to report energy consumption levels at regular intervals help in pinpointing problems by simply reporting that suddenly no electricity is being consumed at specific sites. Phasor Measurement Units (PMUs) located at interconnection points in the grid and that are synchronized in time using the Global Positioning System (GPS) can report phase angle differences that are associated with unstable conditions. The precise timing of the units allows for wide-area sensing and monitoring at rates on the order of 60 samples per second. A Phasor Data Concentrator (PDC) can receive data from multiple PMUs, correlate, store, and trigger response actions to abnormal conditions. Networks of smart meters and PMUs, along with other sensing and reporting devices, make it possible to quickly and accurately determine the details of problems occurring in the grid.

Data Storage. At least 33 states in United States have smart meters installed in place in 2010, and an estimate by the Edison Electric Institute indicates that 60 million will be installed by 2019. The large volume of data from the grid sensing devices underscores the need for large-scale distributed database systems in the Smart Grid. The nodes storing the data may be capable of carrying out data mining to predict future problems, send alerts, and make resource allocation decisions autonomously.

Energy storage, by itself and in combination with distributed generation (termed ES-DER), is a new and emerging technology that has been identified by FERC as a key functionality of the smart grid, and standards related to storage should be treated as a key priority by the Institute and industry in the interoperability standards development process, subject to certain reservations. Coupled with inverter-based technology, these systems can be used to improve EPS performance. Due to the infancy of the use of storage and inverter technologies as a grid-integrated operational asset, there are few standards that exist to capture how it could or should be utilized on the legacy grid and Smart Grid. For example, to date there exist no guidance or standards to address grid-specific aspects of aggregating large or small mobile storage, such as Plug-in Hybrid Electric Vehicles (PHEVs). ES-DER is treated as a distributed energy resource in some standards, but there may be distinctions between electric storage and connected generation. In particular, storage-based systems may function as a load more than 50% of the time.

At the same time, we are moving towards large penetration of renewables into the Grid, which could be destabilizing, but should, in the context of the Smart Grid, allow these renewables to be true utility assets. The potential for instability is twofold; first, due to the intermittent nature of renewables and therefore their unsuitability to be dispatchable resources, and second, due to the interconnection regulations themselves that can lead the electronic interconnection interface (the inverter) to trip off in response to minor variations in grid voltage or frequency. As low frequency is the result of insufficient generation, tripping a high level of inverter-based systems would contribute to the problem and cause possible stability issues in response to a relatively minor disturbance. Appropriate interconnection standards, smart grid devices, and storage are all key elements of the solution. Storage allocation can be done using Linear Programming approach.

Utility Function. Utility is a concept in economics used to express the most basic element of well-being for a representative consumer. Utility functions are generally used to show logical inferences about the trade-offs that consumers make and their likely outcome. The concept of using utility functions to represent consumer *choice* as reflecting likely consumer *behavior* has long been attacked, partly because of it appears (wrongly) to assume economic actors are rational, and partly because it is *tautological*.

A utility is a numerical rating assigned to every possible outcome a decision maker may be faced with. In a choice between several alternative prospects, the one with the highest utility is always preferred. To qualify as a true *utility* scale however, the rating must be such that the utility of any uncertain prospect is equal to the expected value (the mathematical expectation) of the utilities of all its possible outcomes, which could be either "final" outcomes or uncertain prospects themselves. When decisions are made by a so-called *rational agent* (if A is preferred to B and B to C, then A must be preferred to C), it should be clear that *some* numerical scale can be devised to rate any possible outcome "simply" by comparing and ranking these. Determining equivalence in money terms may be helpful in such a systematic process but it is not theoretically indispensable. What may be less clear, however, is how to devise such a rating system so that it would possess the above fundamental property required of a *utility* scale.

Measuring the benefit that is obtained through a management action is the well-studied role of the utility function. A common convention is to normalize the utility function into the range of 0 to 1, with the 0 end of range modeling minimum benefit.

Application Conflicts We presume that distributed software agents will be running applications at various nodes in the grid to summarize and provide information, mine data, and invoke certain control actions. Some applications will compete in their polling actions to acquire necessary data and bandwidth with a resulting degradation in each other's performance. To maintain Quality of Service (QoS) standards, the applications will require allocations of resources.

Large Scale Resource Optimization: There is significant need and benefit to addressing (1) large scale supply chain optimization problems, (2) product pipeline management (here we refer them as '*agents*') decision-making, (3) detailed optimization of process models, (4) design and optimization of nodes in the network and (5) optimization of micro-grid models such as small scale or RTU's designs at strategic, tactical and operational levels. These static and dynamic optimization models, which are commonly in the hundreds of thousands of variables, include non-linear as well as mixed-integer programming problems. These models can be posed in equation form or through black box simulation software. Rigorously handling non-convexities in nonlinear models has emerged as a major challenge that will involve global optimization methods that are potentially very expensive to solve.

RELATED RESEARCH

Mathematical programming has enjoyed a burgeoning presence in theoretical computer science both as a framework for developing algorithms and, increasingly, as a bonafide model of computation. The limits are expressed in terms of sizes of formulations and integrality gaps of formulations as the one described in Arora (2002). Linear formulations are an appealing model of computation because both optimization and decision problems fit naturally into the framework, and both theoretically tractable and efficient practical algorithms exist for solving linear program as defined in R. E. Moore (1991).

Many practical problems in operations research can be expressed as linear programming problems. Certain special cases of linear programming, such as *network flow* problems and *multi-commodity flow* problems are considered important enough to have generated much research on specialized algorithms for their solution. A number of algorithms for other types of optimization problems work by solving LP problems as sub-problems. Hanssmann and Hess (1960) developed a more comprehensive linear programming approach for the production planning problem. This approach became a model for several further research efforts and is the foundation for the simple illustrative model presented in this chapter. A briefer overview of this model can be found in McLeavey and Narasimhan (1985).

Though the existence of optimization techniques can possibly be traced to the era of Newton, Euler, Lagrange, and Cauchy, it was the development of simplex method for linear programming by Dantzig in the mid 40's which, in a sense, started the subject of mathematical optimization the way we understand it today. Another major development was due to Kuhn and Tucker in 1951 who gave necessary and sufficient optimality conditions for nonlinear optimization or non-linear programming problems. Historically, ideas from linear programming have inspired many of the central concepts of optimization theory, such as *duality, decomposition,* and the importance of *convexity* and its generalizations by Hillar and Gerald (2001). Likewise, linear programming is heavily used in microeconomics and company management, such as planning, production, transportation, technology and other issues. Although the modern management issues are ever changing, most companies would like to maximize profits or minimize costs with limited resources. Therefore, many issues can be characterized as linear programming. Although, there may be studies using LP in many applications, we believe no literatures have studied grid resource allocation or assignment problem using LP in the electric grid context.This book chapter uses LP in the context of Smart Grid application.

Collection of Generalized LP Models

Linear programming is the name of a branch of applied mathematics that deals with solving optimization problems of a particular form. Linear programming problems consist of a linear *cost* function (consisting of a certain number of variables) which is to be minimized or maximized subject to a certain number of constraints. The constraints are linear inequalities of the variables used in the cost function. The cost function is also sometimes called the objective function. Linear programming is closely related to linear algebra; the most noticeable difference is that linear programming often uses inequalities in the problem statement rather than equalities. Linear programming is a considerable field of optimization for several reasons.

Linear programming is a problem solving technique involving optimization of a decision, subject to one or more constraints. The basic logic of the approach is displayed in four steps: (1) Recognizing the LP Problem; (2) Formulating the LP Problem; (3) Solving the LP Problem; and (3) Interpreting the LP solution. In recognizing an LP Problem, several conditions must exist before a problem can be addressed by linear programming. (1) There must be a well-defined single objective that can be stated mathematically as either a Minimization or maximization function. (2) There must be a set of decision variables that allow for alternative courses of action. (3) There must be constraints on the achievement of the objective that are imposed by the availability of scarce resources or other restraints such as product demand. (4) The objective and all constraints must be able to be expressed as linear functions.

In Formulating the LP Problem, the simple formulation of the example is provided in Section 2. In Solving the LP Problem, depending on the structure of the linear program formulation, one of several techniques can be applied to solve it. The most common method is the simplex method, or we can apply Branch and Bound, discussed later in the chapter. For interpreting the LP Solution, the interpretation detail of the example is provided later in the chapter.

We discuss generalizations to LP models applicable to electric grid in section 1 and specifically discuss Binary Integer Linear Programming approach in Section 2 with an example of a DER assignment problem and conclude with an analysis of versatility of Linear Programming and the types of problems and constraints which can be handled linearly, as well as some brief comments about its generalizations to handle situations with multiple constraints. The following are collection of LP models that can be applied to electric grid loads and supply resources.

SECTION 1

LP Model by Maximizing Utility

Multi objective programming involves recognition that the decision maker is responding to multiple objectives. Generally, objectives are conflicting, so that not all objectives can simultaneously arrive at their optimal levels. An assumed utility function is used to choose appropriate solutions. Several fundamentally different utility function forms have been used in multi objective models. These may be divided into three classes: lexicographic, multi-attribute utility, and unknown utility.

The lexicographic utility function specification assumes the decision maker has a strictly ordered preemptive preference system among objectives with fixed target levels. For example, a lexicographic system could have its first priority goal as power of not less than 10,000MW; the second priority as leisure of no less than 20 hours a week; the third as power of no less than 12,000MW, etc. This formulation is typical of "goal programming models." (Charnes & Cooper, 1961). The various goals are dealt with in strict sequential order - higher goals before lower order goals. Once a goal has been dealt with (meeting or failing to meet the target level), its satisfaction remains fixed and the next lower order goal is considered. Consideration of the lower level goals does not alter the satisfaction of higher-level goals and cannot damage the higher level goals with respect to target level attainment.

Multi-attribute utility approaches allow tradeoffs between objectives in the attainment of maximum utility. The most common form involves maximization of the sum of linearly weighted objectives. This type of formulation has been used by Candler and Boeljhe and Barnett, Blake, and McCarl. The third utility approach involves an unknown utility function assumption. Here the entire Pareto efficient (non-dominated)

Figure 1. Load demand levels

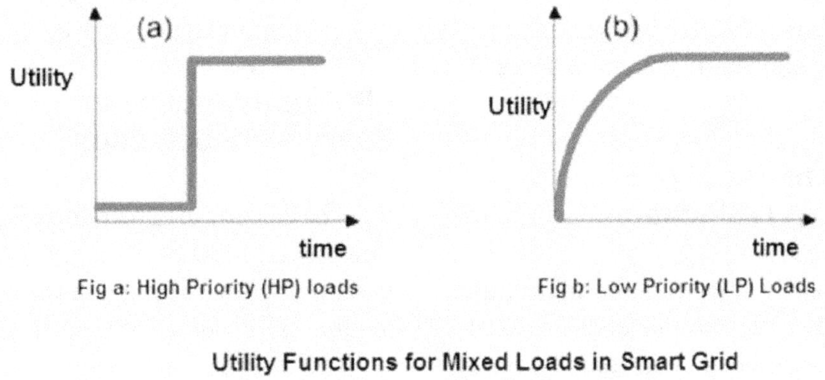

solution set is generated so that every solution is reported wherein one of the multiple objectives is as satisfied as it possibly can be without making some other objective worse off (Steur, 1968). We would follow Candler's approach of Utility functions in this chapter.

We explain LP using utility functions with a simple example as follows. In this Linear Programming model that maximize utility we consider two basic type of loads. They are named as HP loads which are High Priority loads with hard demand requirement i.e, some demand units expect high degree of QoS sensitive to time and LP Loads which are Low Priority Loads with flexible demand requirements. The load demand levels in terms of utility for these two types of loads over time is shown in Figure 1.

Assumptions

Here the model makes following assumptions:

There exists M loads in the electric grid and it has high preference for High priority loads over low priority loads and HP loads more than LP.

Priority: HP Loads first, then LP loads
Pricing: HP loads will pay more.

Two Step Approach for Mixed Demand Load Power Grid

1. Satisfy Quality of Service (QoS) requirement for HP loads (as many as possible)
    a. Admission control
    b. Resource allocation
2. Remaining resources: Distribute among LP loads fairly

$$\max_{Z_{i,k},\, t_i} \sum_{i \in \{LP\ loads\}} V_{LP}(t_i)$$

Subject to:

$t_i \geq Q_i\,,\ \ i \in \{HP\ loads\}$ (HP Load Constraints)

$Qi - QoS\ demand\ of\ i - th\ HP\ loads$

$$t_i = \sum_{l=1}^{L} C_{i,l} Z_{i,l} \qquad i \in \{loads\}, l \in \{clusters\ of\ mixed\ loads\}$$

$C_{i,l}$ $- Capacity\ of\ the\ lth\ cluster\ on\ load\ i$

$$\sum_{i=1}^{Mloads} Z_{i,l} = 1\ \ , Z_{i,l} \in [0,1]$$

Note: Here we define clusters by grouping of adjacent loads in the grid. One reason for using cluster approximation is that it reduces the number of variables during allocation. Qi – QoS demand of i-th HP user.

Time Dependent Supply-Demand (Energy) Allocation

Each supply source in this resource allocation model is said to have two parameters: a) a fixed pricing cost that is associated for employing a selected supply source and b) a demand meeting time period.

The fixed rental charge represents the cost of renting a resource and is paid once if a resource is used in one or more time periods. The following formulation mathematically characterizes the objective function and constraints that are required to identify the resource supply that minimizes the sum of all costs. So, we assume that the model has m time periods for which the demand must be met by a selected supply and n demand-fighting resources.

$$Min\, z = \sum_{j=1}^{m}\sum_{i=1}^{n} c_i H_j D_{ij}$$

Subject to

$$\forall \sum^{m} D_{i,j} \leq Z_i$$

Decision variables:

- D_{ij}: Binary variable that takes on a value of 1 for the time period during which demand is achieved for resources employed.
- H_j: Time period counter.
- C_i: Cost employed for choosing a resource i.e., a power supply.
- Z_i: Binary variable defining whether the i-th resource has been dispatched. Dispatched = 1, not dispatched = 0

LP Formulation with Fuzzy Rule Selection

Let I be a set of m constrained resources, and let R and $Y \cong (Y1;\ Y2;\ldots\ldots\ldots, Y_m)$ be m+1 random variables, with R being a reward and Y_i being the amount of resources of type i consumed. The

joint distribution of R and Y depends on which fuzzy rule is chosen. We make no assumptions about R and Y other than that each has a known expected value for every fuzzy rule selected.

The object is to choose a fuzzy rule within some finite feasible set S, possibly at random, to maximize $E(R)$ subject to $E(Y) \leq b$, where b is an appropriately dimensioned vector of resources and $E()$is the expected value operator. Denote $E(R)$ and $E(Yi)$ when using rule s as ER_s and EY_{si}, respectively. With this notation, we can express our problem as a linear program LP(S) where the variable x_s represents the probability of choosing fuzzy rule s:

$$\text{LP(S): maximize} \sum_{sS} E\, R_s\, x_s$$

subject to

$$\sum_{s \in S} E\, Y_{si} x_s \leq b_i \qquad \forall i \in I$$

$$\sum_{s \in S} x_s = 1, \quad and\, x_s \geq 0 \qquad \forall s \in S$$

The sums in the objective function and resource constraints are $E(R)$ and $E(Y_i)$, respectively, by the conditional expectation theorem. We assume that $b \geq 0$ and that there is a no fuzzy rule exist in S that consumes no resources, so LP(S) has a feasible solution. The value of LP(S) will in general be greater than the value of the restricted problem in which the variables are required to be 0 or 1; that is, it is significant that the decision maker is permitted to choose a strategy at random as long as expected consumption of resources meets the constraints.

SECTION 2

DER Optimization: Formulation of a Resource Allocation Problem as an Example

Optimization constitutes a very important branch of modern applied mathematics. A variety of problems arising in the areas of engineering design, operations research, management science, computer science, financial engineering can be modeled as optimization problems. There are three major aspects of optimization, which are identified as theory, algorithms, and applications. It has been possible to use optimization in real life applications because of availability of efficient algorithms, and these algorithms have been developed because of certain interesting research in Optimization theory. Thus, there is a very strong linkage between theory and algorithms on one hand, and algorithms and applications on the other hand. We should also make a note that development of good software for wide variety of algorithms is too important. Time has proven that practical value of an algorithm is largely determined by its numerical performance, robustness, and ease of computer implementation. We focus our application restricted to smart grid domain.

In this book chapter, a problem on resource allocation for complex power system is taken for illustration and we see how a Branch and Bound algorithm can be robust with respect to variations in demand and fluctuations in power levels. The amount of additional power that DERs can generate and be effectively utilized in power network is a measure of robustness. Hence, we argue that a capacity based the Iterative Binary Integer Linear Programming (C-IBILP) model is inherently a robust resource allocation.

The book chapter focuses on a resource allocation problem for DERs by adapting a binary integer programming approach. For handling power outages from the main distribution circuit, the intelligent agents in the Smart grid have to utilize and negotiate with DER (distributed energy resource) agents that act on behalf of the local generators in the grid, to negotiate power supply purchases to satisfy shortages. We develop an optimization procedure, which can assign these DERs optimally to available multiple Regional Utility Areas or units (RUA) that they can be assigned to operate to meet their power shortages. This is a resource assignment problem. The DERs in our model depict the behavior of power generated through Wind Turbine, Solar powered generation or other renewable power generation units and the region or area refers to a centralized distribution unit.

For example, suppose our simulation study is charged with a need to optimally assign 6 DERs, DER1, DER 2, DER3, DER 4, DER 5, DER 6, to 7 Regional Utility Areas (RUA) based on criteria such as capacity in power levels that these DER are able to generate and preferences in the area that these DER wish to operate. For simplicity in our optimization procedure, we also assume that each RUA can have no more than one DER, and each DER gets exactly one RUA.

The Distributed Energy Resources (DER) can have a preference for the area that they wish to join, and their preferences are considered based on their capacity, i.e., the more power they have been able to generate the power (kW), the higher the capacity. We weigh the preferences based on capacity power level of DER'S through a preference weight matrix (pwm), so that the more power that the DER's can generate, the more their preferences count. Also, we impose multiple constraints such as some RUA have demand, some do not, and some demands are higher than others; DER 3 and DER 4 often work together, so we would like them to be no more than one RUA away, and DER 5 and DER 6 often work together so they also should be no more than one RUA away.

Our approach to solve the assignment problem is to formulate it as a capacity based Iterative Binary Integer Linear Programming (C-IBILP) model and relax the integrality constraints. Our overall objective is to maximize the satisfaction of the preferences weighted by capacity, which will optimally allocate these DERs to optimal areas. This is done through a binary integer programming model by defining a linear objective function.

Our algorithm uses a branch-and-bound procedure with linear programming bounds with 'minimum integer infeasibility' as the branch strategy and 'depth first search for the node search strategy. All simulation results are carried out using optimization tool box in MATLAB. For instance, state-of-the-art approaches to exactly solving large-scale instances of many NP hard problems rely on integer programming approaches that require the repeated solution of integer programs representing problems in Barnhart and Johnson Ellis (1998). We refer to our model for DER assignment as the Capacity based Iterative Binary Integer Linear Programming (C-IBILP) model. There has been little attention given to this type of approach in smart electrical grid analyses.

To illustrate our problem formulation, we assume that there are 7 areas (RUAs) and 6 DER units with the demand and preference levels shown in Figure 2. We define a Regional Utility Area (RUA) as the local distribution power utilities within the micro grid that distributes the power within their network for its loads. For simplicity, we name them Area 1, Area 2....Area 7 as illustrated in Figure 3. The power demand and the preferences in Figure 2 depict a demand driven DER assignment problem that also accommodates preference information. The parameters in the figure are for illustration purposes.

A simple allocation 'text' script on MATLAB would be as follows: text (0.1, .73, 'area1'); text (.35, .73, 'area2'); text (.60, .73, 'area3'); text (.82, .73, 'area4'); text (.35, .42, 'area5'); text (.60, .42, 'area6'); text (.82, .42, 'area 7').

Figure 2. DER vs. RUA assignment problem

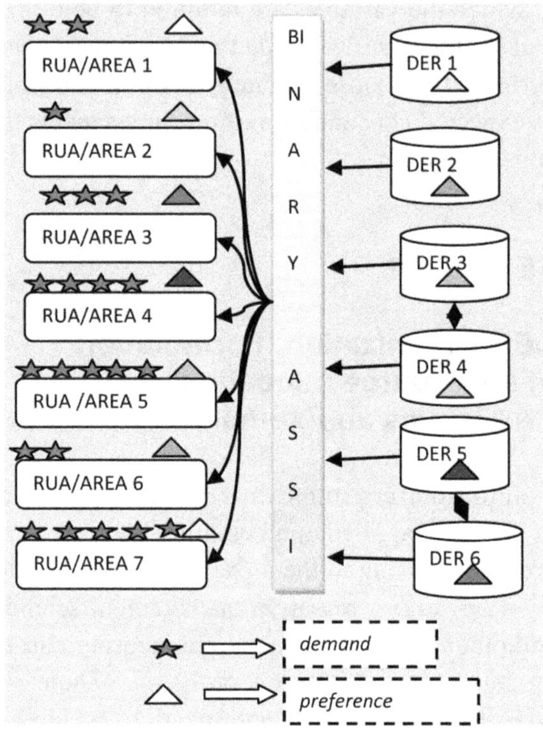

DER-ASSIGNMENT PROBLEM

To develop our problem formulation, the first step is to choose what each element of our solution vector |x| represents. We use binary integer

Figure 3. RUA (area) layout

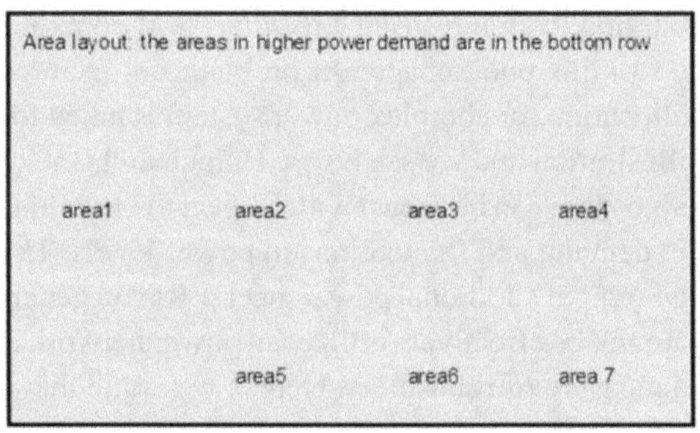

variables, which represents the specific assignments of DERS to RUAs. If the DER is assigned to a RUA, the variable takes the value 1 and if not assigned, the variable takes the value 0. We consider the DER's in sequential order as DER 1, DER 2, DER 3, DER 4, DER 5, DER 6, and DER 7. The nth sequence of elements in vector |x| stores the assignment variables for DER n. In our example |x (1)| to |x (7)| correspond to DER1 being assigned to Area 1, Area 2, etc., up to Area 7. In all, our vector |x| has 6 sequences of 7 elements each or 42 elements in all. Each sequence has a single binary variable set to 1, enforcing a multiple choice condition for each DER.

DER Capacities

We impose constraints based upon DER preference level in their area of operation driven by their capability to generate power. The concept is that the more power that a DER can generate, the higher preference level. For example consider the randomly set power levels given in kiloWatts (kW) below.

a. DER 1→9 kW
b. DER 2→10 kW
c. DER 3→5 kW
d. DER 4→3 kW
e. DER 5→1.5 kW and
f. DER 6→ 2 kW

We create a normalized weight vector based on capacity and also assume that certain DERs should be used in some preferred region or area, such as a DER with higher power generation capability being be used in large demand areas. This normalized weigh vector can be obtained on MATLAB as follows:

```
capacity = [9 10 5 3 1.5 2];
weight vector = capacity/sum (capac-
ity);
>> capacity
Capacity = 9.00   10.00   5.00 3.00
1.500  2.00
>>weight vector
Weight vector = 0.2951     0.3279
0.1639      0.0984  0.049  0.0656
```

RUA Preferences

We set up a preference weight matrix (pwm or prefmatrix) where the rows correspond to AREAS and the columns correspond to DERS. We assume that each DER will give values for each area so that the sum of all their choices, (i.e., their columns), sums to 100. A higher number means the DER prefers the area.

An example of DER preferences is shown below:

```
DER1 = [0; 0; 0; 0; 10; 40; 50];
DER2 = [0; 0; 0; 0; 20; 40; 40];
DER3 = [0; 0; 0; 0; 30; 40; 30];
DER4 = [1; 3; 3; 3; 10; 40; 40];
DER5 = [3; 4; 1; 2; 10; 40; 40];
DER6 = [10; 10; 10; 10; 20; 20; 20];
```

The ith element of a DER's preference vector is the value the ith RUA. Thus, the combined preference matrix is expressed as 'prefmatrix':

prefmatrix = [DER1 DER2 DER3 DER4 DER5 DER6];

```
>> prefmatrix
prefmatrix =
 0   0   0   1   3  10
 0   0   0   3   4  10
 0   0   0   3   1  10
 0   0   0   3   2  10
10  20  30  10  10  20
40  40  40  40  40  20
50  40  30  40  40  20
```

Case 1: We treat the above 'prefmatrix' arrangement as case 1 for analysis. We then weigh the preferences matrix by the |weightvector| to scale the columns by capacity. We also reshape this matrix as a vector in column-order so that it corresponds to our |x| vector. This is achieved in MATLAB script as follows:

PM = prefmatrix * diag (weightvector);

```
>> diag(weightvector)
ans =
0.2951   0       0       0       0       0
0        0.3279  0       0       0       0
0        0       0.1639  0       0       0
0        0       0       0.0984  0       0
0        0       0       0       0.0492  0
0        0       0       0       0       0.0656
```

c = PM (:);

```
>> PM
PM =
0        0        0       0.0984  0.1475  0.6557
0        0        0       0.2951  0.1967  0.6557
0        0        0       0.2951  0.0492  0.6557
0        0        0       0.2951  0.0984  0.6557
2.9508   6.5574   4.9180  0.9836  0.4918  1.3115
11.8033  13.1148  6.5574  3.9344  1.9672  1.3115
14.7541  13.1148  4.9180  3.9344  1.9672  1.3115
```

OBJECTIVE FUNCTION AND ADDED CONSTRAINTS

Objective Function

Our objective is to maximize the total preferences measure weighted by capacity. This is a linear objective function max c'*x or equivalently min -c'*x. We use the BINTPROG script of MATLAB software tool to run our model that is defined as:

$$\min_x f^T x : \quad \begin{cases} A.x \leq b, \\ Aeq.x = beq, \\ x.binary \end{cases}$$

where 'f', 'b', and 'beq' are vectors, A and Aeq are matrices, and the solution *x* is required to be a binary integer vector—that is, its entries can only take on the values 0 or 1.

Constraints

The first set of constraints requires that each DER is assigned to exactly one area. For example, since DER2 is the second DER, we enforce the condition that |sum(x (8:14)) =1|. We represent these linear constraints in an equality matrix Aeq and right hand side vector beq, where |Aeq*x = beq|, by building the appropriate matrices. The matrix |Aeq| consists of ones and zeros. For example, the second row of |Aeq| corresponds to DER2 getting exactly one RUA, so the row pattern is the following as shown in Box 1.

These conditions are implemented in MATLAB code as follows:

```
|Aeq (2,:)*x = 1| is equivalent to
|sum(x (8:14)) = 1|.
numAREAS = 7;
numDERS = 6;
numDim = numAREAS * numDERS;
onesvector = ones (1, numAREAS);
Each row of Aeq corresponds to one
DER.
Aeq =blkdiag (onesvector, onesvector,
```

Box 1.

0 0 0 0 0 0 0 1 1 1 1 1 1 1 0 0 0 0 0 0 0 0 0 0 0 0 ... 0 0 0

Box 2.

1 0 0 0 0 0 0 1 0 0 0 0 0 0 1 0 0 0 0 0 0 ... 1 0 0 0 0 0 0

```
onesvector, onesvector, onesvector,
onesvector);
beq = ones (numDERS, 1);
view the structure of Aeq, that is,
where there are nonzeros (ones)
figure;
```

The second sets of constraints are inequalities. These constraints specify that each AREA has no more than one DER in it, i.e., each AREA has one DER in it, or is empty. We build the matrix |A| and the vector |b| such that |A*x <= b| to capture these constraints. Each row of |A| and |b| corresponds to a RUA and so row 1 corresponds to the DER assigned to RUA 1. In this case, the rows have the type of pattern shown in Box 2 for row 1:

Each subsequent row is similar but is shifted (circularly) to the right by one spot by one position. For example, row 3 corresponds to RUA 3 and enforces that |A(3,:)*x <= 1|, so that AREA 3 cannot have more than one DER. Figures 4 and 5 illustrate equality and inequality constraints that are explained above.

Figure 4. Equality constraints

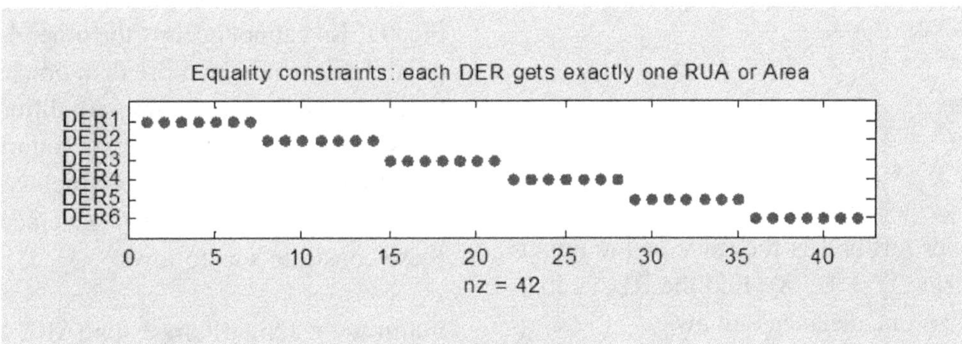

Figure 5. Inequality constraints

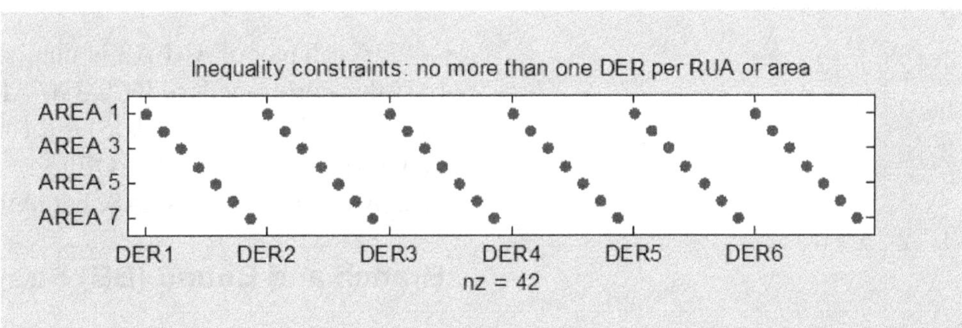

A = repmat(eye(numAREAS),1,numDERS);

b = ones(numAREAS,1);

where '*repmat*' represents the replicate and tile array. The elements of next set of constraints are also inequalities, so they are added to the matrix |A| and vector |b|, that already contain the inequalities from above. We wish to enforce that DER3 and DER4 are no more than one AREA (RUA) from each other, and similarly for DER5 and DER6. First the symmetric distance matrix for the RUAs is built using physical locations and Manhattan (i.e., the "taxicab" metric).

D = zeros(numAREAS); // generates a 7 x 7 Zero matrix

Setting up the top right half of the matrix

```
D(1,2:end)  = [1 2 3 2 3 4];
D(2,3:end)  = [1 2 1 2 3];
D(3,4:end)  = [1 2 1 2];
D(4,5:end)  = [3 2 1];
D(5,6:end)  = [1 2];
D(6,end)    = 1;
```

The lower left half is the same as the upper right D = triu(D)' + D; We find the RUA's that are more than one distance unit away.

D =

```
0 1 2 3 2 3 4
1 0 1 2 1 2 3
2 1 0 1 2 1 2
3 2 1 0 3 2 1
2 1 2 3 0 1 2
3 2 1 2 1 0 1
4 3 2 1 2 1 0
```

```
>> D = triu(D)' + D;
>>D

D =

0 1 2 3 2 3 4
2 0 1 2 1 2 3
4 2 0 1 2 1 2
6 4 2 0 3 2 1
4 2 4 6 0 1 2
6 4 2 4 2 0 1
8 6 4 2 4 2 0
```

[AREAA,AREAB] = find(D > 1);

numPairs = length(AREAA);

This finds |numPairs| pairs of AREAS. For example, if DER3 occupies one AREA in the pair, then DER4 cannot occupy the other AREA in the pair, else it would be more than one unit away in terms of AREA. The same condition holds for DER5 and DER6. This gives |2*numPairs| additional inequality constraints which we add to |A| and |b|. By adding rows to A, we accommodate these constraints as follows:

numrows = 2*numPairs + numAREAS;

A((numAREAS+1):numrows, 1:numDim) = zeros(2*numPairs,numDim);

For each pair of AREAS in numPairs, for the |x(i)| that corresponds to DER 3 in |AREAA| and for the |x(j)| that corresponds to DER4 in |AREAB|, x(i) + x(j) <= 1 i.e., either DER3 or DER4 can occupy one of these AREAS, but not both.

Branch and Bound (BB) Strategy

A large number of real-world planning problems called combinatorial optimization problems share the following properties: They are optimization problems, are easy to state, and have a finite but usually very large number of feasible solutions. While some of these as e.g. the Shortest Path problem and the Minimum Spanning Tree problem have polynomial algorithms, the majority of the problems in addition share the property that no polynomial method for their solution is known. Examples here are vehicle routing, crew scheduling, and production planning. All of these problems are *NP*-hard. Branch and Bound (B&B) is by far the most widely used tool for solving large-scale *NP*-hard combinatorial optimization problems. B&B is, however, an algorithm paradigm, which has to be filled out for each specific problem type, and numerous choices for each of the components exist. Even then, principles for the design of efficient B&B algorithms have emerged over the years. The branch and bound algorithm is a well-known optimal solution method. Branch and Bound (B&B) algorithms are methods for solving non-convex global optimization problems initially proposed by Balakrishnan (1991). They are exact (non-heuristic), in the sense that they calculate a provable upper and lower bounds on the globally optimal objective value and they terminate when all suboptimal feasible solutions have been eliminated.

Branch and Bound (B&B) algorithms can be computationally slow. In the worst case, they require effort that grows exponentially with problem size. We achieved fast convergence in our problems. We do note that due to total unimodularity of the basic A matrix that a network-based customized linear programming solver could be used to provide the lower bounds very quickly in large problems. The BB algorithm is a well-known algorithm in the research community.

We will make use of this algorithm in the context of resource allocation related to distributed energy resources. An example run of the Branch and Bound algorithm is shown in Figure 6 fol-

Figure 6. Branch and bound (BB) search algorithm with inequality constraints

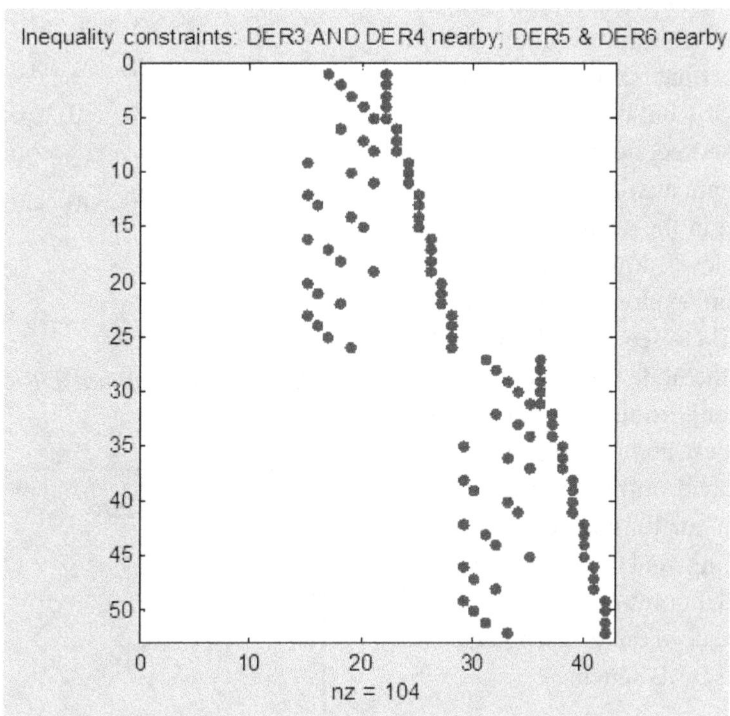

lowed by a snippet of MATLAB code showing the iterative output for each node displayed in the branch and bound algorithm. We let the BINTPROG choose the start point.

```
x0 = [];
f = -c;
options = optimset('Display','iter','
NodeDisplayInterval',1);
[x,fval,exitflag,output] = bintprog(f
,A,b,Aeq,beq,x0,options);
fval
exitflag
output
```

To reduce the number of nodes explored, the time, or number of iterations taken, there are alternative options available. BINTPROG use the options to adjust the algorithm with differing node and branching variable strategies. For example, the default branching strategy is |'maxinfeas'|, which chooses the variable with the maximum integer infeasibility for the next branch, that is, the variable whose value is closest to 0.5. Running the problem again with the branching strategy set to |'mininfeas'|, the variable the minimum integer infeasibility is chosen (that is, the variable whose value is closest to 0 or 1 but not equal to either).

For structuring the tree, depth-first and best-node search strategy are available. For example, in 'df', at each node in the search tree, if there is a child node one level down in the tree that has not already been explored, the algorithm chooses one such child to search. Otherwise, the algorithm moves to the node one level up in the tree and chooses a child node one level down from that node. In best-node (bn) strategy, the node with lowest bound on the objective function is the default. In our limited computational experience, convincing and acceptable results were quickly reached. For future work, we would plan to increase the scale of our test problems and investigate improved B&B schemes.

SOLUTIONS AND RECOMMENDATIONS

The simulation is carried out in a MATLAB platform. The results show that the optimal value is reached after 163 iterations with 54 nodes participation in 1.22 seconds (case1) using the capacity based Iterative Binary Integer Linear Programming (C-IBILP) based branch and bound method which maximizes the satisfaction of the DER preferences weighted by its capacities. The final output shown in Figure 7 presents the DER allocation with the RUA 1 or area 1 treated as empty for optimal assignment.

Case 2: If we change the preferences of DER's according to the matrix shown below, then the optimal solution is reached with 13 iterations, 1 node in 0.047 seconds with default node and branch strategies (see Figure 8).

```
>> prefmatrix
prefmatrix =
 0   0   0   1   3  70
 0   0   0   3  40  10
 0   0   0   3   1  10
 0   0  70   3   2  10
10  90  30  10  10  20
40  40  40  60  40  20
50  40  30  40  40  20
```

Figure 7. An optimal DER assignment solution for case 1

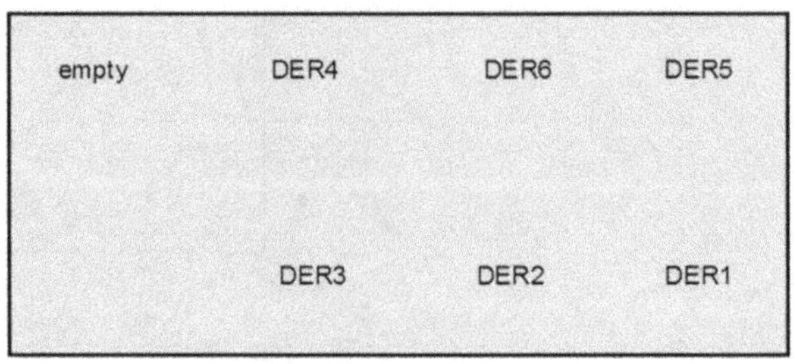

Figure 8. An optimal DER assignment solution to case 2

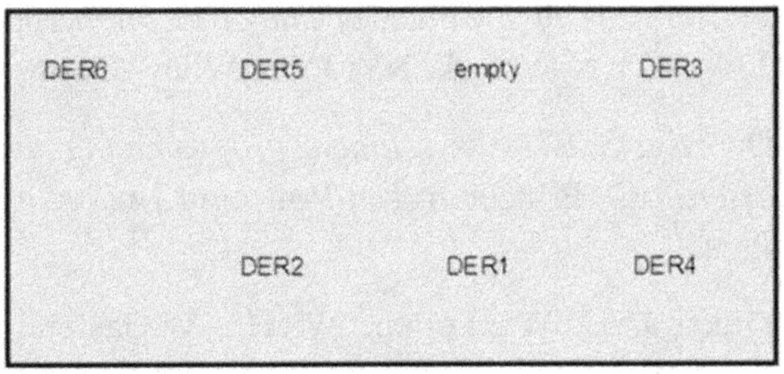

CONCLUSION

We have illustrated the potential value of linear programming in an actual electric grid environment. The book chapter presents a series of LP models for resource assignment problem and proposes a method called a Capacity based Iterative Binary Integer Linear Programming (C-IBILP) model which helps in assigning the distributed energy resources (DER's) during power outage periods to satisfy shortages. Computational results show that our C-IBILP algorithm exhibits very good performance for problem instances tested. Our current effort involves extending the model to a generalized assignment formulation for which multiple number of DER's can serve the RUA. Cyber-Physical Systems lie in the heart of the emerging Smart Grid. Their abilities on providing the "coupling" between the physical world and the business side make them indispensable. However, several issues remain open and significant effort will be needed to understand how cyber physical systems can operate within a complex system of systems such as the Smart Grid as well as the impact of their interactions and cooperation in a network of systems. We believe LP based approach has huge potential in solving optimal resource allocation problems for such architecture.

REFERENCES

Alekhnovich, M., & Arora, S. (2005). Towards strong nonapproximability results in the Lovasz-Schrijver hierarchy. In *Proceedings of the 37th Annual ACM Symposium on Theory of Computing*, (pp. 294–303). ACM Press.

Amin, M. (2004). Balancing market priorities with security issues: Interconnected system operations and control under the restructured electricity enterprise. *IEEE Power and Energy Magazine*, *2*(4), 30–38. doi:10.1109/MPAE.2004.1310871

Amin, M., & Stringer, J. (2008). The electric power grid: Today and tomorrow. *MRS Bulletin*, *33*, 399–407. doi:10.1557/mrs2008.80

Amin, M., & Wollenberg, B. (2005). Toward a smart grid. *IEEE Power and Energy Magazine*, *3*(5), 34–38. doi:10.1109/MPAE.2005.1507024

Arora, S. (2002). Proving integrality gaps without knowing the linear program. In *Proceedings of the 43rd Symposium on Foundations of Computer Science*, (pp. 313–322). IEEE.

Balakrishnan, V., Boyd, S., & Balemi, S. (1991). Branch and bound algorithm for computing the minimum stability degree of parameter-dependent linear systems. *International Journal of Robust and Nonlinear Control*, *1*(4), 295–317. doi:10.1002/rnc.4590010404

Barnhart, C., & Johnson, E. (1998). Branch-and-price: Column generation for solving huge integer programs. *Operations Research*, *46*(3), 316–329. doi:10.1287/opre.46.3.316

Bridge. (2010). Securing the electricity grid. *The Bridge, 40*, 13-20.

Hanssmann, F., & Hess, S. W. (1960). A linear programming approach to production and employment scheduling. *Management Technology*, *1*, 46–52.

Hillier, F. S., & Lieberman, G. J. (2001). *Introduction to operations research*. New York, NY: McGraw-Hill.

Lawler, E. L., & Wood, D. E. (1966). Branch-and-bound methods: A survey. *Operations Research, 14*, 699–719. doi:10.1287/opre.14.4.699

McLeavey, D. W., & Narasimhan, S. L. (1985). *Production planning and inventory control*. Boston, MA: Prentice Hall.

Moore, R. E. (1991). Global optimization to prescribed accuracy. *Computers & Mathematics with Applications (Oxford, England), 21*(6), 25–39. doi:10.1016/0898-1221(91)90158-Z

Nemhauser, G. L., & Laurence, A. (1988). *Integer and combinatorial optimization*. New York, NY: John Wiley & Sons.

Wolsey, L. A. (1998). *Integer programming*. New York, NY: John Wiley & Sons.

ADDITIONAL READING

Alevras, D., & Padberg, M. (2001). *Linear optimization and extensions: Problems and extensions*. Berlin, Germany: Springer-Verlag.

Bland, R. G. (1977). New finite pivoting rules for the simplex method. *Mathematics of Operations Research, 2*, 103–107. doi:10.1287/moor.2.2.103

Clausen, J. (1991). Implementation of parallel branch-and-bound algorithms experiences with the graph partitioning problem. *Annals of Operations Research, 33*, 331–349. doi:10.1007/BF02073939

Clausen, J. L. (1994). Do inherently sequential branch-and-bound algorithms exist? *Parallel Processing Letters, 4*, 3–13. doi:10.1142/S0129626494000028

Dantzig, G. B. (1951). Maximization of a linear function subject to linear inequalities. In *Activity Analysis of Production and Allocation* (pp. 339–347). New York, NY: John, Wiley & Sons.

Dantzig, G. B. (1963). *Linear programming and extensions*. Princeton, NJ: Princeton University Press.

Dijkstra, E. W., Feijen, W. H., & Gasteren, V. (1983). Derivation of a termination detection algorithm for distributed computations. *Information Processing Letters, 16*, 217–219. doi:10.1016/0020-0190(83)90092-3

EPRI. (2002). *Comparing the benefits and impacts of hybrid electric vehicle options for compact sedan and sport utility vehicles*. Palo Alto, CA: Electric Power Research Institute.

Franklin, J. (1980). *Mathematical methods of economics: Linear and nonlinear programming, fixed-point theorem*. New York, NY: Springer-Verlag.

Gale, D. H., Kuhn, W., & Tucker, A. W. (1951). Linear programming and the theory of games. In Koopmans, T. C. (Ed.), *Activity Analysis of Production and Allocation* (pp. 317–329). New York, NY: John Wiley & Sons. doi:10.1002/nav.3800030306

Garey, M., & Johnson, D. (1979). *Computers and intractability: A guide to the theory of NP-Completeness*. New York, NY: W.H. Freeman.

Gärtner, B. (2006). *Understanding and using linear programming*. Berlin, Germany: Springer.

Hoffman, A. J. (1953). *Cycling in the simplex algorithm*. Washington, DC: National Bureau of Standards.

Kempton, W., & Tomic, J. (2005). Vehicle to grid fundamentals: Calculating capacity and net revenue. *Journal of Power Sources*, *144*(1), 268–279. doi:10.1016/j.jpowsour.2004.12.025

Klee, V., & Minty, G. J. (1972). How good is the simplex algorithm? In Shisha, O. (Ed.), *Inequalities III* (pp. 159–175). New York, NY: Academic Press.

Labatt, S., & White, R. (2007). *Carbon finance: The financial implications of climate change*. Hoboken, NJ: Wiley.

Madawala, U., Schweizer, P., & Haerri, V. (2008). Living and mobility A novel multipurpose in-house grid interface with plug-in hybrid BlueAngle. In *Proceedings of the IEEE International Conference on Sustainable Energy Technologies (ICSET)*, (pp. 531–536). Singapore: IEEE Press.

Neumann, J. (1963). Discussion of a maximum problem. In Taub, A. (Ed.), *John von Neumann: Collected Works* (pp. 89–95). Oxford, UK: Pergamon Press.

Parks, K., Denholm, P., & Markel, T. (2007). *Costs and emissions associated with plug-in hybrid electric vehicle charging in the Xcel energy Colorado service territory*. Tech. Rep.NREL/TP-640-41410. Retrieved from http://www.nrel.gov/docs/fy07osti/41410.pdf.

Rahman, S., & Kroposki, B. D. (1993). Photovoltaics – A demand side management performance analysis at a university building. *IEEE Transactions on Energy Conservation*, *8*(3), 491–497. doi:10.1109/60.257064

Roos, C., Terlaky, T., & Jean-Philippe, V. (2005). *Interior point methods for linear optimization* (2nd ed.). Berlin, Germany: Springer-Verlag.

Schrijver, A. (1998). *Combinatorial optimization: Polyhedra and efficiency*. Berlin, Germany: Springer.

Schrijver, A. (1998). *Theory of linear and integer programming*. New York, NY: John Wiley & Sons.

Stigler, G. (1945). The cost of subsistence. *Journal of Farm Economics*, *27*(2), 303–314. doi:10.2307/1231810

Tomic, J., & Kempton, W. (2007). Using fleets of electric-drive vehicles for grid support. *Journal of Power Sources*, *168*(2), 459–468. doi:10.1016/j.jpowsour.2007.03.010

Vanderbei, R. (2008). *Linear programming: Foundations and extensions* (3rd ed.). Berlin, Germany: Springer-Verlag.

Venayagamoorthy, G. K., & Braband, G. (2008). Carbon reduction potential with intelligent control of power systems. In *Proceedings of the 17th World Congress, International Federation of Automatic Control*. Seoul, Korea: Automatic Control.

KEY TERMS AND DEFINITIONS

BB: Branch and Bound algorithm is a well-known optimal solution method. BB algorithms are methods for solving non-convex global optimization problems.

BILP: Binary Integer Linear Programming is a linear program where the decision variables can take on only two values, 0 and 1.

DER: Distributed Energy resources are local generators in the grid to negotiate power supply purchases to satisfy shortages. These include of energy resources such as Wind, Solar.

EPRI: Electric Power Research Institute

GPS: Global Positioning System provides location information about PMU or other related sensor in the grid.

kW: Kilo Watts refer to a unit of power.

LP: Linear Programming is the name of a branch of applied mathematics that deals with solving optimization problems of a particular form.

MW: Mega Watts refer to unit of power.

NERC: North American Electric Reliability Corporation

PDC: Phasor Data Concentrator can receive data from multiple PMUs, correlate, store, and trigger response actions to abnormal conditions

PMU: Phasor Measurement Unit is a sensor that provides phasor information about voltage and current.

QoS: Quality of Service

RUA: Regional Utility Area serves certain number of customers like a local grid.

Section 2
Management and Manufacturing

Chapter 6
Green Computing as an Ecological Aid in Industry

Oliver Avram
Ecole Polytechnique Fédérale de Lausanne, Switzerland

Ian Stroud
Ecole Polytechnique Fédérale de Lausanne, Switzerland

Paul Xirouchakis
Ecole Polytechnique Fédérale de Lausanne, Switzerland

ABSTRACT

This chapter concerns the use of computing for ecological evaluation in the manufacturing industry. Here, ecological evaluation means identification and quantisation of various manufacturing process characteristics from the point of view of the environment. Manufacturing is a complex process with many different interactions between the parameters controlling the manufacturing machine tools. In the past, manufacturing planners and operators have set these parameters without understanding the consequences, leading to waste of energy, cutting fluid, and so on. This chapter presents a computer tool for evaluating and quantifying the effects of different manufacturing choices using chosen criteria. The tool was implemented as part of the work for a European project. It is based on an extensive analysis of machine tools to provide a way of handling the complexities of understanding the use phase of products.

INTRODUCTION AND BACKGROUND

The aim of this chapter is to explain the use and practice of one aspect of green computing in the manufacturing sector. The contents of the chapter concerns work done for a doctoral dissertation (Avram, 2010) performed as part of a European project on the next generation of machine tools (the NEXT project). The work is important for manufacturers to evaluate the consequence of different manufacturing strategies and to quantify the effects of these on the environment. Manufacturing is a very complex task with many interactions making it difficult or impossible to understand the consequences of manufacturing decisions.

The major environmental aspects of machine tool use come during the use phase of the machine tool. At the same time, existing evaluation tools

DOI: 10.4018/978-1-4666-1839-8.ch006

are not sensitive enough to cope with the multiple manufacturing variants. Some work has been reported in the literature (Munoz & Sheng, 1995; Dornkudwar, et al., 1998; Srinivasan & Sheng, 1999a, 1999b; Akbari, et al., 2001; Dahmus & Gutowski, 2004; Jayal & Balaji, 2007; Narita, et al., 2006; Zhigang, et al., 2008, etc.) considering green machining issues; however, in most cases the focus has been limited at the process and process planning level.

Munoz and Sheng developed a model with the three dimensions of material, energy, and time as a basis. The material inputs are the primary material and secondary catalyst material, with output of primary material, waste primary material, and waste catalyst. For the energy dimension, Munoz and Sheng identified thermal, electrical, and chemical as input energy with thermal, chemical, and radiative energy as possible outputs. For the time, there is the initiation time as input and the processing time as output. They found that the geometry of the part, the workpiece material and cutting fluid were the dominant factors affecting process energy.

Domkundwar et al. used features as a basis for analysis for environmental impact assessment. The features were treated on a global basis with additional process or process parameter constraints.

Srinivasan and Sheng worked on quantifying health hazards in wet and dry machining of a simple part with one drilling operation and two end-milling operations. No account was taken of solid particles released in completely dry machining.

Akbari et al. worked with energy consumption in the machining process. An interesting element introduced in this work was the notion of fixed energy consumption versus dynamic energy consumption. This energy is needed for production but is not connected with the production itself. In addition, there is the energy needed for the production itself which can be reduced by improving manufacturing strategies.

Dahmus and Gutowski also worked with energy analysis of the machining process. They categorised the energy consumed into three main types: "Constant start-up operations," "Run-time operations," and "Material removal operations." Based on a set of usage scenarios, which they defined, they compared the energy needs for machining both steel and aluminium on modern and conventional machines.

Jayal, Balaji et al. compared drilling of aluminium with both dry and wet cutting conditions. They concluded that flood application for drilling was needed for high accuracy.

Narita et al. developed a method for estimating the environmental burden of machining in terms of contribution to the global warming potential. They converted different elements, like the electrical consumption, the coolant, lubricants and so on, into equivalent CO_2 emissions and used the total as a measure of the environmental effects.

Zhi et al. investigated the environmental impacts of cutting fluids for cutting processes. They also studied some design aspects for machine tools and thermal deformations of both workpieces and machine tools during cutting.

Informatics is needed as a tool for industry, this can be thought of as 'green applied computing.' The work mentioned above needs to model manufacturing processes explicitly or implicitly and to calculate and present a multitude of diverse factors. This is an application of computing for improving environmental awareness, hence the term "green applied computing." Industry is doing a great deal to improve the environmental impact of manufacturing and computing tools are needed to understand and quantify the effects of manufacturing options. In addition, by exposing students to the topic, future engineers are made aware of their social responsibilities and how they can be managed. This research, on measurement and quantification, is important and needs to be extended so as to gain a better picture of environmental loading. The chapter will describe how

Figure 1. The main subsystems of a machine tool system (MTS)

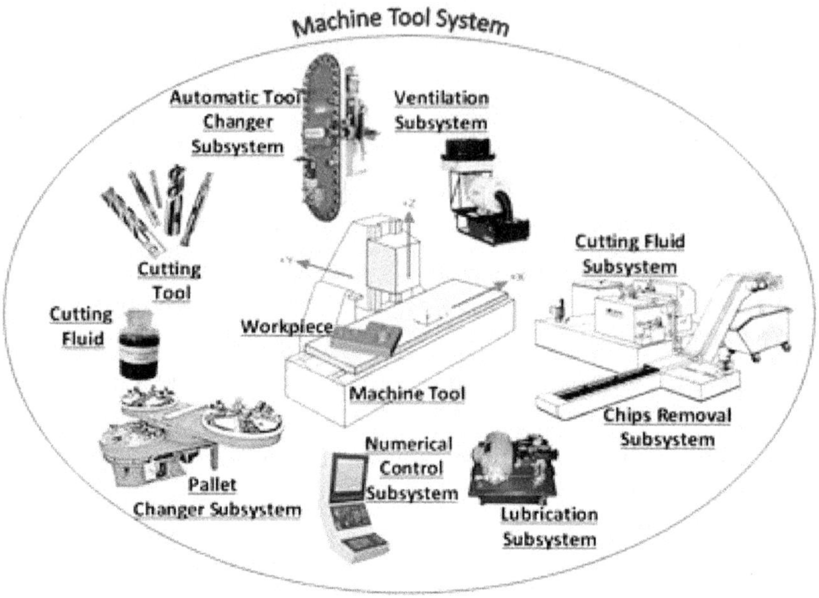

green computing can be applied in an industrial sector, which is an important user of resources.

Machine Tool Systems

Machine tool systems are described as "mother machines" by Suh et al (2008, p. 3). in the sense that it is a machine that makes machines. There are, now, a great many types of machine tool, milling machines, turning machines, grinding machines, EDM machines as well as complex machines capable of performing multiple functions. Machine tools are complex systems with many elements to handle. In addition, the application area of machine tools involves many other variables concerning physical properties of materials, thermal and chemical effects. This chapter focuses on the use of machine tools for so-called "green machining." There are, too, many different manufacturing processes and machine tool types. The main manufacturing processes are milling and turning, but there are other processes, which are used as well, such as EDM, additive processes, laser cutting, stamping, pressing or welding. For the purposes of the research, the milling process was analysed in order to establish the principles for eco-evaluation. However, the analysis method is extensible.

Figure 1 shows a stylised milling machine tool to illustrate the elements. There is a spindle, which holds a rotating cutting tool. The tool is moved in the Y- and Z-directions relative to the workpiece to be machined, which can be moved along X axis. Depending on the architecture of the machine the tool and the workpiece can be moved separetely or in combination.

In addition to the direct cutting processes, there are a number of subsystems for chip removal, lubrication of various components, pumps to supply cutting fluids or fans for cooling, which needed to be taken into account.

Life Cycle Costs of Machine Tools

To understand the ecological effects of any product, including machine tools, it is necessary to consider the life cycle of the product and divide it up into phases. In the design phase, it is possible to consider a number of product variants and optimise these according to different criteria. During manufacturing, the environmental effects of different manufacturing options can be evaluated for the part. The major environmental impact of many manufactured products, including machine tools, comes during the use phase. Finally, there is an End-Of-Life (EOL) phase when there are reuse, recycling, disposal options. The lifecycle cost of a machine tool is the sum of all these multiple effects.

As stated above, the most important environmental effects come during the use phase of the machine tool. This is also the most complicated phase, with an interaction between the ecological effects of the part and the environmental effects from the machine tool due to the way it is used for the manufacturing process. This is illustrated in Figure 2.

Use processes are performed by executing functions of the MTS, but the environmental impact does not result just from deliberated functions. Rather, they result from the behavior of the MTS subsystem responsible for that function. Therefore, it is important to describe the MTS's behavior not just its functions.

The evaluation of the use phase of the product can be simulated by determining settings of different activities within its utilisation such that the environmental impact can be reduced (i.e. finding the optimum combination of cutting parameters and their optimal sequencing among several possible combinations for the same process). The overall goal would be to propose the best-practice solution for each sub-function in order to improve the performance of the overall system.

There are different environmental aspects, which need to be estimated. One important aspect is energy use. Another aspect is the resource use of potential noxious elements as waste, like cutting fluids, or in fumes, particles, etc. released into the air during machining.

EVALUATION OF THE USE PHASE

The methodology developed (Avram, et al., 2011) was based on extensive analysis and measurements of several machines. The experimental data was recorded in a structured relational database. The experiments were carried out to machine different manufacturing features under different conditions and on different machine tools. With the result data, it has been possible to evaluate and quantify the effects of manufacturing decisions. An experimental software tool was developed to show the research results.

Figure 2. Life cycle crossing point of a machine tool system

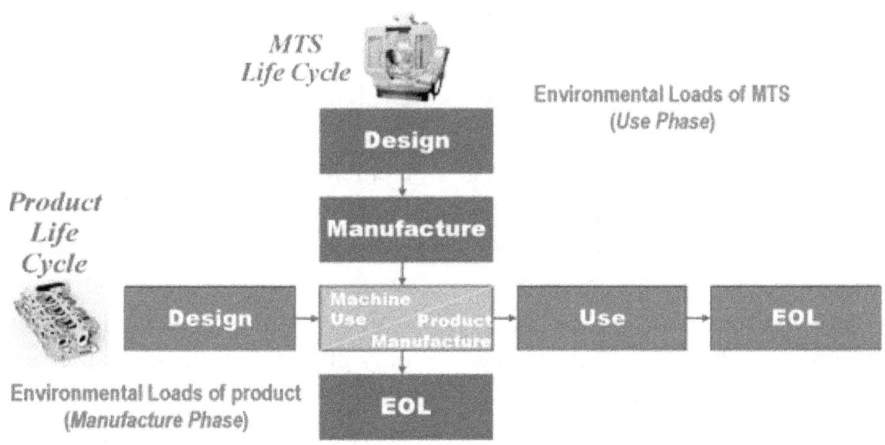

The software tool, GREEM (Global Reasoning for Evaluation of Eco-Machining), worked on two levels. On the first level the user was asked to identify the manufacturing features to be machined, to identify the machining conditions and finally to identify the environmental aspects of interest. This data was then matched to experimental results from the database in order to retrieve relevant measurements. The user is also expected to provide weights for the different environmental aspects as a way of adapting the results to local conditions. For example, if a company uses renewable energy but has waste disposal concerns, then lubrication fluid might be set to have a higher priority than energy use. In this way the tool can be tailored to specific company needs. The results of the analysis are presented both in graphical form and as a textual form, which can be included in a manufacturing report.

While the first level is mainly concerned with aspects related to machining a specific part by an MTS, the second level assesses the machining capabilities of the MTS in terms of component specifications and activities, during its useful life when various shapes and materials have to be machined. Based on the specifications given by the MTS manufacturer for its various components and an energy consumption predictor module (Avram & Xirouchakis, 2011), the goal at the second level is to find the best performing machine tools system to complete a specific machining job.

In addition to the theoretical verification, this model has also been verified using experimental measurement of known machine performance under known conditions. The machines were run both with and without actual cutting so that the effects of the machine tool itself and the effects due to cutting could be measured. The experiments were run under different cutting conditions and with the machine tool moving at different speeds. As well as the main cutting, energy is also used by subsidiary systems and these, too, were measured for quantification. Using the machine tool model, it is possible, also, to estimate the future energy use of machines under development and to optimize the design at an early stage. For industrial users, energy minimization is important both in economic terms as well as for production optimisation.

GREEM Level 1

The first level consists of the assessment of the performance of a specific machining process with respect to a set of criteria. The machining process is defined at an elementary level, that is to say, with a unique tool class, a unique tool path, a set of cutting conditions and a cooling/lubrication alternative. The assessment can be extended at part level by gathering together under one score the performances of the machining process components carried out in order to complete the part.

The steps to be carried out for the analysis are:

- Input/Select the machine tool system
- Input/Select the machining features
- Input/Select the machining alternatives
- Input/Select the measured criteria
- Retrieve the experimental data

On this level, GREEM evaluates the effects of machining with particular options according to selected criteria. The machining description is given as a set of machining features. This method of giving machining instructions is termed "micro process planning" and derives from a new standard for machine tool control, ISO 14649 (2001). The standard allows direct use of Computer-Aided Design (CAD) data, another applied computing area, in machine tool control. The 'features' are elements like holes, pockets, face milling.

The user is first asked to define the workpiece size and material for mass property calculations for example. The user then defines the manufacturing operations in terms of manufacturing features, but only giving their gross geometric characteristics rather than the exact geometry. The user is also asked to define tool characteristics for more precise

Table 1. Feature types for manufacturing

Counterbore	Pattern (Rectangular)	Slot (full open)
Countersink	Spherical cap	Slot (loop)
Planar facing	Step	Slot (closed round
Hole (Round)	Thread	Slot (closed square)
Hole (Tapered)	Chamfer	Slot (closed Woodruff)
Pocket (Closed)	Fillet	Slot (open round)
Pocket (Open)	Profile	Slot (open square)
Pattern (Circular)	Region	Slot (open Woodruff)

knowledge of the cutting conditions. The feature types identified for manufacturing are shown in Table 1.

An alternative, as a component of the machining strategy, refers to the choice between several possibilities all intended to accomplish a specific goal. Even though the focus of this research is on the alternatives for cooling/lubrication of the cutting area, the developed methodology is also able to accommodate other alternatives (i.e. HSM—High Speed Machining). The user is also offered a choice of criteria from which to evaluate the machining plan.

Once the machining plan, the alternatives, and the criteria have been selected, it is possible to evaluate the effects of machining a part. The input data are used to select measured data from a database of experimental results. The user then has to set weights for the different criteria indicative of their relative importance. The criteria belong to the three different categories: economical, technical, and environmental. The user can either give the relative importance of the criteria in a pairwise comparison, or set their relative importance on a global scale, depending on user preference.

The software calculates the best options for machining so that users can evaluate the consequences of their machining choices. Figure 3 shows the graphical presentation of the results which can also be presented for inclusion in a machining report. On the top left, there is a presentation of the global performances based on the weights defined by the user. This shows that the best global performance for manufacturing the features chosen is MQL, or minimum quantity lubrication. MQL means that only a small amount of lubrication is delivered to the contact point between the tool, often as a mist, and the workpiece instead of 'flooding' the working area with a continuous stream of lubrication. On the top right there is a presentation of the contribution factors for machining the round hole, with workpiece quality, force and lubrication evaluations. These were the criteria chosen by the user on which to evaluate the manufacturing steps. The chart shows that for the flood option, with a large amount of cutting lubrication, the main negative environmental effect is from the lubrication itself, with a small amount of force needed for cutting but giving the best workpiece quality. For the minimum quantity lubrication there is a small contribution from the lubrication, small force needed and good workpiece quality. For dry machining, more force is needed and the workpiece quality is poorest. On the bottom left of the figure, the equivalent effects for the planar facing operation are shown. This shows that the major negative effect for flood machining is again the lubricant and the workpiece quality is not too bad. For the MQL option, the main effect is the force needed. For dry machining, much more force is needed and the workpiece quality is lower than with MQL.

The computing tools for the first level of GREEM involve mainly database tools with user-

Figure 3. Graphical presentation

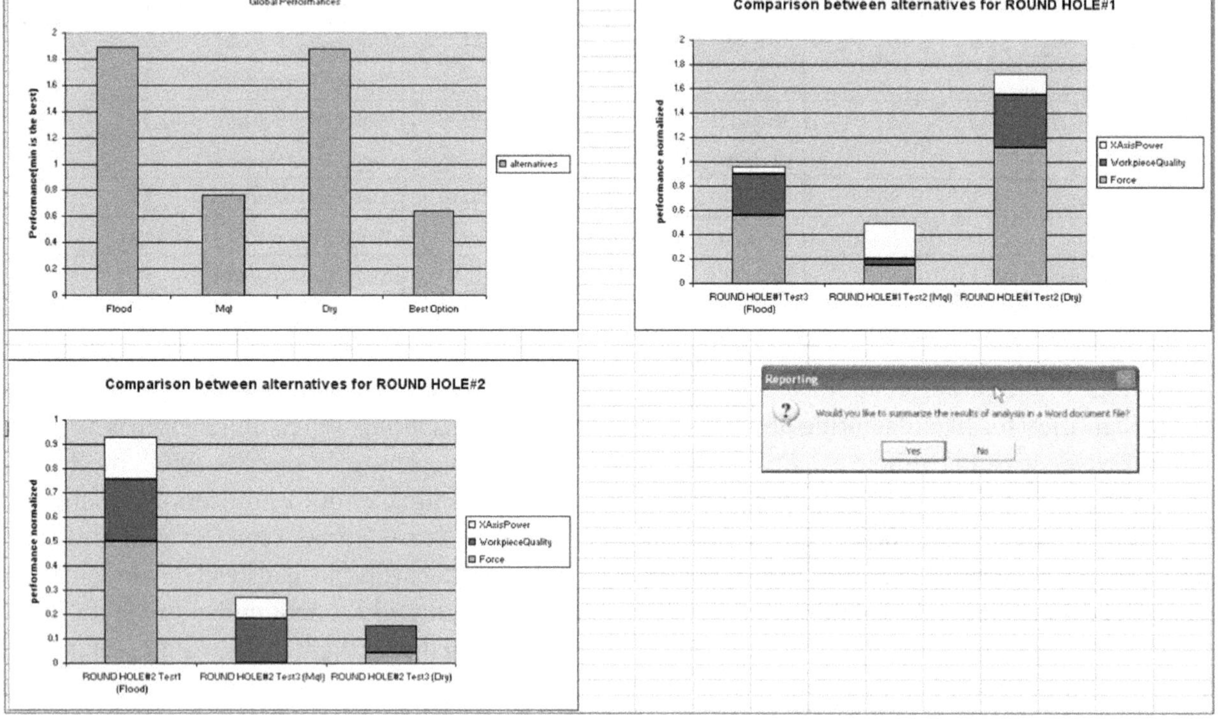

friendly presentation tools. The computer tools are important to make sense of the wealth of information surrounding the manufacturing process.

GREEM Level 2

In addition, similarly to the first level, it is necessary for the user to set priorities. The second level has to address the performance of the machining system as a whole and not just certain aspects. The software makes it possible to compare different machine tool systems with respect to various criteria, e.g. productivity, precision, cost, safety, flexibility/adaptability, and energy. The user has to emphasise the relevant importance of these criteria by using the priority settings, which use the same type of pair wise comparison technique as described earlier. Figure 4 shows the priority setting.

Energy use calculation is complicated and there are effects related to machining strategies. For the research done, a prismatic part was used. In order to calculate the expected energy use for machining a test part, it is necessary to have a model of the machine tool. Such a model has to include the masses of the machine tool elements, spindle and feed axes characteristics, kinematic structure, friction coefficients and so on. Figure 5 shows an example of input data required to estimate the energy used by one element, in this case the feed axis of the machine tool.

In order to calculate the movements of the machine it is also necessary to have the control data. For the purposes of the research, an APT (Automatically Programmed Tools) control data file was used. The APT file contains explicit movement data, which can be used to calculate the power requirements with respect to the chosen tool path. The same file was also used to mill the part while measuring the real power consumption so that the model could be verified.

Figure 4. Priority setting

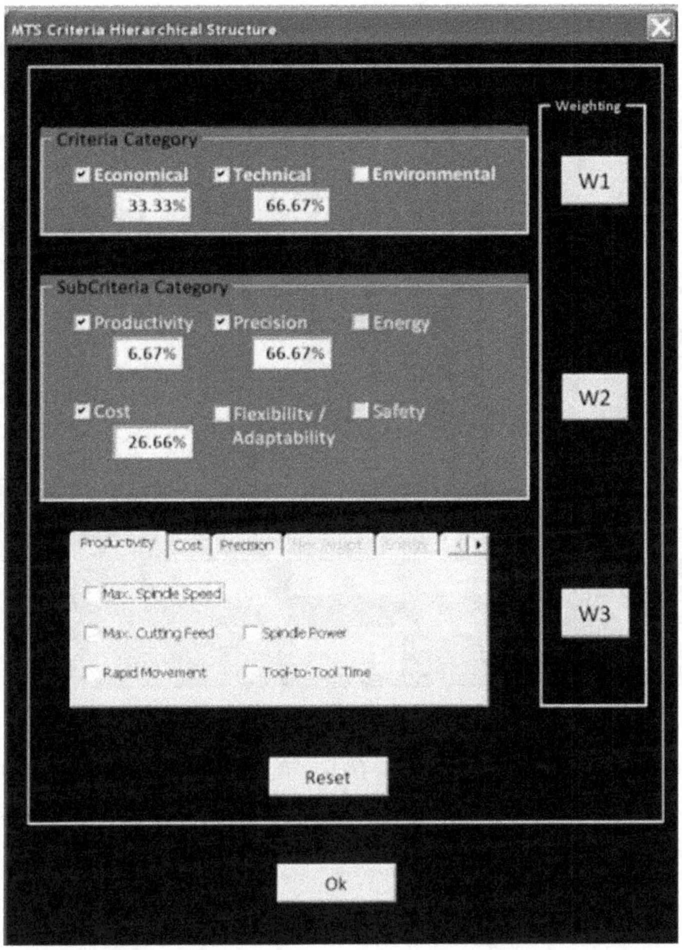

As with level 1 of GREEM, the analysis is divided into machining features. Figure 6 shows some results of the milling energy consumption for the planar face feature of the modelled part, machined under wet conditions and by removing the same amount of material at two different cutting speeds.

The scale on the left shows the number of kilojoules of energy consumed for different operations. On the bottom are the different elements that were measured in the experiment. First there is the energy consumed while the machine is idling, that is, not cutting. Next, for each of the axes of the machine tool (which is a three-axis machine tool in this case), there is the energy measured while the axis was accelerating, moving at constant speed and decelerating. Then there is the contribution to energy consumption by the spindle. The acceleration and deceleration contributions are variable amounts of energy. Also shown is the energy loss due to mechanical effects of the spindle (front bearings, rear bearings, and windage). The energy consumed in actual cutting of material is also shown. The last five columns in the figure show the energy losses due to auxiliary equipment around the machine tool.

This kind of measurement provides a basis for accurate modelling of the machine tool and manufacturing processes. With this data, it is possible to help the user optimise machining parameters.

Figure 5. Feed-axis input for GREEM level 2

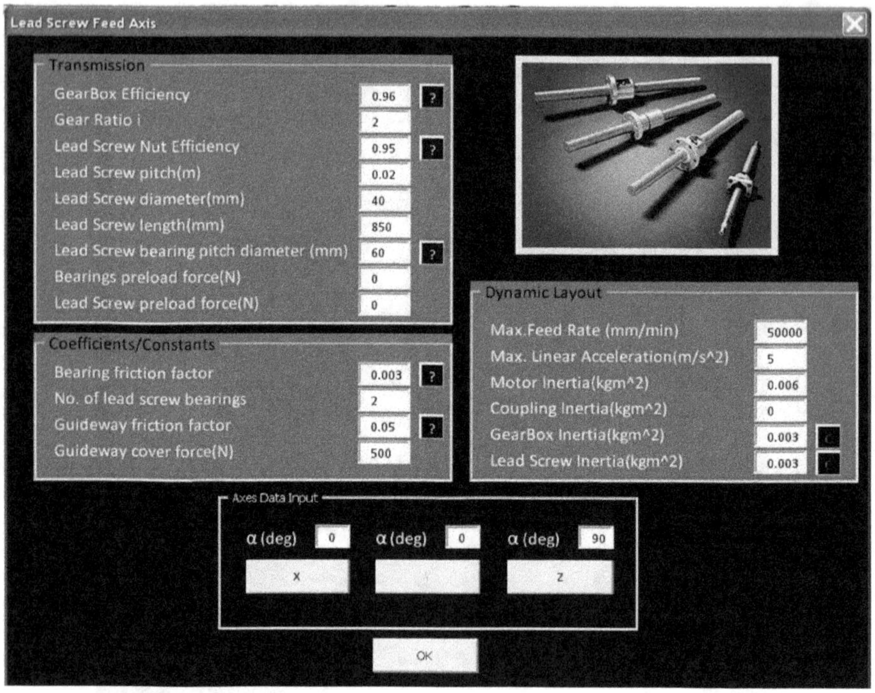

Figure 6. Machine tool system consumed energy (kJ) for planar face milling (low speed vs. high speed)

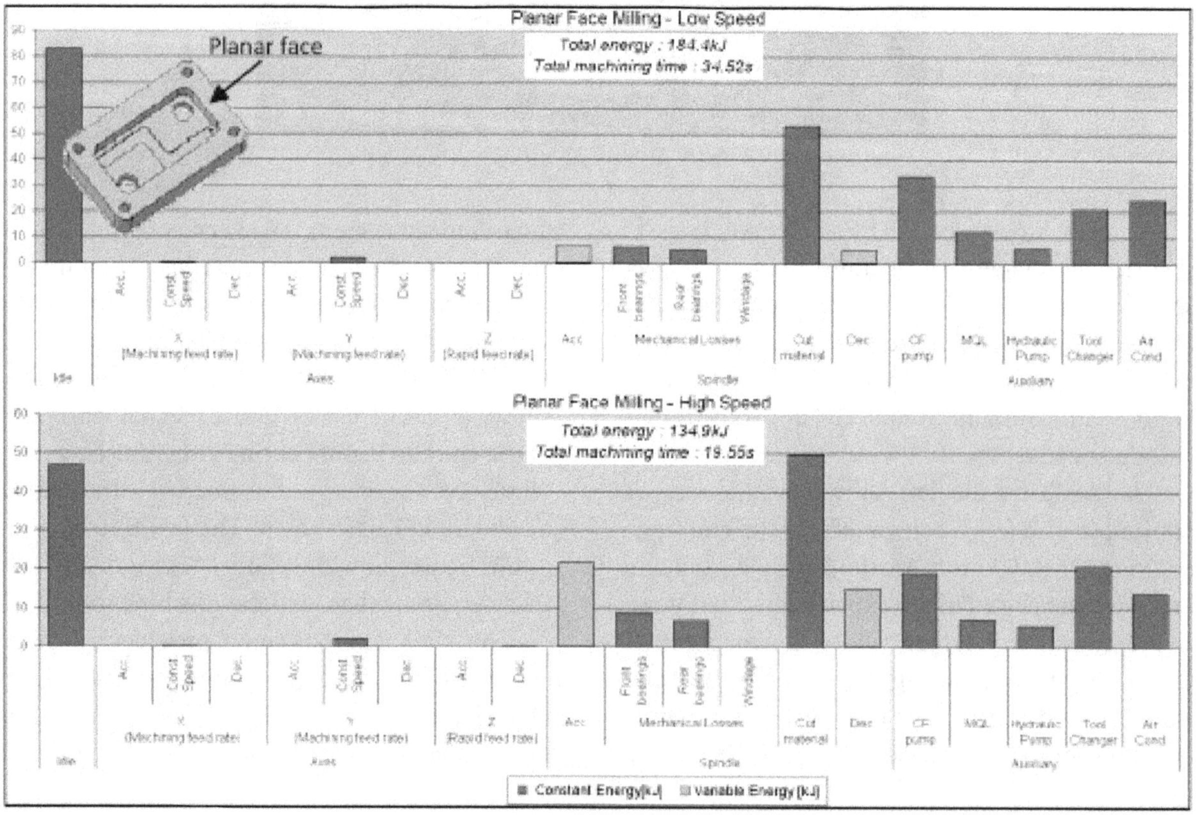

Figure 7. Machine tool system energy consumption (kJ) for part milling (wet vs. dry machining)

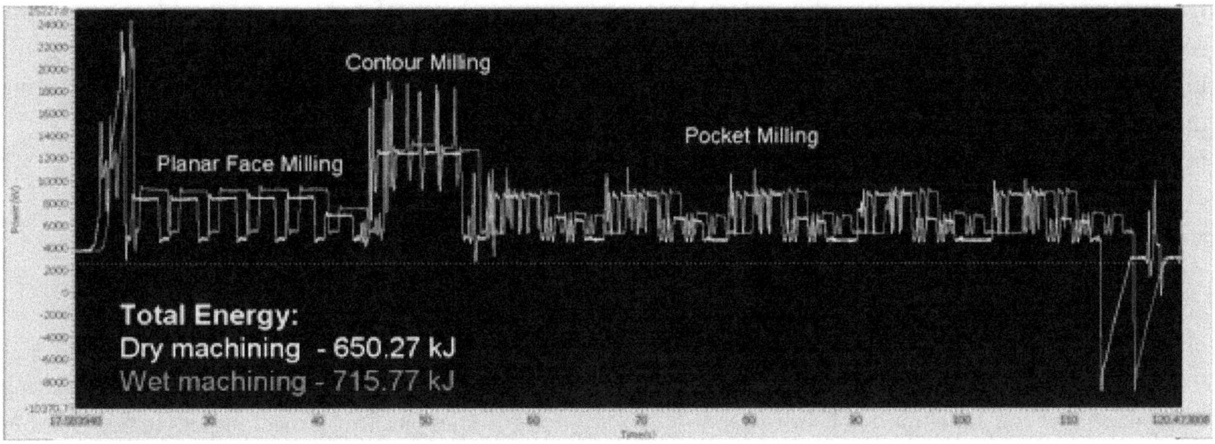

It is also needed for accurate machine tool modelling for advanced machine tool control, to let the machine tool itself optimise its behaviour, which is a current research topic being actively worked on in many places.

The auxiliary equipment concerns other energy consuming subsystems on the machine tool, such as chip removal systems and pumps for providing cutting fluids. The energy consumed by the pump for the delivery of the cutting fluid (CF pump in the Figure 6) is notable because the pump cannot be switched off unless a part is to be machined 'dry.' The power drawn by the pump is highly influenced by the fact that a specific pressure should be maintained and frictions in pipes and fittings as well as elevation differences should be overcome across the length of the process. A more energy efficient alternative is the Minimum Quantity Lubrication (MQL).

Avram (2010) provides an extensive presentation of opportunities for improvement of the environmental impact of machine tools during their use phase. Due to space limitations, only a single example is illustrated here. Figure 7 shows the power profiles acquired during the machining of three milling features by employing identical cutting conditions but different cooling/lubrication strategies. Another interesting conclusion was that high-speed machining, a process where the tool spindle is rotated faster and the axis movement speed increased, can be more economical than conventional machining. Although high-speed machining needs more power, this is applied for shorter periods resulting in a net saving of energy.

Figure 7 shows the power consumption during milling of the example prismatic part used for the experiments. The scale on the left is the power consumption. Note that there is some negative consumption where power is generated in certain phases of the process as the motor becomes a generator under certain conditions. The figure shows two cutting profiles superimposed, in red for wet machining and in white for dry machining. The scale at the bottom shows the time, in seconds, over which the machining took place.

For the test part, there were three machining phases: a planar face milling, a contour milling and pocket milling. The planar face milling removes a layer of material from the workpiece to make it the correct height. The profile milling cuts around the outside of the part to give it the right exterior shape. The pocket-milling phase removes material from the inside of the part.

The face-milling phase consists of a series of passes to cover the whole area of the part; the peaks in the diagram correspond to the different passes. In contour milling, the tool cuts material with the side. This is not done in one pass as cutting too deeply can lead to tool deformation and even breakage. Instead, the tool is passed round the material in a series of steps. For pocket milling, the tool is sunk into the material and then moved around to clear the material. As with profile-milling, this may not be possible in a single step and so the material clearance is done in a set of layers. The diagram shows the peak distributions of these different steps with different axes moving different amounts according to the operational characteristics. Here, only the total energy consumption is shown, not the individual contributions of the systems and subsystems of the machine tool. In order to get an accurate picture it is necessary to measure at high frequency, resulting in a huge amount of data to process.

For the second level of GREEM, database tools are also important as well as simulation, calculation and virtual instrumentation tools.

EXTENDING THE SCOPE OF THE SOFTWARE

The test plan for milling, which has been carried out on a number of machines, is shown as an example for comparison. The other test plans are suggestions for possible extensions to the database of experimental results. The milling tests were carried out, machining the same piece with three options: dry, MQL and flood machining, and with two types of tool: coated and uncoated. This means that six sets of experiments and six sets of tools were required for the tests for each set of machining parameters (feeds and speeds). The more experiments performed the more accurate the results. However, the experiments were performed with new tools for each set of measurements, making the experimental series expensive. The criteria measured were: power consumption of the main power-driven components of the machine, cutting forces and the quantity of cutting fluid applied.

It would certainly be possible to devise an extended set of tests to augment the existing database so that different process plans using different processes could be compared. Four cutting processes are considered here:

1. Milling
2. Turning
3. Wire EDM
4. Sink EDM

Milling is a much used process for creating general shapes using a rotating tool to remove material from a workpiece. Turning is also a conventional process in which the workpiece rotates and the tool is fixed to create rotationally symmetric parts. Wire EDM is a non-conventional process to produce accurate parts from material by cutting parts out from a block of material using a wire. Electric current is passed down the wire resulting in sparks between the wire and the workpiece. The sparks cause enough heat to be generated to vaporise, or erode, a small amount of material. As the wire is moved, it leaves a narrow cut behind it. Sink EDM is another non-conventional process used to create accurate shapes in metal for moulds and dies, for example. It works by slowly lowering a shape towards the block of material to be shape. As different parts of the shape approach the workpiece so sparks occur between the shape being lowered and the workpiece. These sparks erode away the workpiece to leave an accurately shaped cavity.

Figure 8 shows an extended set of test objects devised to provide information about a variety of manufacturing features in the different processes. These features derive from the ISO 14649 specification. For the turned part, there are profiles, chamfers and cut-ins, for example. For the wire EDM process there is a simple profile cutting feature as well as a ‘pocket’ feature where a complete volume of material is vaporised. For the sink EDM process, there are different types

Figure 8. Test parts for process extensions

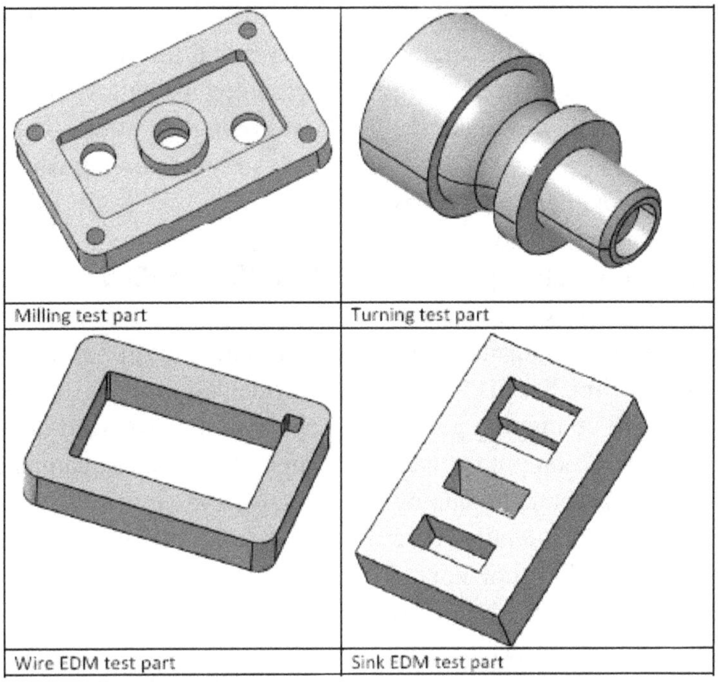

of pockets. Sink EDM is also interesting because it offers an alternative process for mould manufacture and could be compared with conventional and high-speed milling for particular examples.

RESEARCH VISION AND FUTURE RESEARCH

Figure 9 shows an image of the research vision, with an increased understanding of the interaction between different options in manufacturing. This obviously requires a far wider range of analysis

Figure 9. Machine tool system hierarchy of modeling and optimization methods and tools

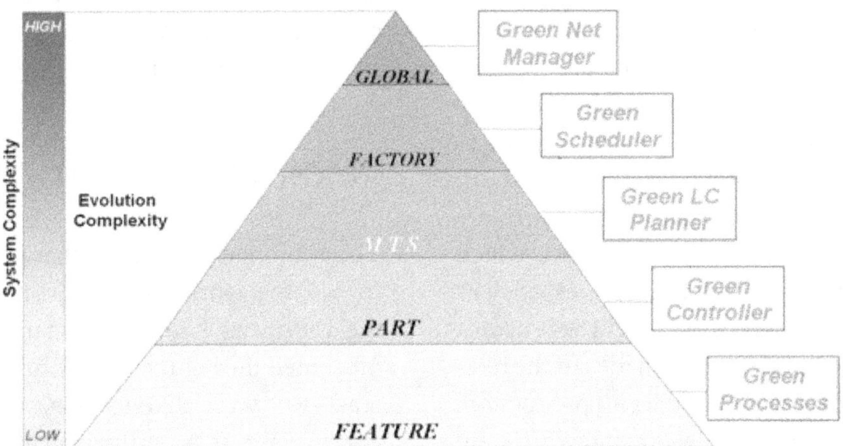

than has yet been performed as well as more time and a greater investment. The milling machine tool analysis is complex enough, but the higher levels are even more complex and need more related research in order to be able to make the whole process more flexible.

An example of the needs for change concerns the modern machine tool control standard, ISO 14649 or STEP-NC. The information provided for control by this new standard allows the machine tool to optimise its own behaviour for its architecture and the part to be machined. This is possible because of the increased computing power now available on the industrial computers used in controllers. Traditionally, movement planning had to be done off-line with neither details of the architecture of the target machine tool nor of workshop loading. An extension to this standard is to have a machine tool model to describe available manufacturing resources as well as to describe machining needs in the control file. Such an extension would allow workshop planners to make last-minute decisions about machine choice.

Another important aspect of improved level of information on the shop floor is the possibility to perform simulations and optimisation of plans. In some processes, such as EDM, the machine operators have extensive knowledge of the process and sometimes perform experiments to improve machining parameters before real production starts.

Concerning global levels, life-cycle analysis tools exist but sometimes need careful application in order to understand the results. An example of a tool, which is easier to use and can be used to compare alternatives is the EcoIndicators99 method. This method uses a scoring system to equate different environmental factors. Transport, machining, part treatments, recycling and so on are all assigned a unit score in a common basis which is then multiplied by part characteristics, such as weight, to arrive at a process score. These scores are then added together to give an idea of the relative environmental impact of the part production. The methods described above keep the different factors separate in order to understand from where the different environmental factors come, a so-called multi-criteria approach. The method presented also allows the user to understand the environmental effects in quantative terms rather than simply knowing which option may be better.

CONCLUSION

Computing at the service of the environment has a wider application than simply improving the environmental aspects of computing itself. Green applied computing can be used to make sense of complex areas where the wealth of data is too great to be comprehensible without processing. In the manufacturing context there is a great deal of complexity already with machining but extending consideration to the factory level and the global supply chain then the complexity far exceeds the level which can be comprehended by users.

For the research presented, the main computing tools used were:

- Database tools
- Data collection and management tools
- Simulation tools
- Calculation

Further research is going on in other projects where advanced simulation and calculation are also being used. Data modelling based on computer science tools also plays a major part in the standardisation work.

ACKNOWLEDGMENT

The authors would like to thank Vincent Capponi for all his work. We would also like to thank the European Union for financing the research presented during the NEXT project. Last but not least, we would like to thank the NEXT project partners for sharing their results and equipment for the experiments that were so necessary for the work.

REFERENCES

Akbari, J., Oyamada, K., & Saito, Y. (2001). LCA of machine tools with regards to their secondary effects on quality of machined parts. In *Proceedings EcoDesign 2001: Second International Symposium on Environmentally Conscious Design and Inverse Manufacturing*, (pp. 347-352). Tokyo, Japan: IEEE.

Avram, O. (2010). *Machine tool use phase: Modeling and analysis with environmental considerations*. Ph.D. Dissertation. Lusanne, Switzerland: EPFL.

Avram, O., Stroud, I., & Xirouchakis, P. (2011). A multi-criteria decision method for sustainability assessment of the use phase of machine tool systems. *International Journal of Advanced Manufacturing Technology*, *53*, 811–828. doi:10.1007/s00170-010-2873-2

Avram, O., & Xirouchakis, P. (2011). Evaluating the use phase energy requirements of a machine tool system. *Journal of Cleaner Production*, *19*, 699–711. doi:10.1016/j.jclepro.2010.10.010

Dahmus, J., & Gutowski, T. (2004). An environmental analysis of machining. In *Proceedings of the 2004 ASME International Mechanical Engineering Congress*. Anaheim, CA: ASME.

Dornkundwar, V., Krishnan, N., Bauer, D., & Sheng, P. (1998). A distributed environmental planning model for machined components. In *Proceedings 5th International Seminar on Life Cycle Engineering*, (pp. 227-239). Stockholm, Sweden: KTH. Eco-indicator-99. (2011). *Website*. Retrieved from http://www.pre.nl/content/eco-indicator-99.

ISO 14649. (2001). *TC184/SC1/WG7, ISO 14649/FDIS, data model for computerized numerical controllers*. New York, NY: International Standards Organization.

Jayal, A. D., & Balaji, A. K. (2007). On a process modeling framework for sustainable manufacturing: A machining perspective. In *Proceedings of the ASME International Mechanical Engineering Congress and Exposition*, (pp. 301-307). ASME.

Munoz, A. A., & Sheng, P. (1995). An analytical approach for determining the environmental impact of machining processes. *Journal of Materials Processing Technology*, *53*, 736–758. doi:10.1016/0924-0136(94)01764-R

Narita, H., Kawamura, H., Norms, T., Lian-Yi, C., Fujimoto, H., & Hasebe, T. (2006). Development of prediction system for environmental burden for machine tool operation. *JSME International Journal Series C*, *49*(4), 1188–1195. doi:10.1299/jsmec.49.1188

Srinivasan, M., & Sheng, P. (1999a). Feature-based process planning for environmentally conscious machining - Part 1: Microplanning. *Robotics and Computer-integrated Manufacturing*, *15*, 257–270. doi:10.1016/S0736-5845(99)00017-4

Srinivasan, M., & Sheng, P. (1999b). Feature based process planning in environmentally conscious machining- Part 2: Macroplanning. *Robotics and Computer-integrated Manufacturing*, *15*, 271–281. doi:10.1016/S0736-5845(99)00018-6

Suh, S.-H., Kang, S.-K., Chung, D.-H., & Stroud, I. A. (2008). *Theory and design of CNC systems*. Berlin, Germany: Springer.

Zhi, G. J., Hua, Z., & Xiao, L. (2008). Study on the cutting technology based on green manufacturing. *Key Engineering Materials*, *375-376*, 158–162. doi:10.4028/www.scientific.net/KEM.375-376.158

Chapter 7
Green IT Project Management:
Optimizing the Value of Green IT Projects within Organizations

Nathalie Bachour
PRINCE2/MSP Practitioner, France

ABSTRACT

With the constant evolution of technology and the world critical environmental status, all private and public Information Technology (IT) businesses are moving towards sustainability. Faced with influences from government regulations, market competition and constraints, as well as watchdogs, IT decision makers within organizations are forced to ride the green technology wave with a challenging and uncertain approach. This chapter defines methods to optimize Green IT projects for sustainable value creation within organizations. It only focuses on economic viability and environmental impact, but could be stretched out in the future to social aspects. The contributions of this chapter allow the project management community and decision makers to follow a framework for Green IT project success evaluation and performance follow-up throughout the project life cycle and the three levels of the organization: operational, tactical, and strategic. A macro-model is also developed to aid them in successfully selecting, prioritizing, managing, and aligning their Green IT projects with the corporate and environmental strategies.

1. INTRODUCTION

"Welcome to an exhausted world!" This slogan was used to welcome people attending one of the conferences on Green Information Technology (IT) during the "Microsoft Techdays 2010" in Paris. Faced to this apocalyptic view, the audience was intrigued on such a different approach to Green IT. Most of Green IT conferences would start with the same cliché: the problematic environmental impact of technology reaching an equivalent of 2% of total worldwide CO_2 emissions, and the important role it could play to reduce the remaining 98%. Looking from a broader perspective, the threat is not limited to the climate change and the 293km^3 of ice lost each year from Greenland. It involves the whole planet's ecosystem from

DOI: 10.4018/978-1-4666-1839-8.ch007

people's health and living trends to biodiversity, raw material and energy resources availability and waste management. The humanity's problems are deeply rooted in its economic structure dictating individuals and organizations' behavior, in addition to the double-edged power of politics and lobbies.

1.1. Economy and Ecology: A Historical Misunderstanding

By going back to etymological roots, the economy is the science of managing resources within a living space. It "*deals with optimizing the production, distribution and consumption of goods and services within this space*" (Wordreference.com). Ecology, is the science of the living space, *it deals with studying the relation between organisms* (in this case, humans) and their environment. Here comes out the absurdity of the economic system, humans are living in. By overusing resources provided by nature at a very cheap price, they destroyed their key provider equilibrium. Two words with the same root were completely misaligned.

To illustrate this contradiction, consider the numerous times a basic consumer replaced a whole electronic device instead of trying to repair it. Maintenance services are very expensive while production is very cheap due to economies of scale. This cycle is being followed at the expense of more resources overuse, pollution, and hazardous wastes generation. Of course, such scenario is not always applicable; sometimes repair services can have a higher environmental footprint than simple replacement. The point is - does the current economic system allow space for evaluation?

Recent observations show signs of tangible change and evolution in the marketplace. Indeed, many business leaders agree that economy and ecology should be aligned, if not on the short term then at least on the long term. "*Green is Green*," declared General Electric CEO, Jeffrey Immelt, while advocating the cooperation between corporations and environmentalists (Coburn, 2008). In this greening wave, some business leaders seem only interested in green technologies from a financial perspective, the same way investors were carried by the Internet bubble in the 1990s. However, this cognitive change in the will to align economy and ecology is gaining popularity in organizations worldwide.

1.2. Integrating Sustainability in Organizations

Do businesses only exist to generate cash or do they have another role in the local, national, and even international society? Companies throughout the world are being asked to act responsibly not only on the environmental dimension but also on social aspects such as poverty, education and healthcare (Esty & Winston, 2006). Which comes first then, shareholder value or corporate social responsibility?

Neither! Sustainability is about finding the right balance between people, the planet, and the profit (3Ps). The will to align business and environmental sustainability is well illustrated in numerous environmental projects launched on the market, covering the fields of IT, construction, energy production, food industry, clothing, etc.

1.3. Why Should Organizations Consider Green IT?

In this changing trend, IT plays a double role. First, it is polluting, so it should be enhanced and re-engineered. Second, it represents a powerful tool that can be used to reduce the environmental impact of other products, operations, or businesses. As an enterprise or IT decision maker, one would be pressured by government regulations, incentives, or market competition, to start that kind of initiative. However, the challenge would be to optimize a selection from the myriad of solutions available on the market, align them with current projects, and evaluate the effective sustainability value created within the enterprise. Many projects are being implemented in this scope in IT departments by following standard processes for project management, quality and service management. IT projects are usually defined in a top-down approach to implement strategies. They may also emerge from a bottom-up initiative to improve operational performance. Through different levels of capability maturity, most IT projects are based on international or customized project management standards. With this green wave of change completely modifying the criteria for success, three questions arise for IT decision-makers and Green IT project managers:

1. Do organizations limit Green IT project success and performance evaluation to the project level and the environmental quality of deliverables?
2. How should organizations follow up and evaluate the environmental performance of Green IT projects, and assess their contribution to sustainable value creation?
3. How can organizations optimize Green IT projects for sustainable value creation?

On a first level, a framework is set for evaluating and optimizing Green IT projects' success within organizations. It is built on organizational project management, integrating the environmental dimension in a coherent cascade going through the three levels of the organization (operational, tactical, and strategic) and covering both the project management and the deliverable implementation processes.

On a second level, a model is built to illustrate the integration of Green IT project management with respect to other quality and systems development models. The whole is further integrated in a macro model showing the interactions between business as a change (project) and business as usual (production). These frameworks can aid IT decision makers and the project management community, in conducting Green IT projects while keeping a holistic view on the business, the corporate environmental impact and external market pressures and opportunities.

2. BACKGROUND

Green IT, also known as eco-computing or green computing, is increasingly implemented in the field of IT infrastructure. It defines a set of best practices in order to optimally use computing resources. Green practices in technology can cover several phases of the product or service life cycle: from acquisition to recycling and final disposal. As a decision maker within an organization or even a project manager involved in Green IT projects, one would have to evaluate the value creation of the project, the project performance and its alignment with other concurrent projects.

Green IT projects can be divided into two categories:

1. *Green for IT*, projects aiming to reduce the environmental impact of IT, also known as, Green IT 1.0. Such projects are mainly shaped around the below topics:
    - Resources usage efficiency
    - Electronic waste management
    - Reduction of hazardous material use
2. *IT for Green*, projects aiming to reduce the environmental impact of organizational & business operations using IT, also known as Green IT 2.0. Such projects are mainly shaped around the below topics:
    - Environmental performance measurement & reporting
    - Operational optimization
    - Operational re-engineering

2.1. Positioning Green IT within Organizations

Academic references analyzing the role of Green IT within organizations are almost inexistent in the field of computer or management sciences.

Within their research agenda, Elliot and Binney (2009) refer to Green IT as EICT and were able to classify practitioner literature around IT and environmental sustainability as seen from the corporate level, into five categories:

1. Source of problems
2. Potential solutions and opportunities
3. Rationale for action
4. Actions
5. Evaluation

The subject of this chapter can be found in two categories: Actions and Evaluation, i.e. what can be done to optimize Green IT projects within organizations and how to evaluate them? In addition to this classification, Elliot and Binney (2009) were able to define a framework for measuring Green IT capability within organizations. The framework was set, based on several academic articles analyzing strategy and sustainability in organizations. Below is a brief overview of the framework:

1. *Awareness* of environmental issues related to carbon footprint and e-waste with unclear relevance to ICT.
2. *Investigation:* IT departments start to understand the link between ICT and business activities, and its contribution to pollution throughout the organization.
3. *Determination:* IT departments seek to measure and monitor ICT environmental footprint, and define potential solutions.
4. *Action:* Solutions are implemented to reduce energy consumption and prevent pollution. Measurement platforms are also put in place to allow smart monitoring.
5. *Evaluation:* Define the appropriate metrics to evaluate Green IT solutions contribution to reduce the organization environmental footprint.
6. *Proliferation:* Using the metrics defined earlier, possible solutions for the business activities are initiated, and procurement criteria are defined.
7. *Integration:* Green thinking becomes integrated on all organizational levels, an EMS and EIA are implemented within the organization.
8. *Business Opportunity:* The advanced Green IT 2.0 is reached. Ventures are created to exploit sustainable business opportunities.

Such framework aids organizations to situate themselves relative to Green IT maturity, and the way ICT can build corporate capabilities. This framework could correspond to Green IT maturity models found on the market such as Accenture Green Maturity Assessment (2008) and IBM Green IT Capabilities Assessment (2009a).

Now that corporate capabilities and dimensions of Green IT were defined from a corporate perspective, it would be interesting to take a look at literature analyzing the alignment of Green IT projects with corporate and environmental strategies.

2.2. Matching Environmental Strategies to Green IT Solutions

Environmental strategies are initiated due to different motivating forces: Technological, organizational, regulatory-market factors and Sociocultural (Jenkin, Webster, & McShane, 2009) If the internal and external contexts shape strategies, environmental strategies should be referenced to define Green IT programs and initiate corresponding projects. In their research agenda, Jenkin et al. (2009) identified four types of environmental strategies with their corresponding impacts and motivating forces. Below is a mapping of these strategies with Green IT concepts found in the marketplace:

- *Type 0:* Image oriented only. What is also known as "*green washing*," where companies try to communicate about their concern for the environment, but no action is really implemented yet.

- *Type 1:* Prevent, Control ⇔ Eco-efficiency. This corresponds to quick wins in *Green IT 1.0*, as well as a portion of Green IT 2.0 where some business activities are enhanced with small changes.
- *Type 2:* Product stewardship ⇔ Eco-equity. Here the product being developed is green throughout its complete life cycle, from manufacturing until disposal. This can be matched to the eco-conception part of Green IT 1.0. Also, IT is used to reduce the environmental footprint of the whole life cycle of other products.
- *Type 3:* Sustainable development ⇔ Eco-Effectiveness. At this level, radical changes are performed on organizational processes. Here, process re-engineering is performed and new means are introduced. Also, IT is used for measurement and reporting of environmental performance and impact of business activities. This is an interesting differentiation from Green IT 2.0, where business operations are not only improved but also re-engineered.

To each environmental strategy formulated, the level of possible impact should be identified at these different levels: technical, individual, group, organizational and societal. Understanding the level of impact is essential for successful management of cognitive and behavioral environmental changes at each level.

2.3. Aligning Green IT Strategies with Corporate Strategies

In their Green IT research agenda, Jenkin et al. (2009) identify a number of authors for their focus on the organizational impact of the environmental strategy and its alignment with the business strategy. Although no one related to organizational project management, there are key methodologies that can be useful for strategic alignment of Green IT projects with corporate and environmental strategies.

There are several theories underlying the behaviors and strategic decisions of corporations (Aravind & Christmann, 2008; Berrone, et al., 2008).

In their article, Berrone et al. (2008), integrate the institutional theory and Natural Resource-Based View (NRBV) to create environmental innovation. Indeed, by complying with legal and environmental constraints and following up with the environment needs, a culture of innovation can be fostered in organizations.

As green initiatives in organizations are being integrated into Corporate Social Responsibility (CSR) aspects, it is important to check if there is literature integrating the two concepts. This complementary strategic setup directly and indirectly adds up to the triple bottom line (People, Planet, and Profit). In this respect, several articles do mention environmental sustainability in organizations and strategic alignment, but no analysis or framework is discussed to include and create a leading thread between CSR and Green IT. In a research done recently on Green IT and Systems in organizations (Jenkin & McShane, 2009), only 1/3 of interviewees related IT to CSR. This shows either a lack of communication and strategic mapping top-down, or a lack of IT and sustainability governance where communication is not facilitated bottom-up in order to identify emergent program.

Fortunately, there are already sustainability management tools available on the market to solve such kinds of misalignment. These tools can bridge sustainability governance to IT governance, linking them back to corporate governance.

2.3.1. Tools for Strategic Alignment

The balanced scorecard and the strategy map are two of the most famous strategic alignment tools used in management. In their article, Moller and

Schaltegger (2005) explain the importance of representing financial as well as non-financial figures, more specifically environmental sustainability performance, for efficient managerial decision-making regarding CSR. In this respect, they enhance the traditional BSC by either adding a new non-market component (representing the environmental sustainability) to the market components (Financial, Customer, Internal, Organization Learning), or adding environmental indicators to each of the existing perspectives. This approach is interesting for this research as it similarly tries to integrate the environmental dimension into project management.

It is now essential to take a look at methodologies used to successfully implement Green IT solutions within organizations. Can general project management standards answer such needs? A market analysis on the integration of project management and green thinking is required. The analysis will not be limited to ICT, as benchmarking other industries is crucial to develop a robust framework.

2.4. Green Project Management

Green PM (Project Management) was introduced in the market by two active PMI members: Tom Mochal and Andrea Krasnoff of TenStep, Inc. The concept is still new and aims at integrating a company's environmental policies into project management processes, irrespective of the type of project being conducted. Authors do not force project managers to consider the environment as a prime stakeholder. They are just suggesting taking into account the environmental dimension instead of ignoring it during the decision-making process.

2.4.1. Green PM Limitations

Green PM concept is quite interesting but it is still vaguely presented. There is no clear framework to follow. Further, as green projects require in general a lot of measurement and monitoring, Green PM does not offer techniques or tools in quantifying baselines and improvements performed against benchmarks or legal environmental standards. For Krasnoff and Mochal, there is no need to put in place new project management processes, "Green thinking" will be integrated in some areas of project management, in order to align with the organizational EMS. Green PM is presented as a scenario analysis (What If), not necessarily leading to a change in decision-making but pushes the project management community to be aware of environmental concerns.

2.5. Green Sigma

IBM Consulting services customized Lean Sigma to become Green Sigma, "*Green improved ROI*" (IBM, 2007). Green Sigma macro-project life cycle is represented in a five-phase macro level framework, which looks like the DMAIC (Define, Measure, Analyze, Improve, and Control) but is focusing on environmental concerns. According to IBM, these principles are applied whenever energy and water are used throughout a client's operations—transportation systems, data centers and IT systems, manufacturing and distribution centers, office facilities, retail space, research and development sites, etc. (IBM, 2008).

Green Sigma can be summarized as follows:

Green Sigma = Lean Six Sigma principles + CSR Consulting + Energy/Water/Carbon Measurement and Management Dashboards.

2.5.1. Green Sigma Limitations

Lean Sigma or its green version Green Sigma, cannot replace project management standards such as PMI or PRINCE2. As Lean complements Six Sigma, Six Sigma complements project quality Management. Therefore, Green Sigma complements the environmental quality management within project quality management. Organizations cannot rely on Six/Lean/Green sigma projects alone, as they do not offer a framework to control/manage the project life cycle; they can however control the improvement of processes within this cycle.

2.6. Environmental Construction Project Management

2.6.1. HQE (France)

"Haute Qualité Environnementale," or High Environmental Quality is an initiative started back in the 1990s, under the sponsorship of PCA (Plan Construction Architecture). HQE is a project management initiative aiming at the reduction of environmental impacts of construction projects such as new buildings and refurbishments. It also targets internal building conditions for health and comfort (ADEME, 2009).

2.6.2. LEED

LEED (Leadership in Energy and Environmental Design) is the equivalent of the French HQE in the States, with some variations. LEED is a green building certification system that applies to any building type and at any building life cycle phase (Design, Construction, Refurbishment, and Operation), that provides third-party verification for buildings and a framework for identifying and implementing practical and measurable deliverables. (US Green Building Council, 2010)

2.6.3. Strengths of LEED and HQE

LEED and HQE both offer a framework of construction recommendations during all the project life cycle. Such checklists could be inspiring for Green IT projects if adapted accordingly. For instance, the HQE auditing process run by a third party after the plan, design and construction could be adapted to the systems development life cycle. It may seem that standards focus more on the products and operations to build the products than on management processes. However going through other references (Alberta Infrastructure, 2007), shows the integration of Green Thinking in several project management processes. HQE and LEED certifications function like a PMO (Project Management Office) where documentation, templates, performance measurements, and recommendations are provided and controlled.

2.7. SSDLC

A sustainable SDLC (Systems Development Life Cycle) model based on the five phases (Planning, Analysis, Design, Implementation, and Maintenance) was introduced by Huang (2008). He included environmental aspects in every phase and added a sixth phase "Disposal." Huang (2008) argues that, although the disposal phase is already defined in the SDLC, it is not used in all models on the markets.

2.7.1. SSDLC Limitations

SSDLC gets very close to the subject of this chapter, since it focuses on information systems. However, it is limited to one aspect of Green IT 1.0, i.e., reducing the environmental impact of new information systems. Parts of the SSDLC can be considered as project life cycles, but the author does not clearly structure the integration of the SSDLC within a project management framework. After reviewing the corporate perspective of Green IT and benchmarking current practices of project management with green thinking, the current practices found in the marketplace for evaluating Green IT projects success and performance will be discussed.

3. GREEN IT PROJECT SUCCESS EVALUATION WITHIN ORGANIZATIONS

Current practices for evaluating Green IT project success were identified during a survey run by Bachour (2010a) as part of a Master's thesis. The survey entitled "Managing Green IT projects in Organizations" assessed companies on:

- Green IT maturity
- Ability to measure the success of Green IT projects, after completion.
- Ability to measure and follow-up the performance of Green IT projects, during execution.

3.1. Green IT Maturity within Organizations

Green IT projects are identified on the market under two categories: 1.0 and 2.0. Such levels do not necessarily refer to a level of maturity. These categories can be implemented simultaneously within the same department.

3.1.1. Green IT 1.0 Projects

As expected from other studies and surveys on the market, most of respondents were focused on IT energy efficiency, one of the main problems of IT. Solutions selected by respondents include measurement, power management, datacenter consolidation, cooling solutions, virtualization, and user education. However, respondents did not limit themselves to efficiency. Indeed, 47% were concerned with IT asset end of life management, another problem with high environmental impact, and 35% were aware of some environmental procurement policy within their company. It is important to mention at this level that the current market trend focuses on energy or systems efficiency but will slowly extend to procurement, design, and disposal phases due to emerging regulations and environmentally demanding business ecosystems.

3.1.2. Green IT 2.0 Projects

Referring back to the survey done in Bachour (2010a), companies seem to be mostly focused on reducing the enterprise environmental impact caused by travelling and commuting using distant collaboration solutions: Remote Work (64%), Videoconferencing (52%), Instant messaging (50%).

Such impact can be measured using financial and CO_2 emissions metrics.

Another focus point was done around dematerialization (52%) contributing directly to lean processes, and reduction of resource overuse such as paper, power, and travelling. Dematerialization highly shapes internal procedures by contributing to reduced operational costs, optimized quality, and operational efficiency. However, respondents seem to miss this link between Green IT solutions and process efficiency since 52% were concerned with dematerialization but only 35% were working on process efficiency. Finally, only 25% were involved in measuring their enterprise environmental impact by focusing on water, air, or ground pollution. The survey shows a lack of concern in EMS (only 11%). This demonstrates a weakness, in either the respondents' awareness, or a lack of the company's capability. Indeed, such aspect could demonstrate the absence of a holistic view of the enterprise and the inexistence of an environmental strategy. This is in line with a Forrester (2009) survey, mentioning that 53% of worldwide enterprises still did not define any environmental strategy.

3.2. Green IT Project Management Success Criteria

The project success criteria defined within Bachour (2010a) survey were based on Dinsmore and Cooke-Davies (2006). Figure 1 shows the Green IT perspective was integrated in the set of classical

Figure 1. Project management success criteria (Bachour, 2010b). Adapted from Dinsmore and Cooke-Davies (2006).

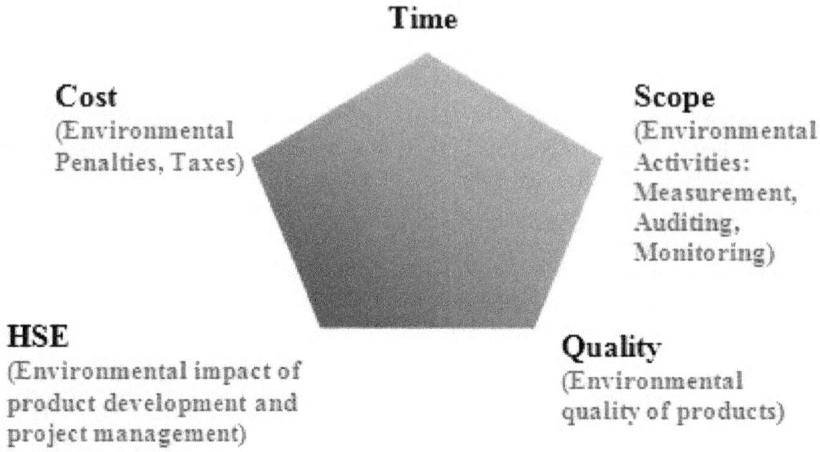

project success criteria: scope, time, cost, quality, and HSE (Health, Safety, and Environment). The environmental dimension was split into two components: the Environmental Impact of project management and product development processes, and the Environmental Quality of deliverables.

Of course such models take into account other technical and security criteria concerned with IT projects. It does not necessarily apply to measuring every Green IT project success. However, it aids the project manager in evaluating a project's success through an environmental framework.

3.2.1. Practitioners' Feedback on Green IT Project Management Success Criteria

In Bachour (2010a) survey, standard project management constraints were not considered by all respondents. However, the environmental impact was considered for both the product and the project (39%, 31%). Deliverables environmental quality only accounted for 25%. Thus, the suggested theoretical model seemed quite applicable and accepted. A first conclusion was formulated as follows:

Green IT project success and performance evaluation could use classical project management success/performance criteria but the environmental dimension cannot be limited to quality. Another criterion should be added to extend the evaluation to the environmental impact of both the project management and the product development processes.

3.3. Green IT Organizational Project Success Framework

Whether green IT projects are implemented within the 1.0 or the 2.0 level, they generate value throughout the enterprise value chain. Results can be observed at the operational, tactical, and strategic levels. As a decision maker within an organization or even a project manager involved in Green IT projects, one would have to evaluate the value creation of the project within an organization, the project performance, and its alignment with other concurrent projects in order to sustain competitive edge.

Bachour (2010a) described how a green IT project aiming to reduce operational costs can cause a positive domino effect throughout the organizational pyramid.

In this respect, classical project success showed to be limited and focus was brought to organizational success defined by Dinsmore and Cook-Davies (2006): "*when overall strategy is implemented effectively, when the productivity of scarce resources is improved and when projects are delivered successfully.*" Applying this perspective to Green IT, if a project is well managed, but does not contribute to a business benefit aligned with the corporate and environmental strategy, it cannot be considered successful. Another representation of organizational project management success can be found in PMI Inc. (2008).

Since such evaluation cannot be achieved at operational levels by project managers alone, it extends to a three-level chain: Project, Program, and Portfolio. In such a governance structure, all potential stakeholders should be identified and put into relation through reliable communication lines.

3.3.1. Green IT Project Benefits Realization

In order to assess an organization's capability to implement the right projects, success evaluation should be run on benefits expected from Green IT projects. If Green IT projects did deliver the final outcome successfully by respecting all success criteria mentioned above, are they contributing to expected benefits?

When defining the expected benefits, they should be SMART (Specific, Measurable, Attainable, Realistic, Timely) and cover multiple perspectives. The balanced scorecard and strategy map are a good example for such multi-dimensional analysis techniques on multiple organizational levels. Indeed, Green IT projects can create benefits for people, operations, customers, finance, and the environmental dimensions throughout the organization's value chain.

Consider the example of a virtualization solution implemented within an IT department. Table 1 is a list of benefits to which such a project can contribute. Each defined benefit should be associated with a metric. Without metrics, no factual evaluation can be achieved, thus decision-making will be based on intuition.

3.3.1.1. Practitioners' Feedback on Green IT Project Benefits Realization

In Bachour (2010a) survey, almost half of respondents (49%) were keen on energy efficiency. Projects being selected in order to reduce energy bills, shows to be well aligned with expected benefits. Indeed, 80% were into virtualization and energy consumption measurement. However, 20% did not seem to have separated IT energy bills from their corporate energy bills. Thus, sponsors can be from two different departments and special care should be taken during stakeholder identification and project governance structure.

Respondents' expectations were almost balanced among operational, environmental, technical and financial efficiency. Only 8% of respondents mentioned no assessment was done at a sponsor level. This demonstrates an alignment

Table 1. Multi-dimension benefits measurement

Dimension	Benefit
People	New competency, culture of innovation
Increased systems use efficiency	Reduced administrative overhead, reduced time to market, improved quality of service, reduce Increase systems use efficiency
Customer	Customer satisfaction, customer retention
Financial	Reduced energy bills, reduced total cost of ownership, reduced maintenance costs, increased ROI
Environment	Reduced CO_2 emissions, reduced resource wasting

among the IT departments' strategy and other business units' strategies or even the business as a whole. Further, this clearly shows that Green IT projects' evaluation within organizations is not limited to the project level.

The survey showed a special focus on expected outcomes creating short to mid-term financial benefits within the enterprise. This is in line with a Forrester survey showing that the first driving force behind Green IT projects is payback (Washburn, 2009). Although the problem of WEEE generation is becoming drastic, organizations will not compromise their business sustainability for the sake of the planet. The target of such evaluation is not to focus exclusively on the environmental dimension but also to align the environment with the economy. Thus, taxes and quotas should not only focus on CO_2 emissions, but also on environmental impact directives such as WEEE or RoHS.

3.3.2. Green IT Project Alignment with Corporate Strategic Objectives

Continuing with the organizational perspective, evaluation should be run on a strategic level. What is the contribution of all the organization's projects to strategic objectives as well as profitability and productivity? This can be conducted by measuring the overall corporate strategic objectives achieved. For instance, did the enterprise wide CO_2 reduction achieve target objectives? If positive, the combination of projects and investment expenditures was done successfully, otherwise combination and prioritization of projects should be reviewed.

From an environmental perspective, strategic alignment should be checked between Green IT and other environmental projects towards the corporate environmental strategy, usually formulated within the CSR.

Again, metrics should be defined for each strategic goal to be achieved. Such metrics can result from consolidated measurement done on the program or project levels.

3.3.2.1. Practitioners' Feedback on Green IT Project Alignment with Corporate Strategic Objectives

The survey in Bachour (2010a) thesis showed a relatively high interest for reducing corporate environmental impact (55%). For companies aiming to reduce their environmental impact, their focus was almost balanced between environmental and financial benefits. Lower interest in "electronic waste generation (46%)" again demonstrates the need for business sustainability.

Respondents focusing on productivity at a strategic level did focus on reducing administrative overhead and improving process efficiency. Those who focused on improving their profitability mostly focused on reducing energy bills (65%), and equally on reducing resource wasting, administrative overhead, and increasing ROI (59%). In most cases, strategic alignment looked consistent when comparing respondents' percentages among different strategies, but seemed less coherent when alignment evaluation was done for a specific strategy. This demonstrates a lack of rigorous methodology followed within the organization to initiate and follow up alignment on the three levels (Project, Sponsor, and Corporate). Only 6% of organizations limited Green IT project success evaluation to the project management level. The survey led to the second conclusion below:

As all projects implemented within organizations, Green IT projects success evaluation extends from classical project management to Organizational Project Management (OPM) where the environmental dimension is included in project performance, benefits realization, and strategic alignment, and evaluated from a project, sponsor, and corporate level perspective respectively.

4. GREEN IT PROJECT PERFORMANCE EVALUATION AND FOLLOW-UP WITHIN ORGANIZATIONS

The first part of the survey assessed organizations' capability to evaluate Green IT projects after completion. Next, they were assessed on their capability to measure the performance of their projects and their alignment with the expected benefits and strategic objectives. How are Green IT projects performing during their execution and are they still consistent with the corporate strategy? As mentioned earlier in each of the theoretical analyses, metrics should be defined at each level of the evaluation. Project, Program and Portfolio performance levels can be matched to operational, tactical, and strategic metrics, elaborated into performance indicators.

4.1. Practitioners' Feedback on Green IT Project Performance Measurement

Although most organizations did mention their evaluation for a Green IT project on the three levels, they showed misalignment between their will to evaluate success or performance and their capability to do it. 96% evaluated their project on the operational level but only 42% defined metrics to do so. 92% evaluated their project on the sponsor level, but only 14% defined metrics to do so. 86% evaluated their project on the strategic level, but only 42% defined metrics.

Only 10% of organizations considered metrics at the three levels. 18% limited their metrics, either to the operational or strategic levels, and 4% to tactical levels.

These observations led to the following conclusions:

Most of organizations evaluate their project performance at least on the operational and strategic levels. The changing corporate environment in which Green IT projects are implemented requires performance alignment with tactical and strategic directions.

At a first level, evaluation should be run by integrating and adding the environmental perspective in the set of success/performance criteria defined on the three levels of OPM. Each success criterion should be associated with a set of metrics and performance/success indicators.

Table 2. Green IT project level success/performance evaluation framework. Adapted from Dinsmore and Cooke-Davies (2006, p. 4).

Success Level	Responsible Parties	Success Criteria	Success Factors
Green IT Projects are done Right	Project manager Project team	• Time, Cost, Quality, Scope • Technical Performance: Availability, Scalability, Efficiency • Systems security standards • Environmental impact: o throughout the product development life cycle o due to the project management process • Environmental quality: Future product/ service environmental specifications • Environmental cost: Penalties, taxes, labeling, certification • Health & safety standards • Social/Ethical norms	Green IT Project management capability • Green IT Project life cycle • SMART goals • Clear technical requirements • Capable and effective team • Adequate resources • Effective planning and control/audit • Adequate product/project metrics • Baseline measurement • Reliable measurement • Organization's standards (HSE, Social, Ethical) awareness • Risk management

So far, success criteria were identified for reliable evaluation of Green IT projects. This chapter will continue on identifying factors for optimizing the path to success.

4.2. Green IT Project Management Success Factors

Green IT project success factors were identified in Bachour (2010b) based on Dinsmore and Cooke-Davies (2006, p. 4).

4.2.1. Green IT Project Life Cycle

In Bachour (2010b), the Green IT project life cycle is built based on PRINCE2 (Specify, Design, Develop, Test, Change Over) and process quality standards Lean Sigma. The association of Green IT projects with quality projects is done based on the measurement efforts involved in Green IT and the pillars of quality improvement: "If you cannot measure it, you cannot manage it." (Kaplan, Deming)

Many quality standards are available in the management field: TQM, Six Sigma, Lean Management, Lean Sigma. Green IT projects get similar to a mix of standards for improving quality and efficiency, such as Lean Sigma. However, the focus is on the environment rather than the customer.

4.2.1.1. Six Sigma and Lean Management

Both methodologies developed by two giant manufacturers, Motorola and Toyota respectively, aim at customer satisfaction by improving quality and eliminating waste. In 1997, TBM Consulting combined Six Sigma and Lean Management based on complementarities of quality and speed in process improvement, by introducing Lean Sigma. Although such methodologies were initially targeting manufacturing processes, they were soon adopted within services and most of business activities including project management. Note that a green IT project life cycle can follow the DMAIC cycle (Define, Measure, Analyze, Improve and Control), in case of green improvement (Green IT 1.0 or 2.0) or the DMADV cycle (Define, Measure, Analyze, Design and Verify), what is also known as "Design for Green," in case of a new green product or service development.

4.2.1.2. Green Sigma Model

The association of Six Sigma principles with green project management was done in Bachour (2008). Another study from Gartner mentions the implementation of "Greening Strategies" within organizations through continuous improvement initiatives like Six Sigma, TQM or lean management but stresses on the fact that "Going green" should not be considered as a project, rather "*a long term endeavor, applying a systematic approach*" (Rowsell-Jones, 2008). In August 2008, IBM introduced Green Sigma, as mentioned earlier in the background section.

4.2.2. Green IT Project Management Framework

Bachour (2010b) identifies the subtle need for a project management framework not provided by Green sigma. A project management framework provides the processes and components required to successfully manage a project through its life cycle. A project follows a path from its start until release into production, whereas project management follows a path allowing initiation and closure as well as effective stage gate reviews, governance and control throughout the project. In order to succeed in managing Green IT projects, Bachour (2010b) recommends the integration of the green perspective in all components of two famous project management methodologies: PRINCE2 and PMI. Table 3 and Table 4 establish an overview of these standards' components with a green perspective and corresponding examples.

4.3. Green IT Organizational Project Management (OPM) Framework

After detailing Green IT project management capability, a brief overview is done through the three-level chain of organizational project management capabilities.

The second level of success criteria was defined at sponsor level and involved the evaluation of benefits realized, showing the right selection of projects. How to ensure the right Green IT projects are implemented? Such capability is perfectly found in program management standards and identified in Bachour (2010b) based on Dinsmore and Cooke-Davies (2006, p. 4).

4.3.1. Sponsor Level Green IT Project Success Factors

According to MSP (Managing Successful Programs) Guide, "*a program is a temporary, flexible organization created to coordinate, direct, and oversee the implementation of a set of related projects in order to deliver outcomes and benefits related to the organization's strategic objectives*" (OGC, 2007). In order to understand how

Table 3. Greening PRINCE2 components (Bachour, 2010a; OGC, 2005)

PRINCE 2 Components	Sections	Green Dimension	Example (When applicable)
Business Case *Defined top-down through project mandate or identified through the project manager investigation.*	Reasons (Why?)	Alignment with environmental strategy	Reduce the corporate CO_2 emissions
		Compliance	ISO14000, WEEE, RoHS, BASEL
	Benefits *Direct & Indirect, non-benefit* *On the 3 levels*	Environmental performance on the operational & corporate levels.	- Energy Savings - Reduced WEEE generation - Reduced CO_2 emissions of IT systems or operations - Resource Usage Savings - Fines for non-compliance with environmental regulations
	Risks	Environmental Impact caused by possible incidents	Toxic Substance Spill
		New environmental regulations	
	Costs	Cost of respecting environmental standards	- Clean-up costs Treatment & disposal of Waste - Environmental auditing costs
	Timeline	Environmental activities	Environmental Auditing
	Investment appraisal	Short-term: Fast ROI due to efficiency & waste reduction	Energy efficiency leads to reduced operational costs
		Negative appraisal unless analyzed with a holistic view of the program, portfolio, business & corporate strategy.	A Green IT 2.0 project taken alone may not represent financial benefits for the IT department but for other functional departments.
		Long-Term: Financial & non-financial benefits	Brand Image, Environmental Stewardship
Organization	Senior User	Green ambassador or champion	
	Project Board	Representation of the environmental department	
	Sponsors	Executives with holistic view able to evaluate strategic benefits on the long-term across departments.	Able to evaluate strategic alignments and benefits on the long-term.

continued on following page

Table 3. Continued

PRINCE 2 Components	Sections	Green Dimension	Example (When applicable)
	Tolerances		
	Scope	Environmental activities	Environmental Auditing, Control, Expertise
	Quality	Environmental norms & standards	Energy Efficiency allowed by EPEAT
	Costs	Non-financial costs expressed in environmental impact thresholds related to product development	- Resource usage: Water, Paper, energy - Power efficiency - Pollution: Air (CO_2, GHGs), Water, and Soil. - WEEE generation - Transport
		Non-financial costs expressed in environmental impact thresholds related to project management processes	- Resource usage: Water, Paper, energy - Traveling/commuting distance - Pollution: Air (CO_2, GHGs)
	Risk	Acceptable Spill Risk	
	Benefit	Expected environmental performance enhancement	25% reduction of IT Energy Consumption
	Product Description	Environmental Specifications	Labels such as EPEAT or Energy STAR
	Work Packages	Environmental constraints	Environmental performance, labels & standards to be respected during task execution communicated to technical manager, contractor or supplier
	Quality control	Environmental Quality control during quality review	Level of waste generated (GHG, CO_2 emissions, toxic waste) Level of resources used (Water, Energy, raw material)
Controls *Before, During & After project execution* *Work Package, Stage, Project & Program Levels.*	Metrics	Based on the environmental reporting framework provided by the Global Reporting Initiative (GRI) for Sustainability. (GRI, 2006b), (GRI, 2006a)	- Resource usage: Water (Volume), Paper (Weight or number/paper per user) & IT energy (Consumption or Savings) - % Renewable Energy, % Recycled material - Raw & Toxic Material Input (Weight) - Power Efficiency
			- Emissions/Waste: GHGs (Weight), Water Discharge (Volume & Quality), Waste by type & Disposal Method (Weight) - Transport: Goods & workforce (Type of transport mean, distance, energy used)
		Relative vs. absolute (Esty & Winston, 2006, p.175)	- CO_2/KWh, CO_2/Trip vs. Total CO_2 emissions - Paper Consumption/meeting or individual vs. Total Paper Consumption
		Captured at multiple levels within project & organization. This leads to the identification of problem areas & best practices.	Measure the impact of product/service development or improvement, i.e., during the development of the product, from conception until release into production.
			Measure the environmental impact of project management activities.
		Same KPI collected throughout the value chain (including customers, suppliers & contactors).	
		Baseline measurement for current environmental performance	Environmental performance can be evaluated with tools available on the market: SAP, Prodware, TerraBytes. (All based on GRI)
		Target Performance figures based on initial metrics	
Quality	Project quality	Environmental quality of project management activities	
	Product quality	Environmental quality of delivered products	
	Acceptance Criteria	Environmental acceptance criteria of products	

continued on following page

Table 3. Continued

PRINCE 2 Components	Sections	Green Dimension	Example (When applicable)
	External to project quality components		
	ISO	Environmental Norms	ISO14000, WEEE, RoHS, BASEL
	Quality Policy	Environmental quality	- Environmental labels: EPEAT, Energy STAR, 80 Plus - Best Practices / Code of Conduct: EU Green Datacenter, Think Green, Greenpeace Ranking
	QMS	EMS	ISO 14000
	Quality Assurance	Internal or External Environmental Auditors/Experts	
Risk	Risk analysis & monitoring	Possible environmental impact & disaster, internal or external to the project	
		Follow-up of future regulations, best practices, norms & market trends	- France: "Grenelle de l'Environnement" - ISO15000

Table 4. Greening PMI (Bachour, 2010a; PMI, Inc., 2008)

PMI Knowledge Areas	Section	Green Dimension	Example
Cost Management		Environmental taxes & penalties	CO_2 taxes & penalties
		Environmental protection expenditure	GRI Protocol Set
		Environmental labeling costs	EPEAT, 80 Plus, Energy STAR
		Green Funds	Provided by banks, governmental programs, e.g., EU Environmental Commission
Communication Management	External Actors	Identification of Green External Actors	Municipalities Chambers of Commerce Ministries NGOs Watchdogs
	Internal Actors	Identification of functional managers getting profit out of the project.	A project targeting energy efficiency: - The IT department pays for the investment - The facilities department will see the payoff in the energy bills - The financial department calculates cost savings Teaming up the three managers enhances chances to get the project approved. (Gale, 2009a), (Gale, 2009b)
	Monitoring & Reporting	Monitoring & Reporting of environmental performance through consolidated published dashboards	Environmental risk, quality, impact, cost & procurement
		Extension of reporting to CSR Strategists	- Communicate environmental performance of corporate IT systems - Communicate contribution of IT systems in reducing corporate environmental impact
Procurement Management		Selection of contractors & suppliers based on a defined list of environmental criteria	- Responsible purchasing of labeled products - Prioritization of contracts offering solutions for products end of life (Reuse, recycling, maintenance support extension)
Human Resources Management		Communication or training (if needed) on corporate environmental initiatives to the project team	
		Training program for identifying the environmental impact & quality in project management disciplines	
		Innovation program for encouraging corporate environmental opportunities in IT departments	A well rooted corporate sustainability culture initiates projects in a two-way, top-down and bottom-up aligned flow

Figure 2. From strategy to projects. Adapted from OGC (2007, p.5) and Pernick and Wilder (2008)

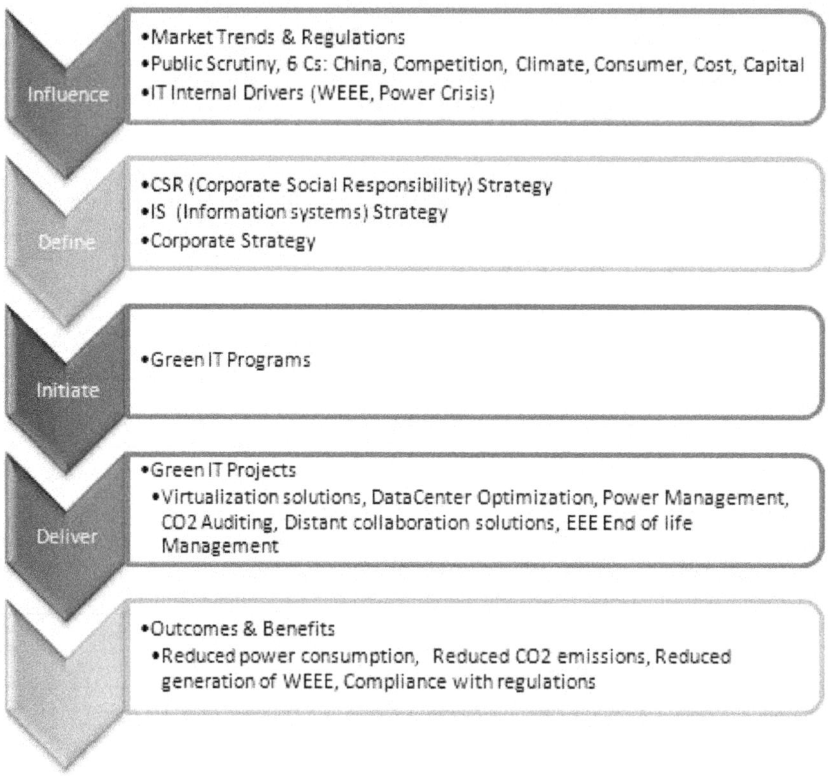

environmental strategies can be linked to Green IT projects, Figure 2 illustrates the chain, from PESTLE influences to project implementation, within organizations.

Managing programs is about coordinating the organization, directing, implementing the group of projects and transferring procedures to achieve outcomes, and realize benefits aligned with the corporate strategy. Selecting the right project and following up its contribution through the project execution as well as after delivery in the business requires both project and program management capabilities.

On one hand, project management techniques help in managing expected benefits and their threats during the whole project life cycle through business case reviews and risk control. Referring back to PRINCE2, benefits were fully integrated in the business case and controls were defined to follow up the project viability against the business case. On the other hand, program management techniques help in defining projects and managing benefits realization, after the project is delivered. MSP benefits realization management theme helps to:

- Ensure benefits and threats are well defined and identified with a clear link to strategic objectives; this can be done using Outcome Relationship (OCR) models and benefits mapping. OCR is very helpful in defining projects and linking them to the expected final outcome and corresponding benefit. Such models and map should integrate the environmental dimension during benefits identification.
- Ensure business lines commitment by designating business change managers within the program management governance. Business change managers can represent

Figure 3. Survey, green IT performance follow-up capability maturity (Bachour, 2010a)

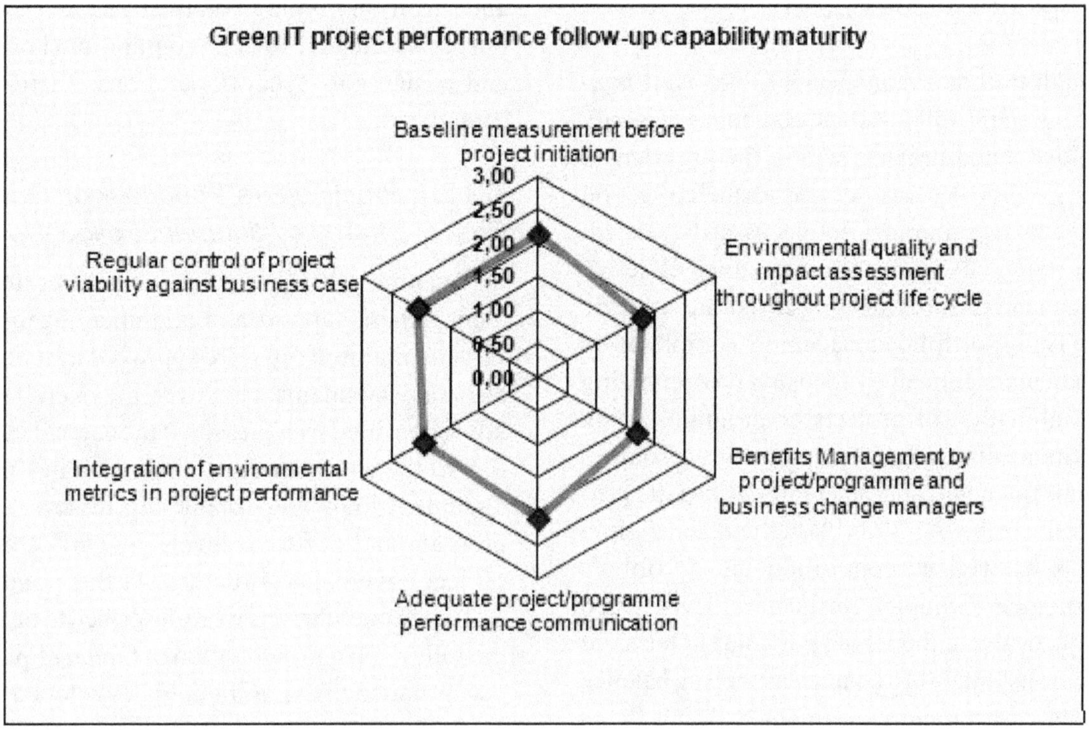

several functional entities impacted by the Green IT project/program change.

- Implement the realization benefit process: measure, track, compare, throughout the defined benefit realization review milestones.
- Ensure alignment between the program and strategic objectives through a well defined governance structure and stage gates reviews (OGC, 2007, p. 61).

4.3.1.1. Practitioners' Feedback on Green IT Project Performance Follow-Up

In Bachour (2010a) survey, most of practices used in Green IT project follow-up on the operational & tactical levels were approved by 50% of respondents. However, the overall maturity evaluation identified weaknesses in benefits realization management and environmental performance follow-up through the project life cycle.

The third level of success criteria was defined at corporate level, and involved the evaluation of Green IT projects alignment with corporate strategic objectives. At corporate level, evaluation is not restricted to Green IT projects/programs, rather to the whole portfolio of related and unrelated projects. Further, not only Green IT project contribution should be aligned but the combination of projects sharing the same resources should be also be optimized against corporate strategic objectives.

What are the success factors towards such performance? Here, all success levels reached earlier are consolidated, i.e., the right projects were managed right. How can the right combination of Green IT projects with other projects in the organization be achieved?

This capability was also identified in Bachour (2010b) based on Dinsmore and Cooke-Davies (2006, p. 4).

4.3.2. Corporate Level Green IT Project Success Factors

Although program management does well integrate alignment with strategic objectives, it is only done for related projects within the program. A portfolio of projects is defined as the collection of projects or programs not necessarily related, but grouping them results in a more effective management towards meeting corporate strategic goals. Thus, portfolio management complements program management by focusing on optimizing the combination of projects or programs using the ultimate allocation of a company's resources (human, financial, and other). It defines the processes for selecting, prioritizing and scheduling projects as well as controlling the portfolio's performance through heartbeat reviews. Such controls evaluate the whole portfolio in terms of risk and sustainability balance as well as balance between cash-in and investment cash-out flows (Dinsmore & Cooke-Davies, 2006, p. 158). In the context of Green IT projects, the environmental perspective should be integrated in all portfolio management disciplines: selection, prioritization, and combination of projects while making sure the CSR objectives are considered in strategic alignment evaluation and follow-up.

In order to maintain alignment with the corporate strategy on the tactical and operational levels, the project management community can use strategic alignment tools such as the SBSC developed by Moller and Schaltegger (2005). SBSC can be adapted to different levels of the project and the business, from senior management to technical teams receiving work packages. An efficient project & Green IT governance should allow the consolidation and communication of targeted reports during stage gate reviews, run in order to check the project business case viability, and heartbeat reviews, run in order to check the portfolio of projects against corporate risk, business & environmental sustainability. Such multi-perspective evaluation leads to a three-dimensional sustainability (People, Planet, and Profit) highlighted on the three levels of the business change (Project, Program, and Portfolio) and business as usual (Operations, Tactics, and Strategies).

4.3.2.1. Practitioners' Feedback on Green IT Project Strategic Alignment Follow-Up

In Bachour (2010a) survey, practices for top management support and alignment against environmental strategy were approved by more than 50% of respondents. However, the overall evaluation identified weaknesses in the use and integration of the environmental dimension in strategic alignment tools and conducting reviews on the program and portfolio levels, as part of a solid project governance structure. At this point, several recommendations were described to optimize the follow-up and evaluation of Green IT projects performance. Next, a model is developed based on the 3Ps capability identified in Green IT organizational project success factors.

5. MODEL FOR GREEN IT PROJECT MANAGEMENT

Bachour (2010b) built a model integrating the following life cycles:

1. Project life cycle: SDLC (Systems Development)
2. Quality Project life cycles: DMAIC or DMADV
3. Project Management life cycle based on PRINCE2 processes
4. Product/Service life cycle

From a perspective spanning beyond project closure, Bachour (2010b) was able to show a macro view of how business change activities disrupt and interact with business as usual on the operational, tactical, and strategic levels.

Based on this model adapted to a Green IT 1.0 project (Virtualization solution) in Bachour (2010b), the below example is built for a Green IT 2.0 project.

5.1. Feasibility Study Project or Stage

This project or phase aims at selecting, prioritizing and initiating Green IT projects. Before selecting the solution to be implemented, the expected benefit should be defined and the current situation should be measured and evaluated.

1. *Define.* Whether Green IT projects are initiated at the operational level or defined within a program, this step focuses on the problem area to be improved. What is the business case? Consider a company seeking to reduce the total CO_2 emissions due to traveling and commuting. At this level, project management processes adapted to a green perspective in Tables 3 and 4 can be used in order to build the project team, identify stakeholders (project board, customers, sponsors, end-users, suppliers, internal functional managers, green responsible) and agree on the project scope.
2. *Measure.* As seen in list of recommendations earlier. The success of a Green IT project relies on performance indicators associated with metrics, defined at the three levels of an organization (Strategic, Tactical and operational). These can be qualitative or quantitative. In this phase, metrics are defined to establish a baseline:
   - Total distance travelled/year for business trips.

Figure 4. Survey, green IT project strategic alignment follow-up capability maturity (Bachour, 2010a)

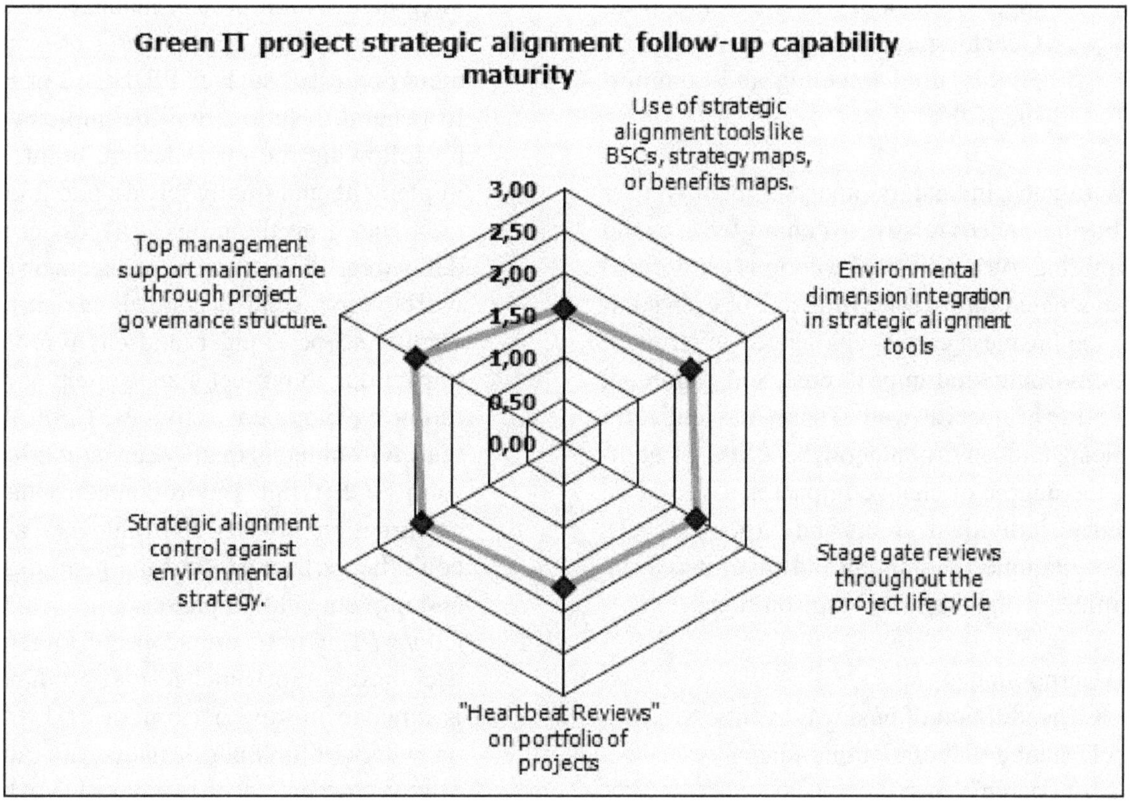

- ◦ Total distance travelled/year for commuting internally.
- ◦ Means of transportation (Subway, Bus, Train, Plane, Car, Bike)
- ◦ Number of car passengers

Although metrics are focused on the problem area, the project success criteria and associated indicators defined in Table 3 should be also defined in order to allow progress reporting and reliable decision making.

3. *Analyze.* In this phase, analysts work on making numbers speak. Trends, behaviors, quick wins, and improvement areas are identified. This can be done with the aid of quality management tools such as cause/effect and Pareto analyses:
    - ◦ Total CO_2 emissions relative to transportation type
    - ◦ Peak travelling during the year
    - ◦ Aim of the trip (Face to face meeting, Workshop, Business Proposal, Coaching, etc.)
    - ◦ Highly used traveling and commuting paths

With such indicators, analysts identify user and business needs relative to committed trips and commuting. As a next step, solutions are defined and evaluated for their feasibility. For each solution, the business case is challenged in terms of risk, environmental impact, cost, and alignment with other projects as well as short and long term financial and environmental benefits. In addition, the degree of change impact is assessed on technical, individual, group, and corporate levels.

For example, commuting and travelling can be optimized using the following solutions:

- Carpooling
- Consolidation of business events
- Distant collaboration platform
- Videoconference
- Web conference
- Telepresence
- Instant Messaging
- Remote Working
- Training on distant management

Internal and external benchmarking can aid companies identifying innovative opportunities and on the contrary avoid unsuccessful solutions. When the solutions are selected and prioritized, other projects or subprojects can be started and project or team managers assigned.

5.2. SOLUTION DEVELOPMENT PROJECT OR STAGE

4. *Improve/Design.* On this level, the solution project life cycle is initiated. Standard project life cycles found in the information systems field can be used such as software or Systems Development Life Cycle (SDLC), integrated with a project management life and orchestrated by standard project management processes such as PRINCE2 or PMI. In general, solutions will be implemented by following the plan, design, build, test, implement, and finally release into production. Based on the project success criteria defined earlier, the project management team will be responsible for controlling risk, time, quality, scope, budget, and environmental impact due to project management and solution development activities. Controlling the environmental quality can be conducted as in HQE (High Environmental Quality) construction projects, environmental audits could be performed after the plan, design, and implementation phases.
5. *Control/Verify.* In the scope of Green IT, each new setup launched into production, is aimed to improve performance (DMAIC) or to deliver high performance and quality relative to new specifications (DMADV).

Table 5. Green IT sponsor level success/performance evaluation framework. Adapted from Dinsmore and Cooke-Davies (2006, p. 4)

Success Level	Responsible Parties	Success Criteria	Success Factors
Right Green IT Projects are done	Project governance, executive sponsor, "client," owner, end-user, SDD responsible, Other functional managers	• Benefits are realized e.g.: o Reduced IT energy bills o Reduced WEEE generation o Reduced CO_2 emissions of IT systems or operations o Reduced TCO of IT systems o Reduced resource wasting o Improved quality of service o Reduced time to market IT services • Stakeholders satisfied	Program management capability • Clear and attainable goals • Stakeholder commitment/Top management Support • Benefits realization management • Project Strategy • Financial & environmental value stream analysis • Adequate program, benefit measurement metrics • Baseline measurement • Business case viability follow-up

Table 6. Green IT portfolio level success/performance evaluation framework. Adapted from Dinsmore and Cooke-Davies (2006, p. 4).

Success Level	Responsible Parties	Success Criteria	Success Factors
Green IT Projects are combined & selected Right	Senior management, CSR strategists	• Corporate Strategy implemented e.g. o Total operational costs reduced o Product/Service differentiated o Innovation culture fostered o Corporate environmental impact reduced o Brand image improved o Employee satisfaction improved o Customer satisfaction improved o Organization agility improved o Competitive edge sustained • Productivity improved • Profitability increased	Portfolio management capability • Continual improvement • Comprehensive and reliable metrics • Baseline measurement • Top-down Vision/strategy communication • Effective decision-making based on reliable bottom-up performance reporting • Defined CSR strategy • Strategic alignment with environmental & corporate strategy follow-up

During the project, regular control or verification is implemented on the project and product (system) performance. Progress reports are communicated to corresponding stakeholders for reliable decision-making. These actions should be performed using project management processes, e.g., direct project, control stage, and manage product delivery, etc. until release into production. As mentioned earlier, the advantage of Six Sigma is the long-term view for quality, where monitoring should continue as part of production, i.e. even after the project ends. This ensures the desired benefits are realized and the new setup sustains the initial targeted performance. General program management standards such as "Managing Successful Programs" provided by APM Group, allow such evaluation through benefits realization management. However, the Six Sigma frame goes beyond the project and program. It establishes continuity between the project scope of a business change and business as usual, until the decommissioning phase.

With the model in Figure 5 and the framework in Table 2, Table 5, and Table 6, the following recommendations can be summarized for optimizing Green IT projects within organizations:

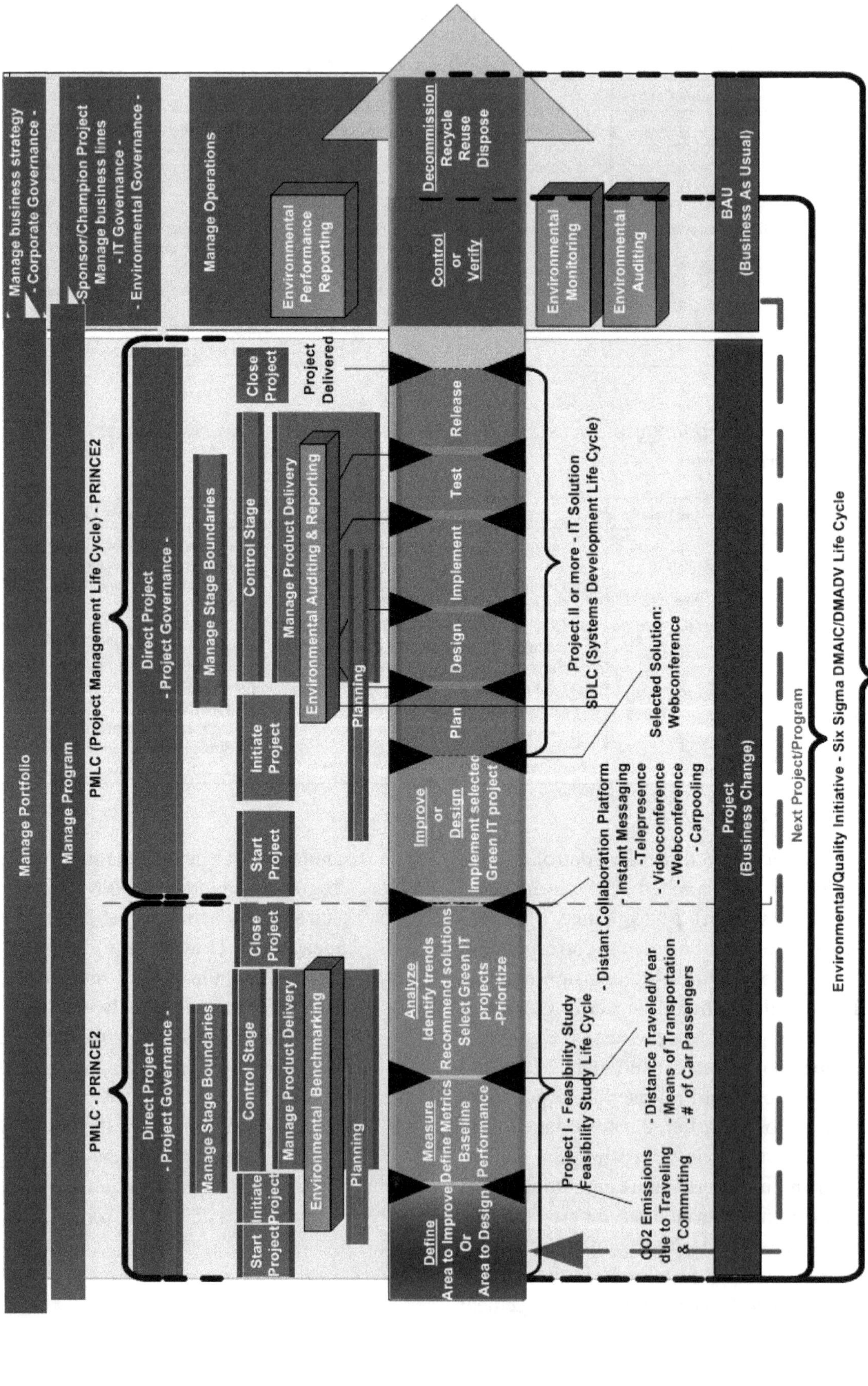

Figure 5. Model for green IT project management: integration of project, quality, systems, and operations life cycle (Bachour, 2010b)

5.3. Green IT Project Recommendations

Sustainable value creation within organizations relies on OPM capabilities and the alignment among the 3Ps levels. In order to optimize Green IT projects within organizations, the following main points should be considered on each level of the 3Ps in addition to standard OPM success factors:

Project

- Follow the Lean Sigma Project life cycle as a leading roadmap (DMADV, DMAIC).
- Start with a feasibility study where the most appropriate solution is selected before implementation.
- Define metrics for all identified environmental areas (cost, risk, quality, scope), and impact of both the project management and the product development processes.
- Baseline current operational environmental performance.
- Ensure project environmental quality is aligned with organizational standards for HSE and Ethics.
- Where possible, include environmental auditing throughout the project execution life cycle for the plan, design, and final deliverable.
- Follow up project performance against the business case.

Program

- Identify internal and external stakeholders who can support or benefit from all aspects (financial and environmental) of the project.
- Engage top management and maintain their commitment through continuous reporting.
- Engage business change managers for benefits realization management.
- Define projects based on strategic alignment tools and benefits mapping.
- Define metrics for environmental benefits measurement.
- Establish a baseline for tactical environmental performance.
- Follow up business case viability against strategic objectives.
- Draw continuity between the project and the business worlds through environmental policies and best practices.

Portfolio

- Select, prioritize, and allocate budget based on alignment with environmental and corporate strategies and organizational resources availability.
- Define comprehensive and strategic metrics.
- Baseline current corporate environmental performance.
- Communicate the CSR or environmental vision and strategy.
- Perform decision-making based on bottom-up performance reporting through a well defined project governance.
- Perform strategic alignment follow-up using strategic alignment tools such as SBSC.
- Link Green IT project governance with IT and Environmental governances.

6. SUGGESTIONS FOR FURTHER DEVELOPMENT

This chapter focused the detailed analysis on adapting project management standards for Green IT projects and linking the 3Ps levels. In order to be coherent on all levels, detailed analysis should be also conducted to examine the feasibility of adapting program and portfolio management disciplines for Green IT governance. At this level of analysis, sustainability should not be limited to the environmental perspective but should also consider social aspects within and outside the organization.

Additional development could be done from a more practical perspective, by building a case study based on the suggested research frameworks, and providing practical enhancements. In parallel to the case analysis, a list of environmental requirements for project management and information systems development can be defined with corresponding performance indicators and metrics. Such list should cover the system's "Cradle to Cradle" life cycle and could be used to provide ranking criteria in the field of Green IT project management.

From the perspective of IT Service management, the green dimension can be integrated in environmental performance monitoring at the operational and business levels. Further research could be done on greening standards such as ITILv3 or COBIT. This would allow a perfect alignment between the transition phase and the run phase of every IT product or service.

7. CONCLUSION

Green IT projects have proven to be quite beneficial for organizations by creating financial and environmental value throughout the value chain. However, environmental management is quite tricky and one can easily fall in the trap of green washing, where the will to be green is neither supported by a reliable evaluation system nor implemented through tangible initiatives. Facing challenging environmental thinking, market competition, and regulatory pressures, IT project managers and decision makers appear to lack the maturity in defining, evaluating, and optimizing a Green IT project or program coherent with other concurrent projects and aligned with the environmental and corporate strategy of their organization.

Despite the rising need on the market, academic and professional literature does not show any thorough analysis for integrating green thinking and organizational project management for Green IT initiatives. Current articles referring to Green IT within organizations are mostly professional and are limited to a list of tips for some areas of project management.

This chapter established a framework at optimizing Green IT projects for sustainable value creation through a framework for Green IT organizational project management and success evaluation, and a macro model for Green IT project management. In this respect, the structure was built around 3 questions:

1. Do organizations limit Green IT project success and performance evaluation to the project level and environmental quality of deliverables?
2. How should organizations follow up and evaluate the environmental performance of Green IT projects, and assess their contribution to sustainable value creation?
3. How can organizations optimize Green IT projects for sustainable value creation?

The first question was answered based on a Master's thesis survey (Bachour, 2010a) distributed to professionals, project managers and decision makers concerned with Green IT and project management. Project Management criteria shown in Figure 1 summarize project management success criteria to be considered during evaluation. It can guide the project management community and decision makers in looking at other dimensions during project follow-up and assessment processes. The survey showed that most respondents considered environmental impact for both the product development and the project management processes before environmental quality. Furthermore, organizations did not limit the success evaluation of Green IT projects to the project level Only 6% did not assess benefits realization and only 13% did not assess strategic alignment of their Green IT projects. Project performance evaluation also extended to upper levels where only 18% of respondents limited the definition of metrics to the operational level.

The second question, regarding success evaluation methodology, was addressed by analyzing the types of success criteria and control practices that are best suited for Green IT projects, at each level of the organization: operational, tactical and strategic. These were selected based on OPM (Organizational Project Management) disciplines, project governance structure for strategic alignment follow-up, and sustainability management. Table 2, Table 5, and Table 6 illustrate a framework for Green IT project success evaluation including a list of success criteria with a summary of corresponding best practices for performance follow-up and sustainable value creation. These mainly include establishing a current baseline, defining metrics with performance indicators, benchmarking and communicating consolidated reports to concerned stakeholders. Environmental indicators were integrated into the framework based on international standards for environmental sustainability reporting such as GRI (Global Reporting Initiative). All these steps should be performed at each of the 3Ps (Project, Program, Portfolio) levels, allowing reliable decision-making throughout the project governance structure. Feedback from practitioners showed a misalignment between the will to evaluate and follow-up performance of Green IT projects and the capability to do it. Only 10% defined metrics on the three levels, and showed weak maturity in benefits realization management and environmental performance follow-up through the project life cycle. Weaknesses were also identified in the use of strategic alignment tools integrating the environmental dimension and conducting reviews on the program and portfolio levels, as part of a solid project governance structure.

The third question, aiming optimization of Green IT projects within organizations, was answered through the qualitative analysis of organizational project management success factors. This detailed analysis in which project management standards were "greened" was summarized and illustrated in a macro model (Figure 5), integrating the project and the business dimensions at different levels of the organization. Within the model, the project management life cycle, the software development life cycle, and the quality project life cycle are matched to create a holistic roadmap for Green IT projects optimization.

Managing Green IT projects goes beyond standard project management methods found on the market. It is more specific in the methodology used, involves a challenging multi-perspective thinking, and seeks alignment with the environmental and corporate strategy. Green IT Project Management is the convergence point of several project, quality, and general management standards, adapted for a technology project with a sustainability perspective.

REFERENCES

1E2008). *Case study dell: Focus on energy efficiency.* Retrieved from http://www.1e.com/download/Whitepapers/Dell_Case%20Study_US.pdf.

Accenture. (2008). *Green maturity assessment.* Retrieved from http://nstore.accenture.com/global/flash/gme/greenmaturity.htm.

ADEME. (2009). *Bâtiments et démarches HQE.* Retrieved from http://www.ecoresponsabilite.environnement.gouv.fr/IMG/BROCHURE_HQE.pdf.

ADEME. (2011). *ADEME, PDE, plan de déplacements d'entreprise.* Retrieved from http://www2.ademe.fr/servlet/KBaseShow?sort=-1&cid=96&m=3&catid=14263.

Alberta Infrastructure. (2007). *LEED project delivery process manual.* Retrieved from http://www.infrastructure.alberta.ca/Content/docType486/Production/LEED_PD_Manual.pdf.

Alliance-TIC. (2009). *L'eco-utilisateur, guide des bonnes pratiques de l'utilisateur informatique et telecom.* Retrieved from http://www.alliance-tics.org/dossiers/environnement/Guide%20éco-utilisateur/GUIDE%20Eco-24p%20BD.pdf.

Aravind, D., & Christmann, P. (2008). Institutional and resource-base determinants of substantive implementation of ISO 14001. In *Proceedings of the Academy of Management.* Anaheim, CA: Academy of Management.

Bachour, N. (2008). *Green project management for green IT.* Unpublished.

Bachour, N. (2010a). Green IT project management: *How to optimize green IT projects for sustainable value creation within organizations?* Unpublished MBA Thesis. Paris, France: SKEMA Business School.

Bachour, N. (2010b). Optimizing the value of green IT projects within organizations. In *Proceedings of the Green Technologies Conference,* (pp. 1-10). Grapevine, TX: IEEE Press.

Barney, J. (1991). Company resources and sustained competitive advantage. *Journal of Management, 17*(1), 99–120. doi:10.1177/014920639101700108

Berrone, P., Gelabert, L., Fosfuri, A., & Gomez-Mejia, L. (2008). Can institutional forces create competitive advantage? An empirical investigation of environmental innovation. In *Proceedings of the Academy of Management.* Anaheim, CA: Academy of Management.

Blue Angel. (2009). *Computer monitors standards.* Retrieved from http://www.blauer-engel.de/en/products_brands/search_products/produkttyp.php?id=486.

Boudreau, M., Chen, A., & Huber, M. (2008). *Green IS: Building sustainable business practices.* Retrieved from http://globaltext.terry.uga.edu/userfiles/pdf/Green.pdf.

Brodkin, J. (2008). Economy driving green IT initiatives. *New World (New Orleans, La.), 25*(49), 16.

Brown, D., Dillard, J., & Marshall, R. (2005). Strategically informed, environmentally conscious information requirements for accounting information systems. *Journal of Information Systems, 19*(2), 79–103. doi:10.2308/jis.2005.19.2.79

Brueil, H., Burette, D., Flury-Hérard, B., Cueugniet, J., & Vignolles, D. (2008). *Rapport TIC et développement durable.* Conseil général de l'environnement et du développement durable, Conseil général des Technologies de l'Information.

Coburn, M. (2008, August 15). GE CEO declares: "Green is green". *The Dartmouth Staff.*

Dimaggio, P., & Powell, W. (1983). The iron cage revisited: Institutional isomorphism and collective rationality in organisational fields. *American Sociological Review, 48*(2), 147–160. doi:10.2307/2095101

Dinsmore, P., & Cooke-Davies, T. (2006). *The right projects done right.* San Francisco, CA: Jossey-Bass.

Elliot, S., & Binney, D. (2009). *Environmentally sustainable ICT: Developing corporate capabilities & an industry-relevant IS research agenda.* Retrieved from http://www.pacis-net.org/file/2008/PACIS2008_Camera-Ready_Paper_209.pdf.

Energy Star. (2008). *Programme requirement for computers, version 5.0.* Retrieved from http://www.energystar.gov/index.cfm?c=about.ab_history.

Environmental Protection Agency. (2011). *Lean manufacturing & the environment.* Retrieved from http://www.epa.gov/lean/thinking/sixsigma.htm.

EPEAT. (2006). *EPEAT background.* Retrieved from http://www.epeat.net/AboutEPEAT.aspx.

Esty, D., & Winston, A. (2006). *Green to gold.* New Haven, CT: Yale.

EU. (1994). *Directive 94/62/EC on P&PW.* Retrieved from http://www.gidos.net/_library/EC.94.62.EC.on.Packaging.and.Packaging.Waste.pdf.

EU. (2002a). *Directive 2002/96/EC on WEEE.* Retrieved from http://ec.europa.eu/environment/waste/weee/index_en.htm.

EU. (2002b). *Directive 2002/95/EC on ROHS.* Retrieved from http://www.rohs.eu/english/legislation/rohs-legislation.html.

EU. (2006). *Regulation EC 1907/2006 on REACH.* Retrieved from http://ec.europa.eu/environment/chemicals/reach/reach_intro.htm.

EU. (2009). *Code of conduct for data centres version 2*. Retrieved from http://re.jrc.ec.europa.eu/energyefficiency/html/standby_initiative_data_centers.htm.

EU. (2011). *Energy star programme*. Retrieved from http://www.eu-energystar.org/en/index.htm.

EU Information Society. (2011). *ICT for sustainable growth.* Retrieved from http://ec.europa.eu/information_society/activities/sustainable_growth/index_en.htm.

Forrester. (2009). *Market overview: The state of enterprise green IT adoption*. Forrester.

Gable, G. (2011, April 23). *The story of e-waste: What happens to tech once it's trash.* Retrieved from http://www.gizmodo.com.au/2011/04/the-story-of-e-waste-what-happens-to-tech-once-its-trash/.

Gale, S. (2009a). Green out. *PM Network, 23*(11).

Gale, S. (2009b). The real deal. *PM Network, 23*(12).

Green Grid. (2009). *The green grid library & tools*. Retrieved from http://www.thegreengrid.org/library-and-tools.aspx.

GRI. (2006a). *Sustainability reporting guidelines.* Retrieved from http://www.globalreporting.org/ReportingFramework/ReportingFramework-Downloads/.

GRI. (2006b). *Indicator protocols set for the environment.* Retrieved from http://www.globalreporting.org/ReportingFramework/Reporting-FrameworkDownloads/.

Hart, S. (1995). A natural-resource-based view of the company. *Academy of Management Review, 20*(4), 986–1014.

Huang, A. (2008). *A sustainable systems development life cycle.* Retrieved from http://www.pacis-net.org/file/2008/PACIS2008_Camera-Ready_Paper_081.pdf.

IBM. (2007). *How to optimise your carbon management through green sigma.* Retrieved from http://www-05.ibm.com/ie/emerging_business_centre/pdf/Green_Sigma_final_Strategy.pdf.

IBM. (2008, August 18). *A measured approach to going green: IBM "green sigma (TM)" consulting offering to help clients reduce energy and water usage.* IBM.

IBM. (2009a). *Green IT assessment and ROI calculator*. Retrieved from http://www-304.ibm.com/businesscenter/cpe/html0/181575.html.

IBM. (2011). *Energy, the environment & IBM.* Retrieved from http://www.ibm.com/ibm/green/index5.shtml.

ISO. (2002). *ISO 14000/14001 environmental management standards.* Retrieved from http://www.iso14000-iso14001-environmental-management.com/index.htm.

Jenkin, T., & McShane, L. (2009). *Green information technologies and systems in organisations: The state of practice.* Retrieved from http://www.albany.edu/business/csr_Jenkin_SpecialIssue_BlindPaper_GreenITS.pdf.

Jenkin, T., Webster, J., & McShane, L. (2011). An agenda for green information technology and system research. *Information and Organization*, *21*(1), 17–40. doi:10.1016/j.infoandorg.2010.09.003

Meyer, J., & Rowan, B. (1977). Institutional organisations: Formal structure as myth and ceremony. *American Journal of Sociology*, *83*(2), 340–363. doi:10.1086/226550

Microsoft. (2011). *Partners Show case*. Retrieved from http://www.microsoft.com/Environment/products-and-solutions/partner-showcase.aspx.

Moller, A., & Schaltegger, S. (2005). The sustainability balanced scorecard as a rramework for eco-efficiency analysis. *Journal of Industrial Ecology*. Retrieved from http://nbis.org/nbisresources/metrics/sustainability_balanced_scorecard.pdf.

OGC. (2005). *Managing successful projects with PRINCE2*. TSO.

OGC. (2007). *Managing successful programs*. TSO.

PMI, Inc. (2008). *A guide to the project management body of knowledge (PMBOK)* (4th ed.). Newtown Square, PA: PMI.

Rikhardsson, P. (1998). *Information systems for corporate environmental management accounting and performance measurement*. Retrieved from http://www.highbeam.com/doc/1G1-55308837.html.

Rowsell-Jones, A. (2008). *Going green: The CIO's role in enterprisewide environmental sustainability*. Gartner.

Scott, W. (1995). *Institutions and organisations*. Thousand Oaks, CA: Sage.

TCO. (2006). *Displays 5.0*. Retrieved from http://www.tcodevelopment.com/.

TenStep, Inc. (2010). *Green project management articles*. Retrieved from http://www.green-pm.com/greenarticles.html.

Terrachoice. (2009). *A marketer's guide to avoid the sins of green washing*. Retrieved from http://sinsofgreenwashing.org/solutions/marketers-guide/.

US Green Building Council. (2010). *What is LEED?* Retrieved from http://www.usgbc.org/DisplayPage.aspx?CMSPageID=1988.

Washburn, D. (2009). *Q&A: The economics of green IT, what are the financial benefits of green IT*. Forrester.

Wernerfelt, B. (1984). A resource based view of the company. *Strategic Management Journal*, *5*, 171–180. doi:10.1002/smj.4250050207

ADDITIONAL READING

1E2008). *Case study dell: Focus on energy efficiency.* Retrieved from http://www.1e.com/download/Whitepapers/Dell_Case%20Study_US.pdf.

Accenture. (2008). *Green maturity assessment*. Retrieved from http://nstore.accenture.com/global/flash/gme/greenmaturity.htm.

ADEME. (2009). *Bâtiments et démarches HQE*. Retrieved from http://www.ecoresponsabilite.environnement.gouv.fr/IMG/BROCHURE_HQE.pdf.

ADEME. (2011). *ADEME, PDE, plan de déplacements d'entreprise*. Retrieved from http://www2.ademe.fr/servlet/KBaseShow?sort=-1&cid=96&m=3&catid=14263.

Alberta Infrastructure. (2007). *LEED project delivery process manual.* Retrieved from http://www.infrastructure.alberta.ca/Content/docType486/Production/LEED_PD_Manual.pdf.

Alliance-TIC. (2009). *L'eco-utilisateur, guide des bonnes pratiques de l'utilisateur informatique et telecom.* Retrieved from http://www.alliance-tics.org/dossiers/environnement/Guide%20éco-utilisateur/GUIDE%20Eco-24p%20BD.pdf.

Aravind, D., & Christmann, P. (2008). Institutional and resource-base determinants of substantive implementation of ISO 14001. In *Proceedings of the Academy of Management.* Anaheim, CA: Academy of Management.

Bachour, N. (2008). *Green project management for green IT.* Unpublished.

Bachour, N. (2010a). Green IT project management: *How to optimize green IT projects for sustainable value creation within organizations?* Unpublished MBA Thesis. SKEMA Business School.

Bachour, N. (2010b). Optimizing the value of green IT projects within organizations. In *Proceedings of the Green Technologies Conference,* (pp. 1-10). Grapevine, TX: IEEE Press.

Barney, J. (1991). Company resources and sustained competitive advantage. *Journal of Management, 17*(1), 99–120. doi:10.1177/014920639101700108

Berrone, P., Gelabert, L., Fosfuri, A., & Gomez-Mejia, L. (2008). Can institutional forces create competitive advantage? An empirical investigation of environmental innovation. In *Proceedings of the Academy of Management.* Anaheim, CA: Academy of Management.

Blue Angel. (2009). *Computer monitors standards.* Retrieved from http://www.blauer-engel.de/en/products_brands/search_products/produkttyp.php?id=486.

Boudreau, M., Chen, A., & Huber, M. (2008). *Green IS: Building sustainable business practices.* Retrieved from http://globaltext.terry.uga.edu/userfiles/pdf/Green.pdf.

Brodkin, J. (2008). Economy driving green IT initiatives. *New World (New Orleans, La.), 25*(49), 16.

Brown, D., Dillard, J., & Marshall, R. (2005). Strategically informed, environmentally conscious information requirements for accounting information systems. *Journal of Information Systems, 19*(2), 79–103. doi:10.2308/jis.2005.19.2.79

Brueil, H., Burette, D., Flury-Hérard, B., Cueugniet, J., & Vignolles, D. (2008). *Rapport TIC et développement durable.* Conseil général de l'environnement et du développement durable, Conseil général des Technologies de l'Information.

Coburn, M. (2008, August 15). GE CEO declares: "Green is green". *The Dartmouth Staff.*

Dimaggio, P., & Powell, W. (1983). The iron cage revisited: Institutional isomorphism and collective rationality in organisational fields. *American Sociological Review, 48*(2), 147–160. doi:10.2307/2095101

Dinsmore, P., & Cooke-Davies, T. (2006). *The right projects done right.* San Francisco, CA: Jossey-Bass.

Elliot, S., & Binney, D. (2009). *Environmentally sustainable ICT: Developing corporate capabilities & an industry-relevant IS research agenda.* Retrieved from http://www.pacis-net.org/file/2008/PACIS2008_Camera-Ready_Paper_209.pdf.

Energy Star. (2008). *Programme requirement for computers, version 5.0.* Retrieved from http://www.energystar.gov/index.cfm?c=about.ab_history.

Environmental Protection Agency. (2011). *Lean manufacturing & the environment.* Retrieved from http://www.epa.gov/lean/thinking/sixsigma.htm.

EPEAT. (2006). *EPEAT background.* Retrieved from http://www.epeat.net/AboutEPEAT.aspx.

Esty, D., & Winston, A. (2006). *Green to gold.* New Haven, CT: Yale.

EU. (1994). *Directive 94/62/EC on P&PW.* Retrieved from http://www.gidos.net/_library/EC.94.62.EC.on.Packaging.and.Packaging.Waste.pdf.

EU. (2002a). *Directive 2002/96/EC on WEEE.* Retrieved from http://ec.europa.eu/environment/waste/weee/index_en.htm.

EU. (2002b). *Directive 2002/95/EC on ROHS.* Retrieved from http://www.rohs.eu/english/legislation/rohs-legislation.html.

EU. (2006). *Regulation EC 1907/2006 on REACH.* Retrieved from http://ec.europa.eu/environment/chemicals/reach/reach_intro.htm.

EU. (2009, December 31). *Code of conduct for data centres version 2.* Retrieved from http://re.jrc.ec.europa.eu/energyefficiency/html/standby_initiative_data_centers.htm.

EU. (2011). *Energy star programme.* Retrieved from http://www.eu-energystar.org/en/index.htm.

EU Information Society. (2011). *ICT for sustainable growth.* Retrieved from http://ec.europa.eu/information_society/activities/sustainable_growth/index_en.htm.

Forrester. (2009). *Market overview: The state of enterprise green IT adoption.* Forrester.

Gable, G. (2011, April 23). *The story of e-waste: What happens to tech once it's trash.* Retrieved from http://www.gizmodo.com.au/2011/04/the-story-of-e-waste-what-happens-to-tech-once-its-trash/.

Gale, S. (2009a). Green out. *PM Network, 23*(11).

Gale, S. (2009b). The real deal. *PM Network, 23*(12).

Green Grid. (2009). *The green grid library & tools.* Retrieved from http://www.thegreengrid.org/library-and-tools.aspx.

GRI. (2006a). *Sustainability reporting guidelines.* Retrieved from http://www.globalreporting.org/ReportingFramework/ReportingFramework-Downloads/.

GRI. (2006b). *Indicator protocols set for the environment.* Retrieved from http://www.globalreporting.org/ReportingFramework/ReportingFrameworkDownloads/.

Hart, S. (1995). A natural-resource-based view of the company. *Academy of Management Review, 20*(4), 986–1014.

Huang, A. (2008). *A sustainable systems development life cycle.* Retrieved from http://www.pacis-net.org/file/2008/PACIS2008_Camera-Ready_Paper_081.pdf.

IBM. (2007). *How to optimise your carbon management through green sigma.* Retrieved from http://www-05.ibm.com/ie/emerging_business_centre/pdf/Green_Sigma_final_Strategy.pdf.

IBM. (2008, August 18). *A measured approach to going green: IBM "green sigma (TM)" consulting offering to help clients reduce energy and water usage.* IBM.

IBM. (2009a). *Green IT assessment and ROI calculator.* Retrieved from http://www-304.ibm.com/businesscenter/cpe/html0/181575.html.

IBM. (2011). *Energy, the environment & IBM.* Retrieved from http://www.ibm.com/ibm/green/index5.shtml.

ISO. (2002). *ISO 14000/14001 environmental management standards.* Retrieved from http://www.iso14000-iso14001-environmental-management.com/index.htm.

Jenkin, T., & McShane, L. (2009). *Green information technologies and systems in organisations: The state of practice.* Retrieved from http://www.albany.edu/business/csr_Jenkin_SpecialIssue_BlindPaper_GreenITS.pdf.

Jenkin, T., Webster, J., & McShane, L. (2009). An agenda for green information technology and system research. *Information and Organization, 21*(1), 17–40. doi:10.1016/j.infoandorg.2010.09.003

Meyer, J., & Rowan, B. (1977). Institutional organisations: Formal structure as myth and ceremony. *American Journal of Sociology, 83*(2), 340–363. doi:10.1086/226550

Microsoft. (2011). *Partners show case.* Retrieved from http://www.microsoft.com/Environment/products-and-solutions/partner-showcase.aspx.

Moller, A., & Schaltegger, S. (2005). The sustainability balanced scorecard as a rramework for eco-efficiency analysis. *Journal of Industrial Ecology.* Retrieved from http://nbis.org/nbisresources/metrics/sustainability_balanced_scorecard.pdf.

OGC. (2005). *Managing successful projects with PRINCE2.* TSO.

OGC. (2007). *Managing successful programs.* TSO.

PMI, Inc. (2008). *A guide to the project management body of knowledge (PMBOK)* (4th ed.). Newtown Square, PA: PMI.

Rikhardsson, P. (1998). *Information systems for corporate environmental management accounting and performance measurement.* Retrieved from http://www.highbeam.com/doc/1G1-55308837.html.

Rowsell-Jones, A. (2008). *Going green: The CIO's role in enterprisewide environmental sustainability.* Gartner.

Scott, W. (1995). *Institutions and organisations.* Thousand Oaks, CA: Sage.

TCO. (2006). *Displays 5.0.* Retrieved from http://www.tcodevelopment.com/.

TenStep, Inc. (2010). *Green project management articles.* Retrieved from http://www.green-pm.com/greenarticles.html.

Terrachoice. (2009). *A marketer's guide to avoid the sins of green washing.* Retrieved from http://sinsofgreenwashing.org/solutions/marketers-guide/.

US Green Building Council. (2010). *What is LEED?* Retrieved from http://www.usgbc.org/DisplayPage.aspx?CMSPageID=1988.

Washburn, D. (2009, February 10). *Q&A: The economics of green IT, what are the financial benefits of green IT.* Forrester.

Wernerfelt, B. (1984). A resource based view of the company. *Strategic Management Journal, 5*, 171–180. doi:10.1002/smj.4250050207

KEY TERMS AND DEFINITIONS

Corporate Social Responsibility: The responsibility of enterprises for their impacts on society. These include social, environmental, ethical, human rights and consumer concerns with respect to business operations and strategies (EU, 2011).

Governance (Corporate, IT, Sustainability, Project): The set of best practices, structure, and processes that allow transparency and control for optimized strategic decision making within organizations.

Green IT: The set of Information Technology solutions aiming to reduce the environmental impact of IT itself (know as Green for IT), and the environmental impact of organizational and business operations (Know as IT for Green).

Green IT Project Management: The convergence point of organizational project management, quality and general management standards, adapted for a technology project with a sustainability perspective.

Green Project Management: The integration of environmental concepts to general project management processes.

Organizational Project Management: The execution of an organization's strategies through projects. This involves the alignment of three capabilities: project management, program management and portfolio management, with defined strategic goals.

Sustainability Management: The management of impacts due to human or corporate activities on vital capitals: Social, Environmental & Economic, what is also known as triple bottom line.

Strategic Alignment: Aligning operations and projects within an organization across functions and resources, with corporate strategic goals and the broader context of the market.

Chapter 8
Towards a Structured Cloud ROI:
The University of Southampton Cost-Saving and User Satisfaction Case Studies

Victor Chang
University of Southampton, UK & University of Greenwich, UK

Gary Wills
University of Southampton, UK

Robert John Walters
University of Southampton, UK

Wendy Currie
University of Greenwich, UK

ABSTRACT

Organisational Sustainability Modelling (OSM) is a new way to measure Cloud business performance quantitatively and accurately, and is a key area offered by Cloud Computing Business Framework (CCBF). OSM combines statistical computation and 3D Visualisation to present the Return on Investment (ROI) arising from the adoption of Cloud Computing by organisations, and makes use of a highly structured and organised process to review and evaluate Cloud business performance. The School of Electronics and Computer Science (ECS), University of Southampton, focusing on cost-savings, is the case study used to illustrate. In addition, i-Solutions and Corporate Planning of the University of Southampton, focusing on user confidence level and service improvement, are another two case studies to support. Data measurements have been taken in the past three years and quantitative analysis has been carefully checked and calculated by OSM to measure ROI. The University of Southampton has achieved cost saving and user confidence with service improvement offered by Cloud adoption and services, which have been deployed by several universities in the adoption of CCBF.

DOI: 10.4018/978-1-4666-1839-8.ch008

1. INTRODUCTION

Cloud Computing (CC) provides added value for organisations; saving costs in operations, resources and staff as well as new business opportunities for service-oriented models (Briscoe & Marinos, 2009; Schubert, Jeffery, & Neidecker-Lutz 2010; Chen, et al., 2010). In addition, it is likely that cloud computing focusing on operational savings and green technology will be at the centre of attention. Achieving long-term sustainability is an important success factor for organisations (Chang, Mills, & Newhouse, 2007), particularly in an economic downturn. This makes cost saving a common organisational goal across different sectors. Cost saving offered by CC is a key benefit acknowledged by academia (Buyya, et al., 2009, 2010; Celik; Holliday, & Hurst; 2009; Khajeh-Hosseini, Greenwood, & Sommerville, 2010; Schubert, Jeffery, & Neidecker-Lutz, 2010) and industrialists (Creeger, 2009; Dunn, 2010; Oracle, 2009, 2010).

The definition and deployment of ROI varies in different sectors and research institutes. Our ROI measurement is a systematic and innovative methodology based on:

1. Nobel-prize models such as the Capital Asset Pricing Models, CAPM (Sharp, 1990);
2. the use of economic and statistical computation for data analysis (Chang, et al., 2010b, 2011b, 2011c);
3. the use of 3D visualisation to present cloud business performance (Chang, et al., 2010b, 2011b, 2011c) and finally;
4. a unique way to use Quality Assurance (QA) to improve the quality of data and research outputs (Chang, et al., 2011b, 2011c).

This leads to the development of Organisational Sustainability Modelling (OSM), which is designed to measure cloud business performance (Chang, et al., 2011b, 2011c). Using OSM offers the following two advantages: (1) it allows performance reviews at any time; and (2) it provides strategic directions and added-values for adopting the right types of cloud business for sustainability.

The structure for this chapter is as follows. Section 2 presents a completed framework and how it can be used to measure ROI. Section 3 describes the University of Southampton case study and our methodology to analyse data. Section 4 presents another two case studies from other two departments in the University focusing on User Satisfaction, explaining how data analysis via OSM and 3D Visualisation can be presented. Section 5 lists seven topics of in-depth discussions. Section 6 describes our Conclusions and proposes future work.

1.1. Cost Saving offered by Cloud and Its Influence to Operations Management

Cost saving offered by Cloud Computing is a key benefit acknowledged by academia (Buyya, et al., 2009, 2010; Celik, Holliday, & Hurst, 2009; Khajeh-Hosseini, Greenwood, & Sommerville, 2010; Schubert, Jeffery, & Neidecker-Lutz, 2010) and industrialists (Creeger, 2009; Dunn, 2010; Oracle, 2009, 2010). It is one of the reasons for its popularity and organizational adoption in economic downturn.

From the academia point of view, Buyya et al. (2009) introduced Service Level Agreement (SLA) led cost saving models and explain how to calculate in detail. Further to their work, Buyya et al. (2010) introduce a Return on Investment (ROI) power model, which can calculate power cost-saving and also present it using 3D visualisation. Celik, Holliday, and Hurst (2009) introduce their Broadcast Clouds techniques which allow communications and cost-savings. They use simulations to support their proposal. Khajeh-Hosseini, Greenwood, and Sommerville (2010) use qualitative research methods to explain how industry can save costs. They present case studies of two companies and demonstrate cost-saving in infrastructure costs, and support and maintenance costs. Schubert, Jeffery, and Neidecker-Lutz (2010) present an overview and opportunities including cost-saving for the EU, and presents cost-saving as an added value offered by Cloud Computing. In the industrial practices, CA Technologies (a global IT firm) use cloud for cost-saving including: US $6.5 million for labour costs; and US $2.4 millions of operational costs in 5 years; and closure of 19 server sites. This allows CA Technologies to consolidate their infrastructure and remove maintenance costs such as staffing and resource expenses (Dunn, 2010). Similarly, Oracle faced a similar challenge after acquiring Sun Microsystems, and they consolidated their infrastructure, resources and migration with Cloud Computing. After spending about six months of transition period, both Oracle and Sun Microsystems are able to share and use the similar level of IT resources and data centres before acquisition, instead of doubling its size. This is largely due to virtualization. Many of their servers and services are in clusters of Virtual Machine (VM) farms, facilitating effective management from architects and management (Oracle, 2009, 2010).

1.2. Selective Models in Green IT

More organisations have used Green IT to manage their resource consolidation, cost-saving and operational management. There are innovative approaches as follows. Firstly, Butler (2011) introduce his Compliance-to-Product (C2P) application model, allowing regulations, policy, technologies and processes to be maintained and managed between manufacturers, organisations and external collaborators. Garg and Buyya (2011) describe a comprehensive review about Green Cloud Computing, and explain a typical usage model for how it can be achieved. They also present a Green Cloud Architecture to explain their rationale, and they use a Case Study to support their Architecture. Thirdly, Zhang, Liu, and Li (2011) propose a Goal Requirements Language (GRL) model to review and present efficiencies in Green IT. Despite their presentation is focused on conceptual model and workflow, they also show that calculations for carbon footprint and cost saving can be done in a quantitative way by using their GRL model.

GRL model is a way to demonstrate how Return on Investment (ROI) can be presented quantitatively. However, the Zhang, Liu, and Li (2011) do not show detailed example for how these data is measured and calculated, which is a core area that need be explained, where Buyya et al. (2010) demonstrate how they define and compute their calculations. Chang et al. (2010b, 2011b, 2010c) demonstrate that their Organisational Sustainability Modelling (OSM) can help organisations to measure its ROI which include a highly structured methodology and the use of statistical computing and 3D Visualisation. These advanced techniques are described in their proposed Cloud Computing Business Framework (CCBF).

2. THE ROLE OF CLOUD COMPUTING BUSINESS FRAMEWORK (CCBF)

Chang et al. (2011a, 2011c, 2011d) demonstrate three business challenges while deploying and migrating organisational infrastructures, applications and services over Cloud. This is the main motivation of proposing Cloud Computing Business Framework (CCBF), which aims to help organisations achieve good Cloud design, deployment and services. The core concept of CCBF is an improved version from Weinhardt's et al. (2009a) Cloud Business Model Framework (CBMF) where they demonstrate how technical solutions and Business Models fit into their CBMF.

The Cloud Computing Business Framework is a highly structured conceptual and architectural framework to allow a series of conceptual methodologies to apply and fit into Cloud Architecture and Business Models. The CCBF can provide the most suitable approaches and methodologies. It has the following objectives.

- Classification of business models to offer Cloud-adopting organisations right strategies and business cases.
- Offer a framework to review cloud business performance accurately.
- Deal with communications between desktops and clouds, and between different clouds offered by different vendors, which focus on enterprise portability.
- Provide linkage and relationship between different cloud research methodologies, and between IaaS, PaaS, SaaS, and Business Models.

Chang et al. (2011a, 2011b, 2011c, 2011d) demonstrate that CCBF has four key areas to work with: (1) Classification; (2) Organisational Sustainability; (3) Portability; and (4) Linkage. This chapter focuses on the second key area, Organisational Sustainability. This includes modelling to review and evaluate cloud project performance in the past and present, and enables forecasting for cloud businesses in the future. Organisational Sustainability Modelling (OSM) is suitable for all IaaS, PaaS, and SaaS. CCBF fits well in our research investigation in relations to Green IT and cost- saving, which fit for OSM.

2.1. Capital Asset Pricing Model (CAPM)

The Capital Asset Pricing Model (CAPM) is a model to calculate investment risks and to determine what the expected return on investment is. In the context of cloud computing, it is a quantitative model for organisational sustainability. CAPM was introduced by Jack Treynor in 1961, William Sharpe in 1964, John Lintner in 1965 and Jan Mossin in 1966, based on Harry Markowitz' work on diversification and modern portfolio theory. CAPM divides risk into two groups. The first group is Systematic Risk (also known as beta), the market of which cannot be diversified away, including recessions and interest rates. The second group is unsystematic risk, the risk of which is specific to individual stocks and can be diversified and managed by investors (Hull, 2009). In CAPM, beta is the only relevant measure of a stock's risk and measures a stock's volatility.

In some interpretations, the Security Market Line (SML) is used to calculate the reward-to-risk ratio. When the expected rate of return for any security is deflated by its beta coefficient, the reward-to-risk ratio for any individual security in the market is equal to the market reward-to-risk ratio, thus:

$$(r - rf)/\beta = rm - rf \quad (1)$$

$(r - rf) = \beta\ (rm - rf)$ [the security market line (SML)].

Finally, to best represent CAPM, the formula is given as:

$$r = rf + (\beta \times (rm - rf)) \quad (2)$$

where r is the expected return of a capital asset
rf is the risk free rate
rm is the expected return on the market and
β is the beta of the cash flows or security being valued.

The term $rm - rf$ is the market risk premium, which is usually considered implicitly rather than explicitly. Therefore, the term $\beta \times (rm - rf)$ is the risk premium on the cash flows (or security) being valued. CAPM example: If the risk-free rate is 1.0%, the beta (risk measure) of the firm is 1.5 and the expected market return over the period is 3%, the stock is expected to return (1.0%+1.5(3%-1.0%)) = 4.0%.

2.2. The 3D Organisational Sustainability Modelling and Other Systems

The CAPM modelling is represented by statistical computation, which needs advanced techniques to present results in visualisation. Although more data can be analysed, a drawback with statistical computing is that more data are generated and often this requires statisticians to understand and further analyse the outcome, and more results and more data are generated. The 3D visualisation can simplify the data analysis process, and it becomes more common to present data in visualisation format in some Web Services, Grid and Cloud research (Pajorova & Hluchy, 2010). Selected results computed by CAPM can be used for 3D Visualisation, which is enabled by Mathematica. While referring back to the market standard for business performance, the stock market is widely accepted and presented business performance in 2D format. Despite stock market is an indication for business performance, it is not a fair system as stock markets are subjective to speculations and a great extent of fluctuations in particular to volatile and uncertain economic periods (Prechter & Parker, 2007). On the other hand, Service Level Agreements (SLA) are often used to present cloud business performance. A drawback is that SLA tends to review cloud business at operational level in terms of usage per hour (Buyya, 2009, 2010), which lacks of strategic directions for achieving cloud sustainability. This means SLA approach permits calculation of a periodic income over time from usage scenarios, however, if the business models are not proposed and executed according to the winning strategy, income over time can be low or below investors' expectations.

To present cloud business performance best, a graphical and dynamic system independent of human-oriented speculations is ideal, and this also provides the best correlation between the organisational focus, strategies and data related to each organisation's cloud computing business models.

2.3. Organisational Sustainability Modelling

Organisational Sustainability Modelling (OSM) is based on the CAPM which is the analysis of return and risks for organisations or projects. Chang et al. (2011a, 2011b, 2011c) demonstrate how OSM can be used to measure Cloud business performance for SAP, Vodafone/Apple and two projects in National Health Services, UK. The proposed approach is to divide return and risk in three areas: Technical, Costs (Financial), and Users (or clients) before and after deploying cloud solutions or products/services. In some context, it can be defined as expected return and actual return. The data to be collected are dependent on organisational focus, which is flexible dependent on different characteristics for any type of technical or business cloud solutions. In this chapter, the focus is on Costs and Users from three different departments at the University of Southampton, and they are as follows:

- Technical: This can be improvements in performance, or improvement in reliability, or any added values or technical gains supported by experiments. This type of data is easier to obtain as experiments can be performed by researcher or collaborators. Risks can be time reduction or percentage of break down or relevant technical risks.
- Costs (Financial): This can be profits, or cost-saving gains, or any fund related. Risks can be loss, or sharp rise in operational or electricity costs.
- Users (or clients): This may mean increases in user confidence, or user community growth or user related area. Risks include reductions in user confidence or numbers or community growth due to factors such as funding, or quality of software, etc.

Figure 1. The cloud computing business framework (CCBF) highlighting organisational sustainability

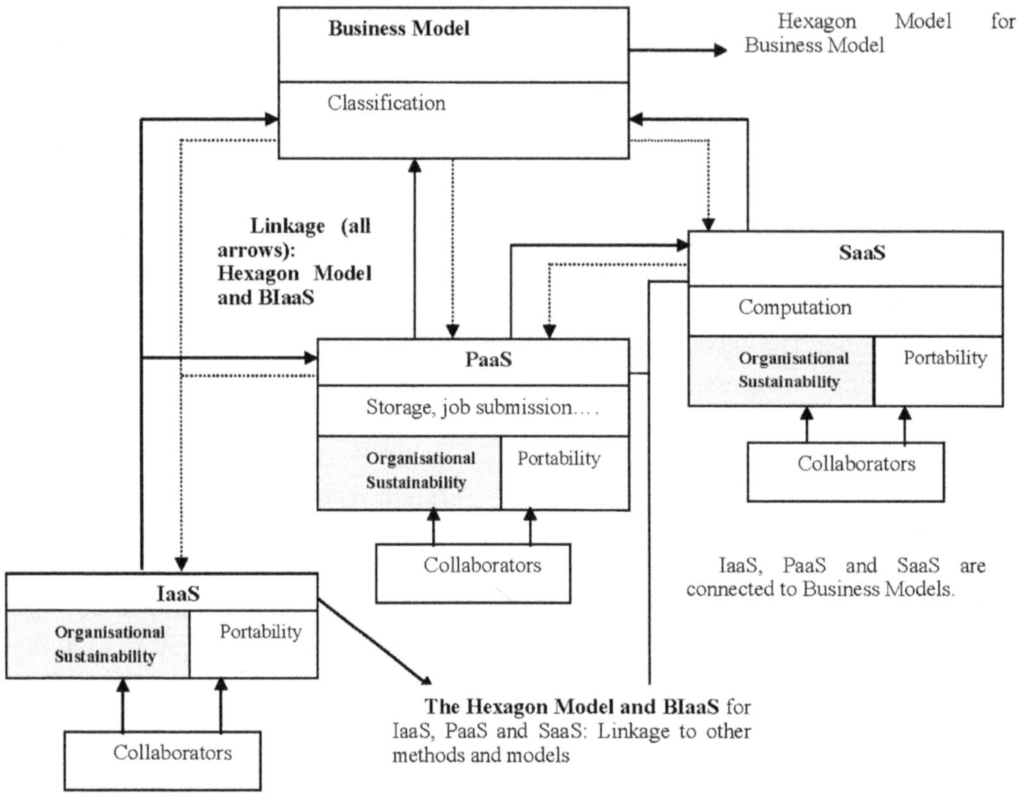

2.4. Organisational Sustainability: How Does It Fit into the CCBF

Figure 1 show the simplified CCBF diagram and explains how Organisational Sustainability fits into the CCBF. Organisational Sustainability is relevant for all organisations adopting or selling Clouds, whether they are IaaS, PaaS, and SaaS. Refer to yellow boxes in Figure 1.

3. THE UNIVERSITY OF SOUTHAMPTON: COST SAVING OFFERED BY GREEN CLOUD COMPUTING

The School of Electronics of Computer Science (ECS), University of Southampton, has used VM-ware and virtualisation since 2007 for different research projects. This fits with the University's long-term Green IT strategy. Some infrastructure consolidation took place in 2008. The final implementation started in late 2008 and was fully completed prior to Easter 2009. Cloud IaaS services commenced in April 2009. The initial focus was to reduce number of servers running continuously to meet green technology requirements. It is the University's aim to save costs and also consolidate number of servers required. By using virtualisation, email servers, storage servers, School Web servers, student record servers and database servers can be used as Virtual Machines (VM) running on private clouds.

This project has two case studies. The first case study is cost saving, which is the focus in this chapter. The second case study is technical added values such as efficiency improvement and time reduction to complete tasks, is in the process of data analysis.

3.1. Our Approach

Measuring ROI requires a structured process, which is useful to begin with during data collection and data analysis, to ensure the quality of presentation. Our ROI approach is a structured process and is divided into three stages, and is described as follows.

- Stage 1: This stage is to identify the cloud project focus of ECS and to define what to measure over a period of time. The focus is cost saving, which mainly includes cost saving from reduced electricity consumption (and bill). No account is taken of any cost saving due to reduced staffing costs. Three years data is obtained detailing the electricity consumption and bills for ECS facilities. This requires analysing electricity consumption (and bills) for ECS servers and storage. The data between November 2007 and June 2010 is used for analysis. Joint effort with the ECS IT and Facilities Teams is made to ensure the range and quality of the data is good prior to analysis in Stage 2. In addition, collaboration with the i-Solutions Group (providing IT services for the University) has been in place to obtain the two years user data about rating of Cloud and HPC services for different users. Data has been carefully examined and studied.
- Stage 2: This involves using OSM for quantitative analysis. The data is firstly used in CAPM statistical computing, which provides a summary. In the process of statistical computation, the Durbin-Watson statistical test is used to establish the accuracy of the output. CAPM statistics then computes the actual and predicted values of analysis in linear regression. The predicted values can be used as an expected result in the next phase. In addition, the data is then used by Mathematica to create a 3D visualisation of the results. 3D Visualisation simplifies the process of analysis; all analysis can be visually presented, enlarged, and rotated in 3D format. The Quality Assurance (QA) process is based on advanced statistical methods to verify data presented in 3D Visualisation has reliable and high quality and accuracy (Chang, et al., 2011b). Similarly, the user data from the i-Solutions Group is used for statistical analysis to interpret the impact and benefits offered by Cloud and HPC services. The most effective data ranges from October 2008 and September 2010.
- Stage 3: The data between June 2010 and December 2010 are compared with the expected values calculated in Stage 2, and to identify any similarities and differences. A minimum of six months data is required for Stage 3. QA is used to improve the quality of actual and predicted analysis. For ISS, the data used for comparison was from September 2010 and January 2011.

The collected data covers November 2007 to July 2010. CAPM can be modelled by statistical languages, in which SAS is a statistical computing language more suitable than others since it can compute more in-depth analysis (Chang, et al., 2010b). SAS code is written to predict the Risk Premiums of an organisation, ECS versus the Market (expected values).

The data is carefully calculated and examined with data consistency and coding algorithms. Thirty-two months of in-depth data can best represent sustainability from the initial phase to establishment. The SAS program for the CAPM is coded to plot required data with a suitable regression method. Table 1 shows a summary of this statistical computing.

Table 1. Regression summary of ECS cost saving

ECS cost-saving CAPM: The AUTOREG Procedure					
Dependent Variable r_ecssave					
Risk Premium for ECS cost saving					
Ordinary Least Squares Estimates					
SSE	5.9374272		**DFE**	40	
MSE	0.14844		**Root MSE**	0.38527	
SBC	44.4976402		**AIC**	41.022301	
Regress R-Square	0.0077		**Total R-Square**	0.0077	
Durbin-Watson	1.2838		**Pr < DW**	0.0079	
Pr > DW	0.9921				
Variable	**DF Estimate**	**Standard Error**	**t Value**	**Approx Pr > |t|**	**Variable label**
Intercept	1	18.6120	2.8950	6.43	<.0001
R_mkt	1	-0.0783	0.1410	-0.56	0.5817 ECS cost-saving Risk Premium for Market

Market is referred as the Expected values. The risk-free rate in this case study means the minimum operational costs in staffing and IT resources. ECS confirms their risk-free rate is reliable, and thus the risk premium is the difference between the expected values and risk-free rate. Table 1 presents the result of auto linear regression summary with Ordinary Least Squares used (OLS), which calculates accurate estimates. The lower the Mean Square Error, the more accurate the regression result. Durbin-Watson is also used to test auto-regression and accuracy of the output. CAPM statistics then computes the actual and predicted values of analysis in linear regression. The predicted values can be used as an expected result in the next phase.

3.2. 3D Visualisation for ECS Cost Saving Model

Further statistical analysis can be computed. This requires advanced statistical skills and also needs to analyse more data from results of statistics. Our major contribution in this project is to present complex statistical analysis using 3D Visualisation, so that no data can be missed for analysis, and also those without advanced statistical backgrounds can understand. This is useful for many decision-makers and directors who need to know business analytic results quickly but do not wish spend too much time to understand them. Chang et al. (2010b, 2010c, 2011b, 2011c) have described how 3D Visualisation techniques are used to present Cloud business performance and quantitative ROI for their collaborators, which include National Health Service (NHS) UK in its two healthcare cloud projects, and organisations such as Vodafone/Apple, SAP, and OMII-UK.

Data from statistical analysis is used for computation, where data is computed in Mathematica. The 3D visualisation models are presented in Figure 2, which is the default 3D model that indicates a high return of cost- saving between 21% and 22% on the y-axis, which is significant reduction in operational costs. It also shows the expected cost-saving between 22 and 26% on

Figure 2. 3D visualisation for ECS cost saving

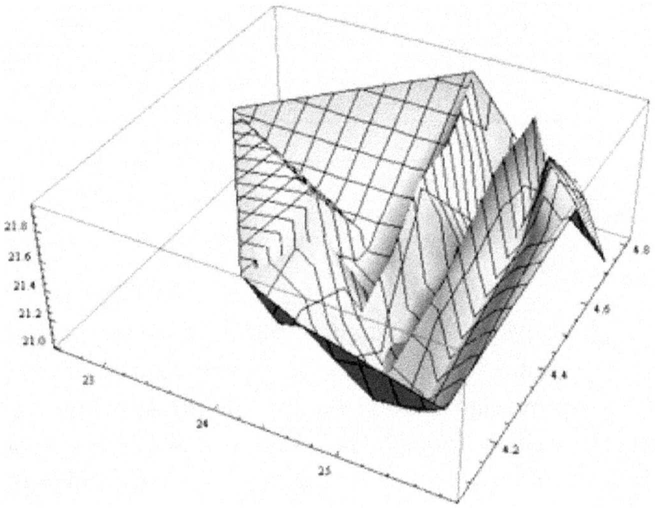

x-axis: Expected return of cost saving (22.5% - 26%)
y-axis: Actual return of cost saving (21.0% - 22.0%)
z-axis: Risk-free rate (4.0% - 4.8%)

x-axis. The z-axis presents risk-free rate (4.0-5.0%), which means minimum expenses to keep operation running. This percentage range can guarantee cost-savings.

3D visualisation takes minimum operational costs (risk-free rate), which include minimum staffing costs, are presented as z-axis. With Cloud Computing, statistics can analyse the cost-saving from consumption and resources required. But the 3D calculation takes hidden areas such as staffing costs into consideration, which means fewer people are required to do the same amount of work. Similarly, Buyya et al. (2010) and Pajorova and Hluchy (2010) use similar 3D Visualisation techniques to present Cloud Computing analysis and challenges.

4. USER SATISFACTION: AN IMPORTANT ASPECT

User satisfaction is an important aspect in organisational Cloud computing adoption. Chowhan and Saxena (2010) explain the role of Customer Relationship Management (CRM) in their Customer Life Cycle (CLC) perspective. They believe using CRM in CLO and Business Strategy perspective can improve customer satisfaction, and eventually exceed their expectations. Nick, Cohen, and Kaliski (2010) explain the development of virtual private clouds also require good user satisfaction to guarantee the delivery of Cloud projects.

4.1. i-Solution Group: University of Southampton User Data and Analysis

The i-Solutions Group (providing Information Systems and Services) of the University of Southampton have offered Cloud Computing, Grid Computing and HPC services for students and staff since 2007 through a number of pioneering projects, which have proven successful and popular. Those services were officially resumed in 2008, serving thousands of users across different campuses for different needs and challenges. The user feedback and rating have been recorded since 2008. We have obtained three years of data between 2008 and 2011 to study the impact of the Cloud Computing and HPC services for University staff and students, particularly more comprehensive data is available for 2009-2010 and 2010-2011 rating. The main objective of this investigation is to identify the level of user satisfaction, as it is an

Table 2. i-Solutions: University of Southampton data overview between 2009 and 2010 (817 valid entries)

	Excellent	Good	Adequate	Poor
Service Rating	16%	44%	27%	13%
	Complete	Some	Little	None
User Confidence	31%	51%	14%	4%
	Better	Same	Worse	
Overall performance	44%	48%	8%	

important factor for achieving good Cloud design, deployment and services (Chowhan & Saxena, 2010). The approach is identical to Section 3.1 with Stage 1 and 2 in place, with the difference in the Stage 3, which collected data between April 2009 and March 2010 by i-Solutions, which used surveys for metrics collection with focus in three aspects: (1) Service Rating; (2) User Confidence, and (3) Overall performance. Out of all responding surveys, there are 817 valid entries, see Table 2. There are additional questions in each section to measure detailed extent of performance comparing to the previous year.

In terms of Service Rating, 87% of respondents have positive comments about the use of Cloud computing for resources and services, which include 16% for 'Excellence,' 44% for 'Good,' and 27% for 'Adequate.' User Confidence has obtained 82% of support, which include 31% for 'Complete,' and 51% for 'Some.' Overall performance for Cloud adoption is encouraging, as it indicates of 44% of better performance, although 48% of respondents still feel the same. Some survey questions also record the rationale of choosing those answers, and details of these data analyses are presented in Section 5: Discussions.

4.2. Corporate Planning: University of Southampton Data and Analysis

Another user survey led by the Corporate Planning of Southampton University took place in March 2011 to measure the user satisfaction rating between 2010 and 2011. Corporate Planning overviews the University's business strategy where Cloud Computing is a core area, in particular providing a good quality of infrastructure and related services. The research method is based on survey with one month spent for data collation. There are 914 valid responses and at least 908 respondents answered all the questions. Table 3 is the summary of some data analysis agreed and supported by Corporate Planning.

Table 3 refers to the user satisfaction rating between Year 2010 and 2011, and there are four areas of focus. The first area is accessibility, which 93.32% of all users have rated a score of 3 and above (scale of 1 to 5). There is also enough financial support for Cloud or computing related research events, where 75.88% of respondents have rated a score of 3 and above. Availability is the third focus, where 89.11% of them, with score of 3 and above, feel there is always adequate provision of computing resources and facilities. Support is the fourth focus, where 89.53% of respondents, with a score of 3 and above, are satisfied with services they have received.

4.3. Statistical Computational Analysis

Organisational Sustainability Modelling (OSM) helps large scale data analysis and presentation of complex data in the form of statistical computation and 3D Visualisation, the latter of which simplifies the process of analysis. In addition, no hidden area can be missed in visualisation and all the key data can be found and analysed in the

Table 3. User satisfaction data between 2010 and 2011 by corporate planning, University of Southampton

a. I have adequate access to the equipment necessary for my research. **Service rating: 5 is the highest and 1 is the lowest score.** **Accessibility (3 out of 5 and above): 93.32%**	**1**	**2**	**3**	**4**	**5**	**NA**
Percent	1.64%	5.03%	13.58%	37.68%	38.23%	3.83%
Number of respondents	15	46	124	344	349	35
b There is appropriate financial support for research activities Enough financial support (3 out of 5 and above): 75.88%						
Percent	6.50%	10.13%	15.64%	29.62%	30.62%	7.49%
Number of respondents	59	92	142	269	278	68
c There is adequate provision of computing resources and facilities Availability (3 out of 5 and above): 89.11%						
Percent	1.43%	6.59%	11.98%	32.08%	45.05%	2.86%
Number of respondents	13	60	109	292	410	26
d I have the technical support I need Sufficient support (3 out of 5 and above): 89.53%						
Percent	1.54%	4.96%	16.52%	37.00%	36.01%	3.96%
Number of respondents	14	45	150	336	327	36

3D domain. Our 3D Visualisation also supports 360 degrees rotation so it assists in any aspects of analysis and further investigation. The data from i-Solution is for 2009-2010 data and the data from Corporate Planning is for 2010-2011 data. These datasets are complementary to each other, since they are to measure how Cloud Computing services are used and rated by users. The objective is to extract and present data, which is not directly and explicitly interpreted from the survey. Therefore, both datasets are jointly used for statistical analysis, where the key data is computed by 3D Visualisation. User satisfaction is a core strategy and a crucial factor for the University of Southampton to succeed with Cloud and IT projects. See Table 4 for details.

The next step is to identify the level of service improvement, which provides useful feedback to organisations adopting and providing IT services such as Helpdesk (Office of Government Commerce, 2007; Hanna, et al., 2009). Understanding the extent of improvement is essential for continuous service improvement and is a good indication for cloud business performance measurement, the second key area in the CCBF. Some of data are from Table 4. Data has been carefully examined and analysed. They are computed in Mathematica to present the result as 3D Visualisation. Based on our data, user satisfaction is largely dependent on user confidence and service improvement, which is a key indicator in our analysis. Modelling based on CAPM is often used as actual return versus expected return along with the risk-free rate. Figure 3 shows the 3D analysis and Figure 4 shows its 90 degrees rotation. The x-axis reflects actual rate of return, user confidence and service improvement between 3.4% and 8.4%. The y-axis reflects expected rate of return, user confidence and service improvement between 1.2% and 5.8%. The z-axis shows risk-free rate between 0.8% and 4.4%. In this case, risk free rate refers to the areas of services that the University will always receive highest rating, and often this refers to response

Table 4. Regression summary of Southampton University user satisfaction

Southampton User Confidence and service improvement CAPM: The AUTOREG Procedure					
Dependent Variable r_sotonuser					
Risk Premium for user confidence and service improvement					
Ordinary Least Squares Estimates					
SSE	15.0740045		**DFE**	55	
MSE	0.27407		**Root MSE**	0.52352	
SBC	94.0305602		**AIC**	89.9444577	
Regress R-Square	0.5500		**Total R-Square**	0.5500	
Durbin-Watson	1.3394		**Pr < DW**	0.0040	
Pr > DW	0.9960				
NOTE: Pr<DW is the p-value for testing positive autocorrelation, and Pr>DW is the p-value for testing negative autocorrelation.					
Variable	**DF**	**Standard Estimate**	**Error t**	**Approx Value Pr > |t|**	**Variable Label**
Intercept	1	-0.8571	0.2017	-4.25	<.0001
r_mkt	1	0.5535	0.0675	8.20	<.0001 Southampton Risk Premium for Market

time to incidents or requests within the time frame defined in the service level agreement.

The extent of service improvement is a straight line with an upward movement. This shows the level for service improvement is continuously progressive and heading upwards. Explanations are likely due to two reasons. Firstly, services offered by Cloud is challenging for operational and service management. The incremental improvement already suggests good progress has made. Secondly, some users may feel positive experience about Cloud services, infrastructure and platform, and they have increased their user confidence level, which includes reduction in time and resources spent on the University IT and Cloud services.

5. DISCUSSIONS

There are also seven interesting discussions. The first discussion asserts there is difference between technical and user perspective towards availability. Users may still find if they are uncertain about the procedures, they need prompt consultation, and do not feel this service is available for them to complete their tasks. The response times to deal with user requests and complete tasks are thus important. The second discussion confirms the positive ROI for the University of Southampton and a popular Cloud adoption in Education. The third discussion presents a structured QA process for data analysis and is supported by Continuous Assurance that offers additional benefits. The fourth discussion introduces Business Integration as a Service (BIaaS), which links different processes and functions together into an integrated platform, with the end result of one process used as the input of another. The fifth discussion addresses the greater role of Green Cloud Computing for Education. The sixth discussion summarises case studies and contributions offered by OSM of CCBF. The final discussion is the comparison between OSM/CCBF and other methods and

Figure 3. 3D visualisation for user confidence and service improvement

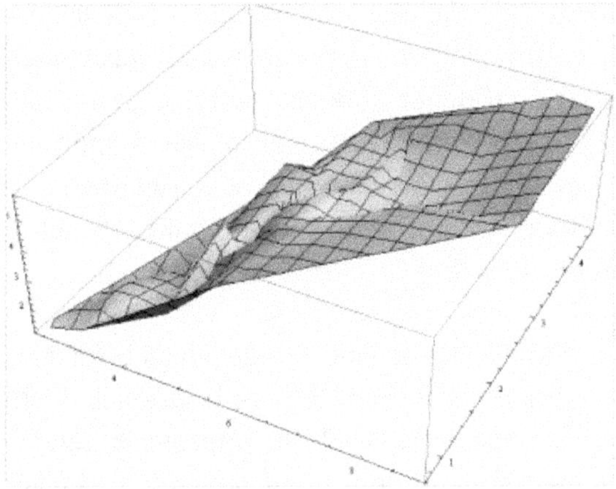

x-axis: Actual Rate of Return - User confidence and service improvement (3.4% - 8.4%)
y-axis: Expected Rate of Return - User confidence and service improvement (1.2% - 5.8%)
z-axis: Risk-free rate (0.8% - 4.4%)

Figure 4. 3D visualisation for user confidence and service improvement, 90 degrees rotation

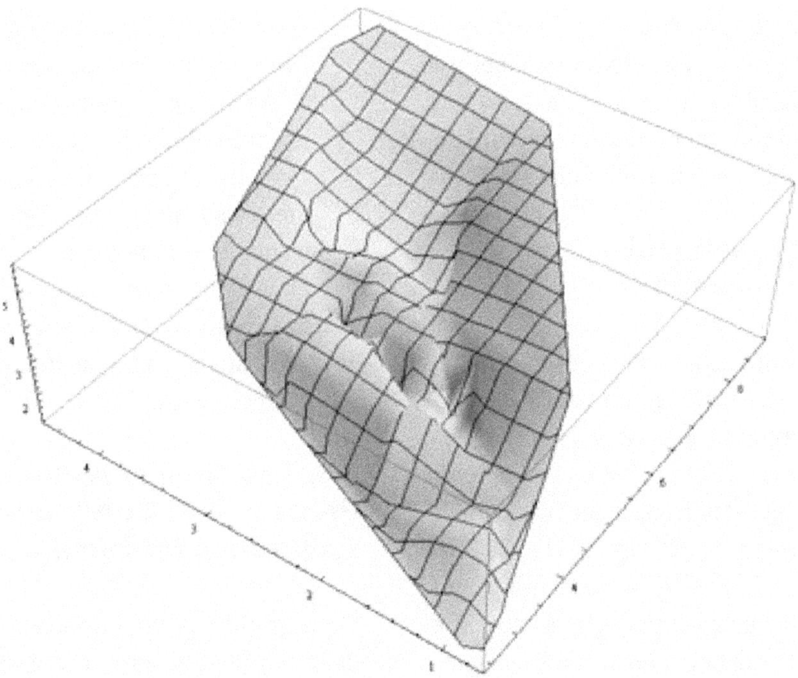

approaches, where OSM and CCBF can provide more added values.

5.1. Availability defined by SLA vs. Availability Perceived by Users

Many papers assert that availability is defined as the up time for services, and should stay close to 100% (Ambrust, et al., 2009; Buyya, et al., 2009, 2010; Brandic, et al., 2009). This is true for the service provider's perspective and all services are available at all times and are accessible by different means. Our data suggests there is only 89.11% availability from users' perspective, and this is considered a good result based on users' feedback. However, the discrepancy between the ideal and users' availability is about how services are handled and managed. Service providers assume a near 100% up time for different types of services. In users' perspective and based on our data, they count the number of incidents that they need further enquiries and support from services. Although numerous services have the ability for automation or features with self-service, a large percentage of users still prefer to consult for help when problems arise, or they have difficulty using self-services. In this case, the response time for each incident or request is important. The University of Southampton is considered above average compared with most of Universities because if the problems happen outside normal office hours, they can respond to the issues and keep users up-to-date. Our data analysis suggests there is a different perception of service availability between technical and users perspectives, and fast response time within acceptable hours to complete the tasks are crucial to retain user satisfaction.

5.2. Cloud Migration and Services for Education

JISC UK has announced Cloud Computing is increasingly attractive for research and education, and they state there are five reasons for University adoption (JISC, 2011). The use of Cloud adoption offers a greater Return on Investment (ROI), where Chang et al. (2011a, 2011b, 2011c) have demonstrated how Cloud Computing and the use of CCBF helps organisations to obtain a good rate of ROI. This helps more Cloud migration and adoption in Education. The University of Southampton, King's College London (KCL) and the University of Greenwich have followed CCBF for Cloud design, deployment, migration and services. Two Southampton case studies have been used for this chapter where 21% – 22% of cost-saving is achieved by ECS, and also 1.2-5.8% increase in user confidence and service improvement is experienced by i-Solution and Corporate Planning, which represent encouraging rate of satisfaction and additional values experienced by students and members of staff. 3D Visualisation also confirms there is a progressive improvement in user confidence and service throughout 2009 and 2011, so that investment in user support and service improvement should be maintained.

Other case studies in educational Cloud adoption are as follows:

- Storage Cloud and self-serviced infrastructure have helped KCL and National Health Service (NHS) UK to improve efficiency and streamline a good process for IT services (Chang, 2009; Chang, et al., 2011b).
- University of Greenwich has three projects that use Cloud resources for development, and this includes Sharepoint development and migration, media server and e-learning development and e-procurement and e-supply chain. These three projects are still in further development with a good progress reported (Chang, et al., 2011d, 2011e).
- The National Grid Services (NGS) UK has deployed Cloud infrastructure and is ready to provide Cloud-related services for scientists and researchers across the UK (Chen, et al., 2010).

5.3. The Quality Assurance Process and Continuous Assurance for Business

Data quality is an important aspect because it checks for data structure and consistency, and rectifies any errors, thus quality of data is improved on an ongoing basis. It ensures statistical analysis, whether in computational or visualisation format, is at a high quality of research output (Chang, et al., 2011b). QA process must be improved and made as efficient as possible. This means time reduction to deliver the same level of services using a tool, and/or automation. Our data (that correlates to 3D Visualisation) have undergone a structured process, and all data are thoroughly tested with

STATA 11 and selected regression techniques to ensure a high quality of data analysis. Chang et al. (2011b) streamline and demonstrate a good process for their QA, which include:

- Review all the data generated by SAS, and determine what need to be further analysed;
- If we are uncertain what needs to be further analysed, then use STATA for testing regression to double check;
- Determine what needs to be further analysed, and make them into a format readable by Mathematica;
- Use Mathematica to compute dynamic 3D Visualisation;
- Use STATA to double check the validity of 3D Visualisation.

Continuous Assurance applies quality assurance across strategy and operations of management and service delivery. Continuous Assurance is critical to Cloud Computing, Services, and Operations in the Enterprise environments, where ISACA (2009), a consulting firm working for an EU project, define the following benefits of adopting it:

- Transparency - Effective and robust security controls are in place, assuring customers that their information is properly secured against unauthorized access, change and destruction.
- Privacy - Privacy controls are in place and demonstrate their ability to prevent, detect, and react to breaches in a timely manner.
- Compliance - Organisations must comply with a litany of laws, regulations and standards, in particular to data protection.
- Trans-border information flow - Data in the Cloud can be anywhere. Physical location dictates jurisdiction and legal obligation.
- Certification - Independent assurance from third-party audits and/or service auditor reports should be a vital part of any assurance program.

Our CCBF can offer both QA and Continuous Assurance for participating organisations, which the University of Southampton that has obtained those benefits.

5.4. Linking all Different Processes, Functions, and Departments Together

Three different departments from the University of Southampton have their Cloud projects and services in place. Each department's work is independent from the others, except i-Solutions and Corporate Planning have the same focus. Even so, it takes some time for communications and establishment of a mutual goal, but an interesting research challenge has surfaced: whether the activities and results from one department can be used for the other department, and whether the end processes of department A can be the used as the input for the processes of department B. This leads to a new and innovative area known as Business Integration as a Service (BIaS), where Chang et al. (2011d, 2011e) have demonstrated how to link different processes and functions together in an integrated platform, and the end result of one process can be used as the input for another process without further translation. BIaS will be useful for universities and enterprises to synchronise all their Cloud services and activities, and can improve the efficiencies without manual work being carried out in each department, and spending much time in communications and getting consensus between different departments.

5.5. The Role of Green IT for Education

Garg and Buyya (2010) demonstrate how Cloud Computing can help organisations to achieve Green IT, and how advanced techniques and technologies can make Green IT ready for enterprise solutions. Case studies from the University of Southampton demonstrate Green Cloud Computing helps the University fulfill two major goals: (1) cost-saving and (2) improved user satisfaction. Green Cloud Computing has also been adopted by the University of Greenwich, where the Business School has received a Sustainability Award. The Business School at the University of Greenwich has a policy to reduce carbon emissions and procurement of environmental-friendly equipment. There are policies and tools working together to deliver benefits such as cost-saving, integration of services, service improvement and user confidence growth offered by Cloud Computing. Current Plan include:

- Streamline departmental printing process and improve recycling rate.
- Digitalise student coursework into electronic format, and improve online marking rate through the use of technologies and policies.
- Collect useful metrics for IT Services and project delivery.
- Collect useful metrics for energy, operational and procurement costs, and analyse data using the use of Organisational Sustainability Modelling (OSM).

Detailed metrics are in the process of collection and analysis. In addition, KCL and National Health Service (NHS) UK have implemented environmental friendly Cloud Storage that not only provides automation and self-services, but also reduces carbon emission with the use of advanced technologies (Chang, 2009). Similarly, the University of Oxford has e-Research projects in collaboration with NGS UK addressing Green Cloud Computing. The role of Green IT has significantly improved and has become more influential for Education, as there are initiatives and pilot projects investigating the use of Green Cloud resources and strategies (Buyya, et al., 2010).

5.6. Summary of Case Studies and Contributions offered by Organisational Sustainability Modelling (OSM) of CCBF

The CCBF has helped several organisations to measure its Cloud business performance including Green IT adoption, and selective results have been presented in several journals, as summarised in Table 5.

Each case study is unique and has either technical, cost or user focus. While leading papers (Amburst, et al., 2009; Brandic, et al., 2010; Buyya, et al., 2009, 2010) explain the technical added values offered by SLA aspects of Cloud Computing, none of them can address user satisfaction and its methodology for measurement and presentation. We would like to emphasise user satisfaction is an important aspect for Cloud ROI measurement and should be managed and monitored by organisations adopting Cloud deployment.

5.7. Comparisons with other Methods and Approaches

Before discussing added values offered by OSM and CCBF, other approaches to define and review business framework or business performance are reviewed. Chang et al. (2011c) have investigated a number of selected frameworks, and compare their strengths and weaknesses with OSM and CCBF, where Table 6 is the summary of all reviews.

6. CONCLUSION AND FUTURE WORK

The CCBF is a relatively new area, and finding the right methods to review business performance can enhance organisational sustainability. Cost saving is particularly important at times of economic downturn, and this is a reason why more organisations are using CC to minimise and control

Table 5. Summary of case studies and contributions offered by OSM, part of CCBF

Organisations	Summary of 3D analysis	Contributions
OMII-UK (Chang, et al., 2010b)	x-axis: The return for OMII-UK (1 and 5%) y-axis: Risk premium for the market (-0.3 and 0.5%) z-axis: Risk-free rate of OMII-UK (0 to 50%)	OMII-UK was a start-up in Cloud services, and experienced a volatile period with uncertainty and competitions. It survived until the end of 2010. OMII-UK is a good case study for start-up's lessons learned.
SAP (Chang, et al., 2011c)	x-axis: The return for SAP (-1 and 1%) y-axis: Risk premium for the market (-1 and 1%) z-axis: Risk-free rate of the market (0.7 to 0.715%)	(1) SME use SAP as a cautious cloud tactics (2) SME prefer to use more predictable or familiar ways to maintain their cloud business sustainability.
Vodafone/Apple (Chang, et al., 2011c)	x-axis: Vodafone's return (21-24%) y-axis: Risk premium of the market return (22-26%) z-axis: Risk-free rate in market (2.0-4.0%)	Their iPhone and iPad strategy has vastly improved profits and business performance.
NHS UK Infrastructure (Chang, et al., 2011b)	x-axis: The return of NHS Infrastructure (6% - 7.2%) y-axis: Risk premium for the market (4.3% - 6%) z-axis: Risk-free rate of the market (3% - 5%)	The use of Cloud infrastructures can improve efficiency. It also results in raising benchmark, the minimum acceptance level to complete concurrent tasks.
NHS UK Bioinformatics (Chang, et al., 2011b)	x-axis: The return of NHS Bioinformatics (1.2% - 7.2%) y-axis: Risk premium for the market (0.2% - 4%) z-axis: Risk-free rate of the market (0.1% - 4.8%)	There are incremental improvements to an NHS Bioinformatics project. The low risk-free rate may imply code development allows reduced time to complete, and objective is clearly met and project delivery is straightforward.
A SME (requests not to reveal identity) in broadband service (Chang, et al., 2011d)	x-axis: The return of SME cost-saving (20% - 23%) y-axis: Risk premium for the market (7.5% - 8.5%) z-axis: Risk-free rate of the market (5% - 5.8%)	CCBF helps them to upgrade their services from IaaS to PaaS. Cost-saving (electricity and operational costs) is achieved with the use of Cloud and Green IT.
ECS, University of Southampton (this chapter)	x-axis: Expected return of cost-saving (22.5% - 26%) y-axis: Actual return of cost-saving (21.0% - 22.0%) z-axis: Risk-free rate (4.0% - 4.8%)	Cost-saving (electricity and operational costs) is achieved with the use of Cloud and Green IT.
i-Solution Group and Corporate Planning, University of Southampton (this chapter)	x-axis: Actual Rate of Return - User confidence and service improvement (3.4% - 8.4%) y-axis: Expected Rate of Return - User confidence and service improvement (1.2% - 5.8%) z-axis: Risk-free rate (0.8% - 4.4%)	The level for service improvement is continuously progressive and heading upwards. Some users feel positive experience about services, infrastructure and platform, and they have increased their user confidence level.

costs. We have demonstrated the case study of the ECS, University of Southampton for cost-saving, and i-Solution and Corporate Planning of the University of Southampton for user confidence and service improvement. These three case studies offer added values. This involves using a highly structured and organised process to review and evaluate. There are three stages and highly structured process involved for data analysis, which have been explained in detail.

Our ROI method introduced by OSM and CCBF provides additional value. Firstly, it includes the use of the CAPM statistics to compute analysis. Secondly, it involves conversion to 3D Visualisation to present cloud business performance. Thirdly, it involves a series of QA tests to ensure the quality of 3D Visualisation is of a high standard. Of the Cloud ROI methods available, ours is the only one demonstrating QA and high quality of analysis among others. The challenge for the UK universities is to improve confidence, particularly in Consumers, which means staff and students can feel the positive impacts of cost-saving offered by CC, which is useful in the economic downturn.

OSM offers the following two advantages: (1) it allows performance reviews at any time; and (2) it provides strategic directions and added-values

Table 6. Review and comparisons between selective frameworks for defining and measuring cloud business and ROI (Chang, et al., 2011c)

Types of framework	CCBF (Chang, et al., 2010b, 2010c, 2011a, 2011b, 2011c, 2011d, 2011e, 2011f)	VBSPN (Zirpins & Emmerich, 2008)	DSGE (Etro, 2008)	CBM (Klems, Nimis, & Tsai, 2009)	SLA-based (Assuncao, Costanzo, Buyya, 2010; Buyya, et al., 2009)	Miscellaneous (Lobo & Arthur, 2005)
Qualitative	Yes. There are proposed business model classification and the use of hexagon model.	Yes. They have a solid qualitative and theoretical framework.	Yes. But only in the EU and make several assumptions.	Yes. They have great descriptions about this.	Might be possible, but take certain assumption on service providers and usage scenarios.	Yes. They provide a series of best practices for framework.
Quantitative	Yes. The use of modelling and simulations to validate.	No, despite they use UML to support theoretical framework.	Yes. He has solid econometrics insights to explain this.	Yes, but not yet done.	Yes. They have numerous ways to measure cloud billing models quantitatively.	Potentially yes, but further work needs to be done.
Linkage (qualitative-quantitative)	Yes. It demonstrates Business Integration as a Service (BIaaS).	No	Yes. But in an early stage. Weak in linkage.	Yes, but not yet done.	No, at least not yet.	No, before they can propose quantitative area.
Cloud Business Performance	Yes. Use Organisational Sustainability Modelling with statistical computing and 3D visualisation. Can offer accurate measurement.	No. This is a generic framework, and is likely to work in their own environment only.	Not yet, although he is working towards it for the EU.	Yes. Use their conceptual framework.	Yes. They are one of pioneers in pay-as-you-go and billing models.	No. Their framework is not designed for cloud, but for generic uses and risk analysis.
Service Level	Yes. Deal with selective IaaS, PaaS and SaaS.	Potentially yes, but in their own setting only.	Potentially yes, but only for the EU.	Yes. Similar to ours, but use a different way to present.	Yes. They have a few papers about this.	Possible, but more work needs to be done.
Orientation of framework	Conceptual and architectural.	Conceptual only	Mathematical Conceptual	Conceptual only	Architectural	Conceptual
Ways to validate	Modelling, simulations, and experiments for quantitative. Hexagon Model, case studies, and interviews for qualitative.	UML and complex business object modelling.	Econometrics and financial modelling based on the data he collects.	Case studies and theoretical hypothesis.	Simulations and calculations.	Theoretical hypothesis and case studies.
What needs to be improved	In progress. More validations and improvements will be presented.	Framework update is required.	Framework update is required.	Need to consider more on quantitative aspects.	Should deal with more strategic levels rather than operational.	Framework update is not published for nearly 5 years.

for adopting the right types of cloud business for sustainability. The ECS, i-Solution and Corporate Planning case studies strongly support the use of Green IT that provides cost-saving benefit and added values for users. 21-22% of cost-saving and 1.2-5.5% increase (on top of high user rating) in user confidence level and service improvement has justified Cloud migration and these services provide a valuable ROI for staff and students. This also fits well into the University of Southampton strategy for cost saving and user satisfaction and our continued collaboration with the University will ensure our work benefits others in addition to the University of Southampton.

There are topics of discussions, among which two topics are worth for further elaboration. The second discussion topic confirms the positive ROI for the University of Southampton and a popular Cloud adoption in Education. This is supported by high percentages of satisfactory survey results (3 out of 5 and above), such as 93.32% for accessibility, 75.88% for enough financial support, 89.11% for availability, and 89.53% for sufficient support for all users involved. User confidence and service improvement have further improved between 3.4% and 8.4%, which are considered satisfactory by users and IT staff. Full support received from the University is also a factor for maintaining good percentages of satisfaction. The sixth discussion topic sums up all 3D analysis and contributions for all collaborators, which include OMII-UK, SAP, Vodafone/Apple, NHS, a broadband SME and University of Southampton (three departments). Those results and analysis have helped collaborators to understand their Cloud business performance, and make the right decisions based on such analysis, which supports that OSM is a unique ROI method with extensive case studies.

CCBF has been extensively used in several organisations such as NHS UK, KCL, Universities of Greenwich, Southampton, Oxford, VMware, Vodafone/Apple, Salesforce, IBM, Commonwealth Bank Australia (CBA) and so on. Some collaborators find it useful for their organisations and contributions from CCBF can positively impact e-Research, Cloud, Grid, Health, Finance and Education Communities.

REFERENCES

Armbrust, M., Fox, A., Griffith, R., Joseph, A. D., Katz, R. H., & Kownwinski, A. … Zaharia, M. (2009). *Above the clouds: A Berkeley view of cloud computing.* Technical Report, No. UCB/EECS-2009-28. Berkeley, CA: UC Berkeley.

Assuncao, M. D. D., Costanzo, A. D., & Buyya, R. (2010). A cost-benefit analysis of using cloud computing to extend the capacity of cluster. *Journal of Cluster Computing*, *13*, 335–347. doi:10.1007/s10586-010-0131-x

Bechini, A., Cimino, M. G. C. A., Marcelloni, F., & Tomasi, A. (2008). Patterns and technologies for enabling supply chain traceability through collaborative e-business. *Information and Software Technology Journal*, *50*, 342–359. doi:10.1016/j.infsof.2007.02.017

Brandic, I., Music, D., Leitner, P., & Dustdar, S. (2009). *VieSLAF framework: Enabling adaptive and versatile SLA-management.* Paper presented at the 6th International Workshop on Grid Economics and Business Models 2009 (Gecon09). Delft, The Netherlands.

Briscoe, G., & Marinos, A. (2009). Digital ecosystems in the clouds: Towards community cloud computing. In *Proceedings of the 3rd IEEE International Conference on Digital Ecosystems and Technologies (DEST 2009)*, (pp. 103-108). New York, NY: IEEE Press.

Butler, T. (2011). Compliance with institutional imperatives on environmental sustainability: Building theory on the role of Green IS. *The Journal of Strategic Information Systems*, *20*, 6–26. doi:10.1016/j.jsis.2010.09.006

Buyya, R., Ranjan, R., & Calheiros, R. N. (2010). InterCloud: Utility-oriented federation of cloud computing environments for scaling of application services. *Lecture Notes in Computer Science*, *6081*, 13–31. doi:10.1007/978-3-642-13119-6_2

Buyya, R., Yeo, C. S., Venugopal, S., Broberg, J., & Brandic, I. (2009). Cloud computing and emerging IT platforms: Vision, hype, and reality for delivering computing as the 5th utility. *Journal of Future Generation Computer Systems*, *25*(6), 559–616. doi:10.1016/j.future.2008.12.001

Celik, A., Holliday, J., & Hurst, Z. (2006). *Data dissemination to a large mobile network: Simulation of broadcast clouds*. Paper presented at the 7th International Conference on Mobile Data Management (MDM 2006). Santa Clara, CA.

Chang, V. (2009). *Cloud storage framework – An integrated technical approach and prototype for breast cancer*. Paper presented at the All Hands Meeting 2009. Oxford, UK.

Chang, V., Bacigalupo, D., Wills, G., & De Roure, D. (2010a). *A categorisation of cloud computing business models*. Paper presented at the CCGrid 2010 IEEE Conference. Melbourne, Australia.

Chang, V., De Roure, D., Wills, G., & Walters, R. (2011c). Case studies and organisational sustainability modelling presented by cloud computing business framework. *International Journal of Web Services Research, 8*(3), 26–53.

Chang, V., De Roure, D., Wills, G., Walters, R., & Barry, T. (2011b). Organisational sustainability modelling for return on investment: Case studies presented by a national health service (NHS) trust UK. *Journal of Computing and Information Technology, 19*(3).

Chang, V., De Roure, D., Wills, G., Walters, R., Li, C. S., Barry, T., & Currie, W. (2011d). The development that leads to the cloud computing business framework. *Information and Management*.

Chang, V., Li, C. S., De Roure, D., Wills, G., Walters, R., & Chee, C. (2011a). The financial clouds review. *International Journal of Cloud Applications and Computing, 1*(2), 41–63. doi:10.4018/ijcac.2011040104

Chang, V., Mills, H., & Newhouse, S. (2007). *From open source to long-term sustainability: Review of business models and case studies*. Paper presented at the UK e-Science All Hands Meeting. Nottingham, UK.

Chang, V., Wills, G., & De Roure, D. (2010b). *A review of cloud business models and sustainability*. Paper presented at the IEEE Cloud 2010, the Third International Conference on Cloud Computing. Miami, FL.

Chang, V., Wills, G., & De Roure, D. (2010c). *Cloud business models and sustainability: Impacts for businesses and e-research*. Paper presented at the e-Science All Hands Meeting 2010, Software Sustainability Workshop. Cardiff, UK.

Chang, V., Wills, G., & Walters, R. (2011e). *Towards business integration as a service 2.0 (BIaaS 2.0)*. Paper presented at the IEEE International Conference on e-Business Engineering, The 3rd International Workshop on Cloud Services - Platform Accelerating e-Business. Beijing, China.

Chen, X., Wills, G. B., Gilbert, L., & Bacigalupo, D. (2010). *Using cloud for research: A technical review*. Retrieved from http://eprints.ecs.soton.ac.uk/21273/.

Chowhan, S. S., & Saxena, R. (2011). Customer relationship management from the business strategy perspective with the application of cloud computing. *Proceedings of DYNAA, 2*(1).

Creeger, M. (2009). CTO roundtable: Cloud computing. *Communications of the ACM, 52*(8).

Croson, R., & Donohue, K. (2003). Impact of POS data sharing on supply chain management: An experimental study. *Production and Operations Management, 12*(1), 1–11. doi:10.1111/j.1937-5956.2003.tb00194.x

Dietmann, A., Helbok, R., Lackner, P., Fischer, M., Reindl, M., & Lell, B. (2009). Pathophysiology and severe malaria – From basic knowledge to new therapeutic approaches. *European Journal of Tropical Medicine and International Health, 14*(2).

Dunn, T. (2010). *Identity management: Cloud and virtualisation.* Paper presented at the European Identity Conference, Munich Cloud 2010. Munich, Germany.

Etro, F. (2009). The economic impact of cloud computing on business creation, employment and output in Europe. *Journal of Review of Business and Economics*, *54*(2), 1–40.

Garg, S. K., & Buyya, R. (2010). *Green cloud computing and environmental sustainability.* Working Paper. Melbourne, Australia: University of Melbourne.

Hanna, A., Windebank, J., Adams, S., Sowerby, S., Rance, S., & Cartlidge, A. (2009). *ITIL V3 foundation handbook: Pocketbook from the official publisher of ITIL*. London, UK: Stationery Office.

Hull, J. C. (2009). *Options, futures, and other derivatives* (7th ed.). Upper Saddle River, NJ: Pearson Prentice Hall.

ISACA. (2009). *Cloud computing: Business benefits with security, governance and assurance perspectives*. White Paper. ISACA.

JISC. (2011). *Cloud computing increasingly attractive to universities, says JISC*. JISC.

Khajeh-Hosseini, A., Greenwood, D., & Sommerville, I. (2010). *Cloud migration: A case study of migrating an enterprise IT system to IaaS*. Paper presented at the 3rd IEEE International Conference on Cloud Computing, Cloud 2010. Miami, FL.

Klems, M., Nimis, J., & Tai, S. (2008). *Do cloud compute? A framework for estimating the value of cloud computing*. Paper presented at Cloud Computing and its Application CCA 08. Chicago, IL.

Lobo, L. O., & Arthur, J. D. (2005). *An objectives-driven process for selecting methods to support requirements engineering activities*. Paper presented at the 29th Annual IEEE / NASA Software Engineering Workshop (SEW-29 2005). Greenbelt, MD.

Msanjila, S. S., Tewoldeberhan, T., Bockstael-Blok, W., Janssen, M., & Verbraeck, A. (2005). *E–supply chain orchestration using web service technologies: A case using BPEL4WS*. Paper presented at IRMA 2005 Conference. San Diego, CA.

Nick, J. M., Cohen, D., & Kaliski, B. S. (2010). Key enabling technologies for virtual private clouds. In *Handbook of Cloud Computing 2010, Part 1* (pp. 47–63). London, UK: Springer. doi:10.1007/978-1-4419-6524-0_3

Office of Government Commerce. (2007). *Official ITIL lifecycle publication suite (5 books)*. TSO Publisher.

Oracle. (2009). *Platform-as-a-service private cloud with Oracle fusion middleware*. White Paper. Oracle.

Oracle. (2010). *Oracle cloud computing*. White Paper. Oracle.

Pajorova, E., & Hluchy, L. (2010). *3D visualization the results of complicated grid and cloud-based applications*. Paper presented at the 14th International Conference on Intelligent Engineering Systems (INES). Las Palmas, Spain.

Papazoglou, M. P., & Georgakopoulos, D. (2003). Service oriented computing. *Communications of the ACM*, *46*(10).

Prechter, R. R. Jr, & Parker, W. D. (2007). The financial and economic dichotomy in social behavioral dynamics: The socionomic perspective. *Journal of Behavioral Finance*, *8*(2), 84–108. doi:10.1080/15427560701381028

Rungtusanatham, M., Salvador, F., Forza, C., & Choi, T. Y. (2003). Supply-chain linkages and operational performance: A resource-based perspective. *International Journal of Operations & Production Management*, *23*, 1084–1099. doi:10.1108/01443570310491783

Schubert, H., Jeffery, K., & Neidecker-Lutz, B. (2010). *The future for cloud computing: Opportunities for European cloud computing beyond 2010*. Retrieved from http://cordis.europa.eu/fp7/ict/ssai/docs/cloud-report-final.pdf.

Sharp, W. F. (1990). *Capital asset prices with and without negative holdings*. Nobel-Prize Economics Lecture. Retrieved from http://www.nobelprize.org/nobel_prizes/economics/laureates/1990/sharpe-lecture.pdf.

Sun Microsystems. (2009). *Introduction to cloud computing architecture*. White Paper. Sun Microsystems.

Weinhardt, C., Anandasivam, A., & Blau, B., & StoBer, J. (2009). Cloud computing – A classification, business models, and research direction. *Journal of Business and Information Systems Engineering*, *1*(5), 391–395. doi:10.1007/s12599-009-0071-2

Zhang, H., Liu, L., & Li, T. (2011). Designing IT systems according to environmental settings: A strategic analysis framework. *The Journal of Strategic Information Systems*, *20*, 80–95. doi:10.1016/j.jsis.2011.01.001

Zirpins, C., & Emmerich, W. (2008). A reference model of virtual service production networks. *International Journal on Service Oriented Computing and Applications*, *2*(2-3), 145–166. doi:10.1007/s11761-008-0025-4

KEY TERMS AND DEFINITIONS

Cloud Computing Business Framework (CCBF): A dynamic framework to help organisations to achieve good Cloud design, deployment, and services with real case studies.

Organisational Sustainability Modelling (OSM): This is a technique and methodology presented in this paper to describe how to measure Cloud business performance or ROI accurately.

3D Visualisation: A technique to present key statistics as 3D to ensure no missing analysis.

Cloud Cost-saving: It is a key benefit of Cloud adoption and this paper presents this aspect in quantitative focus.

Return of Investment (ROI): This is dependent on organisational focus and is defined as (1) cost-saving and (2) service improvement and user satisfaction.

Green IT: Cloud Computing and the use of CCBF can achieve Green IT objectives as demonstrated by the University of Southampton.

User Satisfaction for Cloud Services: This is an important area for Cloud adoption and is essential for service providers.

Chapter 9
Reducing Design Margins by Adaptive Compensation for Thermal and Aging Variations

Zhenyu Qi
University of Virginia, USA

Yan Zhang
University of Virginia, USA

Mircea Stan
University of Virginia, USA

ABSTRACT

Corner-based design and verification are based on worst-case analysis, thus introducing over-pessimism and large area and power overhead and leading to unnecessary energy consumption. Typical case-based design and verification maximize energy efficiency through design margins reduction and adaptive computation, thus helping achieve sustainable computing. Dynamically adapting to manufacturing, environmental, and usage variations is the key to shaving unnecessary design margins, which requires on-chip modules that can sense and configure design parameters both globally and locally to maximize computation efficiency, and maintain this efficiency over the lifetime of the system. This chapter presents an adaptive threshold compensation scheme using a transimpedance amplifier and adaptive body biasing to overcome the effects of temperature variation, reliability degradation, and process variation. The effectiveness and versatility of the scheme are demonstrated with two example applications, one as a temperature aware design to maintain I_{ON} to I_{OFF} current ratio, the other as a reliability sensor for NBTI (Negative Bias Temperature Instability).

DOI: 10.4018/978-1-4666-1839-8.ch009

INTRODUCTION

IT infrastructure has become an essential foundation of human society. Information storage and computation span every aspect of our daily life, from personal entertainment and education, to government and business operation. Moreover, for real-time service providers, data centers and their connections to outside terminals need to be running for a long time without interruption. Examples are banking systems and servers, online shopping and tracking, supply chain management, and multimedia streaming. The unprecedented increasing dependence and requirement on transaction availability and speed call for the deployment of more high performance and reliable computing devices.

However, this leads to at least two problems. At the macroscale level, without careful design and planning the environmental impact due to technology development is unsustainable. For example, the energy cost of IT infrastructure and its carbon footprint doubles in less than five years (Pedram, 2009). Technology-equipped buildings occupy 70% of total electricity usage and emit 40% of greenhouse gases annually in the US (Kleissl & Agarwal, 2010). At the microscale level, high performance and reliability could be at odds for concurrent circuit designs based on the semiconductor technology. For example, high performance designs usually come with high power/temperatures and high stress levels on devices, leading to thermal and reliability problems. In particular, they can cause the instability (including both short time variation and long time shifting) of the threshold voltage of the basic circuit element—CMOS transistors.

The threshold voltage, or V_{TH}, is a critical parameter in circuit design. V_{TH} determines the current flowing through the transistor at least linearly, and in turn affects its equivalent resistance, transconductance, pull-up or pull-down strength and ultimately the speed and power of circuits. Even in advanced sub-micron technologies the dependence of the transistor current on V_{TH} can still be quite super-linear when all short channel effects are considered (Qi & Stan, 2007). The effects in the sub-threshold region are even more prominent since the leakage current grows exponentially as V_{TH} decreases.

Although threshold voltages are specified by technology vendors and treated as fixed and static values in simulations at design time, run-time factors like temperature variability and reliability degradation, as well as process variation, can considerably change V_{TH} both in the short and long term for individual transistors. With the small margins of most circuit designs today, even a relatively small amount of shift can have a significant impact or even lead to circuit failure. On the other hand, preventing these failures by initial overdesigning with large upfront built-in margins can impose large penalties in area and power. For example, an optimized reliability-aware sizing algorithm proposed in (Paul, Kang, Kufluoglu, Ashraful Alam, & Roy, 2006) still incurs an area overhead from 3.31%-13.6% for ISCAS testbench circuits. This conventional method variation by overdesign to tolerate reliability hazards, thermal issues and process leads to unsustainable increase of cost both for the designer and the users.

In this chapter, we propose an adaptive threshold compensation scheme using a transimpedance amplifier and adaptive body biasing to compensate the V_{TH} shift dynamically. The effect of body biasing on modulating transistor threshold voltages has been heavily exploited in recent years to achieve low power and combat process variation (Karnik, Borkar, & De, 2002; Tschanz, et al., 2002). The scheme employed in this chapter works for a range of applications including, but not limited to, manufacturing process variations. Two example applications are presented in this chapter. The first compensates for temperature induced I_{ON} to I_{OFF} mismatch and the second compensates for threshold shifting due to the long time aging/reliability effect.

Figure 1. The proposed transimpedance amplifier as a closed-loop feedback system and its equivalent small signal circuit model

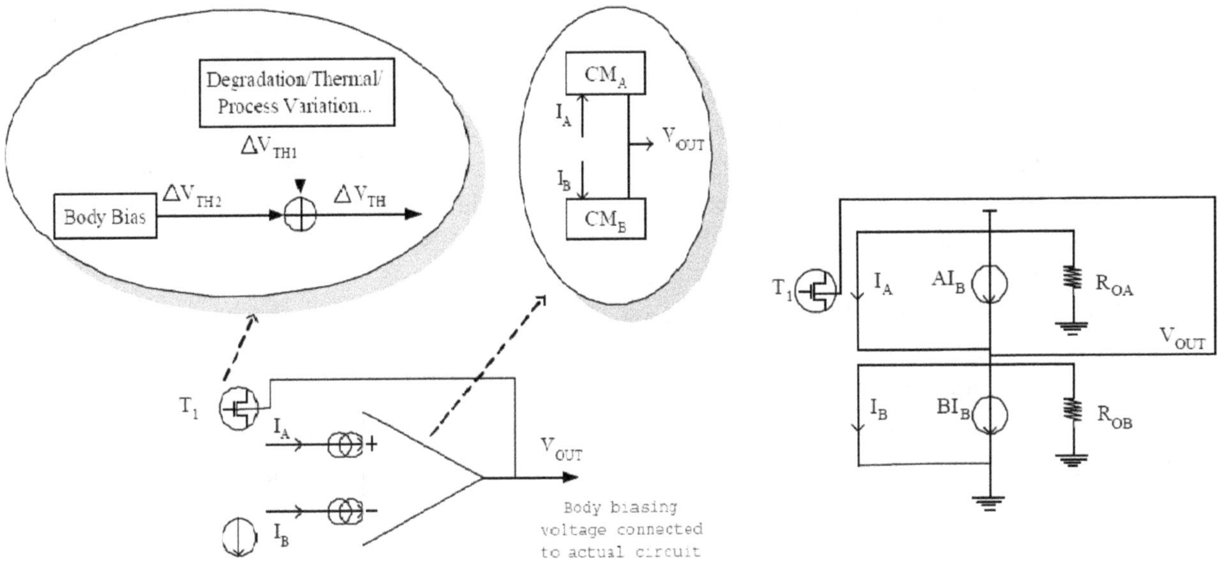

Adaptive compensation of threshold voltage reduces necessary design margins, and enables typical case-based design instead of the traditional worst case-based (or corner based) design. Worst case-based design leads to over-pessimism and unnecessarily large margins/guardbands for the majority of components that do not fall on the tail of the distribution caused by manufacturing or runtime variations. The over-designed components not only take chip real estate and impose non-recurring penalty for designers and manufacturers, but also consume extra power and impose a recurring penalty on the consumer and the environment. The proposed module can also be deployed in large numbers and distributed across the whole chip for adaptive compensation at a finer granularity, and thus achieve more aggressive savings. We believe distribution and adaptation are two key strategies for low-power energy-conserving sustainable computing, and both are embodied in the proposed threshold voltage compensation scheme.

ADAPTIVE THRESHOLD COMPENSATION SCHEME

Figure 1 shows the proposed threshold compensation scheme. It is mainly composed of two parts, the first being the transistor, T1, which could either be a PMOS or an NMOS. T1 serves as a 'proxy' for the transistors in the actual circuit whose threshold voltages need to be compensated due to various reasons including, but not limited to, process and temperature variation and reliability degradation due to aging. The net effect of these factors is the threshold voltage shift noted as ΔV_{TH1} in Figure 1. Ideally, we want to maintain the original V_{TH} value of T1 by offsetting ΔV_{TH1} by ΔV_{TH2}, a voltage shift in the opposite direction of ΔV_{TH1}. In this scheme, ΔV_{TH2} comes from the body biasing effect of T1. Providing the appropriate body bias voltage V_{OUT} to generate ΔV_{TH2} is the goal of the second part of the compensation circuit, shown as a differential transimpedance amplifier on the right.

The transimpedance amplifier mainly comprises two sets of current mirrors, denoted in Figure 1 as CM_A and CM_B. The two differential inputs are the current I_{DS} from the proxy transistor

T1, and a reference current I_B. I_A and I_B are fed into CM_A and CM_B separately, and propagated along to the last stage after being amplified with current gains A and B. Notice that the arrows on I_A and I_B simply mean that they are input variables of the circuit, not necessarily the direction of the currents. The composite current gains A and B can be achieved with multiple stages of current mirrors within CM_A and CM_B. The last stage settles to an equilibrium state only if $A \times I_{ON} = B \times I_{OFF}$. This balance provides the output voltage V_{OUT}. V_{OUT} is the body biasing voltage connected to the actual circuit component that needs threshold compensation. It is also fed back to the bulk of the proxy transistor T1 to complete the closed loop system.

Threshold Compensation Capability Analysis

The threshold compensation system can be analyzed with the equivalent small signal circuit also shown in Figure 1. Since later we will demonstrate an example application of the threshold compensation circuit for NBTI, the following analysis is performed assuming the proxy transistor T1 is a PMOS, since NBTI only affects PMOS transistors. For the ease of notation, all V_{TH} and ΔV_{TH} carry signs within themselves. Therefore, for a PMOS transistor in the saturation mode, according to the alpha-power law model (Sakurai & Newton, 1991).

$$I_A = K\frac{W}{L}(V_{DD} + V_{TH})^{\alpha} \tag{1}$$

Notice that for PMOS transistors V_{TH} is negative. The meanings of the parameters are summarized in Table 1.

Applying Taylor expansion:

$$\Delta I_A = \alpha K\frac{W}{L}(V_{DD} + V_{TH})^{\alpha}\frac{\Delta V_{TH}}{(V_{DD} + V_{TH})} \tag{2}$$

Table 1. Parameters used in the derivation

DEFEINITION/DESCRIPTION OF PARAMETER	
PARAMETERS	DESCRIPTION
W, L	width and length of proxy transistor T1
K	device-transconductance parameter of T1
α	short channel coefficient of T1
V_{DD}	power supply voltage for T1
V_{TH}	actual threshold voltage of T1, signs embedded
V_{TH0}	threshold voltage of T1 without body effect
ΔV_{TH1}	threshold voltage shift of T1 due to non-ideal factors
ΔV_{TH2}	threshold voltage change of T1 from body biasing
ΔV_{TH}	net V_{TH} change, $\Delta V_{TH=}\Delta V_{TH1} + \Delta V_{TH2}$
	body effect coefficient, or bulk threshold parameter of T1
I_A	current from proxy transistor T1, and input to CM_A
I_B	reference current, and input to CM_B
A,B	current gain of CM_A and CM_B
R_{OA}, R_{OB}	small sig. output resistance of the last stage
V_{OUT}	output body biasing voltage to actual circuits

Referring to Figure 1, here:

$$\Delta V_{TH} = \Delta V_{TH1} + \Delta V_{TH2}$$

The output body bias voltage:

$$V_{OUT} = (AI_A - BI_B)\frac{R_{oA}R_{oB}}{R_{oA} + R_{oB}} \tag{3}$$

R_{OA} and R_{OB} are the small signal output resistance of the two current controlled current sources AI_A and BI_B respectively. Assuming the reference, current I_B stays constant:

$$\Delta V_{OUT} = A\frac{R_{oA}R_{oB}}{R_{oA} + R_{oB}}\Delta I_A \tag{4}$$

Equations (2) and (4) can also be obtained by taking derivatives directly. Finally, considering the body biasing effect, the threshold voltage of T1 is (Sakurai & Newton, 1991):

$$V_{TH} = V_{TH0} + \gamma(\sqrt{2\varphi_F - V_{BS}} - \sqrt{2\varphi_F}) \tag{5}$$

This can be further simplified (Sakurai & Newton, 1991) to:

$$V_{TH} = V_{TH0} - \gamma_1 V_{BS} \tag{6}$$

thus:

$$\Delta V_{TH} = -\gamma_1 \Delta V_{BS} \tag{7}$$

Notice that this ΔV_{TH} is essentially the ΔV_{TH2} in Figure 1. Also for PMOS, ΔV_{TH} and V_{BS} have embedded negative and positive signs.

Combining equations (2), (4) and (7), we get:

$$\Delta V_{TH2} \rightarrow = \alpha K\frac{W}{L}(V_{DD} + V_{TH})\alpha\frac{1}{(V_{DD} + V_{TH})}\gamma A\frac{R_{oA}RoB}{R_{oA} + R_{oB}}(\Delta V_{TH1} + \Delta V_{TH2})$$
$$= C(\Delta V_{TH1} + \Delta V_{TH2}) \tag{8}$$

where:

$$C = -aK\frac{W}{L}\frac{(V_{DD} + V_{TH})^{\alpha}}{(V_{DD} + V_{TH})}\gamma_1 A\frac{R_{oA}R_{oB}}{R_{oA} + R_{oB}}$$

and $C>0$.

Now we rearrange equation (8):

$$\frac{\Delta V_{TH2}}{\Delta V_{TH1}} = -\frac{C}{1 + C} \tag{9}$$

Equation (9) is quite interesting. Remember that the ultimate goal of the whole system is to use ΔV_{TH2} to compensate the threshold voltage change ΔV_{TH1} caused by non-ideal factors, therefore ideally we would like $\Delta V_{TH2} = -\Delta V_{TH1}$. However, equation (9) demonstrates that the exact equality is really impossible. However, the larger the constant C is, the more compensation we achieve and the closer the two values can be to each other—C can be viewed as the open loop gain of the feedback system. Increasing the proxy transistor size, the current mirror gain and the last stage output resistance are all possible ways to maximize C. While this opens up optimization opportunities for power and area, these 'knobs' are *not* independent of each other, and the optimization *cannot* be carried out separately, due to the analog nature of the circuit. An important consideration is that the design should always guarantee the correct working range of the current mirrors. How to maximize the gain and quantify the tradeoff of parameters given an area or power budget is our ongoing work.

Finally, notice that although the above analysis assumes a constant reference current I_B, a similar derivation can be carried out in case when I_B is also a variable. We will see an example in the temperature aware design in the next section.

In the following sections, we exploit this adaptive circuit design technique for two example applications, one for reducing temperature sensitivity and the other for mitigating the impact of circuit aging. Both of these examples demonstrate how the proposed adaptive circuit design can help the system achieve its optimum operating point and maximize the system computation efficiency. While temperature- and reliability-aware designs are in themselves important topics for reliable sustainable computing, we mainly choose these two examples to show the ability and versatility of the proposed scheme to adapt to different undesirable environmental factors and mechanisms, thus enabling typical case-based design. One should always remember that *green computing is not necessarily equivalent to low power computing*. The amount of consumed *energy* (or the produced carbon footprint for a given computation load) is the ultimate metric for measuring the 'green' part. Therefore, maintaining a circuit or system at its optimal operating point is the key to exploit every bit of the consumed energy for useful computation.

THRESHOLD COMPENSATION FOR TEMPERATURE VARIABILITY

Now we demonstrate how the proposed circuit technique can be used for maintaining an optimal I_{ON} to I_{OFF} ratio for MOS transistors in the presence of temperature variation. While energy, power, temperature, and temperature variations are all important issues for a system built from semiconductor chips, which in turn are built from transistors, it is important to distinguish among these concepts since the design challenges and issues are not always the same, thus cannot always be tackled together. As we will show in the next section, distinguishing these concepts is critical in identifying and directly targeting the challenge for reducing carbon footprint of computation and bringing forth green computing.

Temperature Variation and Why It Is Important

Thanks to the increase in circuit density and clock speed, power and thermal issues have been upgraded to first tier concerns in modern VLSI design, from chips in high performance servers to low power handheld devices (Chandrakasan, Sheng, & Brodersen, 1992). The first thought that usually comes to mind when thinking about thermal issues is probably power consumption, which is indeed a critical concern especially for battery supported, mobile and handheld consumer electronics and devices. The most popular examples are smart phones, tablets, laptops, portable media players, GPS (global positioning system) devices, etc. This concern about power motivated the whole low power design research initiative in both academia and industry, for good reasons.

However, more precisely speaking, for the above applications what we really need should be called 'low energy' design. Recall that energy is the product of power and time. For the given amount of computation (or device usage), simply lowering power certainly helps in reducing thermal hazards, but does not necessarily reduce energy consumption (or the battery usage). In reality, most low power designs or temperature-aware techniques involve lowering circuit speed and performance, thus increase the operation time for a given user task (Lu, Lach, Stan, & Skadron, 2005). For example, the most common and conceptually simple low power technique that has been long known in academia but has only been slowly adopted by industry in recent years, DVFS (Dynamic Voltage and Frequency Scaling), directly plays with this tradeoff between power and time by dynamically increasing or decreasing voltage and frequency at runtime for different applications. Since lower voltage and frequency immediately reduces power, but also decreases circuit speed (equivalently increasing computation time), therefore applying low power techniques like DVFS blindly does not necessarily reduce the energy consumption and increase battery time, but may actually exhaust their battery faster due the extra area and power overhead from the added DVFS management circuitry, in addition to possibly resulting in an unfriendly user experience due to the slower response time. As another example, parallelism can bring back the performance loss due to voltage scaling in low power design (Chandrakasan, Sheng, & Brodersen, 1992), but with leakage power rising rapidly in modern semiconductor technologies, the total energy consumption may still increase due to the extra transistors and area. Therefore, designers should always double check if they are using the right metric for their purpose, and if the metric is indeed improved after everything is considered. In the circuit design community, the Energy-Delay Product (EDP) is often used to evaluate the optimality, since it takes into consideration both energy consumption and system performance.

In this section, we focus on a less obvious, yet equivalently important, challenge hidden in the 'power wall' (a terminology from Asanovic, et al., 2006) temperature. Temperature is closely related to circuit performance and power. On one hand, the temperature is among the most important environmental factors, which can impact both the performance and power consumption (including dynamic runtime power and static leakage power) of integrated circuits. On the other hand, the temperature can also be affected by the power consumption of the circuit itself since all the power consumed by the circuit ultimately transforms into heat, according to the low of conservation of energy. In fact, since an elevated temperature not only hurts circuit performance but may also lead to a phenomenon called 'thermal runaway' (caused by a positive feedback of temperature and leakage power), which permanently destroy a semiconductor microchip, keeping circuit operational temperature in the specified range has become a critical concern, from integrated circuit packaging to IT facility maintenance, and is very cost-sensitive. At the macroscale level, the cooling facility is a significant portion of IT expense. The old cooling method through ventilation is no longer effective. Nowadays, high performance servers or compute farms need to use sophisticated air-conditioning schemes. At the microscale, the on-chip cooling systems like heat sink and spreader designs are getting fancier and more complicated, at the expense of cost that goes to every single chip, and has become a significant portion of the cost of chip manufacturing. Therefore, it is very important to distinguish between the two related concepts, i.e., low power design and temperature aware design.

While keeping the circuit out of dangerous temperature zones is vital for its health, keeping the circuit at optimal temperature can maximize its efficiency, since the variation of temperature can change the optimal operating point of a circuit. Unfortunately, while it may be possible to maintain the ambient or average temperature, it is almost impossible to maintain the temperature for every transistor on the circuit at a certain value, because of on-chip temperature variations. On-chip temperature variations can happen both spatially across different function blocks of the die, and temporally over time when the system runs different applications. Spatial temperature variations occur when some parts of the circuit consume more power, leading to 'hot spots' on the die while other parts are less active and their temperature is relatively lower. Temporal temperature variations occur when the function of the units on the die are switched on or off, changing the power consumption of the processor, thus changing the temperature of the chip over the time. The sources of temperature variations include ambient temperature and heat generated by circuit itself. Temperature variations are usually defined using 'temperature maps' (temporal and spatial distribution of temperature) which are either directly measured or indirectly derived from 'power maps' (temporal and spatial distribution of power). The calculation of temperature maps from power maps has been investigated by many research groups and various thermal simulators have been proposed for this purpose. On the other hand, temperature variation is also a good indicator of power distribution (Qi, Meyer, Huang, Ribando, Skadron, & Stan, 2010). In practice, circuits on a microprocessor die can experience a wide temperature range of 0-70°C for commercial applications and –55-125°C for military applications.

Due to the importance of temperature and the difficulty to control the temperature for every transistor because of on-chip temperature variations, researchers have proposed different techniques to compensate for temperature variation to reduce its impact (Park, et al., 1995). Most of them focus on scaling and exploiting the "design knobs" of V_{TH} and $V_{DD.}$ For example, using fixed-voltage and scaled-V_{TH} techniques, two sub-1V MOS-based reference current generators were proposed in (Tang, Narendra, & De, 2003) to provide immunity against both temperature and process variations. A dynamic voltage scaling technique based on the temperature variation was proposed in Shakeri and Meindl (2003) to reduce power consumption. In the following, we present the design of a temperature-aware circuit using the proposed threshold compensation technique to dynamically track temperature variations and reduce its impact on circuit performance.

Dependence of I_{ON} and I_{OFF} Current on Temperature

To explain the effect of temperature variation on circuit performance, we first show the dependence of the I_{ON} and I_{OFF} current on temperature. Here we are mainly concerned with digital circuits, which form the dominant portion of modern VLSI system and consume most energy and chip area. The input voltage is either close to V_{DD} (called '1' or ON state) or close to ground (called '0' or OFF state). When in the 'ON' state, a transistor operates in the saturation region. When in the 'OFF' state, a transistor operates in the subthreshold region).

In the ON state, the drain current of a transistor I_{ON} can be expressed by the α power law model (Sakurai & Newton, 1991) as shown in Equation (1). For simplicity we rewrite Equation (1) as:

$$I_{ON} \propto \mu(T)(V_{DD} - V_{TH}(T)) \tag{10}$$

where α is velocity saturation index whose typical value is 2 for long channel MOSFETs and approaches 1 for short channel MOSFET. For current technology α is around 1.3. Notice that the threshold voltage $V_{TH}(T)$ and the mobility $\mu(T)$ themselves also have temperature dependence that can be expressed as follows (Bellaouar, Fridi, Elmasry, & Itoh, 1998):

$$V_{TH}(T) = V_{TH}(T_0) - K(T - T_0) \tag{11}$$

$$\mu(T) = \mu(T_0)\left(\frac{T}{T_0}\right)^{-m} \tag{12}$$

Here T_0 is room temperature (300 Kelvin); K is a threshold voltage temperature coefficient whose typical value is 2.5mV/K and m is mobility temperature exponent whose typical value is 1.5.

According to (11) and (12), it can be seen that both mobility μ and threshold voltage V_{TH} has negative dependence on temperature. Moreover, from (10) we can see that μ and V_{TH} have the opposite impacts on the I_{ON}, thus it is not immediately obvious how I_{ON} changes across a range of temperature values. Figure 2 shows the temperature dependence of I_{ON} for a NMOS transistor with different supply voltage. Based on the analysis in (Park, et al., 1995), if the supply voltage $V_{DD} >> V_{TH}$, I_{ON} is dominantly controlled by the carrier mobility μ and the I_{ON} decreases as temperature increases; However, as V_{DD} approaches V_{TH}, the effect of the V_{TH} shift with temperature becomes stronger and I_{ON} increases with temperature. At a certain V_{DD} value (V_{DD} =0.7 in our chosen technology node as shown in Figure 2), I_{ON} does not

Figure 2. Temperature dependence of I_{ON} with different supply voltages (Zhang & Stan, 2007 ©2007, ACM, Inc. Used with permission)

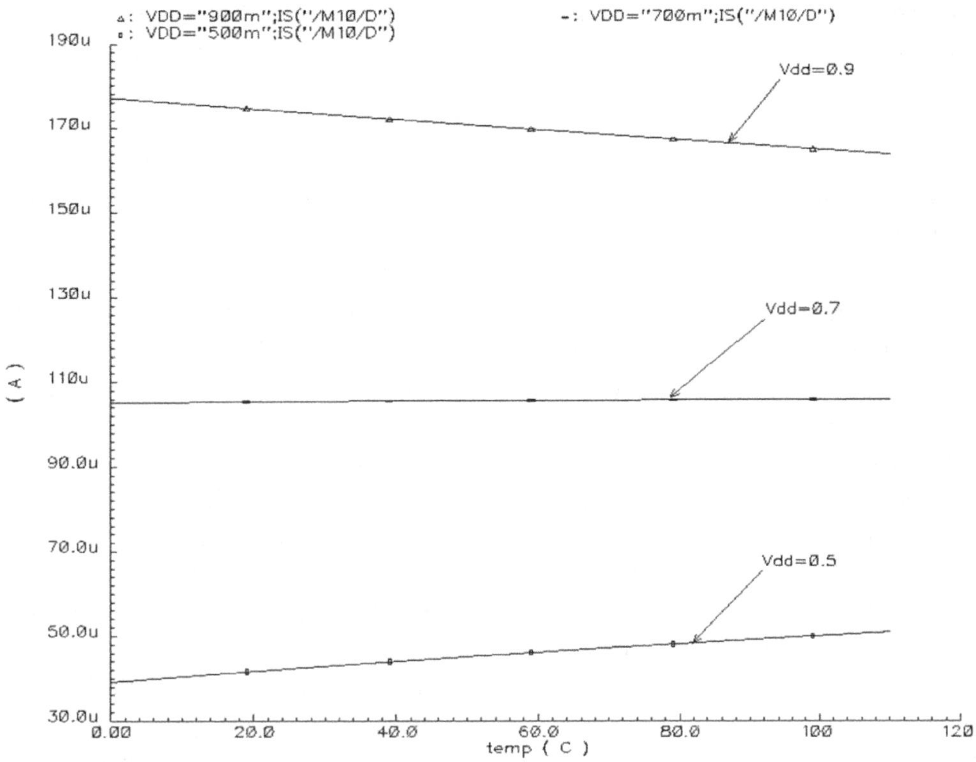

vary with temperature. This voltage is called Zero-Temperature Coefficient (ZTC) supply voltage. In current technologies, V_{DD} is still much larger than V_{TH}, therefore temperature has a negative impact on I_{ON} current, and circuit performance deteriorates at higher temperatures.

When a transistor is turned off, the drain current I_{OFF} is essentially the subthreshold current or leakage current, which can be expressed as:

$$I_{OFF} \propto \mu(T) \exp \frac{V_{GS} - V_{TH}(T)}{nV_t} \left(1 - \exp\left(-\frac{V_{DS}}{V_t}\right)\right) \quad (13)$$

Here n called the subthreshold slope coefficient and V_t=kT/q is the thermal voltage (which is about 26mV at room temperature). Equation (13) shows that I_{OFF} has an exponential dependence on V_t, so I_{OFF} increases exponentially as temperature increases.

Therefore, the effect of temperature variations on I_{OFF} is more significant than its effect on I_{ON}. This leads to a wide range of I_{ON} to I_{OFF} ratio values across the specified temperature range. The I_{ON} to I_{OFF} ratio characterizes the difference between the 'ON' state and the 'OFF' state, thus this value directly affects the robustness of a circuit (for example, a large I_{ON} to I_{OFF} ratio leads to a large noise margin, and a small I_{ON} to I_{OFF} ratio means more susceptibility to power noises and soft errors). This difference must be kept greater than a certain minimum, and the system must be designed to operate at the worst case I_{ON} to I_{OFF} ratio (which in turn implicates the upper bound of the operating temperature range). In a worst case-based design, the circuit will have to be over-designed to accommodate for the worst case. The over-designed circuit will not only incur power and area overhead, but also operate at non-optimal points whenever and wherever the temperature is lower than the upper-bound temperature. For this reason, we adapt the aforementioned threshold compensation scheme to dynamically track the on chip temperature variations and reduce the I_{ON} to I_{OFF} ratio variations across a wide temperature range.

Temperature-Aware Adaptive-Biasing Technique

In order to keep the ratio of I_{ON} to I_{OFF} constant across the temperature range, we can apply the threshold compensation scheme. Recall from Equation (5), positive bias voltage (or forward body bias) decreases the V_{TH} while negative bias voltage (or reverse body bias) increases V_{TH}. The key of the design is to adaptively bias the body voltage so that it tracks the circuit temperature thus produce a relatively constant I_{ON} to I_{OFF} ratio. There are two points we need to consider in the circuit implementation of the general compensation scheme.

First, as we have seen from the analysis in the last section, the temperature variations affect both I_{ON} and I_{OFF}, and their dependence on temperature is quite different Therefore, here it is not enough to simply design the I_A from upper part of the circuit to follow a constant current I_B in the lower part, as shown in Figure 1. Instead, both I_A and I_B are changing with temperature and at different rates, and we need to design for the I_A to I_B ratio (or effectively, I_{ON} to I_{OFF} ratio). Therefore, we need to carefully design the parameters for both the upper and lower portions of the circuit to obtain a dynamic balance of the two, and produce appropriate output body bias voltages to achieve the temperature compensation for the I_{ON} to I_{OFF} ratio.

Second, the most common CMOS (Complementary MOS) based digital circuits have a pull up network comprised of solely PMOS transistors and a pull down network comprised of solely NMOS transistors. The strength of the pull up network affects the rising edge of the output signal, while the pull down network affects the falling edge of the output signal. To reduce the temperature variation induced slew rate changes, both pull up and pull down network need adaptive body biasing to reduce their current change due to the temperature variations. Since NMOS and PMOS transistors have different body bias voltages (the body of NMOS transistor is usually close to ‘0’ or ground, while the body of PMOS transistor is usually close to ‘1’ or power supply voltage), we need to design two sets of body bias circuits to supply different body bias voltages to both transistor types to adjust their drain currents separately.

Figure 3 shows the adaptive body biasing circuit implementation for NMOS transistor temperature compensation. To understand the circuit, we can first view it in two parts, i.e., the upper and lower branches. These two parts monitor *ION* and *IOFF* respectively. In the upper part, N_1 is ON (since its gate is connected to drain) and thus its current is a proxy for I_{ON}. In the lower part, NMOS transistor N_2 is OFF (since its gate is connected to the source) and thus its current is a proxy for I_{OFF}. Then for each part, current mirrors are sized with fixed gains *A* and *B* to separately amplify the *ION* and *IOFF* currents. At the final output the circuit is in equilibrium when $A \times I_{ON} = B \times I_{OFF}$ (again, *A* and *B* are the gains for the ON and OFF branches). The circuit is designed such that when the temperature equals a certain value, the output voltage in the equilibrium state is close to zero, thus the circuit does not provide any bias voltage to the NMOS transistors in the system at this temperature. Then, when the temperature changes due to various reasons including ambient environmental change or on chip power dissipation, the circuit will output a non-zero bias voltage to balance the *ION* and *IOFF* currents and reduce the temperature sensitivity of I_{ON} to I_{OFF} ratio. In this work this zero biasing temperature is set at the middle point of the specified temperature range, but it could also be set to other values, e.g. to the expected working temperature.

Figure 3. Threshold compensation circuit for NMOS to reduce temperature sensitvity of I_{ON} to I_{OFF} ratio (Adatepd from Zhang & Stan, 2007)

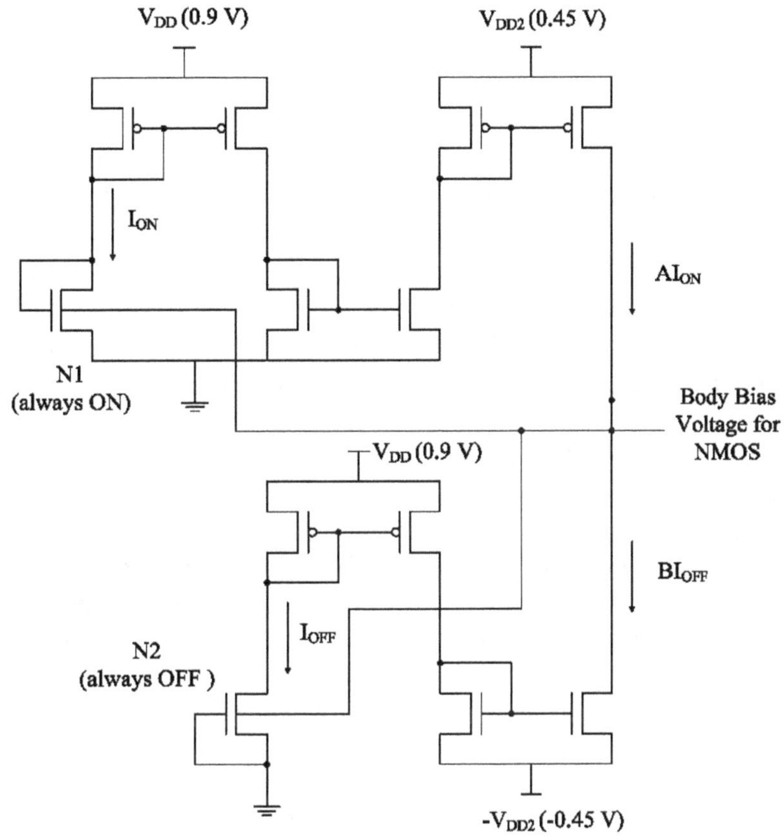

To see how the temperature sensitivity can be reduced, first consider the case when the temperature becomes lower than the zero biasing temperature. Since I_{OFF} decreases much faster than I_{ON}, this leads to an imbalance of $A \times I_{ON} > B \times I_{OFF}$. This imbalance will move the output voltage to be more positive, thus providing a positive bias or forward body bias voltage for NMOS. The forward bias in turn reduces *VTH* and increases I_{OFF} to bring the circuit back to equilibrium. On the other hand, if the temperature is larger than the zero biasing temperature, the faster increase in I_{OFF} leads to an imbalance $A \times ION < B \times I_{OFF}$, which results in a negative output (or reverse body bias for NMOS) which increases *VTH* and decreases I_{OFF} to bring the circuit back to equilibrium.

We note that the threshold compensation circuit requires two additional supply voltages (+0.45V and -0.45V in the implementation shown in Figure 3, but may vary depending on technology and design parameters) to generate the output bias voltage, which would be implemented either through external power pads or internal charge pumps. Moreoever, the value of the output voltage can be used as an indication of the temperature. In other words, each value of V_{OUT} corresponds to a certain temperature value, so with a look up table this circuit can also acts as an on-chip temperature sensor. Finally, implementing the threshold compensation for the PMOS transistor to reduce the temperature sensitivity of its I_{ON} to I_{OFF} ratio is very similar (mirror) to the NMOS one.

Temperature Sensitivity Reduction of I_{ON} to I_{OFF} Ratio

The threshold compensation circuit shown in Figure 3 has been designed in the openly downloadable 70nm Predictive Technology (PTM) (Zhao & Cao, 2006) with the nominal supply voltage of 0.9V. All the simulations in the following sections are also performed with this technology. The output biasing voltage varies from 320mV to –280mV across the temperature range (0 to 110°C). The zero point is set in the middle of the operation temperature range (55°C). We also run Monte Carlo simulations to see the impact of parameter variations. The result shows ~10% standard deviations at high bound and low bound range of output bias voltage.

Figure 4 shows the I_{ON} to I_{OFF} ratios for a single NMOS with and without threshold compensation. The ratio varies from 16600 to 500 without biasing when the temperature varies from 0°C to 110°C. As can be seen, at low temperatures *IOFF* is reduced significantly and circuits operate at a larger *ION/IOFF* ratio than necessary. This leads to sub-optimal performance. By applying forward body biasing to reduce the threshold voltage, the *ION/IOFF* ratio can be brought back to the nominal value. As can be seen from Figure 4, with the proposed compensation scheme the ratio is kept within a much narrower range and only varies from 4500 to 1800. Also, due to the smaller threshold voltage at low temperature, the circuits operate at higher speed. Therefore, in this case the performance can actually be improved with a scheme like Dynamic Voltage and Frequency Scaling (DVFS), which takes advantage of higher circuit speed to reduce computation time with equal or less energy consumption for a given task.

Temperature Compensation for Combinational Circuitry

We first apply the compensation scheme for logic circuitry. A circuit that represents a generic logic

Figure 4. ION/IOFF ratio for a NMOS transistor with and without threshold compensation

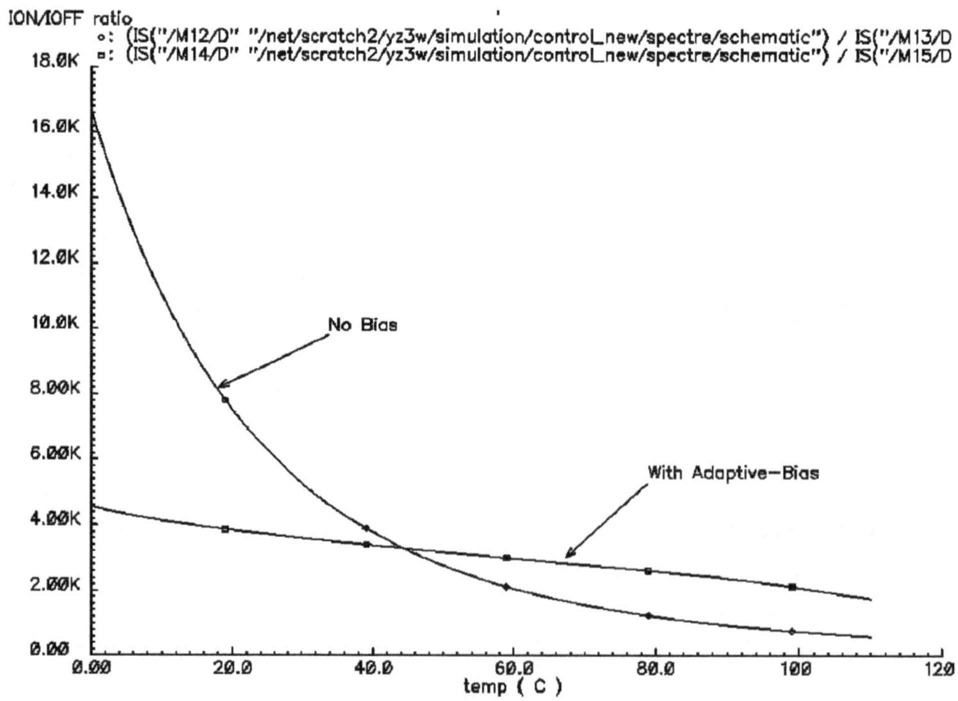

Figure 5. Two tests circuit for temperature variability compensation: (a) represents a combinational circuit and (b) represents a memory circuit (figures from Zhang & Stan, 2007)

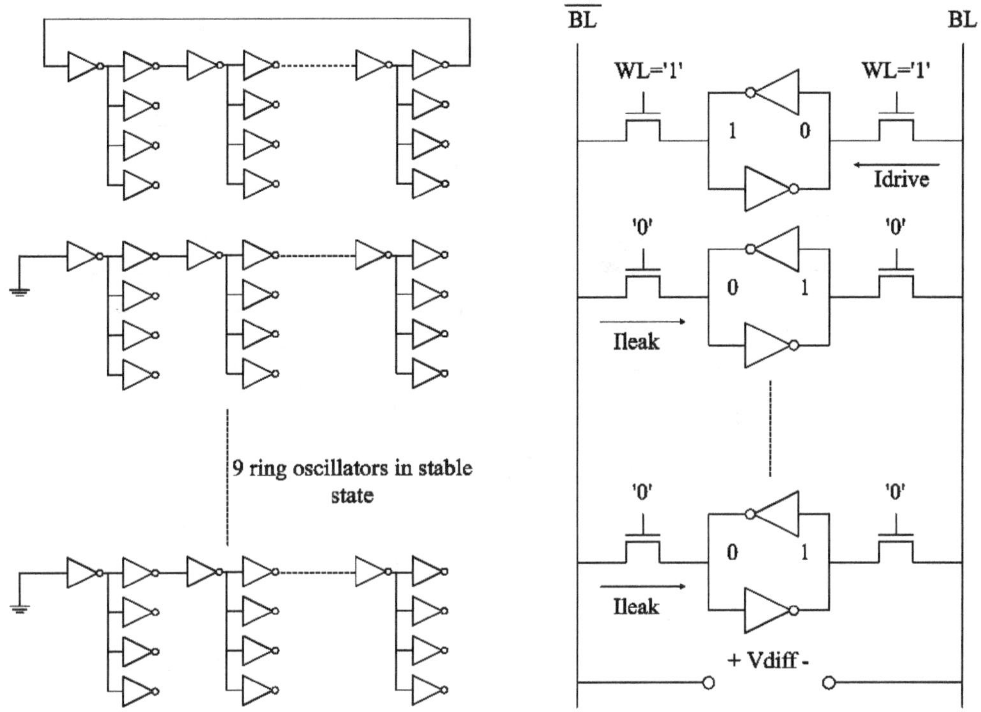

(a) 9-stage ring oscillator with 0.1 switching activity

(b) One column of 6T SRAM cells

circuit is shown in Figure 5(a). The first block in the test circuit is a nine-stage, Fan-Out of 4 (FO4) ring oscillator. Then other nine identical inverter chains are constructed and put in a static state. Together they form a system with an equivalent switching activity factor of 0.1. Since at different supply voltages I_{ON} has a different temperature dependence, two sets of simulations are run at V_{DD}=0.9V (negative dependence) and V_{DD}=0.5V (positive dependence).

The simulation results are shown in Figure 6. There are two other baseline settings in our simulation. "NO BIAS 1" represents test circuit without threshold compensation. "NO BIAS 2" represents a test circuit without threshold compensation but still achieving the same worst-case I_{ON}/I_{OFF} ratio at 110°C (in order to do this we have to increase default threshold voltage to match the actual threshold voltage with threshold compensation at 110°C). Finally, "TAC-ABB" represents the temperature aware circuit with adaptive body biasing circuitry for threshold compensation.

Figure 6a shows the comparison of I_{ON}/I_{OFF} ratio between the three settings for different supply voltages. Here I_{ON} current is the average drain current in the ON state and I_{OFF} the average leakage current in the OFF state. With threshold compensation, the ratio has much less temperature sensitivity and remains almost constant across the temperature range for both supply voltages. Furthermore, "NO BIAS 1" has lower worst-case I_{ON}/I_{OFF} ratio compared to other two settings. "NO BIAS 2" has the same worst-case I_{ON}/I_{OFF} ratio as the "TAC-ABB" but the ratio at lower temperatures is significantly higher than the other two settings.

Figure 6b shows the comparison of delay between three settings for different supply voltages. At low temperatures, the system with threshold

Figure 6. Comparing I_{ON}/I_{OFF} ratio, delay and energy delay product in three scenarios across the temperature range 0 °C - 110 °C. Each coomparison is done for both (figure from Zhang & Stan, 2007).

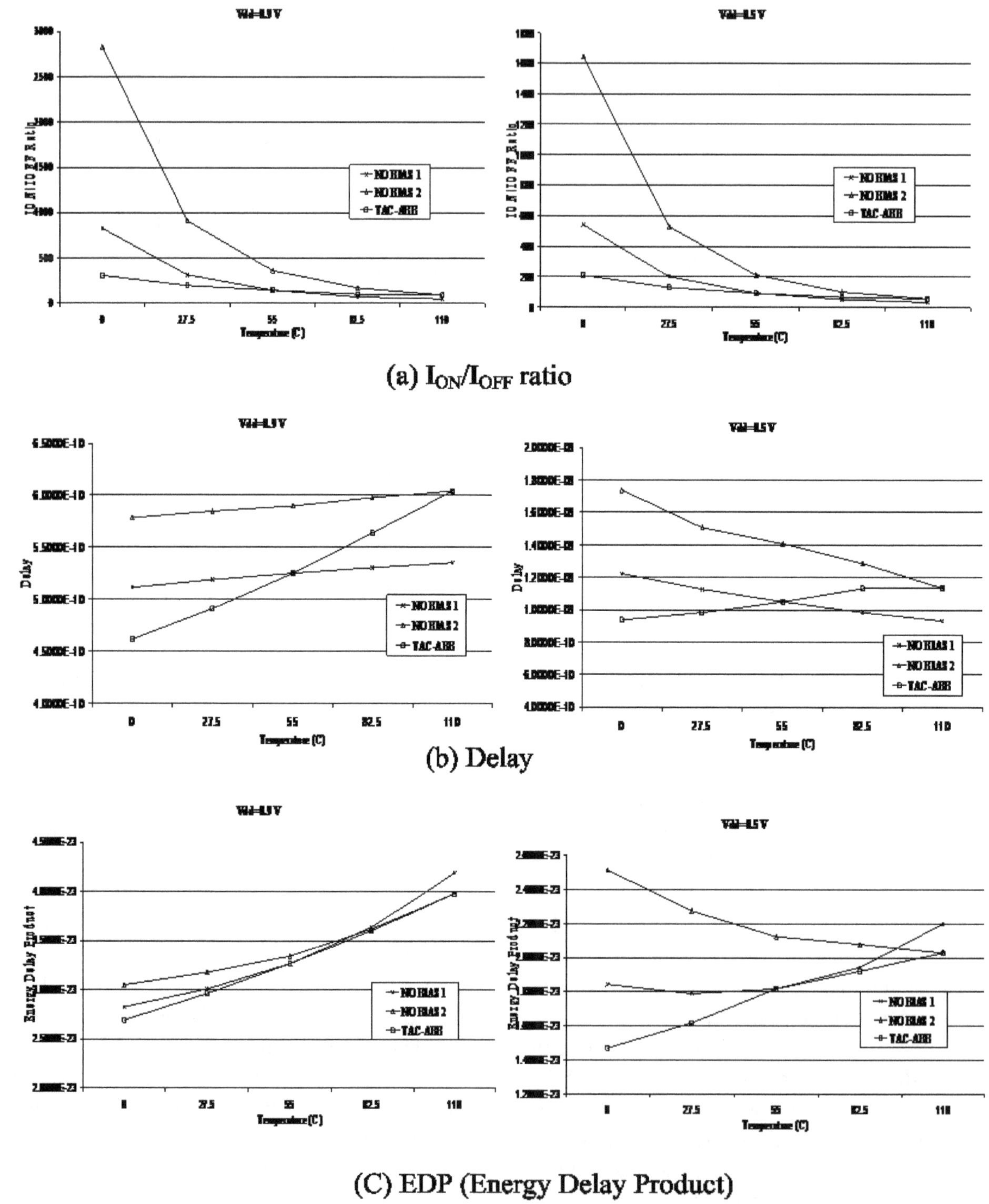

compensation operates at a higher frequency and lower delay thus offering a higher performance compared with both "NO BIAS 1" and "NO BIAS 2." At high temperatures, the worst case-based "NO BIAS 2" design offers the same I_{ON}/I_{OFF} ratio. However, this comes at the expense of longer delay at other temperatures since it has higher V_{TH} value than the other designs. For non-temperature-aware designs like "NO BIAS 1," the leakage current increases drastically, which can cause reduced noise margins, potential thermal hazards and increased leakage power. With the proposed threshold compensation scheme, the system operates at a lower frequency at high temperatures to retain the system's reliability and stability. By operating at lower speed, the circuit also generates less power, which can help cool down the system. Moreover, with reverse bias, the NMOS transistors have higher V_{TH}, which further cuts down the leakage power. Therefore, the threshold compensation scheme can indeed help reliable and green computing.

On the other hand, the right hand side of Figure 6b shows that for lower values (here V_{DD}=0.5 V), the circuit speed could have a positive temperature dependence, i.e., the circuit becomes faster at higher temperature. This reverse temperature dependence usually happens for low power supply voltages (Park, et al., 1995). This phenomenon can lead to a positive feedback, in which the circuit runs faster at higher temperatures, and generates more power to further increase the temperature. This positive feedback loop may cause serious reliability problem. Again, the dynamic threshold compensation technique can throttle this mechanism to prevent potential thermal runaway.

Figure 6c shows the comparison of Energy Delay Product (EDP) between the three settings for different supply voltages. Overall EDP is improved with the threshold voltage for both high and low supply voltages. Compared to "NO BIAS 1," about 5% improvement is achieved at both 0°C and 110°C when V_{DD} = 0.9V; about 20% and 8% improvements are achieved at 0°C and 110°C respectively when V_{DD} =0.5. Compared with "NO BIAS 2" which by design has the same V_{TH} as the threshold compensation circuit, even larger improvements are achieved across all the temperature values. At 0°C EDP is reduced by 11.6% and 41.5% for V_{DD} =0.9V and 0.5V, respectively.

From the above simulation results, it can be observed that by reducing the temperature sensitivity of I_{ON}/I_{OFF} ratio, the threshold compensation scheme offers high performance and reliability. The scheme also achieves a smaller energy-delay product, which means that for the same task, it can reduce the energy without posing a penalty on performance. This is achieved by dynamically tracking the temperature and exploiting higher performance at low temperatures and reducing leakage power at high temperatures. When reverse temperature dependence happens, the same circuit can also throttle the system to prevent potential thermal runaway and unreliable conditions. In summary, the threshold compensation circuit helps the system remain in the optimal and most energy efficient condition.

Threshold Compensation for Memory Circuitry

Due to the large capacity and low switching activity, leakage is a much more critical metric in memory design than in combinational circuit design. However, leakage is important not only because it determines the static power consumption, but it can also affect SRAM stability. In this section, we show how the threshold compensation scheme also helps memory circuits. We implemented a 64-row × 32-column, standard 6T (six-transistor) SRAM array in the 70nm PTM model with 0.9 V supply voltage as an example. Figure 5b shows one column of the SRAM memory.

Assume the top cell which stores a "0" is being read. After both bit lines BL and $\overline{BL}$ are precharged to "1," the top access transistors are selected. BL is discharged while $\overline{BL}$ keeps its pre-charged value. After some time a certain amount of differential voltage (100mV in this example) is built up between the two bit lines and the sense amplifier for this column is activated to poll the stored value. Notice that the above description does not consider the influence of other cells in this column which all share the same bit lines. In the worst case, a "0" is stored in the top cell while all other cells are storing "1," therefore the leakage current from these cells results in a voltage drop on $\overline{\mathrm{BL}}$. This voltage drop not only deteriorates the access time (thus memory performance) but also has an impact on the cell stability and noise margin.

At high temperatures, the leakage current *IOFF* increases faster than *ION*, thus by reverse body biasing and increasing *VTH*, *IOFF* can be reduced, thereby decreasing access time and increasing noise margin. At low temperatures, *IOFF* decreases quickly and the read current (*ION*) becomes the dominant factor of the read speed. Forward body biasing can reduce V_{TH}, increase *ION* and reduce the access time. Therefore threshold compensation can improve SRAM performance across whole temperature range. Notice that this is different from the combinational circuit case, where reverse body bias increases delay and throttles circuit speed.

Figure 7a shows the performance comparison for SRAM. By applying threshold compensation, 5.1% and 18% read time improvement are achieved at low and high temperatures respectively. Write time increases slightly at high temperatures, but write is not a critical operation for memory access since it has a large noise margin, so this penalty is usually acceptable.

We use the method proposed in Hamzaoglu, et al. (2002) to measure the 6T SRAM cell stability. Figure 7(b) shows the stability comparison. At high temperature with threshold compensation, the SRAM cell stability is improved by 15.8% and worst-case cell stability is improved by 10%. This comes with a penalty of less stability at low temperatures. However, this penalty can be easily justified since most reliability problems happen in high temperatures. This increasing stability with rising of the temperature is especially suitable for high performance applications, which usually suffer from degraded performance and stability in elevated temperatures.

THRESHOLD COMPENSATION FOR NBTI

The previous examples have demonstrated how the threshold compensation circuits can be used to improve circuit performance, energy efficiency and thermal reliability by reducing temperature variability for both combinational and memory circuits. In this section, we show another area where the threshold compensation scheme can find applications.

The Circuit Also Ages

Even as it continues unabated, technology scaling hardly comes for free any more. Reliability has become a major concern and poses hefty penalties on future scaling (Borkar, 2005). Reliability is usually measured by MTTF (Mean Time to Failure), or failure rate (Hennessy & Patterson, 2003). However, different reliability hazards impact circuits in very different ways and need different ways of handling. For example, soft errors (also called single event upsets or SEU) occur almost purely randomly and cause random transient failures. Aging mechanisms like NBTI (Negative Bias Temperature Instability), HCI (Hot Carrier Injection), EM (Electro-Migration), TDDB (Time Dependent Dielectric Breakdown), and thermal cycling are more tractable, as they usually appear as a result of a certain amount of circuit functioning, and ultimately limit the system lifetime. Interestingly some aging effects are reversible. For instance, 'healing effects' exist for

Figure 7. Improvement of both performance and stability for SRAM cells by threshold compensation (figure from Zhang & Stan, 2007)

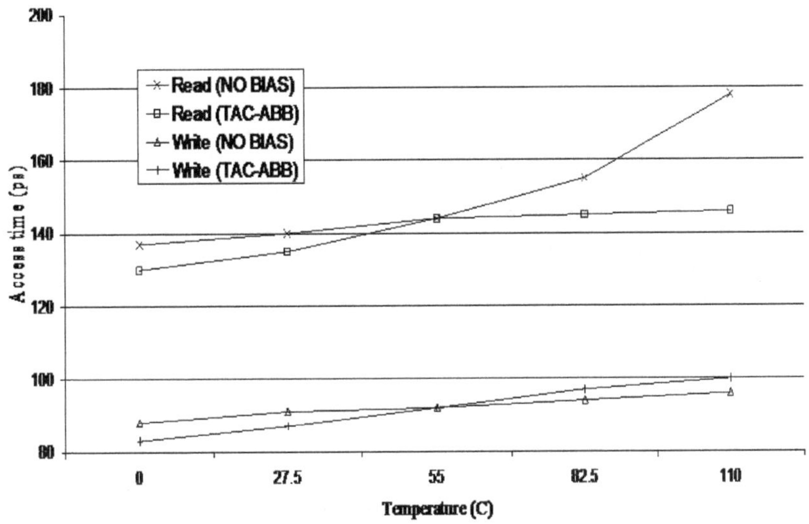

(a) SRAM access time

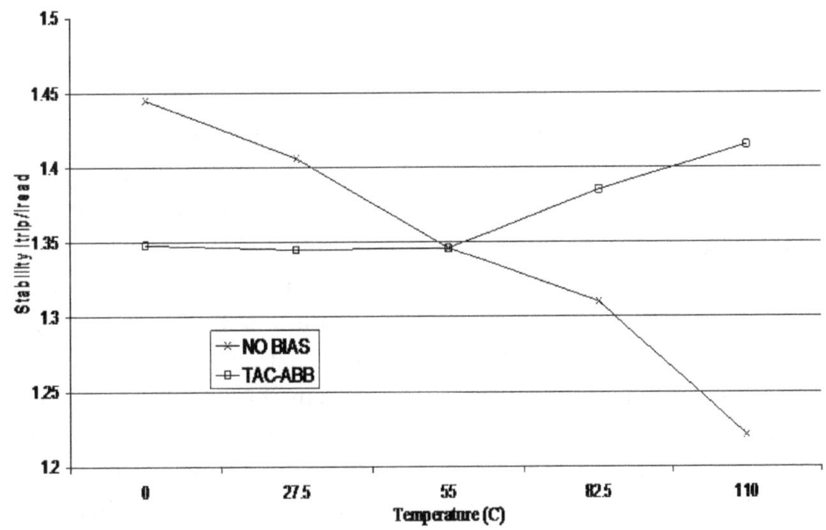

(a) SRAM cell stability

both NBTI (Bhardwaj, Wang, Vattikonda, Cao, & Vrudhula, 2006) and EM (Blaauw, Oh, Zolotov, & Dasgupta, 2003).

At the system level, reliability-aware design schemes, including Dynamic Reliability Management (DRM) and reliability banking, have been proposed (Karl, Blaauw, Sylvester, & Mudge, 2006; Lu, Lach, Stan, & Skadron, 2005). In a general sense, Dynamic Thermal Management (DTM) is also a reliability aware design method that targets temperature cycling (thermal shock). At the circuit level, redundancy based designs like TMR (Triple Modular Redundancy) and Razor (Ernst, et al., 2005) were introduced to

Figure 8. Circuit performance degradation comparison with different reliability aware designs (Qi & Stan, 2008 ©2008, ACM, Inc. Used with permission.)

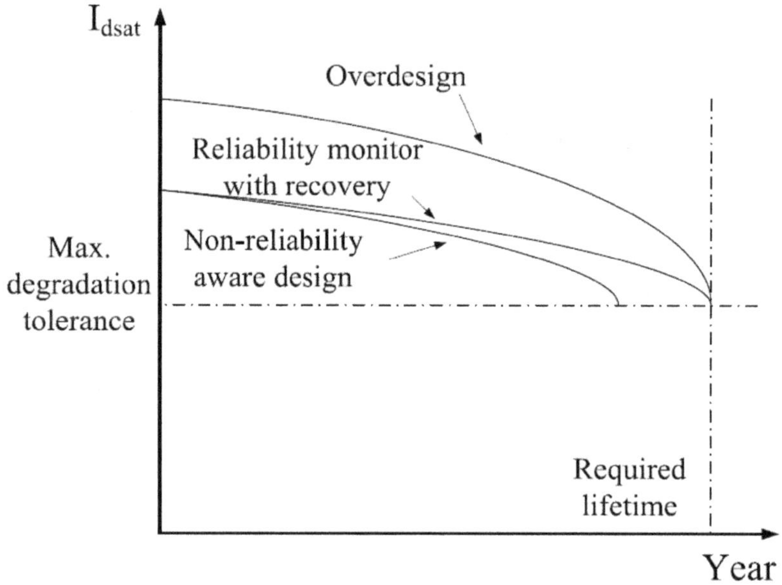

prevent local circuit failures from propagating in the system. 'Circuit hardening' (Zhou & Mohanram, 2006), among other techniques that provide extra margin by overdesigning, is often adopted as well. Redundancy and overdesigning are general methods to deal with most reliability problems, but they could also lead to large area and power overhead by duplicating, triplicating, or upsizing transistors by a significant portion. Accurate reliability models and time consuming reliability simulations, which are not commonly found in current design kits and design tools, may also be needed. In addition, it can be difficult to calculate the required margin for desired lifetime if the realistic input signal pattern is unknown, since the switching activity directly decides the amount of NBTI stress for each PMOS.

It has been suggested that online monitoring with in-situ sensors might be more effective in dealing with reliability problems (Mishra, Pecht, & Goodman, 2002; Mitra, 2007). The idea is to track circuit degradation in real time and issue advance 'warnings' before the system starts to malfunction. A warning indicates either maintenance ('healing,' if possible) or replacement is required. Figure 8 compares the different impact of in-situ sensors and overdesigning on circuit lifetime. In contrast to the above worst-case design with excessive margins, reliability monitors enable typical-case designs. In this chapter, we propose a reliability monitor for NBTI that not only tracks the degradation online, but also *adaptively* heals the degradation and prevents it from harming normal system functioning. In advanced technologies with thin gate oxide, NBTI is predicted to dominate other reliability mechanisms like HCI in system lifetime (Bhardwaj, Wang, Vattikonda, Cao, & Vrudhula, 2006). Although the monitor here is designed specifically for NBTI, it could also be adapted for other reliability hazards as long as they can be monitored or mitigated through transistor threshold voltage, e.g. HCI and thermal hazards.

NBTI Modeling

NBTI becomes a major concern in modern CMOS technologies with thin gate oxide due to its exponential dependence on oxide thickness and temperature (Chen, et al., 2003). NBTI stress occurs in PMOS transistors with negative gate to source voltages, which is often the case in CMOS circuits where PMOS serves as the header device. The most notable aging effect from NBTI is the increase in absolute value of the threshold voltage V_{TH} for PMOS, which contributes to decreased transconductance and weaker driving capability. Although similar mechanisms exist for NMOS, i.e., PBTI (Positive Body Bias Temperature Instability), NBTI imposes a much shorter lifetime compared to PBTI and other major reliability concerns, and has becomes a very active research topic recently (Bhardwaj, Wang, Vattikonda, Cao, & Vrudhula, 2006; Kang, Kim, Islam, Alam, & Roy, 2007; Kang, Kufluoglu, Alam, & Roy, 2007; Paul, Kang, Kufluoglu, Alam, & Roy, 2006; Vattikonda, Wang, & Cao, 2006; Wang, et al., 2007). While this phenomenon has been known for more than 40 years in the device area, the underlying physics is still being studied under various circuit operating conditions (Massey, 2004), and can be more complicated when coupled with process variation (Agostinelli, et al., 2005).

Nonetheless, it is well understood that NBTI stress and recovery phases involve the diffusion and back-diffusion of H and H_2 in the oxide, the Poly-Si and Si. The following equations are from the newest NBTI model from PTM (Zhao & Cao, 2006), also explained in Bhardwaj, Wang, Vattikonda, Cao, and Vrudhula (2006):

$$\Delta V_{th} = \begin{cases} \sqrt{K_v^2 (t-t_0)^{0.5} + \Delta V_{th1}^2} + \delta_v & \text{(stress)} \\ (\Delta V_{th2} - \delta_v)\left[1 - \sqrt{\eta (t-t_0)/t}\right] & \text{(recovery)} \end{cases} \tag{14}$$

where:

$$K_v = A T_{ox} \sqrt{C_{ox}(V_{gs} - V_{th})} \exp(\frac{E_{ox}}{E_0})$$

$$\left[1 - \frac{V_{ds}}{\alpha(V_{gs} - V_{th})}\right] \exp(-\frac{E_{ox}}{E_0})$$

and:

$$E_{ox} = (V_{gs} - V_{th}) / T_{ox} \tag{15}$$

Default values for different PTM technology nodes can be found on the PTM website (PTM Technology). In the above model ΔV_{TH} is proportional to $t^{1/4}$, where t denotes time. Depending on whether H or H_2 dominates the diffusion, the dependence of ΔV_{TH} and t could vary from $t^{1/4}$ to $t^{1/6}$ in Bhardwaj, Wang, Vattikonda, Cao, and Vrudhula (2006), Kang, Kufluoglu, Alam, and Roy (2007), and Kumar, Kim, and Sapatnekar (2006).

Body Bias Circuit

Figure 9 shows the proposed threshold compensation scheme adapted for NBTI monitoring and mitigation. This circuit shares many similarities with the one used for reducing temperature variability shown in Figure 3. The circuit also can also be divided into an upper part and a lower part. In the upper part, transistor $P1$ is acting as a proxy for PMOS transistors experiencing NBTI stress. Transistors $N1$ and $N2$ serve as a current mirror which track the drain current of $P1$ and also provides some current gain. Similarly, $P2$ and $P5$ form the second current mirror that further tracks and amplifies the drain current. The lower part of the circuit also consists of current mirror circuits, except that now the current in the first stage is determined by the NMOS transistor $N3$, whose gate is constantly tied to V_{DD} and is always on. After two stages of current amplification through current mirror pairs $P3/P4$ and $N4/N5$, the amplified drain current of $N3$ eventually becomes the drain current of $N5$. Since the drain currents of $N5$ and $P5$ have to be the same, this balance will decide the output voltage V_{OUT}, which is then used to body-bias PMOS transistors of the actual circuit. The output is also fed back to $P1$ to form a closed loop amplifier. Similar as the circuit used for temperature variability, the magnitude of

Figure 9. Threshold compensation circuit for monitoring and mitigating NBTI. Sizing is shown for PTM 65nm technology with V_{DD} =1V. Feature size (65nm) is taken as unit length (Qi & Stan, 2008 ©2008, ACM, Inc. Used with permission.).

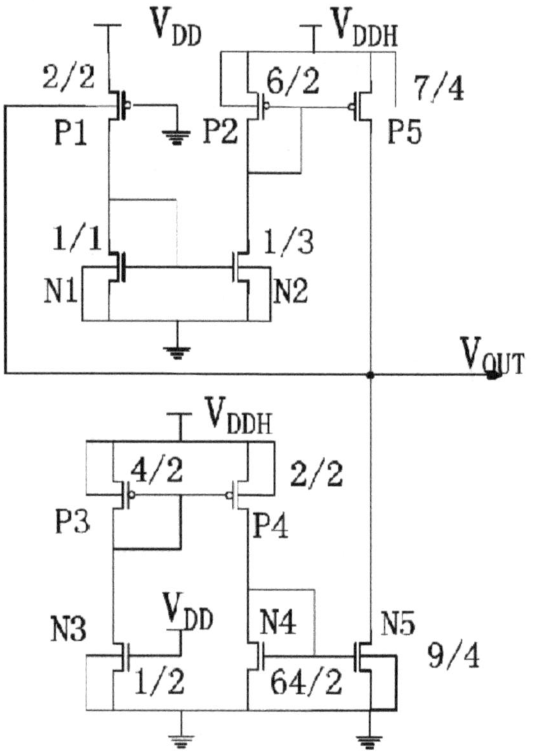

the biasing voltage can also be used as a metric to characterize the NBTI stress amount and the circuit can act as an on chip NBTI monitor.

However, there are some important differences between the design for NBTI in Figure 9 and the design for temperature variability in Figure 3. First, the gate of transistor *P*1 is tied to ground so V_{GS} = - V_{DD} for *P*1 all the time. V_{DD} is a voltage equal to the power supply voltage in the actual circuit to be monitored. Since each PMOS transistor will have its own switching activity, and most of them usually have some periods of stress relaxation when the input gate voltage turns high and shuts off the PMOS, this connection makes *P*1 subjected to the most severe NBTI stress compared to PMOS transistors in the actual circuit. The introduced pessimistic assumption can be dealt with at the circuit level, as will be discussed later. This is different from the case of temperature variability, where temperatures of transistors in the same area are usually quite close, and the proxy transistor could either see a higher or lower temperature than the actual circuit it is monitoring.

Second, there is only one power supply V_{DDH} in the last stage. In the temperature variability case, both positive and negative voltage supplies are needed since the starting point is at the middle and temperature can both rise and fall. On the other hand NBTI degradation only moves PMOS threshold voltage more negative. Therefore, only forward body biasing will be applied to mitigate this change. Since V_{OUT} should equal V_{DD} when no NBTI stress is applied, V_{DDH} needs to be higher than V_{DD} to provide full swing to the output body bias voltage. While this may pose extra design challenges and potential reliability problems, multi- V_{DD} design is not new in today's IC design flows and is well supported by CAD tools. Furthermore, body biasing (both forward and backward) has been adopted as a standard method to systematically reduce the impact of process variation (Tschanz, et al., 2002), which also implies that voltages higher than V_{DD} can be made available.

Finally, it is worth mentioning that at least on the first order ΔV_{TH} is independent of transistor sizing, as can be seen from Eq. (14). On the other hand, the drain voltage will affect NBTI degradation. For PMOS stacks in the pull-up network in CMOS gates, the degradation of each PMOS gate can be different. However, the V_{DS} impact on ΔV_{TH} is rather insignificant, and stacks of PMOS gates are not recommended anyhow due to the slow speed.

Experimental Results

Just as in the temperature-aware design, we start from a simple test case. We first examine the drive current I_{DS} of two PMOS transistors to see how device degradation due to NBTI can be mitigated with the proposed scheme. The following simulation results are obtained with PTM 65nm models, $1V\ V_{DD}$ and $2V\ V_{DDH}$. The NBTI monitor circuit is sized as shown in Figure 9. Figure 10(a) shows how the I_{DS} of each transistor decreases as $|V_{TH}|$ increases. Here we assume that NBTI imposes exactly the same V_{TH} degradation on all PMOS. The reduced I_{DS} amounts to a larger time constant, weaker driving capability and longer charging time, which can eat-up timing margins and eventually lead to timing violations. Although typically NBTI lifetime can be mapped to ΔV_{TH}, what we would like to find out is how lifetime is extended by the reliability monitor. Therefore, we take a 5% I_{DS} reduction as the cut-off point for lifetime estimation and the ΔV_{TH} shown in all the figures does not take body biasing into account in order to isolate ΔV_{TH} caused by NBTI. This is achieved by modifying parameters in the PTM model file during simulation.

From Figure 10b, it can seen that with ABB, I_{DS} reduction in $P2$ is much slower than $P1$, which means P2 tolerates more NBTI stress and has longer lifetime. Indeed, the 5% ΔI_{DS} for $P2$ is reached when V_{TH} degrades almost 100mV from the unstressed value. In contrast I_{DS} in $P1$ drops 5% as soon as V_{TH} changes by only 25mV. Figure 10b compares the sensitivities of I_{DS} in both transistors. It can be seen that body biasing essentially reduces the sensitivity, especially in the range where the maximum reliability tolerance is not violated and the device is supposed to meet the performance requirement set by timing analysis.

Figure 10. Current and output voltage change with threshold voltage drift, as well as their sensitivity to threshold voltage drift (Qi & Stan, 2008 ©2008, ACM, Inc. Used with permission.).

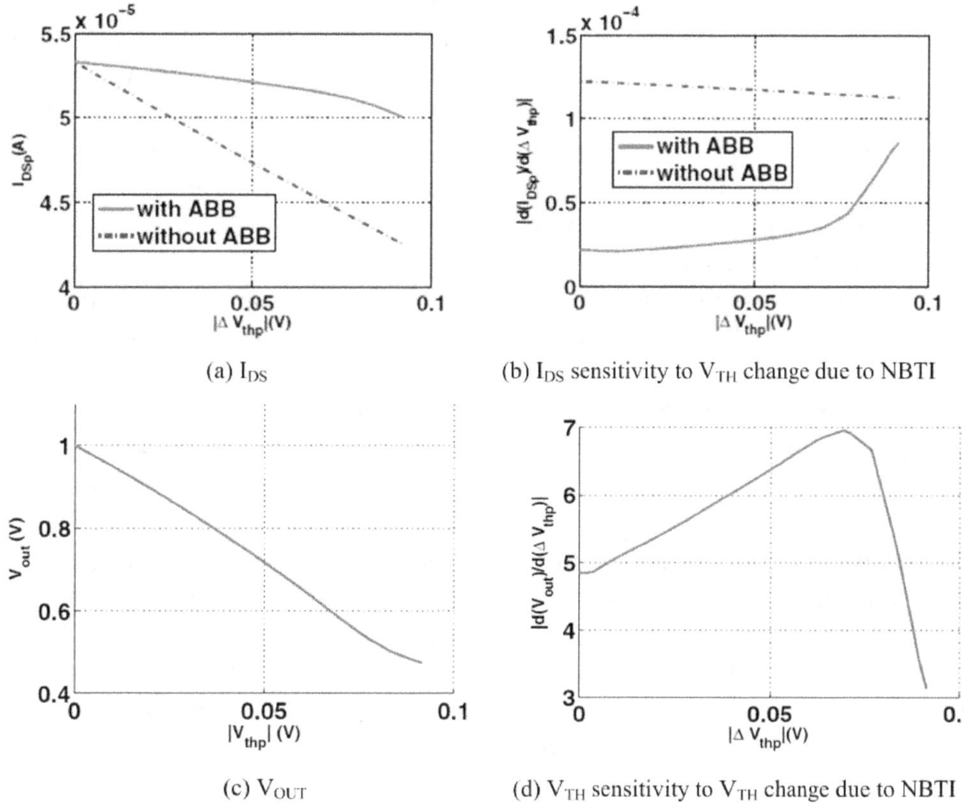

Figure 11. Threshold compensation slows down the frequency degradation of a 21 stage oscillator as PMOS threshold voltage being degraded by NBTI (Qi & Stan, 2008 ©2008, ACM, Inc. Used with permission.).

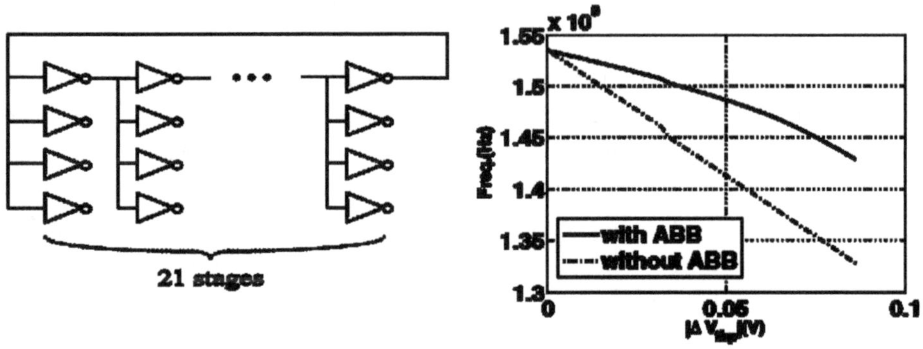

Referring to the NBTI model from earlier sections, ΔV_{TH} depends on $t^{1/n}$ where n varies from 1/4 to 1/6. With the reliability monitor, the tolerance of ΔV_{TH} increases four times from 25mv to 100mv, which in turn is equivalent to a prolonged device lifetime 256 to 1296 times the original depending the effective n value in the NBTI model. However, notice that the above values are *not* representative of the actual circuit lifetime due to the assumed constant static stress. The above analysis is for static stress with no recovery phase during the whole lifetime. In practice very rarely will a PMOS gate be tied constantly to ground, which is the worst-case scenario. Moreover, device lifetime is also limited by temperature and other reliability hazards. Nonetheless, the improvement in lifetime demonstrates the effectiveness of the proposed NBTI monitor.

Figure 10c and 10d show the output voltage and the small signal gain of V_{OUT} to V_{TH}. The circuit is sized such that this gain is maximized within the lifetime requirement. This is an important guideline for this design. Another constraint is that the PMOS body bias voltage must equal V_{DD} at the beginning when any NBTI stress is applied, as it would be if the monitor did not exist. For different technologies, the circuit in Figure 9 needs to be resized accordingly to these requirements.

To show the effectiveness of the NBTI compensation for general circuits, we again resort to the ring oscillator structure. This time we use a twenty-one stage FO4 ring oscillator designed in PTM 65nm technology. Again, with the assumption that all PMOS transistors have the same NBTI stress and thus the same ΔV_{TH}, we monitor the frequency degradation. As shown in Figure 11, with the NBTI monitor, the circuit can tolerate almost twice ΔV_{TH}. However, in this case the lifetime cannot be simply translated from $t^{1/4}$ or $t^{1/6}$, since this is dynamic NBTI stress with relaxation and 'healing' within every clock period.

Actual energy savings from employing the proposed scheme relate to a number of factors, including the system specification, usage model, circuit-switching activity, operating environment, deployment of the proposed circuit, and the alternative approach when the proposed scheme is not adopted, etc. To make our analysis more straightforward, we still resort to the simple but representative ring oscillator in Figure 11. With a specification that states that the frequency cannot fall below 900MHz, this results in about 5% guardband, which translates to 0.08V threshold degradation. Without the proposed scheme, at 0.08V threshold degradation the circuit would be operating at 830MHz, thus violating the specification. Ignoring all second order factors, the maximum frequency is inversely proportional to the lesser of the charging and discharging cycles of the circuit. Since NBTI only affects PMOS, the charging cycle becomes the limiting factor. The charging cycle is proportional to the maximum of the current that a PMOS can supply. Therefore:

$$f \propto \frac{1}{t_{charging}} \propto I_{PMOS} \propto W_{PMOS} \tag{16}$$

It is immediately clear that in order to maintain 900MHz when NBTI degrades the V_{TH} by 0.08V, one needs to add an up-front guarband of (900-830)/830 = 8.4% on the PMOS width, which translates to 8.4% extra energy consumption and area overhead.

Notice that the above is an over-simplistic first order estimation. First, overdesign doesn't need to be applied on every path. However, one cannot overdesign only the critical paths either. Identifying which paths to overdesign is a big challenge and requires extensive low-level circuit simulation based on accurate models from device-level to system-level (Saluja, Vijayakumar, Sootkaneung, & Yang, 2008). Second, the actual energy savings will be offset a little by the energy consumption of the threshold compensation circuit itself. The overhead incurred by the proposed circuit is largely affected by its deployment, which is discussed in the next section.

It should also be emphasized that the benefits of the proposed scheme go far beyond the immediate energy saving from the above guardband reduction. In fact, the more implicit energy savings can actually be more significant. By tolerating more aging degradation, the circuit improves the efficiency of analog circuits which often rely on precise matching and voltage biasing. It also improves noise characteristics and soft error immunity in memory systems. In contrast, without adaptive threshold compensation and merely adding guardband disrupts the normal design. For example, upsized PMOS transistors form a stronger pull-up network than the pull-down network, thus reducing noise margins. With the adaptive compensation scheme, the reduction of excessive noise and probability of failure lead to potential energy savings which would otherwise be incurred from engaging error correction systems or replacing failure part. For example, in a recent paper (Mostafa, et al., 2011) which proposes an alternative body bias scheme, our circuit is shown to improve write margins by 4.6%, and reduces the read and write failure probablity by 0.28% and 0.14% respectively for standard 6T SRAM banks. In fact the proposed circuit is also shown to be 1.2X smaller in area, and leads to 5.5% less leakage power than the ABB circuit proposed in Mostafa, et al. (2011).

DEPLOYMENT OF PROPOSED CIRCUIT

As shown in both applications, the proposed threshold compensation circuit serves dual purposes. It can act as an on chip sensor to monitor temperature or reliability effects, and at the same time reduce their impact on the system by compensating the threshold voltage change via adaptive body biasing. A common issue with on chip sensors is how to let them see the same environmental variables (temperature, NBTI, etc.) of the actual devices that they are monitoring. Each individual device can experience different temperature or stress, and it is impossible to assign one sensor to each device for the whole circuit. Process variations further complicate the problem. A more realistic and practical way is to allocate a sensor to a group of devices in their neighborhood in order to reduce the overhead and the impact of local process variation. The granularity at which the sensors are deployed reflects a tradeoff between overhead and accuracy, and the optimization can be assisted with a-priori knowledge of temperature or failure distribution.

As mentioned above, compensating for aging effects is more challenging than for temperature variability, since the former is based on switching activity which differs from device to device and ideally the compensation should have much finer granularity than for temperature variability. In the following we use the NBTI application as an example and provide some analysis and insight. In Figure 9 *P*1 is under constant NBTI stress, thus only representing the worst stress case. This pessimism may result in a situation where a PMOS is over-biased. However, in this case the PMOS, threshold compensation may provide a larger drive current than the nominal value used in timing analysis. While this may cause matching problems in analog circuits, for digital circuits it actually helps setup mode timing, as long as the hold mode timing margin is not very stringent. This is usually the case for most designs and most paths even in a design with a tight hold time margin. Recall that NBTI becomes a concern because it causes timing problems by slowing down the critical paths and causing timing violations after a certain amount of stress time. For circuits like shift registers, where hold timing is critical, NBTI is most likely not an issue anyway and we simply do not deploy any compensation for them. Furthermore, Figure 10 and Figure 11 both show that even with the NBTI monitor the V_{TH} degradation is still not fully removed, so some extra amount of forward biasing still helps recover the lost margin.

If one really wants to remove the pessimism, there are at least two compatible approaches to choose from, or combine. The first is to identify the most critical paths whose timing analysis result is extremely sensitive to NBTI degradation. This can be achieved with reliability simulations, either by commercial tools supported by foundry-provided device reliability parameters (Liu & McGaughy, 2006) or implementing one's own device reliability models and incorporating them in the simulation (Wang, et al., 2007). An NBTI-aware design scheme that only upsizes those PMOS on critical paths is also presented in Yang and Saluja (2007). The difference is that, for overdesigning, accurate sizing for each transistor directly depends on modeling and simulation precision which may even reply on a correct assumption of input signal switching activity, while for monitor deployment we only need to identify the most vulnerable portions of the circuit.

The second scheme, as illustrated in Figure 12, is to only selectively turn on the monitor when NBTI degradation is large enough and a degradation constraint either on I_{DS} or V_{TH} is violated. The violation detection can be done either off-line in a special testing mode enabled at preset time points, or online with error detection structures as in the

Figure 12. Selectively power on/off the monitor to prevent over-stressing and save power (adapted from Qi & Stan, 2008)

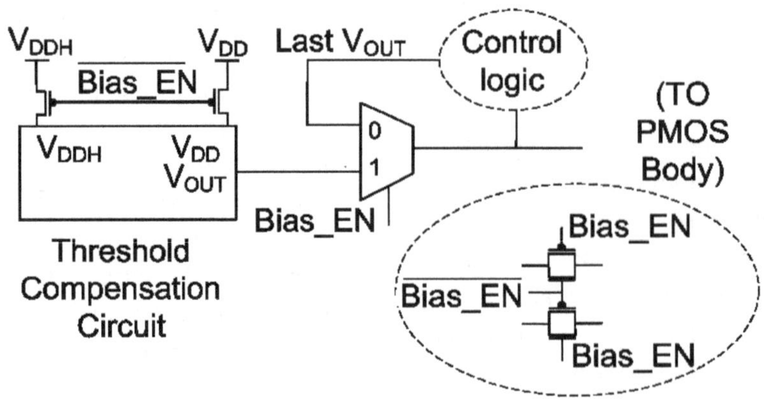

Razor architecture (Ernst, et al., 2005). Notice that when the body bias output is selected by the multiplexer, the reliability monitor should also be turned off to prevent the proxy (*P*1 in Figure 9) from being over-stressed. In addition, the pass transistor logic multiplexer needs to pass an analog output V_{OUT}, and cannot be replaced by a standard cell implementation. Pass transistor logic can be used here because V_{OUT} is connected to the body with small current and feedback; power savings can be achieved by being able to turn off the NBTI monitors.

SUMMARY

In this chapter, we proposed a threshold compensation scheme that can optimize the system energy efficiency. We demonstrated that the design can effectively mitigate the transistor threshold voltage shift due to non-ideal factors, like temperature variability and reliability degradation, and prevent the system drifting far away from its optimal operation condition. A detailed analysis of the proposed scheme shows that although ideal compensation is physically impossible, several design parameters can be tuned to maximize the compensation capability. Two example applications are demonstrated for temperature variation and NBTI-caused threshold shifts. In these examples the proposed scheme is shown to improve circuit performance, reliability and lifetime for both combinational and memory circuitry.

With little modification, the proposed scheme can be adapted for a wide range of applications. For example, compensating the threshold shift by HCI (Hot Carrier Injection) or PBTI (Positive Bias Temperature Instability) can be done very similarly as for NBTI. The scheme can also be applied to adjust shifted threshold voltages due to manufacturing process variation. Intra-die or inter-die variation can be mitigated through an on-chip proxy transistor and an off-chip reference current (I_B in Figure 1). Depending on the actual application, possible simplifications may be made, as in the case of NBTI sensor, which exploits the fact that only forward biasing is needed.

We believe adaptive and distributed computing are two viable approaches for making future computing systems more sustainable. Adaptive and distributed computing can be implemented across all the levels from the bottom circuit level to the top system level. The proposed scheme can be viewed as one such example at the circuit level. The ability to adapt to the runtime usage and environmental variations is critical to maximize the system efficiency. Distributing such compensation blocks across the whole chip can achieve more aggressive savings at a finer granularity. By adaptively compensating for both short time variation and long time shifting of transistor threshold voltages, the proposed scheme enables typical case-based design, thus effectively reducing design margins and associated chip area and power consumption. The two different example applications, i.e., compensating for short time temperature and long time NBTI aging induced threshold voltage instability, demonstrate its versatility.

REFERENCES

Agostinelli, M., Pae, S., Yang, W., Prasad, C., Kenche, D., Ramey, S., et al. (2005). Random charging effects for PMOS NBTI in ultra-small gate area devices. In *Proceedings of Annual International Reliability Physics Symposium.* Reliability Physics Symposium.

Asanovic, K., Bodik, R., Catanzaro, B. C., Gebis, J. J., & Husbands, P. K. (2006). *The Landscape of parallel computing research: A view from Berkeley*. Berkeley, CA: University of California at Berkeley.

Bellaouar, A., Fridi, A., Elmasry, M., & Itoh, K. (1998). Supply voltage scaling for temperature-insensitive CMOS circuit operation. *IEEE Transaction on Circuits and System II*, *45*(3), 415–417. doi:10.1109/82.664253

Bhardwaj, S., Wang, W., Vattikonda, R., Cao, Y., & Vrudhula, S. (2006). Predictive modeling of the NBTI effect for reliable design. In *Proceedings of IEEE Custom Integrated Circuits Conference.* IEEE Press.

Blaauw, D. T., Oh, C., Zolotov, V., & Dasgupta, A. (2003). Static electromigration analysis for on-chip signal interconnects. *IEEE Transactions on Computer-Aided Design of Integrated Circuits and Systems, 22*(1), 39–48. doi:10.1109/TCAD.2002.805728

Borkar, S. (2005). Designing reliable systems from unreliable components: The challenges of transistor variability and degradation. *IEEE Micro, 25*(6), 10–16. doi:10.1109/MM.2005.110

Chandrakasan, A. P., Sheng, S., & Brodersen, R. W. (1992). Low-power digital CMOS design. *IEEE Journal of Solid-state Circuits, 27*(4), 473–484. doi:10.1109/4.126534

Chen, G., Chuah, K. Y., Li, M. F., Chan, D. S., Ang, C. H., Zheng, J. Z., et al. (2003). Dynamic NBTI of PMOS transistors and its impact on device lifetime. In *Proceedings of Annual International Reliability Physics Symposium,* (pp. 196–202). Reliability Physics Symposium.

Ernst, D., Das, S., Lee, S., Blaauw, D., Austin, T., & Mudge, T. (2005). Razor: Circuit-level correction of timing errors for low-power operation. *IEEE Micro, 24*(10), 10–20.

Hamzaoglu, F., Ye, Y., Keshavarzi, A., & Zhang, K., S., N., Borkar, S., et al. (2002). Analysis of dual-VT SRAM cells with full-swing single-ended bit line sensing for on-chip cache. *IEEE Transactions on Very Large Scale Integration (VLSI) Systems, 10*(2), 91–95.

Hennessy, J. L., & Patterson, D. A. (2003). *Computer architecture: A quantitiative approach* (3rd ed.). London, UK: Elsevier Science Ltd.

Kang, K., Kim, K., Islam, A. E., Alam, M. A., & Roy, K. (2007). Characterization and estimation of circuit reliability degradation under NBTI using on-line IDDQ measurement. In *Proceedings of Design Automation Conference,* (pp. 358–363). Design Automation Conference.

Kang, K., Kufluoglu, H., Alam, M. A., & Roy, K. (2007). Efficient transistor-level sizing technique under temporal performance degradation due to NBTI. In *Proceedings of IEEE International Conference on Computer Design,* (pp. 934–939). IEEE Press.

Karl, E., Blaauw, D., Sylvester, D., & Mudge, T. (2006). Reliability modeling and management in dynamic microprocessor-based systems. In *Proceedings of Design Automation Conference,* (pp. 1057–1060). Design Automation Conference.

Karnik, T., Borkar, S., & De, V. (2002). Sub-90nm technologies: Challenges and opportunities for CAD. In *Proceedings of International Conference on Computer Aided Design,* (pp. 203 - 206). Conference on Computer Aided Design.

Kleissl, J., & Agarwal, Y. (2010). Cyber-physical energy systems: Focus on smart buildings. In *Proceedings of Design Automation Conference,* (pp. 749 - 754). Design Automation Conference.

Kumar, S. V., Kim, C. H., & Sapatnekar, S. S. (2006). Impact of NBTI on SRAM read stability and design for reliability. In *Proceedings of International Symposium on Quality Electronic Design.* Quality Electronic Design.

Liu, Z., & McGaughy, B. W. (2006). Design tools for reliability analysis. In *Proceedings of Design Automation Conference,* (pp. 182–187). Design Automation Conference.

Lu, Z., Lach, J., Stan, M. R., & Skadron, K. (2005). Improved thermal management with reliability banking. *IEEE Micro, 25*(6), 40–49. doi:10.1109/MM.2005.114

Massey, J. G. (2004). NBTI: What we know and what we need to know - A tutorial addressing the current understanding and challenges for the future. In *Proceedings of the IEEE International Integrated Reliability Workshop*, (pp. 199–211). IEEE Press.

Mishra, S., Pecht, M., & Goodman, D. L. (2002). In-situ sensors for product reliability monitoring. In *Proceedings of SPIE* (pp. 10–19). SPIE. doi:10.1117/12.462807

Mitra, S. (2007). Circuit failure prediction enables robust system design resilient to aging and wearout. In *Proceedings of the IEEE International On-Line Testing Symposium*, (p. 123). IEEE Press.

Mostafa, H., Anis, M., & Elmasry, M. (2011). Adaptive body bias for reducing the impacts of NBTI and process variations on 6T SRAM cells. *IEEE Transactions on Circuits and Systems. I, Regular Papers*, *58*(12), 2859–2871. doi:10.1109/TCSI.2011.2158708

Park, C., John, J., Klein, K., Teplik, J., Caravella, J., Whitfield, J., et al. (1995). Reversal of temperature dependence of integrated circuits operating at very low voltages. In *Proceedings of the International Electron Devices Meeting*, (pp. 71 - 74). Electron Devices Meeting.

Paul, B. C., Kang, K., Kufluoglu, H., Alam, M. A., & Roy, K. (2006). Temporal performance degradation under NBTI: Estimation and design for improved reliability of nanoscale circuits. In *Proceedings of European Design and Test Conference*, (pp. 780–785). European Design and Test Conference.

Pedram, M. (2009). Green computing: Reducing energy cost and carbon footprint of information processing systems. In *Proceedings of the ACM Great Lakes symposium on VLSI*. ACM Press.

PTM. (2011). *PTM technology*. Retrieved from http://ptm.asu.edu/.

Qi, Z., Meyer, B., Huang, W., Ribando, R., Skadron, K., & Stan, M. (2010). Temperature-to-power mapping. In *Proceedings of IEEE International Conference on Computer Design*, (pp. 384 - 389). IEEE Press.

Qi, Z., & Stan, M. R. (2007). Accurate back-of-the-envelope transistor model for deep sub-micron MOS. In *Proceedings International Conference on Microelectronic Systems Education*. Microelectronic Systems Education.

Qi, Z., & Stan, M. R. (2008). NBTI resilient circuits using adaptive body biasing. In *Proceedings of the ACM Great Lakes symposium on VLSI*, (pp. 285-290). ACM Press.

Sakurai, T., & Newton, A. R. (1991). A simple MOSFET model for circuit analysis. *IEEE Journal of Solid-state Circuits*, *38*(4), 887–893.

Saluja, K., Vijayakumar, S., Sootkaneung, W., & Yang, X. (2008). NBTI degradation: A problem or a scare? In *Proceedings of the VLSI Design, 2008: VLSID 2008 21st International Conference*, (pp. 137 - 142). VLSI.

Shakeri, K., & Meindl, J. D. (2003). Temperature variable supply voltage for power reduction. In *Proceedings of IEEE Computer Society Annual Symposium on VLSI*. IEEE Press.

Stan, M., Cabe, A., Ghosh, S., & Qi, Z. (2007). Teaching top-down ASIC/SoC design vs bottom-up custom VLSI. In *Proceedings of the International Conference on Microelectronic Systems Education (MSE)*. MSE.

Tang, S., Narendra, S., & De, V. (2003). Temperature and process invariant MOS-based reference current generation circuits for sub-1V operation. In *Proceedings of of International Symposium on Low Power Electronics and Design*. Low Power Electronics and Design.

Tschanz, J. W., Kao, J. T., Narendra, S. G., & Nair, R. A. A., D., Chandrakasan, A. P., et al. (2002). Adaptive body bias for reducing impacts of die-to-die and within-die parameter variations on microprocessor frequency and leakage. *IEEE Journal of Solid-state Circuits*, *37*(11), 1396–1402. doi:10.1109/JSSC.2002.803949

Vattikonda, R., Wang, W., & Cao, Y. (2006). Modeling and minimization of PMOS NBTI effect for robust nanometer design. In *Proceedings of Design Automation Conference*, (pp. 1047–1052). Design Automation Conference.

Wang, W., Yang, S., Bhardwaj, S., Vattikonda, R., Vrudula, S., Liu, F., et al. (2007). The impact of NBTI on the performance of combinational and sequential circuits. In *Proceedings of Design Automation Conference*, (pp. 364-369). Design Automation Conference.

Yang, X., & Saluja, K. (2007). Combating NBTI degradation via gate sizing. In *Proceedings of International Symposium on Quality Electronic Design*, (pp. 47–52). Quality Electronic Design.

Zhang, Y., & Stan, M. (2007). Temperature-aware circuit design using adaptive body biasing. In *Proceedings of the ACM Great Lakes Symposium on VLSI*, (pp. 84–89). ACM Press.

Zhao, W., & Cao, Y. (2006). New generation of predictive technology model for sub-45nm early design exploration. *IEEE Transactions on Electron Devices*, *53*(11), 2816–2823. doi:10.1109/TED.2006.884077

Zhou, Q., & Mohanram, K. (2006). Gate sizing to radiation harden combinational logic. *IEEE Transactions on Computer-Aided Design of Integrated Circuits and Systems*, *25*(1), 155–166. doi:10.1109/TCAD.2005.853696

ADDITIONAL READING

Basoglu, M., Orshansky, M., & Erez, M. (2010). NBTI-aware DVFS: A new approach to saving energy and increasing processor lifetime. In *Proceedings of the 16th ACM/IEEE International Symposium on Low Power Electronics and Design*. ACM Press.

Bsoul, A. A., Manjikian, N., & Shang, L. (2010). Reliability and process variation-aware placement for FPGAs. In *Proceedings of the Conference on Design, Automation and Test in Europe*. Design Automation and Test.

Dadgour, H., & Banerjee, K. (2010). Aging-resilient design of pipelined architectures using novel detection and correction circuits. In *Proceedings of the Conference on Design, Automation and Test in Europe (DATE 2010)*. Design Automation and Test.

Firouzi, F., Kiamehr, S., & Tahoori, M. B. (2011). A linear programming approach for minimum NBTI vector selection. In *Proceedings of the Great Lakes Symposium on VLSI*. ACM Press.

Fu, X., Li, T., & Fortes, J. (2008). NBTI tolerant microarchitecture design in the presence of process variation. In *Proceedings of the Microarchitecture IEEE/ACM International Symposium*, (pp. 399-410). IEEE Press.

Guardiani, C., Shibkov, A., Brambilla, A., Gajani, G., Appello, D., Piazza, F., et al. (2009). An I-IP based approach for the monitoring of NBTI effects in SoCs. In *Proceedings of the IEEE International On-Line Testing Symposium*. IEEE Press.

Kothawade, S., Chakraborty, K., & Roy, S. (2011). Analysis and mitigation of NBTI aging in register file: An end-to-end approach. In *Proceedings of the International Symposium on Quality Electronic Design (ISQED)*. ISQED.

Rabaey, J., Chandrakasan, A., & Nikolic, B. (2003). *Digital integrated circuits: A design perspective* (2nd ed.). Upper Saddle River, NJ: Pearson Education, Inc.

Rahaman, M. S., Duan, Q., & Chowdhury, M. H. (2011). Spatial- and temporal-reliability aware design for nano-scale VLSI circuits. In *Proceedings of the IEEE International Symposium on Circuits and Systems*. IEEE Press.

Rakesh, R., & Yadav, K. S., & Jaipal. (2011). Reduction of leakage by input vectors with constrained NBTI degradation. *International Journal of Information Technology and Knowledge Management*, *4*(1), 163–167.

Ramakrishnan, K., Wu, X., Vijaykrishnan, N., & Xie, Y. (2008). Comparative analysis of NBTI effects on low power and high performance flip-flops. In *Proceedings of the IEEE International Conference on Computer Design*, (pp. 200 - 205). IEEE Press.

Salfelder, F., & Hedrich, L. (2011). An NBTI model for efficient transient simulation of analogue circuits. In *Proceedings of edaWorkshop*. EDA.

Siddiqua, S. G. T. (2009). NBTI-aware dynamic instruction scheduling. In *Proceedings of the IEEE Workshop on Silicon Errors in Logic - System Effects*. IEEE Press.

Siddiqua, T., & Gurumurthi, S. (2009). NBTI-aware dynamic instruction scheduling. In *Proceedings of the IEEE Workshop on Silicon Errors in Logic - System Effects*. IEEE Press.

Song, J., Han, Y., Zhang, L., Li, H., Li, X., & Yan, G. (2009). M-IVC: Using multiple input vectors to minimize aging-induced delay. In *Proceedings of the Asian Test Symposium*. Asian Test.

Subramanyan, P., Jangir, R. R., Tudu, J., Larsson, E., & Singh, V. (2009). Generation of minimal leakage input vectors with constrained NBTI degradation. In *Proceedings of the 8th East-West Design and Test Workshop*. East-West Design and Test.

Vazquez, J., Champac, V., Ziesemer, A., Reis, R., Teixeira, I., Santos, M., et al. (2009). Built-in aging monitoring for safety-critical applications. In *Proceedings of the 15th IEEE International On-Line Testing Symposium*. IEEE Press.

Vazquez, J., Champac, V., Ziesemer, A., Reis, R., Teixeira, I., Santos, M., et al. (2010). Low-sensitivity to process variations aging sensor for automotive safety-critical applications. In *Proceedings of the VLSI Test Symposium (VTS)*. VTS.

KEY TERMS AND DEFINITIONS

Transistor/NMOS/PMOS: A transistor (also called MOS or MOSFET) is the basic building element for modern integrated circuits. There are two types of transistors: NMOS (N-type MOS transistor, also called NFET) and PMOS (P-type MOS transistor, also called PFET).

Transimpedance/Transconductance: For a two-port circuit, transimpedance is the ratio of the output voltage change to the input current change, and transconductance is the ratio of the output current change to the input voltage change.

Design Margin/Guardband: The extra capability designed into a circuit system slightly beyond the minimum requirement from design specification in terms of performance, reliability, power, etc, to compensate for all kinds of variations or uncertainties. This can translate to extra circuit area, extra power consumption, etc.

Feedback: Used in the context of control theory, meaning the system output is fed back into as part of the input to control the system. Feedback improves stability of a system.

Open/Closed Loop: Systems with/without feedback mechanism

Variation: Circuit uncertainty and drift from the values designed for in all aspects of design specifications. Variation can come from many sources including manufacturing process (process variation), power supply instability (power variation), circuit operating temperature (thermal variation), aging degradation (aging variation), etc.

Reliability Degradation/Aging: The performance of a transistor can change (usually resulting in inferior performance, e.g., slower transition, reduced noise margin, or even failure to turn on) after it is turned on and passes currents for some time. Mechanisms include NBTI (Negative Bias Temperature Instability), HCI (Hot Carrier Injection), TDDB (Time Dependent Dioxide Breakdown), etc.

Threshold Voltage: The gate voltage needed for forming a low-resistance channel (inversion layer) between the source and drain of a transistor. In oversimplified terms, it can be considered a point beyond which the transistor is "turned on" (passing current) and below which "turned off" (open circuit).

NBTI/PBTI: Negative/Positive Bias Temperature Instability. NBTI happens in PMOS with negative voltage bias between gate and source terminal. The effect is increased threshold voltage and larger transimpedance. The degradation increases with the biasing time and elevated temperature, but can partially recover when the bias is removed. PBTI is the symmetric mechanism for NMOS.

HCI: Hot Carrier Injection, a phenomenon in which high energy ("hot") electrons tunnel through the gate dioxide of a transistor, causing threshold voltage drift and degraded transistor performance.

Chapter 10
Multidimensional Analysis of Supply Chain Environmental Performance

Antti Sirkka
Tieto Finland, Finland

Marko Junkkari
University of Tampere, Finland

ABSTRACT

Monitoring the environmental performance of a product is recognized to be increasingly important. The most common method of measuring the environmental performance is the international standards of Life Cycle Assessment (LCA). Typically, measuring is based on estimations and average values at product category level. In this chapter, the authors present a framework for measuring environmental impact at the item level. Using Traceability Graph, emissions and resources can be monitored from the data management perspective. The model can be mapped to any precision level of physical tracing. At the most precise level, even a single physical object and its components can be analyzed. This, of course, demands that the related objects and their components are identified and mapped to the database. From the opposite perspective, the authors' model also supports rough level analysis of products and their histories. In terms of the Traceability Cube, multidimensional analysis can be applied for traceability data.

INTRODUCTION

Green computing is usually analogized with Green IT as pure technical issues closely related hardware and software solutions following Murugesan's (2008) definition "the study and practice of designing, manufacturing, using, and disposing of computers, servers, and associated subsystems—such as monitors, printers, storage devices, and networking and communications systems—efficiently and effectively with minimal or no impact on the environment." Instead, we agree

DOI: 10.4018/978-1-4666-1839-8.ch010

with those authors (e.g. Donnellan, et al., 2011) who see that green computing not only involves the previous definition for Green IT, but also possesses methods to use ICT in business processes to reduce environmental impact of enterprises.

Our contribution is to serve fine-grain tracing approach for monitoring and analyzing life cycle data of products. As our example (forest industry) illustrates our model is general—not only for manufacturing and tracing of computers and data centers. Any physical product has its own supply history represented as a supply chain. In practice, the history of the product is the history of its parts composed in the supply chain. The precision of traceability of products depends on how detailed the history of the components can be traced. We give a logical framework for tracing and analyzing the emissions and resources of products both at the item and patch levels.

The problem with measuring the environmental impact caused by a product at the item level is that supply chains are dynamic. A manufacturer can use various subcontractors and supply various end manufacturers or retailers in different countries. For example, a product that is transported from another continent to a supermarket is bound to have different environmental impact than another product that is transported to a supermarket from a nearby producer. However, the common method of calculating the environmental impact on a product is to measure the resources used, emissions and production in some time period and calculate the average environmental impact on the product. This does not take the dynamic nature of the supply chains into account.

To be able to track the objects through the dynamic supply chain, the products/patches must be identified at the physical level. The development of an auto identification enables us to identify an object moving in the supply chain. This means that we can connect the physical world objects with their virtual counterparts in databases. With the traceability we can track the relationships among properties of processes, in this case the environmental burden caused by processes, and actual product instances.

Unlike existing methods our model enables analyzing environmental impact on the product level – not only average values. The model supports for monitoring emissions (e.g. CO_2) and resources (e.g. Energy) at any precision level only depending on how precisely physical products and patches can be identified and monitored. Our approach is based on the Traceability Graph (Junkkari & Sirkka, 2011) that enables tracing of products on any level of precision. In practice the method, produce a huge amount of data. Multidimensional methods (e.g. OLAP, online analytic processing) enables viewing this data through several dimensions at different levels of granularity. We design a data cube, called Traceability Cube, for advanced analysis of traceability data.

WORKFLOWS

Generally, workflows are used to model the flow of materials, documents and other pieces of information from one process to another (van der Aalst & van Hee, 2002; Bonner, 1999). Modern software modeling methods such as UML contain activity diagrams for mapping real-world activities to the underlying software solution (or vice versa). There are also a number of commercial applications that have a component for drawing workflow diagrams. The common feature of these applications is that they support the illustration of different types of processes.

The workflows can be divided into two main categories, process- and data-centric. So far, the process-centric workflow modeling focusing on processes and the timing between them has been the dominant approach. However, recently the data-centric workflow modeling has gained popularity. In the data-centric workflow modeling the focus is on the transformation of data sets—initial, intermediate, and final (Akram, et al., 2006). The data sets are used as parameters to services that consume the input data set and create output data sets. The data-centric workflows are most commonly used in scientific problem solving.

In scientific problem solving the primary feature of the scientific workflow methods is to concentrate on functionality of processes based on the underlying data (Curcin & Ghanem, 2008). Analytical steps instructions how to handle the underlying data sets, i.e. data-centric workflows are used to model, design, and execute an analysis, the capturing series of analytical steps.

Scientific workflows formalize the process of the scientific analysis by modeling the data transportation, transformation, and analytical steps between distributed computational steps. Scientific workflows are directed graphs where arcs represent the transitions between places, which represent computation steps. The main components of the scientific workflow model are:

- Workflow engine invokes the services based on the predefined instructions
- Services which form the workflow are accessed through interfaces.
- Applications implement the functionality of the services.
- Tokens of data are consumed and produced by services.

A scientific calculation process is generally geographically distributed over users and resources and involves huge data sets. Scientific workflows play an important role in Grid computing which combines multiple computer resources to solve a single task, such as geophysics, astronomy, or bioinformatics where the amount of data can be petabytes (Foster, et al., 2001; Chervenak, et al., 2001). The main principles in Grid are the non-centralized control of resources, standardized open protocols, and interfaces. A Grid network is a distributed system that enables sharing antonymous resources based on their availability. There normally is a semantic mismatch between the applications in the Grid network, and calculations executed by applications in the Grid network can be time-consuming. These requirements must be taken into account when designing scientific workflow systems (Barker & van Hemert, 2007).

The ability to reproduce and reuse results is one of the most important requirements in scientific analysis. To be able to trace the chain of the analysis we need the provenance information, which enables the users of a scientific workflow system to share the results and reproduce the calculation process (Freire, et al., 2008; Simmhan, et al., 2005). The provenance information can be categorized into two main categories (Clifford, et al., 2008). Prospective provenance means information about the calculation steps in a workflow, i.e. the specification of the scientific workflow that needs to be executed. Information about the environment and executed calculation steps used to derive the data sets is called retrospective provenance. The provenance includes the inputs of a workflow, the outputs of the workflow, and definitions of the calculation steps and data tokens between the steps.

The provenance data that is collected from supply chain is used to connect information on processes to products. The example used in this chapter is the environmental burden caused by processes, and actual product instances. The Traceability Graph is not bound to any existing data or workflow methods. The model can be applied using existing formalisms and systems. First, the supply chain is modeled as a basic workflow model. The workflow model provides a starting point for the Traceability Graph—presented in this chapter—which can be seen as a data-centric workflow where each supply chain process is a calculation step—node—in which the life cycle data are allocated to a product. Instead of concentrating on the process functionality, which is one of the main aspects of data-centric workflows (Curcin & Ghanem, 2006), the Traceability Graph emphasizes the handling of aggregation and movement of data between processes. The life cycle information collected from each process can be shared among supply chain stakeholders using the services as discussed in Section Future Research Discussion.

ENVIRONMENTAL PERFORMANCE MONITORING

The most common method used for measuring environmental performance in organizations is to calculate the total environmental impact for the whole company. The dominant methodologies used are Greenhouse Gas Protocol, which is a guideline for organizations to estimate their greenhouse gas emissions and the Global reporting initiative reporting framework which defines the sustainability reporting framework for the organization. Both methods result in a total environmental burden for a whole company. These resulting values cannot be used to measure an environmental impact for a certain product because the emissions are not allocated to the individual products.

The main approach for assessing the environmental impact for a product is international standard of the Life Cycle Assessment (LCA) ISO 14040 series (International Organization for Standardization, 1997). There are also specifications for the assessment of the greenhouse gas emissions of products. The Publicly Available Specification (PAS) 2050 (Carbon Trust, 2008) that builds on ISO standards for the life cycle assessment by describing the requirements for the assessment of the greenhouse gas emissions and ISO 14067 standard for Quantification and Communication of the Carbon footprint of a product.

Life cycle assessment is a standardized method for calculating the environmental impact caused by a product during its life cycle. The goal of LCA is to compare the environmental impact caused by a product so that the customer can choose the least burdensome one.

In the first phase of LCA, the scope of the life cycle assessment is defined, i.e. which life cycle processes, inputs, and outputs are included in the assessment. In addition, the functional unit for the assessment is selected. The functional unit can be described as a meaningful amount of a product used in the assessment. For example a cubic meter of timber, one mobile phone or one tomato.

The second phase, called life cycle inventory analysis, the input and output flows for each process is defined (see Figure 1). There are two flow types. Elementary flows describe the inputs and outputs for process (e.g. raw materials and energy resources and outputs of waste and emissions respectively). Product flows are used to describe flow of products and by-products through process. If a process produces more than one product, an allocation is also needed. For example, in sawing the main product is a centreboard with certain dimensions. The sawing process also produces sideboards, wood chips, and sawdust that are considered as by-products. In this case, the emissions caused in the sawing process can be allocated to centreboard, sideboards, wood chips, and sawdust using volume or value based allocation method.

In the third phase, called impact assessment, the results of previous phase are assigned to the impact categories, which include (International Organization for Standardization, 2000):

- Climate change
- Depletion of the stratospheric ozone layer
- Acidification of land and water sources
- Eutrophication
- Formation of photochemical oxidants
- Depletion of fossil energy resources
- Depletion of mineral resources

For example, the emissions to air like carbon dioxide and methane are assigned to the Climate Change category. In the final, interpretation phase, the conclusions of the assessment are made. In this chapter, we present the model for tracing and storing the life cycle data about product manufactured in dynamic supply chain. Unlike the existing methods, our model enables analyzing resources

Figure 1. Process and flows

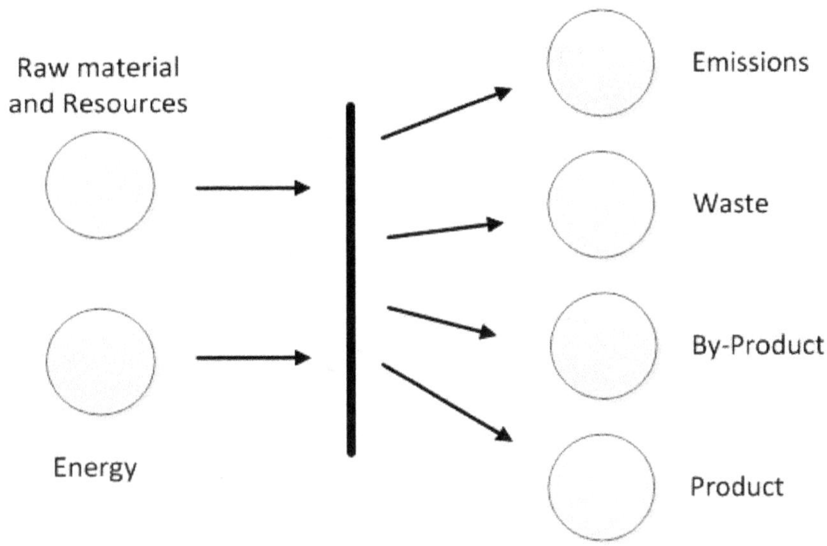

and emissions at the single product level—not only average values. This is achieved by allowing gathering real monitored activity data from supply chain processes.

IMPLEMENTATION OF LCA PHASES

Our approach supports the different phases the life cycle assessment as follows. In LCA Phase 1, the Traceability Graph is used to model the life cycle processes, inputs and outputs are included in the assessment. The traceability data is collected to a database based on the Traceability Graph in LCA phase 2 - Life Cycle Inventory (LCI) analysis. The phases 3 and 4 are implemented in terms of the Traceability Cube. The Traceability Cube provides abilities to use the OLAP (Online Analytic Processing) type operations (Chaudhuri & Dayal, 1997) for analyzing the information of the Traceability Graph. OLAP is a method used for describing the analysis of the complex data in the data warehouses. OLAP Council has defined three main functions that are provided by OLAP systems: multidimensional views of data, ability to perform complex calculations and intelligent handling of time, i.e. time intelligence.

LCA Phase 1: Traceability Graph

The Traceability Graph is used to model the supply processes of physical products and resources and emissions associated with the products and their components. The Traceability Graph has the ability to manipulate products and the transformations of the products. For example, a product may be composed from many parts or a product may be manufactured using masses of raw materials. The Traceability Graph is not bound to any existing data or workflow methods. The model can be applied using existing workflow formalisms and systems.

The Traceability Graph has also the ability to manipulate the properties of processes and to allocate them to products that are handled in that process. The Traceability Graph can be presented using nodes and edges and their properties. A node is used to describe a supply chain process. An edge is used to describe a product flow between processes. The supply chain can be viewed as supply processes following each other in a partial order. A manufacturing process is an event that

Figure 2. A sample traceability graph

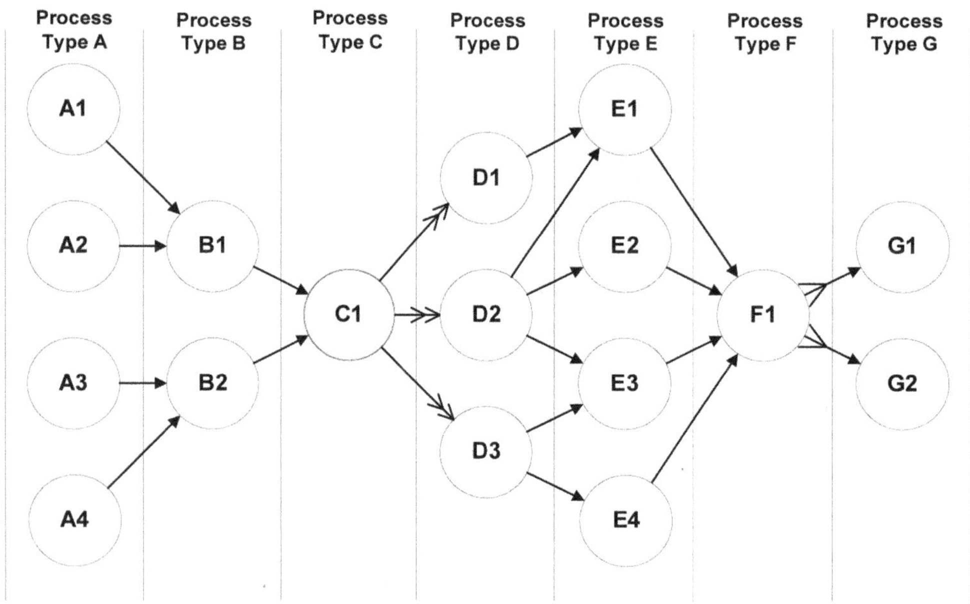

transforms the input elements (raw material, energy) into output elements (product, waste, and emissions).

In a Traceability Graph, processes can be grouped based on their process types, i.e. similar processes are instances of a process type. Within a process type the specific properties of processes, such as timing, placing, etc., may vary.

In Figure 2 there are seven process types (A, B, ..., G). Process type A has four instances. These nodes have no predecessor, which means that the traced objects have been created in these nodes. The objects are transferred forward in the graph. For example, objects from Nodes A1 and A2 are transferred to Node B1. Now objects are not changed but object sets (product portions) of A1 and A2 are unionized in B1. This also means that the resources and emissions of A1 and A2 are cumulated to the new set of objects. The objects from B1 are transferred to the Node C1 where products are classified and sent to one of the D processes.

In the D processes objects are divided into several objects. A double-headed arrow illustrates this. For example a physical object is decomposed or divided into parts. Then, parts may be classified and sent to forthcoming processes. The E nodes receive product portions consisting of these parts. In an E process they are refined and sent to Node F1 which is a shared process for all products. The products of F1 are components for the G processes, i.e. in G1 and G2 objects are composed from the objects that F1 yields. A shared start arrow illustrates this, i.e. several component objects of the start node are needed for a single composition object of the end node.

In a Traceability Graph it is possible to trace the supply chain of an object, i.e. to find all the preceding processes where the object at hand has participated. This also means that all the information related to those processes can be attached to the object. Given the running example, let us assume that we are interested in an object that belongs to Node G1. Then, the processing history of the object is a subgraph of the main graph. In Figure 3 the colored nodes are processes in which the underlying object, its part, or a whole related to the parts has participated. In the example, parts of

Figure 3. The supply chain of an objects belonging to node G1

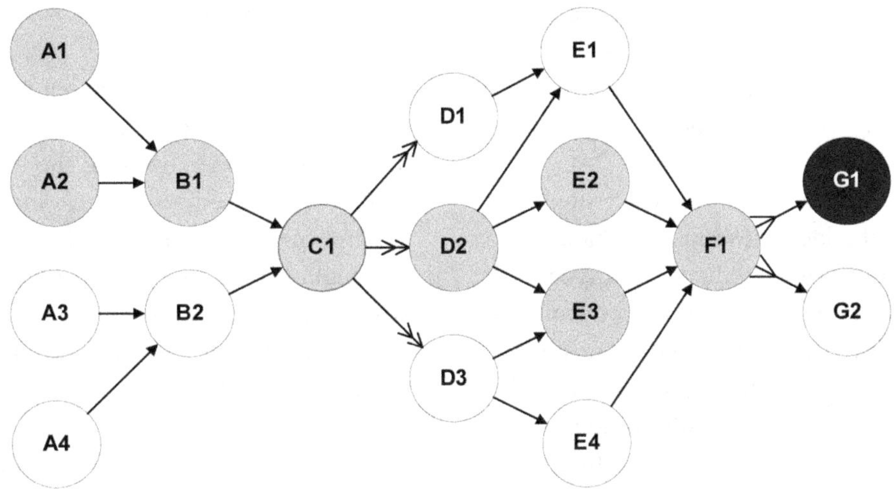

the underlying object have gone through F1, E2, E3, and D2, whereas the larger objects consisting of the parts have gone through C1, B1, A1, and A2. This subgraph is also the supply chain of the underlying object.

The Traceability Graph can also be used for analyzing different aspects on processes. For example a process type can be selected and we can see how much some process causes the environmental burden. Further, this analysis can be done in a supply chain of one object or a set of objects.

Next, we introduce the properties of edges and nodes of the Traceability Graph.

An *object* is a unit of tracing in a phase of the related supply chain. This can be a single product or a patch depending on the precision of tracing in the underlying supply system.

A process node contains the identity of a process, the set of product portions (patches) and the set of attributes associated with the process. A *product portion* involves the quantity of products, the identifiers of objects and the ratio of the emissions and resources compared with the total ones in the process node. The ratio is calculated by an application specific method. It can be based on the portion of mass or used time of machines, for example. Product portions of a process are viewed through the end products of a process.

An *attribute* of a node determines information associated with a process. Input attributes describe the resources of a process whereas output attributes can be used for determining the emissions of a process. Each attribute has two values: one for the underlying process and the other for containing the cumulated values from the previous nodes. A cumulated value is calculated based on the ratios of product portions and quantity that is sifted from the previous nodes via edges.

Via edges, products are sifted from a process to another more precisely from a product portion to another. An edge also determines the mapping of objects between two processes. The mapping can be:

1. Equivalence: Objects from a start node of a product portion are sifted to the related product portion of the end node.
2. Subsetting: Only some objects are sifted to the related product portion of the end node.
3. Supersetting: All the objects are sifted to the related product portion of the end node but the product portion of the end node contains similar objects from another process node.

4. Division: Objects of a start node are divided into smaller objects. If an object represents a single product, this is portioned.
5. Composition: Products of the start nodes are components for the end node.

In 1-3 the objects maintains their identities but in 4 and 5, the identities must be changed. In case 4 the identity of a product is mapped with the identities of parts that are produced from the product. In case 5 several objects are needed for one composition object, i.e. the identities of components are mapped with the identity of the related composition. It is worth noting that a product of an end node may contain components from several start nodes.

Through an edge, the information of sifted products from a node to another node is transferred to an end node following the mapping of objects. An edge involves those objects that are sifted from a start node to the end node (only some products of a product portion may be selected from other processes). This part of the product portion of the start node is called a *sifted product portion*. In transferring products from a process to another, the attributes must be re-calculated for corresponding to the sifted product portion. This is based on the ordinary and *derived attributes*. The derived attribute is associated with an edge and it determines the amount of an ordinary attribute that is related to the sifted product portion.

This information collected throughout supply chain processes—nodes—enables to connect the products and processes using the Traceability Graph that can be thought as a provenance model, where provenance is collected as a set of nodes—discrete activities through supply chain processes—that describe the route of an object traveling through a supply chain. Each node includes the following information: location, the time interval when an object was present in the node and elementary flows connected to the process.

LCA Phase 2: Storing Traceability Data

The Traceability Graph is mapped to the relational database as presented in Figure 4. We selected the relational database because the standard OLAP methods are used to further analyze the huge amount of data that is a result for tracing the individual objects. In Figure 4 PK means primary key and FK means foreign key.

The information of the Traceability Graph is stored into eight relations:

- Node relation is used to store the identities of process nodes.
- Attributes (e.g. raw materials, energy) are stored into Attribute relation.
- Product types are stored into Product relation.
- The relation NodeAttribute is used to store the process (Node) specific attributes. For example □Process#1, Electricity, 100 kWh□ specifies the use of electricity of Process#1.
- The relation NodeProduct is used to store the information about product portions of a specific process (Node). The column Ratio is used to allocate the environmental burden between the portions of products and by-products. For example □Process#1, Product#1, 0.6□ specifies that Product#1 is an end product of Process#1 and the related ratio is 0.6.
- The relation Object is used to store the object specific information like physical code of the object and its volume.
- The relation ObjectRelation is used to store the object mapping when object identities are changed. The column Transformation function is used to calculate the cumulated attribute values.
- The route of the objects through a supply chain is realised by the Route relation. This corresponds to the sifted product portion.

Figure 4. Database schema for the traceability graph

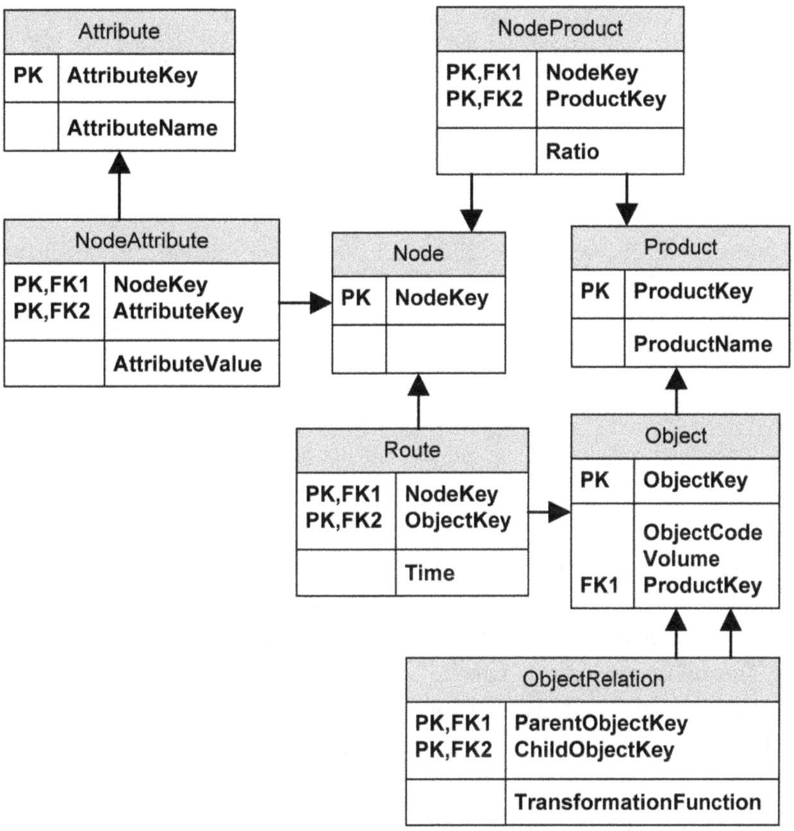

The relation model in Figure 4 can be easily extended to include more product and supply chain specific information. For example, we can implement an organisational hierarchy by creating Process, Site and Organisation relations (Node→Process → Site→ Organisation). This kind of extension enables analysis of environmental data by using the hierarchy as a dimension in multidimensional OLAP model.

LCA Phases 3 and 4: Traceability Cube

There are three different type of OLAP systems, the relational, multidimensional and hybrid OLAP. The data and dimensions are stored as relational tables in the database in the relational OLAP. The multidimensional OLAP stores the data in the optimized multi-dimensional array called OLAP Cube. OLAP Cube is a data structure that allows fast analysis of data from multiple perspectives (dimensions). The cube consists of facts that are called measures and dimensions that categorise the facts. The hybrid OLAP mixes the previous types, i.e. the usage of relational tables and cubes can be selected for each case separately (Chaudhuri, 1997). To be able to use the OLAP type operations for analyzing the information of the Traceability Graph we must combine the previous tables as a data cube. Common operations include:

- *Slicing* and *dicing* the data providing multidimensional view of data based on subsets corresponding to the selected dimensions.
- *Drilling down/up* rises or lowers the level of aggregation. For example we can view

Figure 5. Traceability cube schema

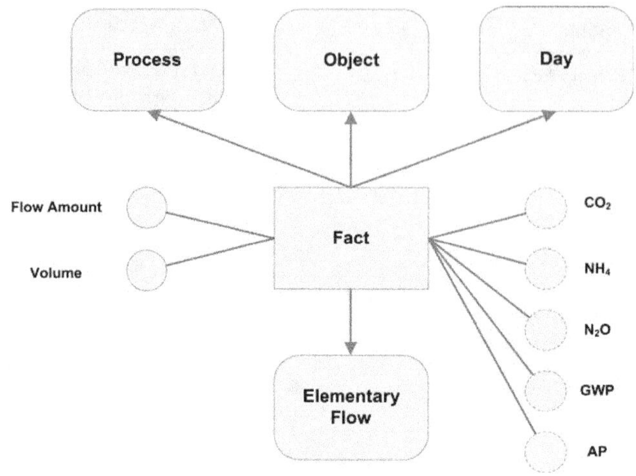

the data in day, week, month or yearly level.

- *Pivot/rotate* operation provides alternative view of data. The pivot operations can be used to change the dimensional orientation of the data cube.

In this work, we use the multidimensional data model "MD" that is presented in Torlone (2003). In figures, the dimensions are presented as round-cornered boxes, the facts are presented as boxes, the description of dimensions are presented as small diamonds, and the measures as circles. The circles drew with dashed line presents calculated measures.

Figure 5 presents the Traceability Cube with some example dimensions. The Process dimension can be used to compare the environmental impact between manufacturing sites and manufacturers. The Object dimension is used to aggregating the environmental data for different product groups. The measure Flow Amount is the amount of elementary flow used in a process in a certain date allocated to specific object. The measure Volume specifies the volume of an object. In Figure 6 some sample instances over the Traceability Cube are

Figure 6. A sample instance over the traceability cube

EPC	Process	Site	Product	Day	Month	Year	Flow	Amount	Volume
1	Kiln Drying	Mill#1	Timber	1	1	2010	Electricity	0,5 kWh	0,30 m^3
2	Kiln Drying	Mill#1	Timber	1	1	2010	Electricity	0,5 kWh	0,35 m^3
3	Kiln Drying	Mill#1	Timber	1	1	2010	Electricity	0,5 kWh	0,33 m^3
1	Transporting	Truck#1	Timber	2	1	2010	Lorry Transport	100 km	0,30 m^3
2	Transporting	Truck#1	Timber	2	1	2010	Lorry Transport	100 km	0,35 m^3
3	Transporting	Truck#1	Timber	2	1	2010	Lorry Transport	100 km	0,33 m^3
4	Kiln Drying	Mill#1	Timber	1	1	2010	Electricity	0,53 kWh	0,40 m^3
5	Kiln Drying	Mill#1	Timber	1	1	2010	Electricity	0,53 kWh	0,42 m^3
6	Kiln Drying	Mill#1	Timber	1	1	2010	Electricity	0,53 kWh	0,38 m^3
4	Transporting	Train#1	Timber	2	1	2010	Rail Transport	200 km	0,40 m^3
5	Transporting	Train#1	Timber	2	1	2010	Rail Transport	200 km	0,42 m^3
6	Transporting	Train#1	Timber	2	1	2010	Rail Transport	200 km	0,38 m^3

Figure 7. Emission and impact calculation: climate change

EPC	Machine	Process	Day	Elementary Flow	Flow Amount	Volume
2	Truck#1	Transport	2.1.2010	Lorry Transport	100 km	0,35 m2
4	Rail#1	Transport	2.1.2010	Rail Transport	200 km	0,40 m2

↓

EPC	Emission	Emission Amount	Unit		Impact Category	Amount	Unit
2	CO_2	19,9102	kg	→	GWP	19,9102	as kg of CO_2 eq
2	CH_4	0,0078544	g	→	GWP	0,19636	as kg of CO_2 eq
2	N_2O	0,000777828	g	→	GWP	0,15401	as kg of CO_2 eq
4	CO_2	22,1193	kg	→	GWP	22,1193	as kg of CO_2 eq
4	CH_4	0,0055732	g	→	GWP	0,13933	as kg of CO_2 eq
4	N_2O	0,043629899	g	→	GWP	8,63872	as kg of CO_2 eq

presented. The EPC column is the unique identity of an object.

The Flow Amount is used for calculating the calculated measures – amount of emissions (e.g. carbon dioxide, methane, nitrous oxides) and amount of key environmental performance indicators (see e.g. Lim & Park, 2009). The emission amount is the amount of emissions caused when using elementary flow (raw material or energy) in some process. For example, carbon dioxide emissions when using electricity from Tampere electricity station in Finland were 194 g / kWh in the year 2008. There are many environmental databases that comprise life cycle inventory data from different supply chain processes. For example, the ELCD core database by European Commission - DG Joint Research Centre - Institute for Environment and Sustainability comprises more than 300 process datasets (e.g. key materials, energy carriers, transport, and waste management).

In Figure 7 some emissions and impact category global warming potential for objects with code 2 and 4 are. Key environmental performance indicators are calculated based on the emissions.

The impact category climate change is calculated based on carbon dioxide, methane, nitrous oxide and several other emissions. The measurement unit for the climate change is kg of carbon dioxide equivalent which means that all the other emissions are converted by using a conversion factor. For example the conversion factor of Methane is 25. Full list of emissions and factors can be found from PAS 2050 (Carbon Trust, 2008).

In Figure 8 the impact category acidification of land and water resources and some of the emissions affecting it are presented. The acidification potential is represented in terms of sulphur dioxide (SO_2) equivalents. Thus, all the other emissions are converted to sulphur dioxide equivalents using conversion factors: NO = 1.07, N_2O = 0.7, NO_x = 0.7, NH_3 = 1.88, HCl = 0.88, HF = 1.6 (Heijungs, et al., 1992).

Process Dimension

The process dimension contains the organisational hierarchy and other process specific information, in this example Machine → Process → Site → Organisation. The machine level includes the

Figure 8. Emission and impact calculation: acidification of land and water resources

EPC	Machine	Process	Day	Elementary Flow	Flow Amount	Volume
2	Truck#1	Transport	2.1.2010	Lorry Transport	100 km	0,35 m2
4	Rail#1	Transport	2.1.2010	Rail Transport	200 km	0,40 m2

↓

EPC	Emission	Emission Amount	Unit		Impact Category	Amount	Unit
2	NO_x	8	g	→	AP	5,6	as kg of SO_2 eq.
2	N_2O	0,0048	g	→	AP	0,00336	as kg of SO_2 eq.
2	NH_3	0,0072	g	→	AP	0,013536	as kg of SO_2 eq.
4	NO_x	64	g	→	AP	44,8	as kg of SO_2 eq.
4	N_2O	0,07	g	→	AP	0,049	as kg of SO_2 eq.

spatial description location that can be used for spatial analysis of environmental burden.

The process dimension can also be used to benchmark between different processes or manufacturers, i.e. it can be used to compact workflow. For example we could merge the transporting nodes of our running example and get the total amount of transporting in our sample supply chain. In an enlarged example, presented inFigure 9, we analyze the effect of different transporting methods by merging the transporting nodes based on the transporting type – the machine type in example dimension.

By comparing the environmental performance of same type of processes companies can see if the performance metrics of their process compare to industry best practices. This allows companies to notice where they could make the biggest improvements in their environmental performance.

Object Dimension

The object dimension is used to describe the properties of object and can be used to benchmark the environmental burden between different products and product types. The total environmental impact caused when manufacturing a single object is a result when slicing the cube using object level. The slices contain environmental impact from the processes that are extensionally connected with the product (see Figure 10).

By using the slice, we can also get the information about the supply chain of a product. This accurate item level information improves the visibility of the supply chain enabling organizations to improve their product life cycle management. For example in forest industry, the companies are certificating their product using PEFC (www.pefc.org) chain-of-custody certification (Forests area: 229 million ha, Forest owners: > 475,675). The PEFC chain-of-custody certification is a method for tracing wood from forest to the final product to ensure the wood or wood-fibre can be traced back to certified forest. The certification has two methods of realizing this:

- Percentage based method – the method allows mixing certified and non-certified raw materials taking into account that the percentage of certified raw material must be known. Company can sell as certified the

Figure 9. Process dimension and benchmarking

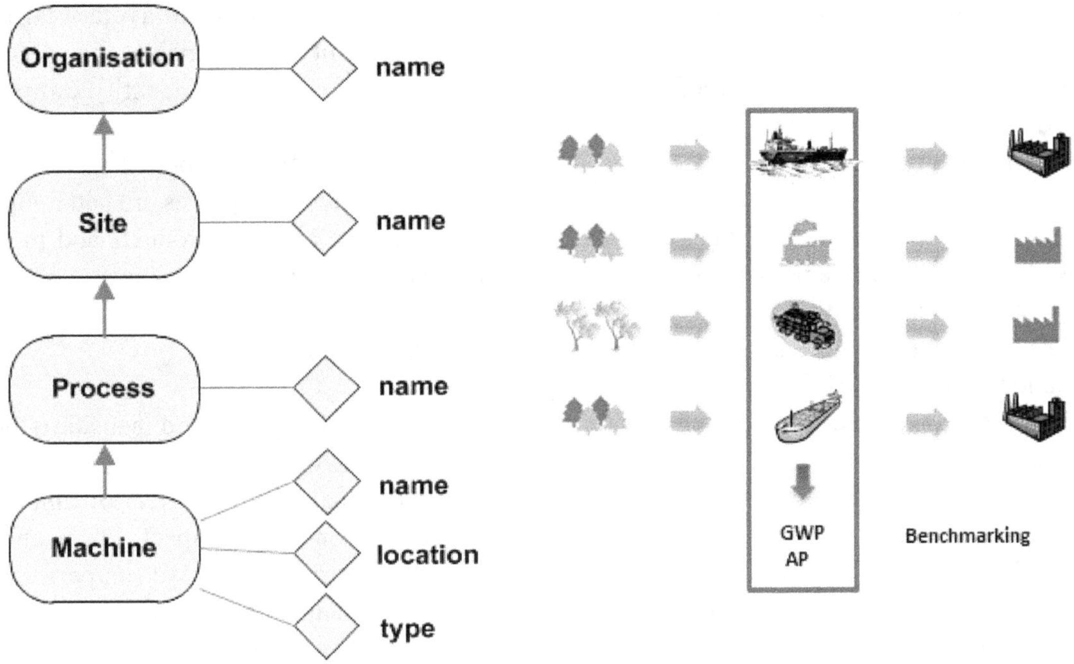

Figure 10. Object dimension and a supply chain of a product

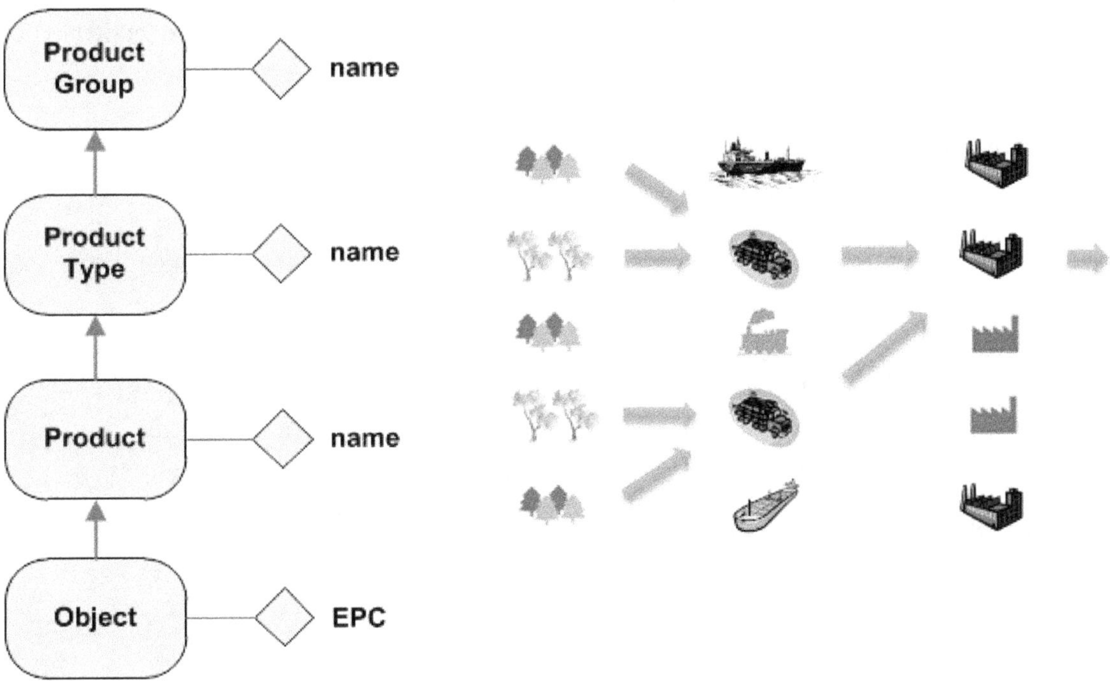

proportion of its production, which equals the proportion of the certified raw material

- Physical separation method – the method requires certified and non-certified raw material to be physically separated throughout the supply chain.

As seen, the methods can be integrated and improved using the Traceability Cube. The accurate item level information of the origin can be used to create a transparent and trustworthy certification system for origin.

Other usage example is recalling unsafe, defective, or hazardous products. The item level traceability allows manufacturers to recall just the particular products that contain the unsafe elements. For example, Toyota recalled approximately 9 million cars in 2009-2010 for three related recalls pedal entrapment/floor mat problem and accelerator pedal problem. If the reason of recall is for example a faulty set of components, a manufacturer could have only recalled those cars that include the faulty components, not the whole set of manufactured cars in some time period.

The item or patch level traceability can also be used in food industry to avoid total recalls. The Traceability Cube enables manufacturers, suppliers and resellers to identify the products where some suspected raw material was used and all customers whom the products were delivered. Using the system companies are better equipped to retrieve the affected products and to protect their reputation and brand value.

Date Dimension

In data warehousing the date dimension is the most frequently used dimension. The date dimension is used for analysing the effect of time regarding the environmental impact. For example by using the date dimension, we can perform trend analysis, for detecting a pattern of behaviour in a data series. Also historical and period comparison analysis can be made.

For example, we can track if the environmental burden of a manufacturing facility is diminishing or growing. The trend information can also reveal if environmental burden is affected by seasons or other time related matters. For ex-

Figure 11. Date dimension and trend report

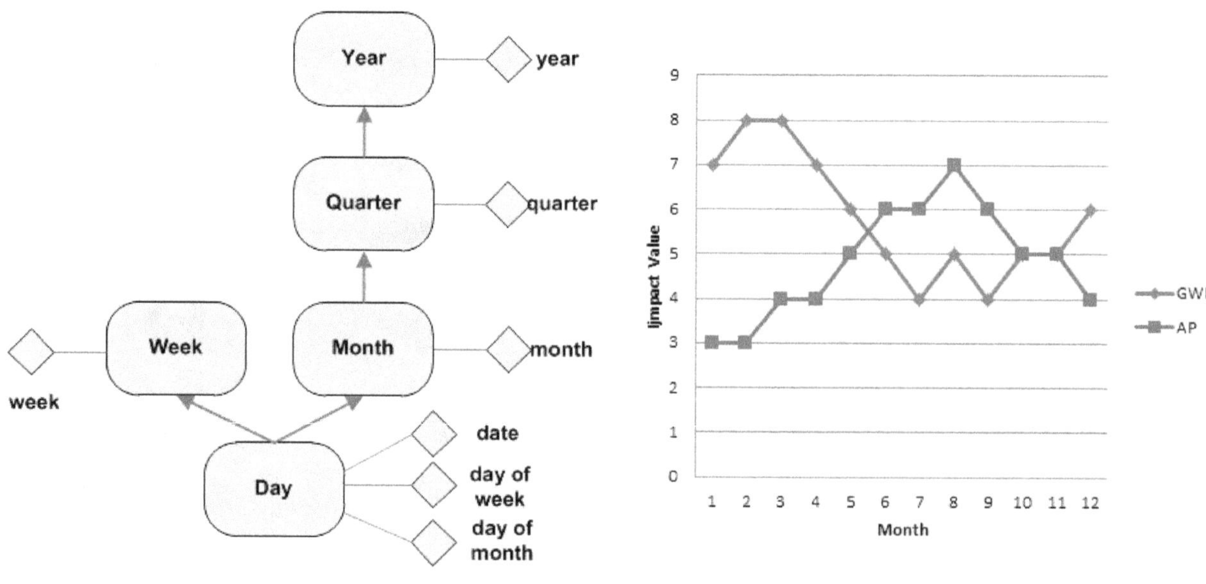

ample the Figure 11presents a situation where acidification potential is higher during summer months and global warming potential is higher during fall, winter, and spring. The trend information can easily be used for forecasting also the future environmental burden using standard algorithms as time series analysis.

Elementary Flow Dimension

The elementary flow dimension describes the resources and raw materials used to manufacture an object in each process (see Figure 12).

In this chapter, we have used the environmental values as an example. The Traceability Cube can be easily extended to include also economic and social information. By adding a price for an elementary flow we can also monitor and analyze the economic performance of the processes in similar methods than presented above. The social elementary flow can also be added to the system. For example work related accidents happened during the process. This enables us to monitor and analyze the products and their manufacturing process in all three viewpoints of sustainability and thus allowing us to choose the most sustainable way of manufacturing.

The analytics capabilities of the Traceability Cube can be used for analyzing the environmental data. For example, environmental data can be summed up to create the total environmental impact for the whole life cycle of the product. The data can also be used for comparing the performance between different manufacturers or manufacturing sites. The possibility to analyze the supply chain on the process and item level allows the end users to select a product, which creates least environmental burden. This creates pressure for the manufacturers to improve the eco-efficiency of their supply chains.

FUTURE RESEARCH DIRECTIONS

The precision of traceability of the resources and emissions depends on the underlying data model and ability how strictly physical products and their components can be identified. Our model can be applied to any granularity of tracing. For applications, it is required physical identity mechanism that can be mapped to their logical counterparts in the database.

One option for marking the objects is Radio-Frequency Identification (RFID) technology, which can be compared to the bar code identification: an identification code is embedded to an object. The RFID system can be divided into three components:

Figure 12. Elementary flow dimension and sustainability performance comparison report

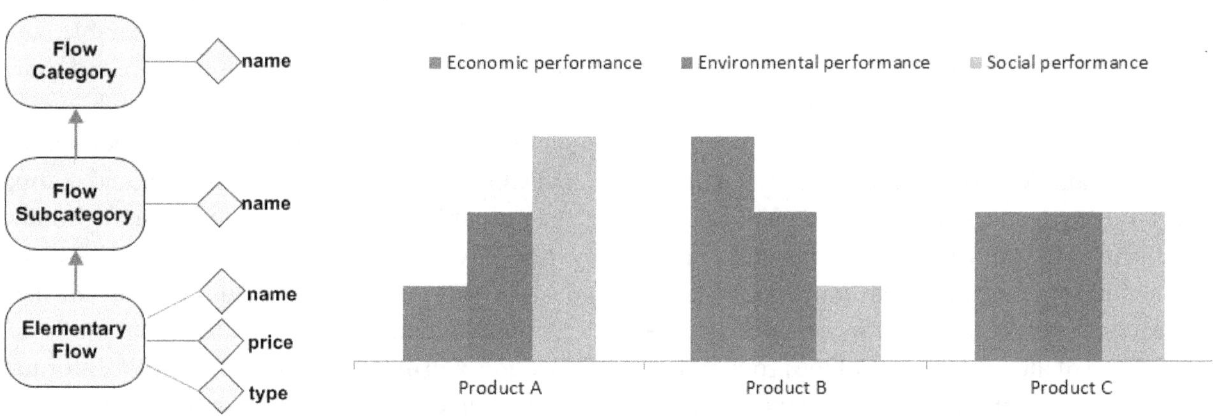

- RFID-tag. The main parts of the tag are a chip, an antenna, and a packaging. The chip holds the individual code attached to the physical object. The antenna transmits the information to the reader and the packaging holds the antenna and the chip.
- RFID-reader. RFID-readers read the object identifier from the RFID-tag without needing a clear line of sight. The RFID-reader can be integrated to a machine or it can be just a hand held device.
- Host. The Host computer hosts the application that maps the RFID-data stream from RFID-readers to more understandable form that is useful for for the manufacturing execution and enterprise resource planning systems.

Unlike bar code identification, the RFID system does not require a line of sight and the orientation of reader and tag is not signification for communication. Moreover, the reading is automatic and each item is individually labeled with a tag that is more difficult to counterfeit than a simple barcode. With RFID technology a vast amount of tags can be read simultaneously. RFID tags can be active, passive, or semi-passive. The RFID technology and data management has been researched extensively during the last decade—e.g. RFID data staging to OLTP- and OLAP-applications (Krompass et al., 2007) provides a method that answers to different kinds of needs that transactional and analytical applications poses to RFID data, and the RFID data warehousing (Gonzalez, et al., 2006). Dada and Staake (2008) describe a method of presenting the carbon footprint of the product instance to the customers.

The example above presented how environmental impact information can be analysed. To be able to cover the whole supply chain the environmental monitoring system must be modular, i.e. each actor is responsible for generating data from a part of the supply chain of the product. The main challenge of analysing the environmental impact of a product is the precision of underlying environmental data. In the optimal situation, all the products in the world would be traced through the full supply chain where all processes have smart metering systems for measuring the emissions generated when manufacturing products and this information would be shared through standard interfaces. However, this is not realistic as a starting point.

As discussed in Usva et al. (2009) there is a need for a modular approach where the starting point is to use the environmental impact values produced by current approaches, varying from expert judgments and partial estimates to the usage of aggregated product group data generated from national input-output tables or data from life cycle assessments that is calculated using averages. These default values would be used when the part of the supply chain does not have environmental monitored traceability data available. These values should be defined so that the stakeholders would be encouraged to produce more accurate data. This is achieved by using somewhat higher values than average of the industry.

The modularity of a supply chain means that each actor is responsible for generating data from a part of the supply chain of the product and to share it with other stakeholders. To be able to share product related information in the complex supply chains the organizations have to agree on a common standard. One of the most promising is EPC global Architecture Framework[1] standards, which are generally accepted methods for sharing product data in supply chains. They enable supply chain stakeholders to capture, store and share product related data. The EPCGlobal architecture includes EPC Information Services specification (EPCGlobal, 2007) that defines storing and sharing the traceability data that is created when a product marked with an RFID-tag passes an RFID-reader in a process in a supply chain. This event data normally contains unique identification code, location, and time. By extending the EPC Information Service specification also environmental data can be included in event data. For example: 'At location X in time Y the object Z was observed with the environmental data [elementary flow #1, elementary flow #2]'. In Björk et al. (2011), we have given more detailed technical description of sharing environmental data and scalability of the system in real world application.

To be able to generate a total carbon footprint for a product the organizations must share environmental information of the products that were handled by them in their part of the supply chain. For example, in our running example, some organizations are responsible for harvesting the timber; sawmill companies handle the sawing and glued laminated timber manufactures are using the boards sawn in sawmills. All these stakeholders own a part of the final product's life cycle information. To be able to share environmental information each stakeholder must implement an EPC Information Service that implements the extension for environmental data. To be able to handle the object transformation (division or composition) in the supply chain, the stakeholder responsible for the transformation part of the supply chain is also responsible for aggregating the environmental information from up to that point. In other words, when a manufacturer is further processing products, the manufacturer is responsible for calling the EPC Information Services of a supplier and to add this (derived attribute in the Traceability Graph) information to the environmental information of the further processed product.

To be able to present the environmental data coherently and reliably there has to be a standard set of rules which are enforced by an external auditing party. The ISO 14025 (International Organization for Standardization, 2006) specifies a method for creating environmental declarations and programmes for creating them. One programme operator is an international EPD system, which maintains a specific set of product category rules that provides specifications for creating an environmental product declaration for products. The rules could be extended so that the best level of the environmental product declaration would be a monitored real-time traceability based environmental data.

When the environmental impact of the whole supply chain is visible to the customers in a reliable and visible way, they can make educated choices of selecting the product with least environmental impact, which would encourage companies to produce data for their part of the supply chain and to optimize their production in a sustainable way.

CONCLUSION

We presented a model how emissions and resources can be monitored from the data management perspective. The model can be mapped to any precision level of physical tracing. At the most precise level, even a single physical object and its components can be analyzed. This, of course, demands that the related objects and their components are identified and mapped to the database. From the opposite perspective, our model also supports rough level analysis of products and their histories. We showed how multidimensional analysis can be applied for OLAP analysis based on the Traceability Graph.

In short, our approach supports the different phases the life cycle assessment as follows. The Traceability Graph is used to model the processes (LCA phase 1). The traceability data is collected to a database based on the graph (LCA phase 2). The Traceability Cube is used to represent the impact assessment and analyzing the results (LCA phases 3 and 4).

We also presented an infrastructure, which allows tracing the environmental burden of an individual product in a complex and dynamic supply chain. The system can be used with any level of physical tracing from using the yearly values to the component level of the product. The model presented can be used for creating a reliable system for measuring an environmental impact of a single product and to present this information to stakeholders, so that they are able to choose the products, which create less environmental burden

REFERENCES

Akram, A., Allan, R., & Kewley, J. (2006). Data centric approach for workflows. In *Proceedings of the Tenth IEEE International Enterprise Distributed Object Computing Conference Workshops (EDOCW 2006)*. IEEE Press.

Barker, A., & van Hemert, J. I. (2007). Scientific workflow: A survey and research directions. In *Proceedings of the Third Grid Applications and Middleware Workshop (GAMW 2007)*, (pp. 746-753). ACM.

Björk, A., Erlandsson, M., Häkli, J., Jaakkola, K., Nilsson, Å., & Nummila, K. (2011). Monitoring environmental performance of the forestry supply chain using RFID. *Computers in Industry*, *62*(8-9), 830–841. doi:10.1016/j.compind.2011.08.001

Bonner, A. J. (1999). Workflow, transactions, and datalog. In *Proceedings of the Eighteenth ACM Symposium on Principles of Database System (PODS 1999)*, (pp. 294-305). ACM Press.

Carbon Trust. (2008). *Specification for the assessment of the life cycle greenhouse gas emissions of goods and services, PAS 2050*. Carbon Trust.

Chaudhuri, S., & Dayal, U. (1997). An overview of data warehousing and OLAP technology. *SIGMOD Record*, *26*(1), 65–74. doi:10.1145/248603.248616

Chervenak, A., Foster, I., Kesselman, C., Salisbury, C., & Tuecke, S. (2001). The data grid: Towards an architecture for the distributed management and analysis of large scientific datasets. *Journal of Network and Computer Applications*, *23*, 187–200. doi:10.1006/jnca.2000.0110

Clifford, B., Foster, I., Hategan, M., Stef-Praun, T., Wilde, M., & Zhao, Y. (2008). Tracking provenance in a virtual data grid. *Concurrency and Computation*, *20*(5), 565–575. doi:10.1002/cpe.1256

Curcin, V., & Ghanem, M. (2008). Scientific workflow systems - Can one size fit all? In *Proceedings of the 5th Biomedical Engineering Conference (CIBEC 2008)*, (pp. 18-20). CIBEC.

Dada, A., & Staake, T. (2008). Carbon footprints from enterprises to product instances: The potential of the EPC network. *GI Jahrestagung*, *134*, 873–878.

Donnellan, B., Sheridan, C., & Curry, E. (2011). A capability maturity framework for sustainable information and communication technology. *IEEE IT Professional*, *13*(1), 33–40. doi:10.1109/MITP.2011.2

EPCglobal. (2007). *EPC information services (EPCIS) version 1.0.1 specification, EPCglobal ratified standard*. Retrieved May 10, 2011, from www.epcglobalinc.org/standards/epcis/epcis_1_0_1-standard-20070921.pdf.

Foster, I., Kesselman, C., & Tuecke, S. (2001). The anatomy of the grid: Enabling scalable virtual organizations. *The International Journal of Supercomputer Applications*, *15*(3), 200–222. doi:10.1177/109434200101500302

Freire, J., Koop, D., Santos, E., & Silva, C. (2008). Provenance for computational tasks: A survey. *Computing in Science & Engineering*, *10*(3), 11–21. doi:10.1109/MCSE.2008.79

Gonzalez, H., Han, J., & Li, X. (2006). FlowCube: Constructing RFID FlowCubes for multidimensional analysis of commodity flows. In *Proceedings of the 32nd International Conference on Very Large Data Bases*, (pp. 834-845). Very Large Data Bases.

Heijungs, R., Guinée, J. B., Huppes, G., Lankreijer, R. M., Udo de Haes, H. A., & Sleeswijk, A. ... de Goede, H. P. (1992). *Environmental life cycle assessment of products*. Leiden, Germany: CML.

International Organization for Standardization. (1997). ISO 14040, environmental management - Life cycle assessment - Principles and framework. *ISO, 14040*, 1997.

International Organization for Standardization. (2000). ISO 14020, Environmental LABELS AND DECLARATIONS -- General principles. *ISO, 14020*, 2000.

International Organization for Standardization. (2006). ISO 14025, environmental labels and declarations -- Type III environmental declarations -- Principles and procedures. *ISO, 14025*, 2006.

Junkkari, M., & Sirkka, A. (2011). Using RFID for tracing cumulated resources and emissions in supply chain. *International Journal of Ad Hoc and Ubiquitous Computing, 8*(4), 220–229. doi:10.1504/IJAHUC.2011.043582

Krompass, S., Aulbach, S., & Kemper, A. (2007). Data staging for OLAP- and OLTP-applications on RFID data. In *Proceedings of Datenbanksysteme in Business, Technologie und Web (BTW 2007)*, (pp. 542-561). BTW.

Lim, S.-R., & Park, J. M. (2009). Environmental indicators for communication of life cycle impact assessment results and their applications. *Journal of Environmental Management, 90*(11), 3305–3312. doi:10.1016/j.jenvman.2009.05.003

Murugesan, S. (2008). Harnessing green IT: Principles and practices. *IEEE IT Professional, 10*(1), 24–33. doi:10.1109/MITP.2008.10

Torlone, R. (2003). Conceptual multidimensional models. In Rafanelli, M. (Ed.), *Multidimensional Databases: Problems and Solutions* (pp. 69–90). Hershey, PA: IGI Global. doi:10.4018/978-1-59140-053-0.ch003

Usva, K., Hongisto, M., Saarinen, M., Katajajuuri, J. M., Nissinen, A., Perrels, A., ... Koskela, S. (2009). Towards certifies carbon footprints of products – A road map for data production. *VATT Research Reports, 143*(2).

Van der Aalst, W., & van Hee, K. (2002). *Workflow management, models, methods, and systems*. Cambridge, MA: The MIT Press.

ADDITIONAL READING

Carbon Trust. (2008). *Guide to PAS 2050 – How to assess the carbon footprint of goods and services*. Carbon Trust.

Dada, A., Rau, A., Konkel, M., Staake, T., & Fleisch, E. (2010). *The potential of the EPC network to monitor and manage the carbon footprint of products: Dynamic carbon footprint demonstrators*. White Paper WP-BIZAPP-054. Auto-ID Labs.

Dada, A., Staake, T., & Fleisch, E. (2009). *The potential of the EPC network to monitor and manage the carbon footprint of products: Carbon accounting*. White Paper WP-BIZAPP-047. Auto-ID Labs.

Defra. (2006). *Environmental key performance indicators: Reporting guidelines for UK business department for environment*. London, UK: Queen's Printer and Controller.

Finkenzeller, K. (2004). *RFID handbook fundamentals and applications in contactless smart cards and identification* (2nd ed.). London, UK: Wiley.

Ilic, A., Staake, T., & Fleisch, E. (2009). Using sensor information to reduce the carbon footprint of perishable goods. *IEEE Pervasive Computing / IEEE Computer Society and IEEE Communications Society, 8*, 22–29. doi:10.1109/MPRV.2009.20

International Organization for Standardization. (2006). *ISO 14064, greenhouse gases -- Part 1: Specification with guidance at the organization level for quantification and reporting of greenhouse gas emissions and removals. ISO 14064-1:2006.*. International Organization for Standardization.

Johansson, O., & Pålsson, H. (2009). The impact on Auto-ID on logistic performance: A benchmarking survey of Swedish manufacturing industries. *Benchmarking International Journal (Toronto, Ont.)*, *16*(4), 504–522.

Junkkari, M., & Sirkka, A. (2011). Using RFID for tracing cumulated resources and emissions in supply chain. *International Journal of Ad Hoc and Ubiquitous Computing*, *8*(4), 220–229. doi:10.1504/IJAHUC.2011.043582

Lim, S.-R., & Park, J. M. (2009). Environmental indicators for communication of life cycle impact assessment results and their applications. *Journal of Environmental Management*, *90*(11), 3305–3312. doi:10.1016/j.jenvman.2009.05.003

Ranganathan, J., Corbier, L., Phatia, P., Schmizt, S., Gage, P., & Oren, K. (2004). *The greenhouse gas protocol – A corporate accounting and reporting standard*. WBCSD and WRI.

Simmhan, Y. L., Plale, B., & Gannon, D. (2005). A survey of data provenance in e-science. *SIGMOD Record*, *34*(3), 31–36. doi:10.1145/1084805.1084812

KEY TERMS AND DEFINITIONS

RFID: Radio-Frequency Identification (RFID) is a technology that is used to identify an object by using radio wave technology to exchange the identity data between the tag attached to an object and RFID-reader

EPCglobal Architecture Framework: EPCglobal Architecture Framework is a collection of standards of hardware, software and data which aim to exchange product level supply chain information transfer through the use of Electronic Product Codes.

EPCIS: Electronic Product Code Information Services standard specifies the EPC data sharing between the supply chain stakeholders. The goal is to enable supply chain stakeholders to gain shared view of the EPC-bearing objects in the supply chain.

LCA: Life cycle assessment is a standardized method for calculating the environmental impact caused by product during its life cycle. The goal of LCA is to compare the environmental impact caused by a product so that the customer can choose the least burdensome one.

LCI: Life cycle inventory is an inventory input and output flows of a certain system. Input flows include input of energy and raw materials. Output flows include releases to land, water and air. The inventory data is related to the functional unit defined in the life cycle assessment.

Environmental Indicator: Environmental Indicator is a measure of performance used to measure the state of the environment

OLAP: Online Analytical Processing is a method used for describing the analysis of the complex data in the data warehouses.

Traceability Graph: A workflow model for tracing an allocating emissions and resources of products in supply chains.

Traceability Cube: A multidimensional model for analyzing life cycle information based on the Traceability Graph.

ENDNOTE

1 See EPC Global (2011).

Section 3
Mobile Handheld Devices

Chapter 11
Mitigating the Environmental Impact of Smartphones with Device Reuse

Xun Li
University of California – Santa Barbara, USA

Pablo Ortiz
University of California – Santa Barbara, USA

Brandon Kuczenski
University of California – Santa Barbara, USA

Diana Franklin
University of California – Santa Barbara, USA

Frederic T. Chong
University of California – Santa Barbara, USA

ABSTRACT

The rapid growth of information technology has not only brought substantial economic and societal benefit but also led to an unsustainable disposable model in which mobile devices are replaced in a matter of months. The environmental impact of this stream of handsets in terms of manufacturing energy, materials, and disposal costs is alarming. This chapter aims at raising today's environmental issues of the increasing smartphone market, as well as providing a quantitative analysis on the environmental impact of different life-cycle stages of the smartphones, including the manufacturing stage, using stage, and recycling. To achieve sustainable computing and best utilize the energy consumed during manufacturing the large number of devices, this chapter demonstrates the methodology and techniques towards reusing smartphones by presenting a case study on reusing smartphones for elementary school education.

DOI: 10.4018/978-1-4666-1839-8.ch011

1. AN INTRODUCTION TO SMARTPHONES

During the last 10 years, there has been a rapid increase in the power and popularity of mobile devices. The consumer enthusiasm generated by the latest devices has led to a dramatic shift in personal computing. This shift is characterized by consumers moving away from traditional desktop computers in favor of their mobile devices. The magnitude of consumer enthusiasm and the degree to which the personal computing shift can be felt are made tangible by observing the sheer volume of mobile devices sold around the world. For example, roughly one million units of the iPhone 3G model were sold worldwide in the first weekend it was being sold (Dalrymple, 2010). Similarly, over three hundred thousand iPads were sold on the first day of its release (Lance, 2009). This excitement inevitably means that the number of these devices is increasing extremely quickly.

1.1. The Cellphone Market

The increasing number of cellular phones in the United States is of particular interest. In accordance with demand, the number of cellular phone shipments to the United States has grown very quickly in the last few decades, and the result is the enormous volume of devices currently observed. As can be seen in Figures 1 and 2, the amount of these devices that have been used and replaced in

Figure 1. This graph depicts the trend in phone sales worldwide from 1997-2009. The units for the Y-axis is millions of smartphones. The X-axis, of course, shows the relevant year (Kairer, 2005).

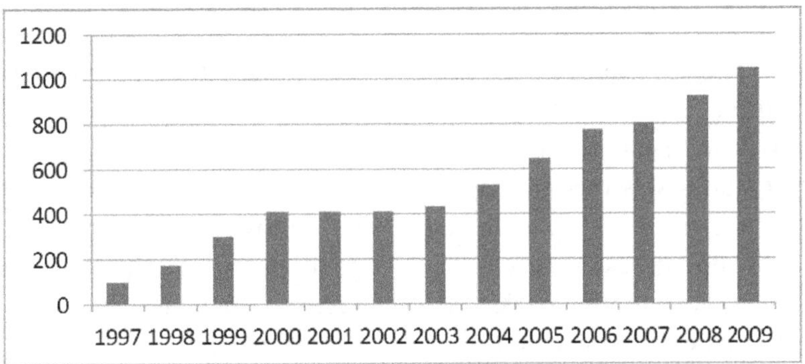

Figure 2. This graph depicts the number of US cell phone subscribers from 2004-2008 and provides a forecast for how the trend will look in later years (Ortiz, 2008).

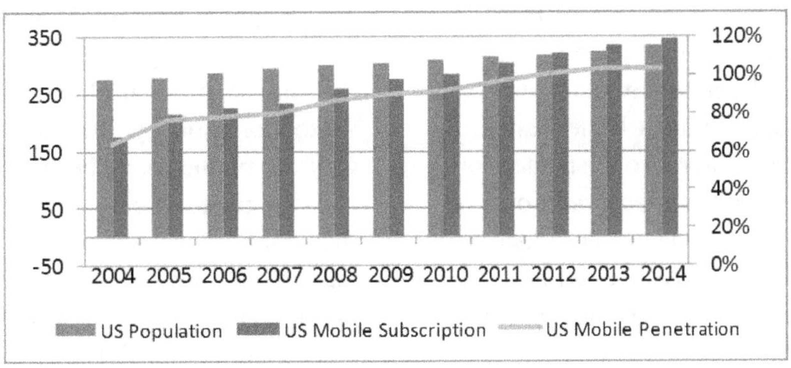

the last few decades is on the order of hundreds of millions, both nationally and in a global scope. It is easy to see that the scale of the problem of what to do with all the discarded cellular phones is growing similarly. It is becoming increasingly important to research a particular subset of cellular phone due to its consumer popularity. An understanding of this type of cellular phone will aid in efforts to improve the way in which cellular phones are handled at the end of their life cycles.

1.2. From Cell Phones to Smartphones

Among cellular phones, smartphones are the fastest growing subset of cellular phones in terms of popularity. A smartphone, in its current manifestation, is a cellular phone with highly advanced telecommunication features and computational power comparable to a desktop computer created in the last decade. They usually run complete operating systems that allow software developers to produce a wide range of applications for the smartphone platform. The large library of advanced applications that consumers can run and install on their smartphones sets this particular type of phone apart from other subsets of cellular phone. Smartphones are not to be confused with feature phones. Feature phones are now considered to be those low-end mobile phones with a few special features tacked on (i.e. a limited browser, a game or two, etc.). Up until phones displayed computational power comparable to a desktop computer and large libraries of downloadable applications were made available to them, the terms smartphone and feature phone were used interchangeably to describe any phone that was more advanced than the most basic design. As smartphones became more sophisticated over time, consumers began to replace traditional cellular phones with smartphones. Smartphones will continue to become more advanced with time and consumers will make use of them more and more. The current state of the overall cellular phone market is indicative of this trend.

The rate of growth of the smartphone market is much greater compared to the overall cellular phone market. According to the *International Data Corporation*, the smartphone market is expected to grow by 49.2% in 2011. This growth rate is four times faster than the expected growth rate of the overall cellular phone market for the same year (IDC, 2011). In fact, the overall cellular phone market has seen a decline in sales to match the state of other markets in the United States. Despite the economic recession, the smartphone market still generates a great deal of activity, as shown in Figure 3. To further illustrate how dramatic this growth is, research from *In-Stat*, a respected group of market researchers, predicts that half of all handset shipments in the United States will be smartphones by the year 2012 (Business Wire, 2011). Forecasts of this type make it clear that traditional cellular phones are being replaced by their more advanced counterparts. Thus, it is likely that these advanced counterparts will dominate the future cellular phone market. Traditional cellular phones that are currently in circulation or will be in circulation in the future will gradually be removed from the market all together. The phenomenon of rapid growth in the smartphone sector is not market behavior that is unique to the United States.

As shown in Figure 4, smartphones have been replacing traditional cellular phones, at a greater or lesser rate than the United States, all over the world. The same sorts of indicators present in the cellular phone market of the United States can be observed in many other countries. For example, the amount of smartphone shipments in Asia is set to quadruple by 2015 (IMS Research, 2010). Cellular phones have been extremely popular in Asia for quite some time. Thus, it is not surprising that smartphones, the next step in mobile phone technology, have been taking over the cellular phone market there, as well. Pyramid Research estimates that over sixty percent of cellular phones in Spain will be smartphones by 2014 as shown in Figure 5. This amount of growth easily rivals that seen in the smartphone market of the United States (Bokun, 2011). Similar data detailing the growth of the smartphone market of a particular region can be found for many more regions than those previously discussed. Thus, it cannot be said that the widespread consumer adoption of smartphones is a phenomenon exclusive to the United States. It is, in fact, a global reality.

1.3. Smartphone Components

Smartphones are composed of many different kinds of hardware components. Though the specifications of these components may differ from model to model, much of the same type of hardware is used. Components common to current smartphones include random access memory, embedded processors, batteries, internal data storage space, display screens, sensors, and cameras. Each one of these components, as found in smartphones,

Figure 3. This graph portrays what percentage of mobile phones being sold between 2008-2009 are smartphones (Myslewski, 2009).

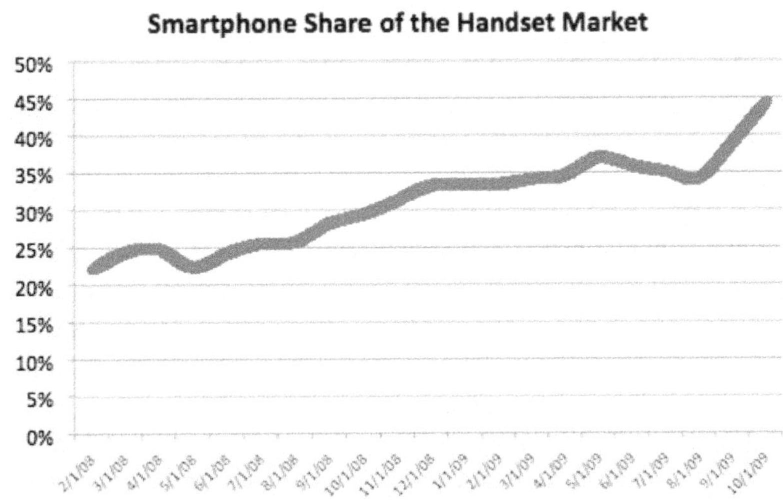

Figure 4. This graph gives both an optimistic and pessimistic forecast for global smartphone shipments between 2009-2013 (CIOL, 2009).

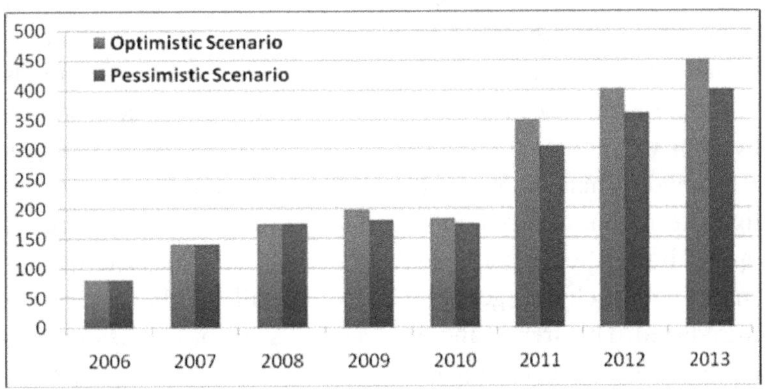

Figure 5. This graph forecasts how the smartphone market will look in Spain up to the year 2014 (Bokun, 2011).

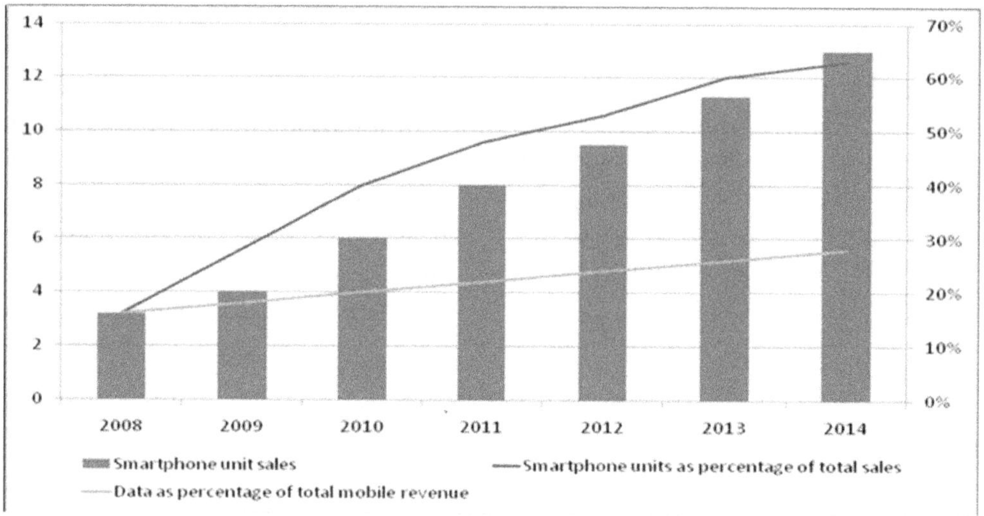

have been steadily evolving since their introduction. The rate at which each component improves in quality can vary significantly depending on the hardware demands of software developers and consumers as well as the technological ability to meet those demands at any given time.

- **Memory**: The amount of memory on smartphones has been growing as smartphones become more complex. More memory on a phone provides the opportunity for better software services. Software services often take the form of useful operating system features or new resource intensive applications. As consumer demand has made the smartphone a popular development platform, the demand for more RAM on phones does not cease. The amount of memory on smartphones has been on the order of hundreds of megabytes for some time now and is steadily making its way towards the gigabyte realm. Smartphones of particular model lines often receive an increase in memory with each new generation that gets introduced. This trend can be observed in Figure 6 where the evolution of memory in two popular smartphone brands is compared. Similar growth patterns have been observed for the clock frequency of smartphone processors.
- **CPU**: The clock frequency of smartphone processors is a rough estimate for the speed of the overall hardware. It is by no means a precise estimate, but it can still be used as an indicator for how processors in smartphones are evolving. The rate of growth of clock frequency in smartphone processors is very similar to the rate of growth of the size of memory in smartphones. The fact that Figures 6 and 7 have a similar shape supports this claim. Since the two figures are a comparison of the hardware components of two of the most popular smartphone hardware brands over time, they can be taken as being representative of the state of market at large. Smartphone processors are currently being produced with clock frequency in the range of many hundreds of megahertz to one gigahertz. They are all still single-core processors, as well. The sorts of processors seen in smartphones are reminiscent of the ones seen in desktop

Figure 6. This graph depicts the evolution of the size of random access memory in the Apple iPhone and HTCs line of smartphones

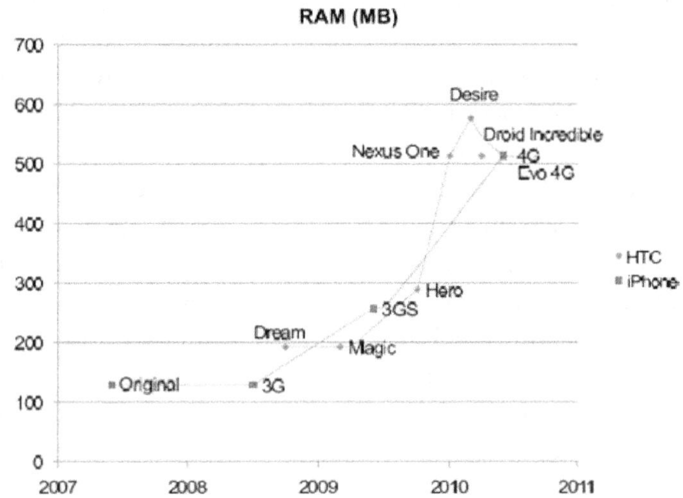

Figure 7. This graph depicts the evolution of processor clock frequency in the Apple iPhone and HTCs line of smartphones

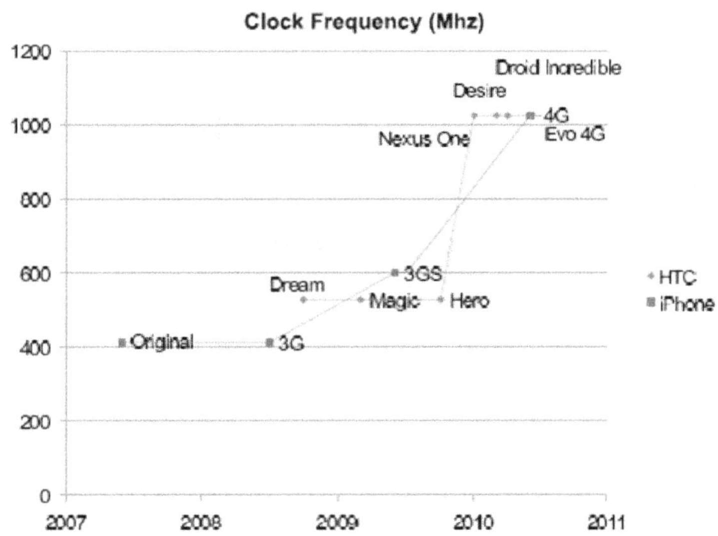

computers at the beginning of the millennium in terms of performance. The evolution of battery capacity diverges from the trends seen thus far.

- **Battery**: The battery capacity of a smartphone determines to some extent how long a phone can function without recharging. To truly determine how long a phone can go without recharging, one must have a model for typical user activity and a hardware profile for a specific phone. In other words, it must be known at what rate power will be consumed. It is not well known how the battery capacity of smartphones will evolve in the future. Advances in battery technology are much slower than in other hardware components. Therefore, smartphone manufacturers are much more likely to design more power efficient phones before incorporating a battery with a longer life. It is interesting to note, however, that, as shown in Figure 8, the battery capacity of smartphones is not as predictable as components in silicon technology.

- **External Storage**: The desire for ever-more storage space in smartphones is similar to the desire for more memory. The amount of external storage on a phone is a tangible limit for the amount of applications, music, videos, etc. a consumer can have on their phone. The amount of storage currently provided on phones is on the order of tens of gigabytes. Space on this scale is enough for smartphones to be used as portable music players and gaming devices. The growth rate of external storage capacity seems to have been following Moore's law on Apple and HTC smartphones. In other words, storage capacity doubles about every two years. This trend is similar to that seen for memory and clock frequency.
- **Screen**: A description of the evolution of display screens on smartphones is more feature-centric than the numeric trends discussed for previous hardware components. The display screens on cellular phones used to be for visual output only. The screens found on current smartphones do much more than that. Now, consumers use their screens as the primary input device for their phones. Smartphones are often set up such that selecting applications, changing settings, typing, etc. are all done via the touch screen. On top of being used for visual output and primary input, touch

Figure 8. This graph depicts the evolution of the battery capacities in the Apple iPhone and HTCs line of smartphones

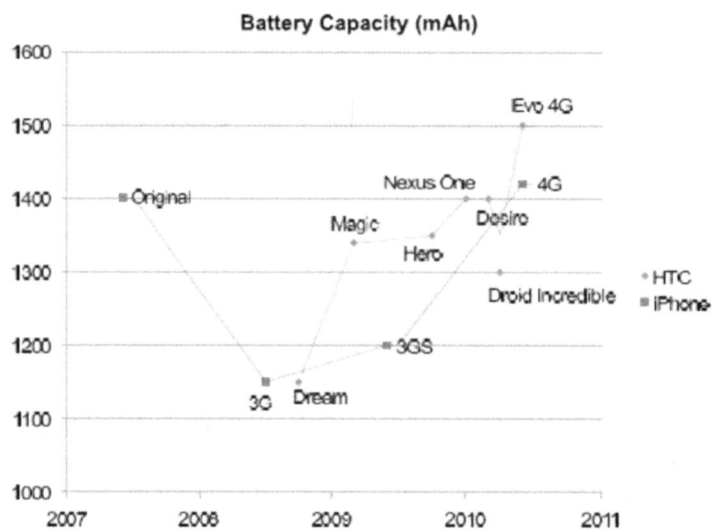

screens are now capable of understanding and responding to different gestures made by users. This feature is called multi touch. Display screens have evolved by way of acquiring new features. These features are centered around improving the I/O experience of a user. It is not clear how display screen features will evolve, however. It is more than likely that display screen technology for smartphones will continue to advance alongside other hardware components.

Smartphone hardware is projected to take a technological leap in the near future. Processor manufacturers are moving towards chips with *multiple cores*. These processors will allow software developers to make certain applications run much faster. Memory and clock frequency should continue to grow in size at the same rate they have been until the power wall for smartphones has been reached. The power wall for smartphone hardware has not been heavily studied and will depend on the state of battery technology. As is typical for batteries, advances in the capacity of smartphone batteries will occur slowly. This is consistent with the trend observed in advances in battery technology of other devices. Phones will also receive a boost in the amount of storage space available to users. As consumers abandon traditional computers in favor of their mobile devices, inevitably, they will want to store more data on their phones. Furthermore, more powerful smartphone hardware could lead to larger applications. This would necessitate the increase in the amount of storage space available on smartphones. The next popular innovation in smartphone displays will likely be a new I/O feature. Perhaps displays will incorporate 3D display technology on top of their multi-touch capabilities much like the Nintendo 3DS. Sensors like accelerometers and gyroscopes will likely be encountered in more devices to enhance the *user interface*. Finally, the cameras found in smartphones should improve steadily over time. At some point, smartphones will have camera resolutions exceeding ten megapixels. Phones with camera resolutions exceeding ten megapixels already exist but they are not widespread.

2. ENVIRONMENTAL IMPACT OF SMARTPHONES

Consumers benefit greatly from the efforts made by hardware developers to follow the predictions laid out by Moore's Law. It is the accuracy of and desire to follow this prediction that has led to the many advances in computer hardware enjoyed today, including smartphones. However, the rush to advance computer hardware further and further requires a tremendous resource commitment and causes considerable environmental stress. Environmental stressors originate in the manufacture of smartphone components; in transportation of components to assembly and of assembled phones to consumers; during the use of the phone and the network; and at the end of the phone's life. Although comprehensive information about the resource requirements and environmental impacts of mobile phone use are not well documented, some key results have emerged:

1. The manufacture of cellular phones is more energy intensive than the use of the phones and generally causes greater environmental impacts. Some key contributors to manufacturing-phase impacts include semiconductors, printed circuit boards, and batteries. Semiconductor manufacturing, in particular, is highly energy-intensive and requires large quantities of water and industrial chemicals, many of which are toxic. Smartphone manufacture also consumes many non-renewable resources such as gold, platinum, and other precious metals. The way these materials are used in electronics manufacture makes them difficult to recover at the end of life.

2. During use, the operation of the radio access network requires dramatically more energy than is consumed by the handset. Some studies have estimated that modern radio network energy usage exceeds handset usage by a factor of 100 or more on a per-subscriber basis.
3. Mobile phones are significant contributors to the large and growing stream of Waste Electrical and Electronic Equipment (WEEE), often generated in the developed world and processed in the developing world, where it leads to considerable environmental and health burdens. However, efforts by both industry and policymakers to promote the wise management of WEEE as well as ongoing projects to re-use, re-purpose, or recycle end-of-life mobile phones are developing and show promise.

A significant driver of all three points of concern is the sheer number of mobile phones in use and the constant growth of this technology segment, coupled with the short life span of mobile phones. Given that smartphones are replacing traditional cellular phones extremely quickly and that a large portion of the consumer population is in possession of one, determining how to minimize the environmental impact of smartphones would significantly reduce the negative effects of mobile phone waste in the future.

2.1. Methodology: Life Cycle Assessment

To understand the impact of the fast increasing market of smartphones to the world, it is critical to quantify the resource been used during different stages of smartphone life cycles. Computing the resource requirements and environmental impacts of a product system is often carried out through Life Cycle Assessment (LCA). LCA is a standardized methodology for computing environmental burdens based on the network of processes that lead to a product being produced, delivered, and disposed at the end of life (Rebitzer, et al., 2004). Often an objective of a life cycle study is to compare the resource and energy requirements of the production phase (raw material extraction, manufacturing, and distribution) to the use phase and end-of-life disposal phase. Other objectives include the identification of "hot spots" in a product life cycle that contribute disproportionate impacts in areas of environmental concern, and the comparison of different product systems that serve similar functions. LCA results which describe the pre-consumer stages of a product can be called "cradle-to-factory-gate" or simply "cradle-to-gate" results, while results which cover the entire life cycle are called "cradle-to-grave."

A life cycle assessment leads to two main types of quantitative results. The first is a Life Cycle Inventory (LCI), which describes resource requirements and environmental emissions due to the product system. Subsequently, a Life Cycle Impact Assessment (LCIA) is then performed to estimate the potential environmental impacts due to the inventoried emissions. LCIA results are generally reported in terms of a quantitative indicator of the potential effects in a category of interest (Guinée, 2002). For instance, potential contribution to climate change is reported in units of the equivalent amount of CO_2 (kg CO_2-eq) emitted to the atmosphere as a consequence of the product system's operation. Other categories of environmental impacts are reported in appropriate equivalencies (atmospheric acidification may be reported in units of SO_2-equivalent; human toxicity may be reported in units of arsenic-equivalent; and so on.) Because the resource requirements and emissions associated with manufacturing processes are proprietary and often closely held by industry participants, it can be difficult to obtain adequate data to perform an LCA, particularly of high-tech products such as smartphones.

2.2. Existing Life Cycle Studies

With the advent of the "3G" Universal Mobile Telecommunications System (UMTS) around the turn of the century, numerous mobile phone manufacturers and research groups undertook life cycle studies of mobile phones. However, these studies have not been updated to reflect recent developments in mobile communication, and no public LCA of smartphones is presently available.

A 2004 EU directive to develop a multi-stakeholder integrated product policy for reducing the environmental impacts from electronic products led to a report describing LCAs of Nokia handsets (Singhal, 2005). The report summarized the results of a number of LCA studies that had been conducted by Nokia and which are not currently published. The study's authors found that the production and use stages used comparable amounts of energy, with the production phase generally dominating the use phase. Production of electronic components such as printed wiring boards and integrated circuits were the largest contributors to life cycle impacts. Different usage modes were found to have a significant but not large effect on total life cycle energy requirements. The cumulative energy requirements of production of a 3G mobile phone were found to be 150 MJ, equivalent to about 4.3 L of gasoline. In contrast, transportation of components and of the finished phone to the first customer required about 30 MJ, while the consumer's use of the phone and charger required 35-50 MJ per year (see Figure 9).

A report from the same time prepared by Ericsson looked at the energy use associated with the production and operation of a pilot-scale 3G network (Malmodin, 2004). They found that, although manufacture of subscriber handsets outweighed handset use, the operation of the network carried far greater impacts than the handsets on a per-subscriber basis (see section 2.4.2). Faist Emmenegger et al. (2006) found that in both GSM and UMTS cases, impacts from production of the handset again outweighed use phase impacts, but the impacts were roughly 45% higher for the UMTS phone. They traced the higher impacts to increased production of electronic components.

Less information is available describing modern smartphones. While the academic literature is very sparse, some industry information is available in the form of environmental performance reports published by manufacturers. Apple Inc. estimates that the life cycle of an iPhone 4 leads to the emis-

Figure 9. (a) Life cycle energy requirements of a 3G Nokia handset in 2003 (Singhal, 2005). (b) Life cycle greenhouse gas emissions of a 3G system, including phone and radio network, on an annualized per-subscriber basis (Malmodin, 2004).

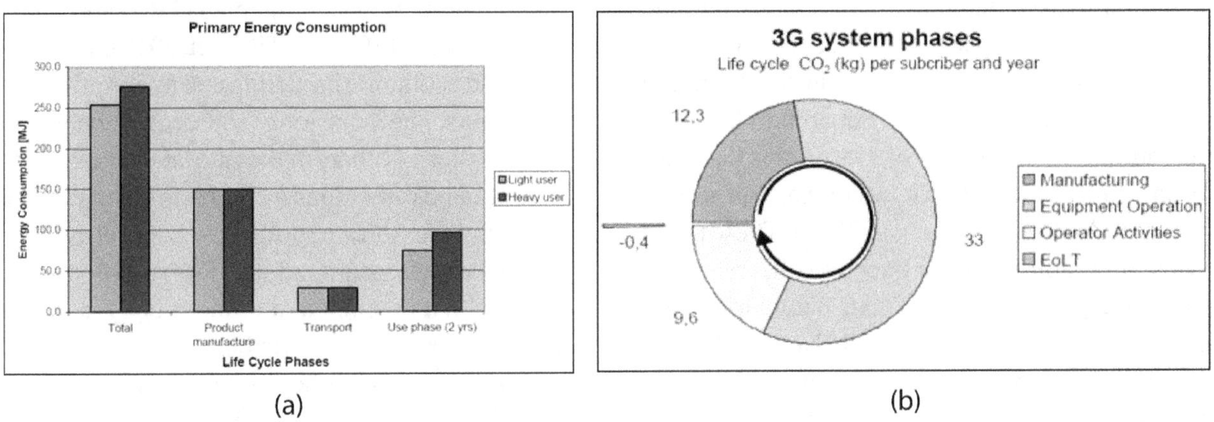

(a) (b)

sions of 45 kg CO2 equivalent, (Apple Inc., 2010). This impact is comparable to combusting 19 gallons of gasoline. According to Apple's estimates, the use phase contributes 34% of greenhouse gas emissions, versus 57% for production, 8% for transportation, and 1% for recycling the phone at the end of its life. The Apple study assumes a 3-year life span for the phone.

2.3. Manufacturing of Smartphone Components

Several of the components fundamental to smartphone function also carry significant environmental impacts to produce. Semiconductor components in particular require extraordinarily pure silicon which undergoes numerous processing stages in large fabrication plants. A 2-gram microchip was found to require 1,700 grams of material input to produce (Williams, et al., 2002). A later study of an 8-layer, 1.2-cm^2 CMOS processor found its manufacture to require 1.87 kWh of facility electrical energy, equivalent to between 18-22 MJ of primary energy, and to release 2.9 kg CO_2-eq (Boyd, et al., 2006). This amount is almost a quarter of the annualized per-subscriber manufacturing impacts found in the Ericsson study (Malmodin, 2004). Modern chip manufacturing technology has become slightly more efficient on a per-die basis but remains energy intensive (Boyd, et al., 2009).

As mentioned in Section 2.1, LCA is the most commonly used technique for evaluating the environmental impacts of products. However, LCA studies of mobile phones in the literature are scarce and out of date (Scharnhorst, 2008). In addition, LCA studies of electronics have not been found to be reliable in comparison to one another (Andrae & Anderson, 2010). These problems of reliability can be traced to differences in modeling assumptions, system boundaries, and life cycle inventory data used. Nonetheless, it is possible to make a lower bound estimate of the environmental impacts due to smartphone manufacture by synthesizing published cradle-to-gate inventory and impact data for various components required to build a smartphone. The accuracy and sensitivity of these individual values can then be independently assessed. In a study of this nature, typical component requirements are combined with published inventory and impact results found in the literature. The impact intensity of manufacturing the various components is first reported, in terms of impacts per functional unit of output (for instance, the energy required to produce a cast and machined aluminum part, reported in terms of the weight of the finished part). These intensity values are then paired with estimates of the functional requirements for each component in a typical smartphone (for instance, the weight of aluminum in a phone with a cast, machined aluminum body may be approximately 20 g). The product of the impact intensity and the functional requirement equals the total impact.

Table 1 shows the results of one such study conducted by the authors for a set of hypothetical smartphones. This estimate is only representative of typical conditions for common components and does not constitute an environmental assessment of any actual phone. In this table, the literature sources for impact intensities are shown for each component. The functional requirements represent an estimate by the authors, based on published specifications for popular phones (such as Apple Inc., 2010), results of "teardown" studies (such as Rassweiler, 2010), the International Technology Roadmap for Semiconductors (ITRS) specification, 2010 update (www.ITRS.net, 2010), and technical and economic modeling and assumptions by the authors. The inclusion of some components and the omission of others makes the tabulated result a lower bound rather than a complete estimate. The study includes three hypothetical handset models: one made with a stamped stainless steel body, one made with a machined aluminum body, and one made with an ABS plastic body.

Energy and greenhouse gas emission impact intensities reported are drawn from the GaBi

Table 1. A list of typical smartphone components combined with estimates of manufacturing energy and greenhouse gas emissions drawn from the literature. See the text for an explanation of modeling assumptions and data quality indicators (DQI).

Phone Body	**Intensity**		**Source (DQI)**	**Functional Requirement**			**Total Impact**	
	Energy	**GHG**		**Steel body**	**Alum. body**	**Plastic body**	**Energy**	**GHG**
Material	MJ/kg	kg CO_2eq /kg		kg	kg	kg	MJ	kg CO_2eq
Plastic, ABS, Injection Molded	109	5.0	GaBi Prof-DB (**B**)	0.003	0.005	0.018		
Steel, Stainless 304	62	4.5	GaBi Prof-DB (**B**)	0.039	0.010	0.010		
Aluminum, cast, machined	165	12.1	GaBi Prof-DB (**B**)		0.020			
Glass, LCD	62	4.4	Ecoinvent 2.2 (**C**)	0.040	0.040	0.040		
Total Impacts (steel body):							**5.2**	**0.4**
Total Impacts (aluminum body):							**6.9**	**0.5**
Total Impacts (plastic body):							**5.1**	**0.3**

Phone Components	Reference Unit (Source, DQI)	Intensity per unit		Functional Requirement MJ	Energy	GHG
		Energy (MJ)	GHG (kg CO_2eq)		MJ	kg CO_2eq
CMOS Logic, 45nm	110 mm² die (Boyd 2009, **A**)	58	3.4	CPU – 50 mm²	26.4	1.55
512 MB DDR SDRAM– 46 mm²					24.3	1.42
Other Logic – 50 mm²					26.4	1.55
NAND Flash, 45nm	8 GB die (Boyd 2011, **A**)	48	2.0	1 die	48.	2.0
Total Impacts (semiconductor, lower bound):					**125.**	**6.5**
Printed Wiring Board	1 m² SMD(Ecoinvent 2.2, **B**)	4800	290	20 cm²	9.6	0.6
LCD Display	1 kg LCD (Ecoinvent 2.2, **C**)	787	59	0.010 kg	7.9	.6
Battery, Li Ion	114 Wh (Notter 2010, **C**)	104	6.0	5.2 Wh	4.7	0.27
Total Impacts (other components):					**22.**	**1.4**
Total for component manufacture, including uncertainty estimates: (Body, semiconductors, and other components, minimum)					**130-176 MJ** **7.0-9.7 kg CO_2eq**	

Professional Database maintained by PE International (PE International, 2011); the Ecoinvent database, version 2.2 (Frischknecht, et al., 2010); contemporary studies of semiconductor manufacturing (Boyd, et al., 2009, 2011), and of lithium ion batteries (Notter, et al., 2010). Each component includes a data quality indicator of A, B, or C, which is a result of the authors' interpretation of the accuracy of the data with respect to smartphone manufacture. A DQI of "A" indicates a report about contemporary technology whose results are well-documented and highly applicable to the study at hand; these data sources are given an uncertainty of +/-10%. A DQI of "B" indicates a general result that is a good technical approximation of the manufacturing process and are given a +/-25% uncertainty estimate. A DQI of "C" indicates an older, less reliable, or less applicable study, which is used in the absence of any superior data. These studies are given an uncertainty of -50%/+100%. In the case of LCD glass and display components, the data are based on a study by the US EPA that is more than 10 years old (Socolof et al, 2005), justifying the "C" rating. In the case of batteries, the data are based on studies for lithium batteries used in automotive transportation (Notter, et al., 2010).

Impact results are based on European production conditions, so results may differ for production in Asia or in other areas. Each cited source provides either directly reported values for energy and GHG intensity, or provides a complete life cycle inventory from which the intensity indicators can be computed. In the latter case, the Global Warming Potential indicator maintained by the Institute of Environmental Sciences Leiden (CML) was used to compute GHG intensity (Guinéee, 2002).

The table shows that *semiconductor manufacturing* dominates the impacts of smartphone production, with a subset of required silicon components generating 80-85% of energy requirements and greenhouse gas emissions for all components considered. In contrast, though there are significant differences in the manufacturing energy for different types of phone bodies, the differences are overwhelmed by the large requirements of semiconductor manufacturing.

Other components, such as digital cameras and high-performance touch displays, have minimal or no information available in the literature and the impacts due to their production are difficult to estimate. These components were omitted from the table.

2.4. Use Phase

The use phase of a smartphone is the stage in the smartphone life cycle where an actual consumer makes use of the device. To calculate the amount of energy a smartphone consumes during this phase, it is imperative to possess a model of user behaviors and to determine how much energy is consumed by a particular hardware component. Once an estimate for the amount of energy consumed in the use phase is obtained, it can be compared with the estimate obtained for the manufacturing phase.

2.4.1. User Behavior

A model that corresponds to how users interact with their smartphones in the real world is quite complex as there is a great deal of variation in how users interact with their phones. In a large study of smartphone users and their behaviors, it was found that different users would interact with their phones 10-200 times a day on average, would have interactions with mean lengths of 10-250 seconds, and receive 1-1000 megabytes of data per day where 10-90% is a result of active use. The reason these numbers are so meaningful is that the study found that user behavior formed a continuum within ranges that spanned several orders of magnitude. There were no significant groupings observed for the data points in these ranges. Thus, it is important when estimating how much energy a phone consumes during its use phase to consider the entire distribution of user behaviors.

One could conceivably study a large group of people and aggregate all the resulting individual distributions to create a general model for the consumer population. The model would be able to predict when a phone is on, how long the phone is on, and what types of applications are likely to be used. The data garnered from such a model would be enough to produce a reasonable estimate of how much energy smartphones consume in the use phase of their life cycles.

Although it was said previously that there is a great deal of variation in user behavior, there are some things that can be said about smartphone user behavior in general. Smartphones are in the idle state the vast majority of the time they are powered on. Parallels to this behavior can be seen in CPU usage which is either at 100% or below 10%. In addition, users do not typically switch between brightness levels on their display screens nor do they install any type of power management software (Shye, 2010). These observations should make the process of estimating the energy consumption in the use phase of a smartphone a bit simpler.

2.4.2. Indirect Energy Consumption

Use of a mobile phone requires the availability of a pervasive radio network, which is always on and provides communication capacity to a large number of users in geographically diverse locations. It is thus unsurprising that the *radio network* is the greatest energy consumer in the mobile phone system. The authors of the Ericsson study found that over 60% of the global warming potential generated by the 3G pilot system they studied originated in the operation of network base stations (Malmodin, 2004).

Subsequent life cycle studies of 3G networks confirmed this fundamental result. Faist Emmenegger et al. (2006) compared older GSM technology to the emerging UMTS network. Overall, the authors found different results depending on the functional unit of output. On a transmitted data basis, the higher-bandwidth UMTS system was a superior performer. The GSM system required 1.14 GJ of primary energy per Gbit of data transmitted between mobile phones, while the UMTS system required 0.8 GJ of energy for the same function. However, on a per-subscriber basis the results were reversed. In this comparison, the UMTS system required 2.1 times primary energy and led to 2.4 times as much greenhouse gas emissions per subscriber as the GSM network. A study of a mature national 3G network in Japan found that the ratio of network infrastructure terminal energy use was 150:1 on a per-subscriber basis, with over 80% of the infrastructure energy requirements attributed to Node B antennas (Etoh, 2008). It is likely that as radio networks continue to advance and data transfer rates increase, the energy requirements of emerging networks will continue to rise.

2.5. End of Cycle

The speed at which consumers are replacing their phones with newer models is an important, contributing factor to the growth of smartphone electronic waste (Jang & Kim, 2010). The problem does not lie just with people replacing traditional cellular phones with smartphones. If this were the case, the speed at which cellular phones are being replaced would eventually slow down to a more manageable level. The problem is perpetuated by *consumers* who replace their old smartphones for the next model. Even though consumers in the United States are keeping their phones slightly longer due to the state of the economy, they still keep their phones, on average, for *less than two years* (JD Power, 2010). Unfortunately for the environment, new models of smartphones are produced very quickly. The old phones are then done away with, often in a less than optimal way. Due to a lack of public awareness and a flawed recycling system, smartphone disposal often consists of either a consumer placing their phone into some kind of storage or inadvertently sending their phone to a landfill. The former case is simply wasteful, the latter is harmful.

The large quantity of waste mobile phones disposed leads them to contribute to the stream of Waste Electrical and Electronic Equipment (WEEE) which is the cause of significant health effects and environmental degradation in the developing world (Robinson, 2009). Mobile phones contain a broad spectrum of hazardous substances, including heavy metals and flame retardants, and are often recycled informally to extract semiprecious metals like copper, leading to leaching of toxicants into groundwater and release into air (Lim, 2010).

One way of reducing the negative environmental impact of smartphones is to have them be recycled at the end of their life cycle (Silveira, 2010). The recycling process begins when a consumer sends their old cellular phone to a recycling organization. From there, recycling typically involves dismantling the hardware in some fashion and separating the components that are still usable as they are from those materials that require further processing to be made useful again. In this way, salvageable components can be garnered for use in some future application, and the effects of hazardous, electronic waste can be reduced. Those parts that are not usable are typically disposed of or processed in a responsible way. Despite considerable environmental benefits, mobile phone recovery for either reuse or recycling is only marginally economical (Geyer, 2010), though careful policy design could provide incentives to improve recovery (Zoeteman, 2009).

Many resources and expenditures can be saved through the recycling of smartphones. It is estimated that if the 130 million cellular phones disposed of every year in the United States were to be recycled, enough energy would be saved to power more than 24,000 homes for an entire year (EPA, 2010). Of that 130 million, the percentage of smartphones is growing rapidly. Many smartphones are recycled before they become useless. Components such as memory and batteries that would be expensive to manufacture can simply be used to refurbish or repair other devices. If the parts are new enough, they could even be used to produce completely new smartphones. In this way, some of the costs and detrimental environmental effects of manufacturing can be avoided. Cellular phones, in general, contain enough precious metals to make their retrieval financially worthwhile. These metals could then be used to make new products instead of wasting away at the bottom of a landfill. It was estimated in 2006 that the value of the precious metals in cellular phones in storage and in use at that time in the United States was in the hundreds of millions of dollars (Sullivan, 2006). Therefore, there is plenty of incentive for industry to recycle smartphones. The plastic in cellular phones can also be recycled for a variety of uses. Despite the many benefits of recycling, consumers are not recycling enough for those benefits to be reaped.

The percentage of consumers that recycle their cellular phones is quite low. According to a 2008 report by Nokia, almost half of all old cellular phones end up packed away somewhere (FierceWireless, 2008). A quarter of the consumer population pass along their old phones to friends and family who may be in need of one. An even smaller percentage sell their phones. Only three percent of consumers recycle their cellular phones and four percent throw very old cellular phones into the garbage (Global, 2008). The most common reason why consumers do not recycle their phones is poor awareness of the issue. Some who know that old cellular phones can be recycled abstain due to an assumption that the recycling process consumes more energy and resources than it is worth. There is some effort being made towards educating the general public of the benefits of recycling their cellular phones, but the scale of the effort is not large enough to effect a monumental change in consumer behavior. Despite the many benefits that can be reaped from recycling smartphones, there are some problems with the way recycling is currently envisioned.

The biggest problem with recycling smartphones is that many smartphones are recycled *while* they are still perfectly functional. This method is not sufficient if the goal is to reduce the environmental impact of smartphones. To recycle such a phone would be a waste of resources that could be better used in other applications. A great deal of energy is saved if a fully functional phone or even a somewhat functional phone is reused by another person until it ceases to be functional. Thus, a good solution would be to have a working smartphone be reused until it reaches the end of its life cycle and then recycle it.

3. TOWARDS GREEN COMPUTING: SMARTPHONE REUSE

As the smartphone industry consumes more and more energy and affects a greater environmental impact, it is critical to think Green for every stage of the smartphone life cycle. Although it is also important to apply new technologies to reduce the energy and materials used during the manufacturing process of smartphones, it is rather beyond the scope of computing technology. Hence, this chapter focuses on how to improve the energy efficiency of the use-phase and post-use-phase of smartphones.

3.1. Reducing Use-Phase Energy Consumption

As we have analyzed in the previous section, the four components that consume the most energy in smartphones are: the *screen*, *CPU*, *GPS* service, and *network* connection. Although the energy consumed by the network connection could vary substantially when the network interface changes from cellular service/3G to Wi-Fi or Bluetooth, the average energy consumption is high in general.

3.1.1. Reduce Energy Consumption for Screens

It is first important to enhance the energy efficiency in display technologies, which is rather the focus of researchers in electrical engineering and materials science. For computer scientists, one popular trend towards reducing the energy consumption of screens is to *adjust the brightness of the screen in a smarter way*. Studies by Shye (Shye, 2009) show that the current screen brightness adjustment strategy does not match with user behavior perfectly. Energy is wasted due to long screen intervals when users are not using their phone. Their new scheme of reducing the brightness of the screen slowly over time demonstrates 10.6% total system energy savings with a minimal impact on user satisfaction.

3.1.2. Reduce Energy Consumption for CPU

Nowadays smartphones are using commodity general purpose processors. Hence, most of the existing techniques to improve the energy efficiency of general purpose processors would also benefit smartphones, such as Dynamic Voltage and Frequency Scalding (DVFS) (Burd, 1995), which has been widely applied to embedded systems and architectures (Henkel, 2007).

Two streams of approaches exist in the context of reducing energy consumption; particularly for smartphone CPUs. One popular technique to reduce the energy needs of mobile devices is *computation offloading*: Applications can take advantage of the resource-rich infrastructure by delegating code execution to remote servers. Such approaches either rely on programmers to partition the program (Flinn, 2001), or rely on full process migration (Osman, 2002). Recent projects (Cuervo, 2010) aim at providing an automatic framework for determining how to partition programs for remote execution.

Another way to optimize the computational energy efficiency is to *utilize the computational power of other components in the smartphone*; specifically, the Graphics Processing Units (GPU). Mapping parallel computation to GPUs is nothing new (Owens, 2007), but it is still challenging for smartphones. The limited graphics memory resource in smartphones prevents the migration of most, existing algorithms into smartphone GPUs. Pioneer work has been proposed to study the collaboration of CPU and graphics units in smartphones to improve performance and reduce energy consumption (Wang, 2010).

3.1.3. Reduce Energy Consumption for GPS Service

GPS has become one of the most important features in smartphones today and most applications rely on it. Frequently updating location information through a GPS service consumes substantial amounts of energy. Researchers have been working on several approaches to address this including: *location approximation,* which uses other information, such as cellular/Wi-Fi triangulation, or other sensors, such as accelerometer and orientation sensors, to estimate the location (Lin, 2010), and *reducing the frequency* of requests for GPS information through different optimization strategies (Zhuang, 2010).

3.1.4. Reduce Energy Consumption for Network Connections

A smartphone today typically uses a cellular (3G) network or Wi-Fi for network connections. It is well known that Wi-Fi consumes much less energy compared to a 3G network. Hence, using Wi-Fi as much as possible is the key to achieving more energy efficient smartphones. The main challenge for this is that a Wi-Fi network is not always available, and the delay could be intolerable if one always tried to wait for Wi-Fi connections. Researchers have been optimizing the energy-delay tradeoff between using Wi-Fi networks and cellular networks (Ra, 2010).

3.2. Smartphone Reuse

Unlike recycling which breaks down used phones into raw materials and parts that are used to make new items, reuse is an attempt to use old smartphones as complete functional units for the same purpose they were intended for or entirely different purposes. Although reusing smartphones for the same function (as a phone) is also technically defined as reusing, it only has a small market, and does not have capture the interest or pose a challenge for computer scientists. Hence, this chapter focuses on techniques of reusing smartphones for different purposes.

3.2.1. Benefits of Reuse

The average lifetime of a smartphone is currently about 1.5 years. This is typically due to the duration of contracts and the high frequency with which new generations of phones come out. However, at the time when those phones are discarded or placed in a drawer forever, they are most likely still fully functional and powerful enough to be reused for various purposes for another 5 years.

To understand the benefit of reusing smartphones, it is important to notice that reuse does not automatically save energy. Instead, the total lifetime energy will increase since the phone will be used for a longer period of time. The key benefit of reuse is that energy efficiency, which is defined as the ratio between manufacture energy and use-phase energy, is improved substantially. Furthermore, the high manufacturing cost of the smartphones does not get wasted since they could benefit the society for a much longer period of time.

Figure 10 shows the energy consumed in different stages of the smartphone life cycle for both normal use and repurposing. Dark portions give the energy consumption in a typical smartphone life cycle: production manufacturing, transportation, and actual using. Total Cost is the sum of Product Manufacturing and Transport. Gray parts give the energy for an extra smartphone reuse stage. As can be seen from the figure, before reuse the cost of a new smartphone was much larger than the energy consumed to use the phone (about 4:1). After 5 years of reuse, the cost of reuse is relatively low as the ratio between the cost and extended use phase energy becomes much more balanced (about 1:1). The vast increase in energy efficiency indicates that reuse is necessary and beneficial.

3.2.2. Reused Smartphones

Smartphones degrade while being used. Although we imagine that typically when a smartphone is discarded after 1 or 2 years of normal use, it remains mostly functional. Even if certain com-

Figure 10. Manufacturing, transport, and user phase energy for regular use and recycling

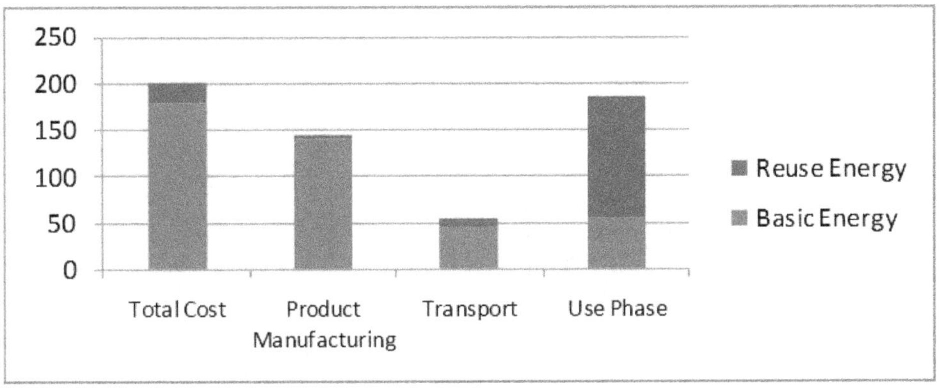

ponents break or wear out, the phone might still be reusable as long as its key components are still working. Yet, what defines "key components" in smartphones is highly dependent on what kinds of applications are the targets of reuse. The following components are generally considered as important: the screen, processing unit, storage (internal and external) and battery. This section will provide the degradation model for the above components as well as discussions about other components in smartphones.

- **Screen**: Screens will play an important role for any user-interactive applications, but will not necessarily be needed for other purposes. Most screens in smartphones are manufactured using Liquid Crystal Display (LCD) technology, possibly under different variations. For example, the screen of the iPhone 4 is made using TFT LCD, a variant of LCD which uses Thin-Film Transistor (TFT) technology to improve image quality. The typical life time of LCD displays under normal brightness conditions and normal operating temperature is on average approximately 40,000 hours (Tamuz, 2007). Even a user is using the phone all the time while awake, i.e. 16 hours a day, the LCD screen will still have a life cycle of almost 7 years. Hence, the screen should remain functional in most cases for a XX-year reuse unless it is broken by the users.
- **CPU**: A smartphone has to retain certain computational ability in order to perform any task. Processing units are simply a collection of transistors, and there are known transistor degradation mechanisms, such as gate-oxide breakdown and hot-electron effects, that slowly change transistor performance. This can slowly degrade the timing for signals across a chip and, eventually, cause the chip to not work. It can also result in a bit flip if degradation happens in the chip's cache. Ultimately, the chip will slowly become less capable, i.e. the clock frequency will decrease slowly until it becomes unable to run at the advertised speed, and act unstably.

Figure 11 shows the different degradation models for a typical CPU in PCs. The middle line models the minimum required voltage needed for a processor to continuously run at 100% load for the period shown along the x-axis. During this time, the processor is subject to its specified maximum core voltage and is never over-clocked, which is often the case for smartphones. Since the working temperature is also lower in smartphones than PCs, it is expected that the degradation model

Figure 11. Typical CPU degradation model: under different models a processor could last for 6-9 years

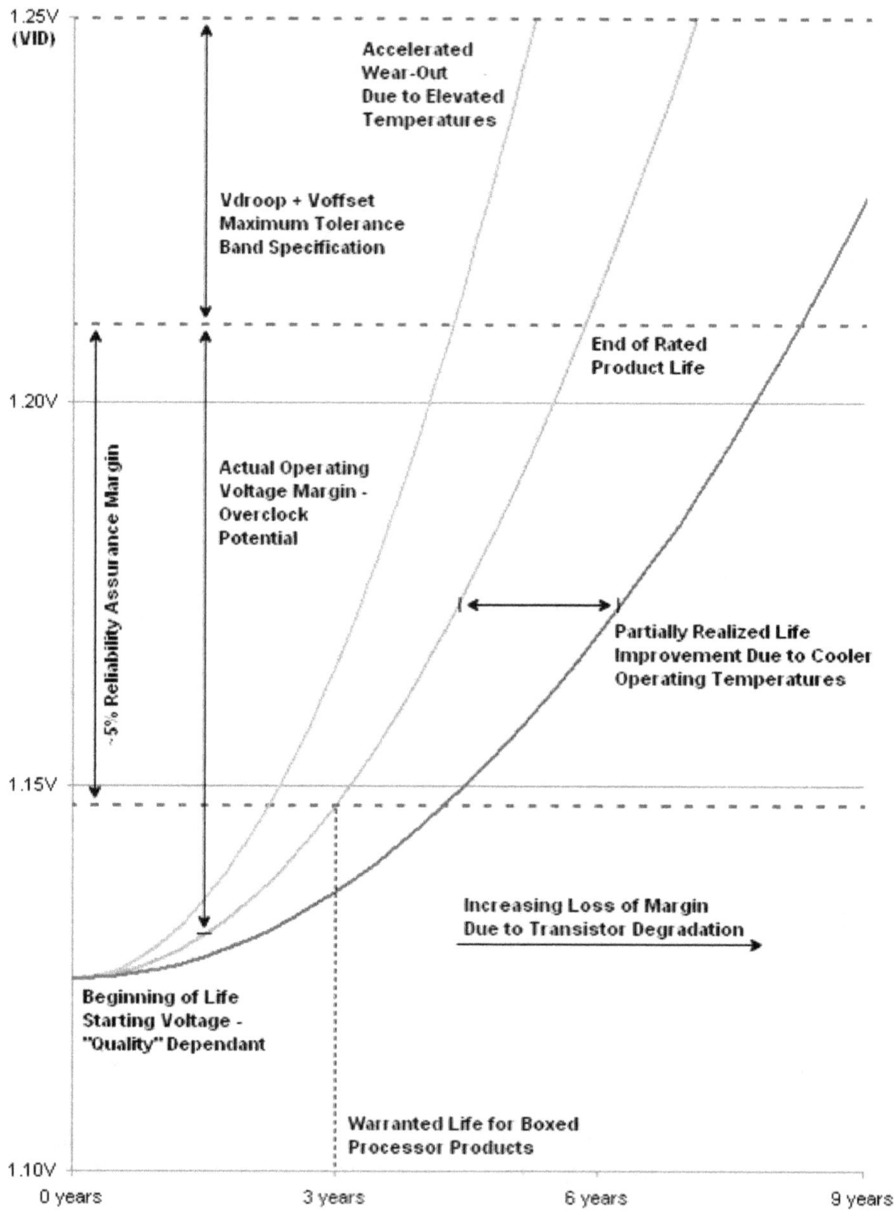

for the CPU in smartphones fits into the rightmost curve, i.e. to have a 9-year lifetime.This is much larger than the typical use time for smartphones including reuse.

- **Storage**: The smartphones of today usually have hundreds of megabytes of internal memory for running the operating system and applications, as well as tens of gigabytes of external memory to store applications and personal data. When they are being reused for purposes that require running rich applications on the device, both internal and external storage will need to be usable. Both internal memory and external storage in smartphones are typically flash memory. However, some smartphones (such as the iPhone) use embedded DRAM as internal memory, which is considered to rarely degrade compared to flash devices.

Flash storage is known to have a limited lifespan due to issues with write endurance. Depending on the usage pattern of the primary user, the flash memory of a repurposed phone may be severely limited. Recall for the SLC NAND flash, which is the present technology used today, Program/Erase cycle endurance is approximately 100,000 cycles, as opposed to MLC which is only 10,000 cycles. The following formula could be used to calculate the lifetime of a flash storage (Olson, 2008):

Parameters

- NAND flash P/E cycle endurance
    - SLC 100,000 cycles, MLC 10,000 cycles
- NAND flash P/E utilization
    - The allowable amount of P/E utilization before flash memory is considered worn out – 95%
- SSD capacity
    - Total capacity of the SSD
- Capacity rate factor – The amount of written overhead accompanying the data
    - Sequential write has a constant value of 1.1
    - Mixed written data has a value that varies depending on the percent mixture of sequential and random written data for MLC.

Equation

$$Lifetime(years) = \frac{(SSD\_Capacity(GB))(P / E)(Percent\_Utilization)}{(Usage / Day)(Capacity\_Rate)(365\ days / year)}$$

Based on the above formula, using 3.2GB/day as the usage rate (which should be much larger than the actual usage rate in smartphones), a 32GB external SSD storage would have a lifetime of 2,366 years! Hence, the degradation of flash devices will not affect the reuse of smartphones. Similar trends can also be seen which would have even less pressure int the context of reuse for flash-based internal memory.

- *Battery:* Batteries are the most important components in smartphones. They are required for any target application of re-use. On the other hand, batteries are also the most notorious components in smartphones for degrading quickly. A typical Lithium-ion battery has a recharging endurance of 400-1200 cycles depending on the concrete technology and the actual usage model. With an average value of 800 cycles, assuming the phone battery is recharged every 1.5 days, the lifetime of the battery is only about 1.5 years. This matches the average life cycle of today's smartphones. When a phone is prepared for reuse, a new battery is required.
- *Other Components:* Other components such as wireless cards, which provides Wi-Fi connectivity, are also important for many different types of applications. However, they are not generally considered to be components that "degrade." These components should either work properly for a long period of time or simply stop working under certain probability.

3.2.3. Applications for Reuse

One natural application for the reuse of mobile handsets would be the formation of a robust and self-maintaining *environmental monitoring infrastructure*. For example, an incarnation of this idea could be a network of distributed nodes that monitor a physical region of interest, where audio or visual data can be gathered, aggregated to increase accuracy and completeness. The data could then be processed to locate events or objects of interest. Specific deployment scenarios include: home alarm systems, wildlife detection and tracking systems, and emergency reporting systems for monitoring earthquakes along fault lines or the spread of wildfires. Leveraging the large number of repurposed devices allows us to mask intermittent and permanent failures with redundancy and transparently failover. The software infrastructure would monitor nodes and perform adaptive load-maintenance on a per-resource basis; dynamically reassigning tasks away from failed components while leveraging resource rich nodes to their fullest.

Another compelling application would involve the reuse of handsets as *GPS trackers* for mobility measurement and trace gathering. Current research on mobile networks such as MANETs or vehicular networks is severely constrained by the lack of representative mobility models (Saha, 2004). While recent work has shown theoretical mobility models such as Random Walk or Random Waypoint to be inaccurate, the measurement of real mobility models is constrained by the size, cost, and availability of wireless computing clients. Repurposed handsets would provide a scalable, low-cost solution that minimizes the inconvenience to participants. Handsets can use GPS-locators or cellular triangulation to compute, track, and record their movements. Measurements can be aggregated, filtered, compressed, and anonymized in-network before being gathered by wired base stations. For vehicular networks, distributed sensors can not only build accurate mobility models, but also provide a distributed monitoring network that measures and predicts traffic patterns. They can even detect and alert authorities to accidents in real-time (Elbatt, 2006). Finally, accurate, large-scale mobility measurements would not only fill a critical need for networking researchers, but also provide invaluable insights into the optimization and design of transportation infrastructures.

The low-cost of repurposed devices also makes them ideal for widespread deployment as *information kiosks* and receivers at both remote locations and highly populated environments. Handsets can store a variety of content into local memory and storage, and provide them to interested users through audio or visual displays. For example, we can replace signs, pamphlets, and human guides at information kiosks with a bank of interactive handsets that responds to voice-driven commands and queries with locally stored data. As in other scenarios, resource intensive tasks such as voice recognition, voice generation, and dedicated storage can be tasked to resource-rich devices. Devices with graphical displays can provide an augmented-reality experience to users by intuitively and graphically presenting location-based information and services. What's more, wireless connectivity allows these devices to pull data updates from relevant sources on a periodic or on-demand basis, providing automated information management (Sherrif 2007).

Furthermore, we can repurpose these wireless devices as *security tokens* that provide their users with an out-of-band authentication mechanism. Secure, one-time activation codes sent via SMS can complement traditional knowledge-based authentication mechanisms for access to online or physical services. Similarly, repurposed handsets can use SMS messages and short calls as secure payment mechanisms for micropayments, replacing coin-operated vending machines and copy machines.

Finally, reusing smartphones for *educational purposes* has been an emerging research area and shown to be more efficient than conventional education. With a cellular phone in each child's hands, the problem sets being taught could be personalized to each student. One advantage this offers is that short math problems can be tweaked so that the collaboration between two students with slightly different problems becomes more of an interaction of, "How did you get your answer?" rather than, "What answer did you get?" Another advantage personalization offers is the ability to match problem difficulty more closely to a student's abilities, making the learning process less daunting (or less trivial) for individual students. Participatory interaction is encouraged by working through distributed devices, since students who are shy in a traditional setting are more inclined to collaborate via text or graphical annotations. In addition, due to the nature of online interaction, every student's input is weighed equally since no single comment will interrupt or "talk over" another comment.

3.2.4. Reusing Smartphones for Education

Although pioneer research has been done in reusing smartphones, they approaches used are rather random and a systematic methodology is still missing. To set up a framework for reusing smartphones for any purpose, we demonstrate the general process of research in reusing smartphones by showing a case study of reusing smartphones for elementary school education.

A general process towards reusing smartphones should contain the following steps: Market Research (Scenario Definition), Requirement Specification, Feasibility Analysis, Deployment, and Design Feedback.

- *Market Research:* The first step of reusing smartphones is to choose the appropriate market that would require a large amount of used smartphones without demanding full functionality, and analyze how well the chosen market and reused smartphones match with each other. In our case study, we chose elementary school education as our target market for smartphone reuse.

The first step is to justify the match between the target market (education in our case) and used smartphones. We ask three questions to satisfy this goal:

a. *Do smartphones benefit the target market?* In our case, the advantage of using smartphones in classrooms has already been discussed in the previous section.
b. *Are "old" smartphones also valuable in the target market?* The answer is not hard to figure out in our case: students do not need brand-new smartphones, which could be used to make phone calls and run quickly. Instead, they simply need a smartphone, which can run various educational applications in a reasonable way.
c. *Could the volume of reused smartphones serve the demand of the target market?* Billions of smartphones are being discarded every year which could easily satisfy the demand of reused phones in school education.

- *Requirement Specification:* During the market matching stage we only vaguely predict that used smartphones should be powerful enough to satisfy the demand of elementary school education. However, the detailed requirements for running educational applications are unclear until we investigate further. It is also possible that certain changes and adaptations are required in order to enable reused smartphones to be used for educational purposes.

To do so one needs to collect a representative set of educational applications that should include most features needed by all such types of applications. Then a detailed analysis will be performed to look at the functional requirements of each application. Static functional requirements (components in the phone used by the application, including wireless, cameras, voice, GPS, compass, and so on) can usually be easily figured out by either looking at their requirement specifications (such as the Android platform) or simply using them. However, dynamic resource requirements (including storage usage, memory usage, CPU utilization, power consumption, etc.) cannot be observed in a straightforward manner.

To obtain the dynamic requirements of those applications, experiments that involve running those applications and measuring the different types of resources used are necessary. Monitoring the memory, storage usage and CPU utilization can be implemented easily by probing system information periodically. However, it is challenging to measure the dynamic power consumption. Several approaches can be considered:

1. If an analytical model is available to estimate the power consumption of each component based on the usage rate obtained from system information, then one can estimate the total power consumption by feeding system probe information into such model;
2. Without an estimation model, one has to measure the real power consumption, by wiring power meters between the battery and the smartphone. Setting this up on smartphones might be very difficult since one has to take the battery off plug meters in-between the battery and the phone, which would make using the phone a lot harder. What could be worse is that it is impossible to take off the battery for certain brands of smartphones.
3. Since it is difficult to measure the power consumption directly on a phone, one can use smartphone-on-chip (such as TI OMAP processors) to perform experiments, which might look "ugly" as a phone but is much more convenient to measure and debug.

- *Feasibility Analysis:* Once different types of requirements for running the target applications are collected, one then needs to compare such requirements with the functions potentially left in a reused smartphone. If a reused smartphone in general can satisfy the requirements of the target application, we can conclude that such reuse is feasible (in our case, a reused smartphone should have all the resources needed by educational applications except the battery). If certain functions are missing or cannot satisfy the demand, one will need to calculate the cost of upgrading such components or functionalities.
- *Development:* Smartphones were manufactured for making phone calls as well as for running entertainment applications. They might not be equipped with applications or functions designed for the target market. In our case, although there are already a substantial number of educational applications, they are neither designed for classroom use nor embedded with teaching and collaborating facilities. The modification of existing educational applications will be necessary. Furthermore, new strategies need to be made in order to maximize the value of reused smartphones in the target market. In our case, new ways of teaching that would incorporate reused smartphones and maximize their value are still open questions.
- *Deployment and Feedback:* Once all the above analyses and planning are finished, it will be possible to deploy reused smartphones in real settings. There shall be either cooperation with cellphone companies or other channels in order to get used smartphones with defected components upgraded. The used smartphones would then be deployed in the target market and feedback would be collected, which can be used to refine the whole process. In our case, surveys and suggestions can be collected from students and parents.

3.2.5. Challenges

The major challenge in repurposing used smartphones is the extreme heterogeneity of smart-

Figure 12. The global smartphone market share in the first quarter of 2010 among different smartphone vendors. The Y-axis corresponds to the percentage of the market a particular smartphone vendor controls.

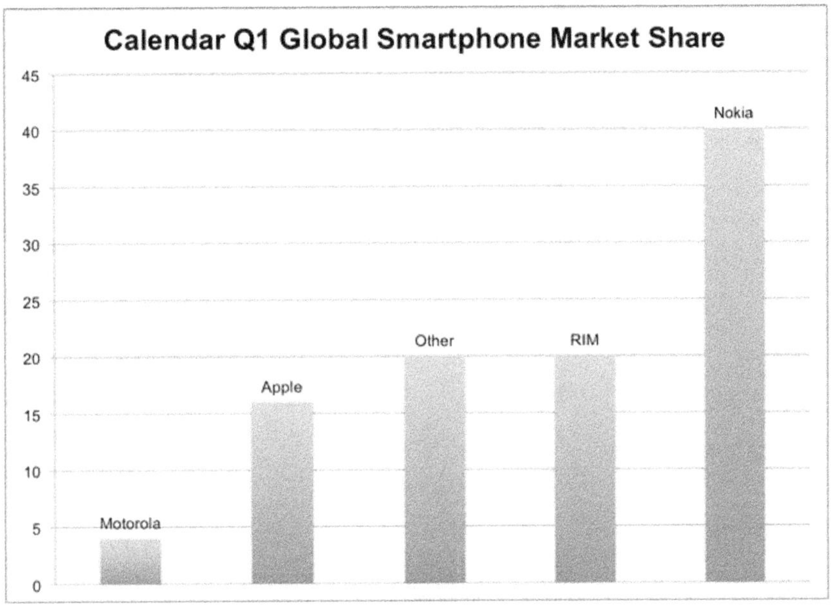

phones. Figure 12 shows the smartphone vendor market share in 2009 and 2010. From the figure, we can see that not only are there already five different major vendors in the market, but there are also hundreds of small vendors that contribute to 20% of the whole market. Devices from different vendors can have different components, different low lever interfaces, and different parameters. Even if one can somehow abstract all these devices in a standard way, there could be different operating systems running on those devices, as shown in Figure 13. Different operating systems will require programmers to write their applications under different APIs and consider different types of system interactions. Even for devices from the same vendor running the same operating system, they can still largely differ from each other if they are different generations. Nowadays most of the major smartphone vendors introduce a new generation of smartphones every year or even more than once a year.

4. CONCLUSION

Smartphones have replaced traditional cellphones and become the most pervasive and popular handset devices. The increasing demand of smartphones leads to the fast expanding of the smartphone market. As a result of the continuing innovation of technology and Moore's Law, new smartphone generations are produced every year with more powerful processing capability and more diversified sensors and functionalities. The dark side, however, is that smartphones become out-of-date so quickly that their average lifetime is only 1.5 years. Billions of used smartphones are discarded every year, among which most devices are still functional.

There are significant energy and resource requirements associated with the manufacture of mobile handsets and the operation of access networks. Device manufacture accounts for a large share of device life-cycle impacts, driven mostly by the high-energy cost of semiconductor technology. Lengthening device life times would

Figure 13. The market share among different smartphone operating systems, Nov 10th to Jan 11th, postpaid mobile subscribers, n = 14,701 (Source: The Nielsen Company)

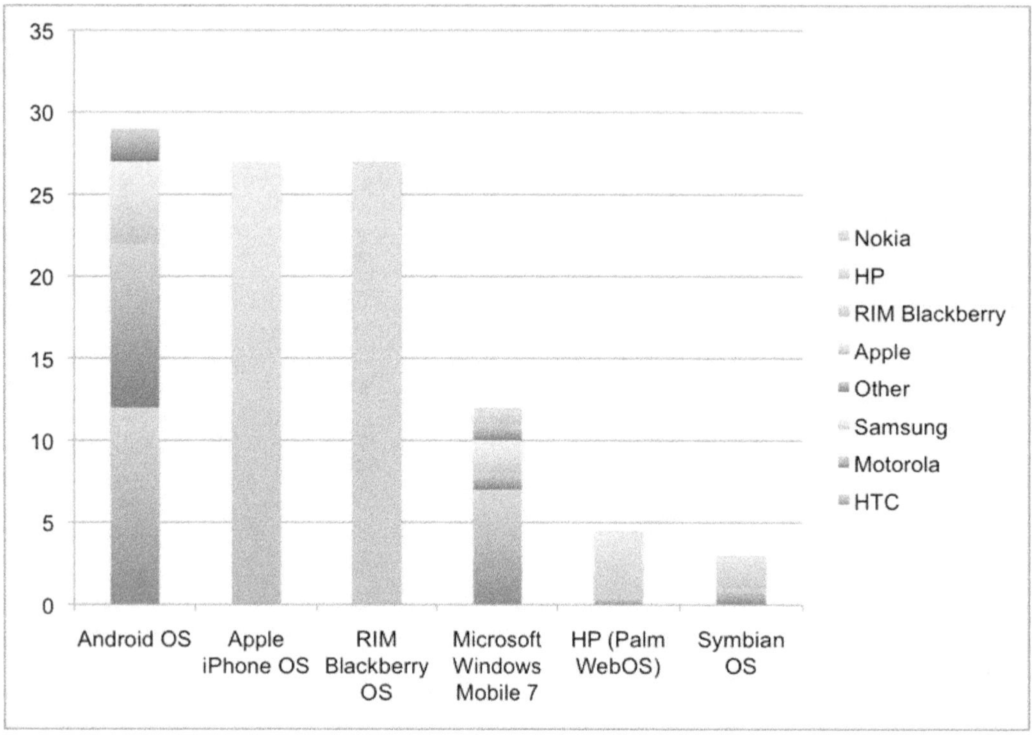

have the effect of spreading manufacturing impacts across a longer useable period. This would reduce the generation of waste as well as the intensity of environmental impacts from smartphone use. Hence reusing those smartphones is a key approach towards green computing. Among many different ways of reusing smartphones, elementary school education is most promising and perfectly matching. Although there are still challenges towards reusing smartphones, the preliminary analysis and study demonstrated in this chapter shall shed light to this line of work for future reference.

REFERENCES

Andrae, A. S. G., & Anderson, O. (2010). Life cycle assessments of consumer electronics – Are they consistent? *International Journal of Life Cycle Assessment*, *15*, 827–836. doi:10.1007/s11367-010-0206-1

Apple Inc. (2010). iPhone 4 environmental report. *Apple Computer*. Retrieved May 7, 2011 from http://images.apple.com/environment/reports/docs/iphone_4_product_environmental_report.pdf.

Bokun, S. (2009). Vodafone kills two birds with one customer loyalty program. *Pyramid Research*. Retrieved May 6, 2011 from http://www.pyr.com/points/item/090820.htm.

Boughton, K. (2008). Intel's 45nm dual-core E8500: The best just got better. *AnandTech*. Retrieved May 10, 2011 from http://www.anandtech.com/show/2468/6.

Boyd, S. B., Dornfeld, D., & Krishnan, N. (2006). Life cycle inventory of a CMOS chip. In *Proceedings of the 2006 International Symposium on Electronics and the Environment*. Scottsdale, AZ: IEEE.

Boyd, S. B., Horvath, A., & Dornfeld, D. (2009). Life-cycle energy demand and global warming potential of computational logic. *Environmental Science & Technology*, *43*, 7303–7309. doi:10.1021/es901514n

Boyd, S. B., Horvath, A., & Dornfeld, D. (2011). Life-cycle assessment of NAND flash memory. *IEEE Transactions on Semiconductor Manufacturing*, *24*(1), 117–124. doi:10.1109/TSM.2010.2087395

Burd, T. D., & Broderson, R. W. (1995). Energy efficient CMOS microprocessor design. In *Proceedings of the 28th Annual Hawaii International Conference on System Sciences.* (pp. 288–297). Los Alamitos, CA: IEEE Press.

Business Wire. (2011). Research and markets: Smartphones: A worldwide view - More than half of US handset shipments will be smartphones by 2012 and will move to 1 billion by 2015 worldwide. *Business Wire: A Berkshire Hathaway Company*. Retrieved May 6, 2011 from http://www.businesswire.com/news/home/20110210005842/en/Research-Markets-Smartphones-Worldwide-View---Handset.

CIOL. (2009). Smartphones ring in healthy growth in 2009. *CIOL: Information Technology News*. Retrieved May 15, 2011 from http://www.ciol.com/Biz-Watch/News-Reports/Smartphones-ring-in-healthy-growth-in-2009/53091168230/.

Cuervo, E., Balasubramanian, A., Cho, D., Wolman, A., Saroiu, S., Chandra, R., & Bahl, P. (2010). MAUI: Making smartphones last longer with code offload. In *Proceedings of the 8th International Conference on Mobile Systems, Applications, and Services,* (pp. 49-62). New York, NY: ACM Press.

Dalrymple, J. (2010). Apple: Over 300,000 iPads sold on first day. *CNET News*. Retrieved May 6, 2011 from http://news.cnet.com/8301-13579_3-10472602-37.html.

Deng, L., Babbitt, C. W., & Williams, E. D. (2011). Economic-balance hybrid LCA extended with uncertainty analysis: Case study of a laptop computer. *Journal of Cleaner Production*, *19*(11), 1198–1206. doi:10.1016/j.jclepro.2011.03.004

Elbatt, T., Goel, S. K., Holland, G., Krishnan, H., & Parikh, J. (2006). Cooperative collision warning using dedicated short range wireless communications. In *Proceedings of the 3rd International Workshop on Vehicular ad hoc Networks,* (pp. 1-9). New York, NY: ACM Press.

EPA. (2010). EPA highlights recycling opportunities during national cell phone recycling week. *US Environmental Protection Agency*. Retrieved May 7, 2011 from http://yosemite.epa.gov/opa/admpress.nsf/e77fdd4f5afd88a3852576b3005a604f/2f0711847e1f1c18852576f80053f5fd!OpenDocument.

Etoh, M., Ohya, T., & Nakayama, Y. (2008). Energy consumption issues on mobile network systems. In *Proceedings of the IEEE International Symposium on Applications and the Internet.* IEEE Press.

Faist Emmenegger, M., Frischknecht, R., Stutz, M., Guggisberg, M., Witschi, R., & Otto, T. (2006). Life cycle assessment of the mobile communication system UMTS. *International Journal of Life Cycle Assessment*, *11*, 265–276. doi:10.1065/lca2004.12.193

Falaki, H., Mahajan, R., Kandula, S., Dimitrios, L., Govindan, R., & Estrin, D. (2010). Diversity in smartphone usage. In *Proceedings of the 8th International Conference on Mobile Systems, Applications, and Services,* (pp. 179-194). New York, NY: ACM Press.

Fierce Wireless. (2008). Global consumer survey reveals that majority of old mobile phones are lying in drawers at home and not being recycled. *FierceWireless.* Retrieved May 7, 2011 from http://www.fiercewireless.com/press-releases/global-consumer-survey-reveals-majority-old-mobile-phones-are-lying-drawers-home-an-0.

Flinn, J., Narayanan, D., & Satyanarayanan, M. (2001). Self-tuned remote execution for pervasive computing. In *Proceedings of the 8th Workshop on Hot Topics in Operating Systems (HotOS),* (pp 61-66). Schloss Elmau, Germany: HotOS.

Frischknecht, R., Jungbluth, N., Althaus, H.-J., Doka, G., & Dones, R. (2007). *Overview and methodology*. Final report ecoinvent data v2.0, No. 1. Dübendorf, Germany: Swiss Center for Life Cycle Inventories.

Geyer, R., & Doctori-Blass, V. (2010). The economics of cell phone reuse and recycling. *International Journal of Advanced Manufacturing Technology*, *47*, 515–525. doi:10.1007/s00170-009-2228-z

Guinée, J. B., Gorrée, M., Heijungs, R., Huppes, G., Kleijn, R., & Koning, A. … Huijbregts, M. A. J. (2002). *Handbook on life cycle assessment: Operational guide to the ISO standards*. Dordrecht, The Netherlands: Kluwer Academic Publishers.

Hagelüken, C., & Buchert, M. (2008). The mine above ground - Opportunities & challenges to recover scarce and valuable metals from EOL electronic devices. In *Proceedings of 7th International Electronics Recycling Congress*. Salzburg, Austria: IEEE.

Henkel, J., & Parameswaran, S. (2007). *Designing embedded processors: A lower persepctive*. Dordrecht, The Netherlands: Springer. doi:10.1007/978-1-4020-5869-1

IDC. (2011). IDC forecasts worldwide smartphone market to grow by nearly 50% in 2011. *IDC Home: The Premier Global Market Intelligence Firm.* Retrieved May 6, 2011 from http://www.idc.com/getdoc.jsp?containerId=prUS22762811.

International, P. E. (2011). *GaBi professional database. GaBi Software, version 4.131, Service Pack 18*. PE International.

International Technology Roadmap for Semiconductors. (2010). *ITRS 2010 update*. Retrieved May 5, 2011 from http://www.itrs.net/links/2010itrs/home2010.htm.

Jang, Y.-C., & Kim, M. (2010). Management of used & end-of-life mobile phones in Korea: A review. *Resources, Conservation and Recycling*, *55*, 11–19. doi:10.1016/j.resconrec.2010.07.003

Kairer, R. (2005). Mobile phone sales to exceed one billion in 2009. *Palm Infocenter*. Retrieved May 6, 2011 from http://www.palminfocenter.com/view_story.asp?ID=7978.

Lim, S. E., & Schoenung, J. M. (2010). Toxicity potentials from waste cellular phones, and a waste management policy integrating consumer, corporate, and government responsibilities. *Waste Management (New York, N.Y.)*, *30*, 1653–1660. doi:10.1016/j.wasman.2010.04.005

Lin, K., Kansal, A., Lymberopoulous, D., & Zhao, F. (2010). Energy-accuracy trade-off for continuous mobile device location. In *Proceedings of the 8th ACM/USENIX Annual International Conference on Mobile Systems, Applications and Services (MobiSys 2010),* (pp. 285-298). New York, NY: ACM Press.

Malmodin, J. (2004). *Summary of the study, "life cycle assessment of a third generation (3G) system at Ericsson*. Ericsson.

Murph, D. (2010). Study: We're keeping our mobiles longer, despite sinking prices. *Engadget*. Retrieved May 6, 2011 from http://www.engadget.com/2010/09/27/study-were-keeping-our-mobiles-longer-despite-sinking-prices/.

Myslewski, R. (2009). iPhone conquers half the (smartphone) world. *The Register: Sci/Tech News for the World*. Retrieved May 6, 2011 from http://www.theregister.co.uk/2009/11/24/admob_mobile_metrics_report/.

Notter, D. A., Gauch, M., Widmer, R., Wagner, P., Stamp, A., Zah, R., & Althaus, H.-J. (2010). Contribution of li-ion batteries to the environmental impact of electric vehicles. *Environmental Science & Technology*, *44*, 6550–6556. doi:10.1021/es903729a

Olson, A. R., & Langois, D. J. (2008). Solid state drives data reliability and lifetime. *Imitation*. Retrieved May 10, 2011 from http://www.imation.com/PageFiles/83/SSD-Reliability-Lifetime-White-Paper.pdf.

Ortiz, C. (2008). Worldwide and US mobile subscriber penetration. *About Mobility*. Retrieved May 15, 2011 from http://weblog.cenriqueortiz.com/mobility/2008/12/29/worldwide-and-us-mobile-subscriber-penetration-dec-2008/.

Osman, S., Subhraveti, D., Su, G., & Nieh, J. (2002). The design and implementation of zap: A system for migrating computing environments. In *Proceedings of the Fifth Symposium on Operating Systems Design and Implementation (OSDI)*, (pp. 361-376). New York, NY: ACM Press.

Owens, J. D., Luebke, D., Govindaraju, N., Harris, M., Krüger, J., Lefohn, A. E., & Purcell, T. J. (2007). A survey of general-purpose computation on graphics hardware. *Computer Graphics Forum*, *26*(1), 80–113. doi:10.1111/j.1467-8659.2007.01012.x

JD Power and Associates Reports. (2010). *Average length of time wireless customers keep their mobile phones increases notably*. JD Power and Associates.

Ra, M., Paek, J., Sharma, A. B., Govindan, R., Krieger, M. H., & Neely, M. J. (2010). Energy-delay tradeoffs in smartphone applications. In *Proceedings of the 8th International Conference on Mobile Systems, Applications, and Services*, (pp. 255-270). New York, NY: ACM Press.

Rassweiler, A. (2010). *Google nexus one carries $174.15 materials cost, iSuppli teardown reveals*. Press Release. Retrieved May 15, 2011 from http://www.isuppli.com/Teardowns/News/Pages/Google-Nexus-One-Carries-$17415-Materials-Cost-iSuppli-Teardown-Reveals.aspx.

Rebitzer, G., Ekvall, T., Frischknecht, R., Hunkeler, D., Norris, G., & Rydberg, T. (2004). Life cycle assessment: Part 1: Framework, goal and scope definition, inventory analysis, and applications. *Environment International*, *30*(5), 701–720. doi:10.1016/j.envint.2003.11.005

Research, I. M. S. (2010). Smartphone shipments to quadruple in Asia by 2015. *IMS Research: Excellence in Market Intelligence*. Retrieved May 6, 2011 From http://imsresearch.com/news-events/press-template.php?pr_id=1350.

Robinson, B. H. (2009). E-waste: An assessment of global production and environmental impacts. *The Science of the Total Environment*, *408*, 183–191. doi:10.1016/j.scitotenv.2009.09.044

Saha, A. K., & Johnson, D. B. (2004). Modeling mobility for vehicular ad-hoc networks. In *Proceedings of the 1st ACM International Workshop on Vehicular ad hoc Networks*, (pp. 91-92). New York, NY: ACM Press.

Scharnhorst, W. (2008). Life cycle assessment in the telecommunication industry: A review. *International Journal of Life Cycle Assessment*, *13*, 75–86.

Sheriff, I., Acharya, P. A. K., Sampath, A., Zhao, B. Y., & Belding, E. M. (2007). Integrated data location in multihop wireless networks. In *Proceedings of the 2nd International Conference on Communication Systems Software and Middleware*, (pp. 1-10). IEEE Press.

Shye, A., Scholbrock, B., & Memik, G. (2009). Into the wild: Studying real user activity patterns to guide power optimizations for mobile architectures. In *Proceedings of the 42nd Annual IEEE/ACM International Symposium on Microarchitecture,* (pp. 168-178). New York, NY: ACM Press.

Shye, A., Scholbrock, B., Memik, G., & Dinda, P. A. (2010). Characterizing and modeling user activity on smartphones: SUMMARY. In *Proceedings of the ACM SIGMETRICS International Conference on Measurement and Modeling of Computer Systems,* (pp. 375-376). New York, NY: ACM Press.

Silveira, G. T. R., & Chang, S.-Y. (2010). Cell phone recycling experiences in the United States and potential recycling options in Brazil. *Waste Management (New York, N.Y.), 30,* 2278–2291. doi:10.1016/j.wasman.2010.05.011

Singhal, P. (2004). *Stage I final report: Life cycle environmental issues of mobile phones. Integrated Product Policy Pilot Project.* Espoo, Finland: Nokia Corporation.

Socolof, M. L., Overly, J. G., & Geibig, J. R. (2005). Desktop computer displays- A life-cycle assessment. *Journal of Cleaner Production, 13,* 13–14. doi:10.1016/j.jclepro.2005.05.014

Sullivan, D. (2006). Recycled cell phones - A treasure trove of valuable metals. *US Geological Survey.* Retrieved May 7, 2011 from http://pubs.usgs.gov/fs/2006/3097/.

Sullivan, J. L., & Gaines, L. (2010). A review of battery life-cycle analysis: State of knowledge and critical needs. *Argonne National Laboratory Technical Report ANL/ESD/10-7.* Retrieved from http://www.ipd.anl.gov/anlpubs/2010/11/68455.pdf.

TAMUZ. (2007). Technical bulletin: General information about TFT-LCD displays. *TAMUZ Broadcast Monitors.* Retrieved May 10, 2011 from http://www.tamuz.tv/manuals/tamuz-tb-lcd-effects.pdf.

Wang, Y., Pang, S., & Cheng, K. (2010). A GPU-accelerated face annotation system for smartphones. In *Proceedings of the International Conference on Multimedia,* (pp. 1667-1668). New York, NY: ACM Press.

Whitney, L. (2009). Jobs: Over 1 million new iPhones sold. *CNET News.* Retrieved May 6, 2011 from http://news.cnet.com/8301-13579_3-10269991-37.html.

Williams, E. D., Ayres, R. U., & Heller, M. (2002). The 1.7 kilogram microchip: Energy and material use in the production of semiconductor devices. *Environmental Science & Technology, 36,* 5504–5510. doi:10.1021/es025643o

Zhuang, Z., Kim, K., & Singh, J. P. (2010). Improving energy efficiency of location sensing on smartphones. In *Proceedings of the 8th International Conference on Mobile Systems, Applications, and Services,* (pp. 315-330). New York, NY: ACM Press.

Zoeteman, B. C. J., Krikke, H. R., & Venselaar, J. (2009). Handling WEEE waste flows: On the effectiveness of producer responsibility in a globalizing world. *International Journal of Advanced Manufacturing Technology, 47,* 415–436. doi:10.1007/s00170-009-2358-3

ADDITIONAL READING

Duan, H., Eugster, M., Hischier, R., Streicher-Porte, M., & Li, J. (2009). Life cycle assessment study of a Chinese desktop personal computer. *The Science of the Total Environment, 407*(5), 1755–1764. doi:10.1016/j.scitotenv.2008.10.063

Finnveden, G., Hauschild, M. Z., Ekvall, T., Guinée, J., Heijungs, R., & Hellweg, S. (2009). Recent developments in life cycle assessment. *Journal of Environmental Management*, *91*(1), 1–21. doi:10.1016/j.jenvman.2009.06.018

Guinée, J. B., Heijungs, R., Huppes, G., Zamagni, A., Masoni, P., & Buonamici, R. (2010). Life cycle assessment: Past, present, and future. *Environmental Science & Technology*, *45*(1), 90–96. doi:10.1021/es101316v

Hischier, R., Classen, M., Lehmann, M., & Scharnhorst, W. (2007). *Life cycle inventories of electric and electronic equipment: Production, use, and disposal*. Dübendorf, Switzerland: Swiss Center for Life Cycle Inventories.

Kahhat, R., Kim, J., Xu, M., Allenby, B., Williams, E., & Zhang, P. (2008). Exploring e-waste management systems in the United States. *Resources, Conservation and Recycling*, *52*(7), 955–964. doi:10.1016/j.resconrec.2008.03.002

Kuehr, R., & Williams, E. (2003). *Computers and the environment: Understanding and managing their impacts*. Dordrecht, The Netherlands: Kluwer Academic Publishers.

Mackay, D. J. (2009). *Sustainable energy - Without the hot air*. Cambridge, UK: UIT Cambridge, Ltd.

Nnorom, I. C., & Osibanjo, O. (2009). Toxicity characterization of waste mobile phone plastics. *Journal of Hazardous Materials*, *161*(1), 183–188. doi:10.1016/j.jhazmat.2008.03.067

Oliver, J., Amirtharajah, R., Akella, V., Geyer, R., & Chong, F. T. (2007). *Life cycle aware computing: Reusing silicon technology*. Washington, DC: IEEE Press.

Scharnhorst, W., Hilty, L. M., & Jolliet, O. (2006). Life cycle assessment of second generation (2G) and third generation (3G) mobile phone networks. *Environment International*, *32*(5), 656–675. doi:10.1016/j.envint.2006.03.001

Singhal, P. (2005a). *Integrated product policy pilot project. Stage I Final Report: Life Cycle Environmental Issues of Mobile Phones*. Espoo, Norway: Nokia.

Singhal, P. (2005b). *Integrated product policy pilot project. Stage II Final Report: Options for Improving Life-Cycle Environmental Performance of Mobile Phones*. Espoo, Norway: Nokia.

Singhal, P. (2006a). *Integrated product policy pilot on mobile phones. Stage III Final Report: Evaluation of Options to Improve the Life-Cycle Environmental Performance of Mobile Phones*. Espoo, Norway: Nokia.

Singhal, P. (2006b). Integrated product policy pilot on mobile phones. *Stage IV Final Report: New Environmental Initiatives & Experiences from the Pilot*. Espoo, Noway: Nokia.

Suh, S., & Huppes, G. (2005). Methods for life cycle inventory of a product. *Journal of Cleaner Production*, *13*(7), 687–697. doi:10.1016/j.jclepro.2003.04.001

Williams, E., Kahhat, R., Allenby, B., Kavazanjian, E., Kim, J., & Xu, M. (2008). Environmental, social, and economic implications of global reuse and recycling of personal computers. *Environmental Science & Technology*, *42*(17), 6446–6454. doi:10.1021/es702255z

KEY TERMS AND DEFINITIONS

Cradle to Gate: The set of upstream processes, including extraction and refinement of materials, production of energy, and supply chain flows, that are together required to create a product (as to the outbound gate of the factory where it is produced).

Cradle to Grave: The complete set of processes, including material extraction, energy, and supply chain flows, required by a product throughout its life cycle until it is finally disposed.

Process Inventory: A record of the inputs (resources, raw materials, energy, and other products) and outputs (products, wastes, and environmental emissions), associated with carrying out an industrial activity.

Life Cycle Inventory: An aggregate record of natural resources required and environmental emissions generated by the set of processes that make up the life cycle of a product. Life cycle inventories are often reported as either cradle-to-gate or cradle-to-grave.

Life Cycle Impact Assessment: An estimate of the environmental impacts incurred as a consequence of a product's life cycle. A life cycle impact assessment is computed from a life cycle inventory in conjunction with a model of the environmental impacts due to different resource requirements and emissions.

Functional Unit: A quantitative measurement of the utility or service provided by a product system. The selection of a functional unit is the basis for a life cycle assessment.

Reference Flow: The material or service outputs from a process that deliver the function expressed by the functional unit.

Chapter 12
Power Management Strategies based on Multi-Agent Systems for Portable Devices Equipped with Renewable Power Sources:
Laptop Case Study

Rafael J. Valdivieso-Sarabia
University of Alicante, Spain

Juan M. Garcia-Chamizo
University of Alicante, Spain

ABSTRACT

Electronic portable devices present some technological challenges like battery life, providing services, and integration of renewable power supplies. These challenges can be improved using power management based solutions. Power management systems for electronic portable devices are ad-hoc solutions for each device, so solutions are not general enough to be applied to others. A model driven architecture approach is defined. It allows defining a general power management systems solution for portable devices. The general solution is based on agent modelling, which is parameterized, so it allows automatic retrieval of the power management multi-agent system for a particular device giving values to the model's parameters. Distributed management strategies are based on giving priority to renewable power sources and reducing power spikes when there is not enough power. A laptop equipped with a photovoltaic charger is introduced for simulating power demand and supply in order to discuss power management strategies for reducing electrical consumption.

INTRODUCTION

There is a significant growth in portable, embedded, and mobile computing devices like mobile phone, smartphones, handsfree, mp3 players, laptops, tablets, global positioning systems, autonomous wheelchair, medical monitoring sensors, etc. This fact may be consequence of miniaturization of electronic components like processors, flash memories, antennas, etc. and the social success of this kind of products. The Achilles heel of this kind of electronic devices is

DOI: 10.4018/978-1-4666-1839-8.ch012

the power consumption-performance ratio. The power supply is determinant for power autonomy in these devices because they should be operative as long as they can, so power leaks could mean an authentic disaster in some critical situations. Electronic portable devices try to minimize power consumption of each intern component, although more and more are composed by a higher number of components in order to provide more and better services. Solutions based on larger batteries for improve power autonomy could make the device bigger and heavy. Moreover, this approach is not environmental sustainable because raw materials used to build batteries are finite and toxic. Sustainable development is defined by Brundtland (1987) as "development that meets the needs of the present without compromising the ability of future generations to meet their own needs." Goodland and Daly (1996) relate social, economic, and environmental aspects as requirements to achieve it. Environmental sustainable development regarding to energy is related to renewable power supplies based on wind, sun, biomass, etc. The use of renewable power sources is usually associated with medium-large energy systems like: grid power plant, houses (Lucia & Gervino, 2006), vehicles (Tsuchiya, 2008; Ray, Lever, Streeter, & Price, 2007), although the use of renewable microgeneration for portable devices is becoming popular (Paradiso & Starner, 2005; Chang, Tran, Wang, Fuh, & Lin, 2010). The use of renewable power sources using smaller batteries seems to be environmental sustainable. The scenario composed by electronic portable devices, which should be operative while there is no grid connection available, equipped with batteries and renewable power sources suggest the use of power management systems in order to maximize the power autonomy. Power management policies are used in electronic devices for energy saving, although the different approaches are ad-hoc solutions, based on the particular features of the electronic device.

This chapter proposes a general power management system for portable devices that can be equipped with renewable power sources. Our approach is based on model driven architecture paradigm (Mellor, Scott, Uhl, & Weise, 2004). We define an agent based model of mobile power systems composed of multiple power supplies, storage elements and consumption components. The model makes it possible to create a power management solution for a particular portable device from the general model, giving appropriate values to the model parameters in order to improve the power autonomy. The power management system is conceived using distributed computational paradigm, specifically multi-agent system. It is automatically generated for JADE multi-agent software development framework (Bellifemine, Poggi, & Rimassa, 2001). Agent's interactions determine the power management strategies to apply in each moment. Strategies optimize energy distribution from power supply units to power consumption components using sustainable criteria and managing power consumption through cutting it down. Global power reduction is carried out by disabling or modifying non-critical features of consumer components in order to reduce the local consumption of each one. We present two strategies for triggering the load management. The first one consist in detecting components that are demanding energy over their average for disabling features of other components to make up for the overconsumption. The other strategy is to reduce power consumption if there is not power enough for all the components. It is determined when any batteries state of charge reaches a critical level, which should to be defined for each particular device. A laptop equipped with a photovoltaic generator and a full charged battery is simulated in order to discuss the power management strategies for this particular case study.

POWER MANAGEMENT SYSTEMS BACKGROUND

Power management systems are used in many application scenes or contexts. Contexts can be classified using criterion based on power handled in the electrical systems. Using this criterion, we determine these contexts: countrywide electrical grids, buildings/houses, high-speed vehicles, low-speed vehicles, and portable electronic devices. Although power management systems aids are common for all the contexts, solution approaches used in each and every one are different because solutions tries to make the most of particular features.

Power management gains importance at large electrical systems due to the raised energy demanded. These kind of distributed and heterogeneous systems involve the use of power management and control systems, putting in a nutshell smart grid concepts (Farhangi, 2009). This scenario introduce some challenging problems like reducing environmental pollution through improving the use of renewable power sources (Carrasco, et al., 2006), optimal distribution satisfying networks power constraints. Power grid markets are based on transactions (Miguelez, et al., 2004) and their power management systems are based on decision-taking, planning algorithms which minimize economical cost, satisfying requirements like stability and voltage regulation (Bevrani, Hiyama, & Mitani, 2008; Pathak, Li, Honavar, & McCalley, 2007). There are ad-hoc solutions for particular electrical utilities, (Zeng, Liu, Ngan, & Wu, 2009) models wind turbine, photovoltaic array, battery, and power grid connection, but it is not able to manage other kind of power sources. Li, Poulton, and James (2007) present an agent based control system, which coordinates distributed energy sources for large-scale distribution networks, using a broker agent for coordination mechanism among all the agents, which are able to generate or request power.

Power management systems are relevant as well for self-powered houses or buildings equipped with renewable power sources, regardless of whether they are grid connected or not. It is a common practice in not grid connected houses, to study the energy required by each one in order to calculate the kind of power sources and batteries to generate and store electricity. There are works related to modeling power consumption and power generation for houses, buildings, or inhabitants. Aydinalp-Koksal and Ugursal (2008) and Aydinalp, Ugursal, and Fung (2003) discuss three models used for determine the power required by a neighbourhood and isolated houses. Xu, Hu, Wu, and Zhou (2008) propose social influence model to evaluate prize policy and habits of power consumption in residential houses. Jian, Longyun, Binggang, and Liuchen (2005) present a study of power management for renewable distributed generation using raw materials like wind, sun and biomass. The generated energy is stored in supercapacitors, chemical batteries and flywheels. The system is controlled by a control unit for each pair of power sources and storage unit. Although this approach introduces a distributed control unit, is not general enough because each power source needs to be related to a storage unit and a control unit.

The next power management context is for high-speed vehicles including all kind of transport like trains, cars, trucks… Indeed, there are many trains and cars, which use regenerative brake (Ogasa, 2008), hybrid technologies (Xiaoling & Hodgson, 2002a; Xiaoling & Hodgson, 2002b; Cerovsky & Mindl, 2005), hydrogen fuel cell (Tsuchiya, 2008). These works introduce the use of different alternative power supplies and a particular control system for store and use energy from each element in order to guarantee power supply.

The next context is for low speed vehicles, robots. Ray et al. (2007) presents an autonomous transport robot, which is four-wheel-drive, solar-powered designed for Antarctica and Greenland. The work defines photovoltaic control algorithm that stores power in a battery and manages power between mobility and own consumption, so when it is not enough power, the robot decreases the cruising speed. Bouquain, Blunier, and Miraoui (2008) propose an autonomous wheelchair equipped with fuel cell, which is the main power supply, and a battery, which is used for satisfying peak loads.

The last context and the most relevant are represented by portable electronic devices. Power management systems use reduction consumption techniques for improving autonomy. These techniques can be classified as statics and dynamics. The static ones include technologies based on components choice in design time like: low consumption logic, physical design, software, etc. The suitable components choice during design time is important at the moment of increasing the autonomy, but it does not exploit load variations that suffer the system during the common usage. In that context, it is where dynamic technologies such as Dynamic Voltage and Frequency Scaling (DVFS) (Kihwan, Soma, & Pedram, 2005) and Dynamic Power Management (DPM) acquire greater importance. The first is used, mainly, in processors for scaling the voltage level enough as be able to guarantee computation time constraints (Gilberto & Margaret, 2005). Dynamic technologies model components consumption management and the works loads, principally as discreet events systems (Irani, Singh, Shukla, & Gupta, 2005). This technology uses models based on finite state machines and transitions between states, which are triggered by events. Ren, Krogh, and Marculescu (2005) define an architecture, which is composed by service requester, service request queue, service provider, and power manager. The service requester asks for a service, which is queued in the service request queue. The service provider will resolve the next request when it achieves the appropriate state. The power manager gets the service request queue and the service provider state and using the selected policy, it throws an event to trigger the state change. DPM use runtime behaviour of the system to reduce power when the system is on idle or serving light workloads. There are three kind of policies for determine when it is appropriate to turn off a particular component. Heuristic policies use a fixed or dynamic timeout; stochastic, which are based on statistical theories; and predictive, which turn off a component when it is known that the inactive period is enough as to amortize the cost of switched off and switch on. Abbasian, Hatami, Afzali-Kusha, and Pedram (2008) propose a DPM predictive policy based on wavelet for predicting service provider inactive time period. After each single prediction, the system is updated with new values, so it is able to learn.

Literature contains power management systems that are context independent, some of them are based on multi-agent paradigm. The potential value of multi-agent systems technology to the power industry is described by McArthur et al. (2007), which introduces concepts and approaches of multi-agent systems that are appropriate to power engineering applications. Agent's advantages regarding other techniques are autonomy, social abilities, proactive nature in order to achieve its goals, etc. Lagorse, Paire, and Miraoui (2010) propose a multi-agent system for management of distributed power sources. This approach is focused on a photovoltaic generator, lead-acid battery, super-capacitor and grid connection. It uses a token passing to decide which agent controls the voltage in the bus. The active loads are not controlled by an agent, because it supposes that the final user will control the loads. This is a passive approach because the overall behaviour of the system depends of the user habits, so this solution is not able to reduce power consumption automatically. Abras et al. (2008) introduce a multi-agent system for power management in home automation where each agent is embedded into a power resource and the negotiation protocol used in cooperation is detailed. The control algorithm is based on two strategies: emergency mechanism and anticipation mechanism. The first one guarantee constraints and the second one determine the best future set-points for production and consumption.

Dependent context solutions given above, specially power management strategies for portable devices, are ad-hoc solutions which tries to make the most using particular features of the device. These solutions are not generalized and cannot be applied directly into other conditions. This fact stresses the absence of model proposals that supports different contexts. Independent context solutions are more general but strategies present some limitations or are too much complex to be applied to portable electronic devices. Our approach is based on developing a distributed and general model, which can be automatically applied into several portable electronic devices. The management strategies are based on giving priority to renewable power sources and using power reduction policies to cut components consumption down to balance the overall consumption, when it is not enough power.

AGENT BASED MODEL FOR POWER MANAGEMENT OF ELECTRONIC PORTABLE DEVICES

Our approach is Model Driven Architecture (MDA) (Mellor, et al., 2004) compliant. MDA express the problem domain and the solution using concepts from a high abstraction level that allows excluding useless details. It is called platform independent model and it is independent of any technology. This general model can be derived via some transformations into a particular platform, so it is called platform specific model. This ad-hoc model for a particular platform, framework or middleware can be used for getting solutions instantiating the independent model for each particular problem. The platform independent model uses mathematical notation in order to represent the agent based model that is powerful enough for represent complex, dynamic and adaptive systems (Macal & North, 2010). The framework used for developing the platform specific model is JADE multi-agent software development framework.

The main advantage of our MDA approach allows to instantiate particular problems from particular problem domain. From the domain of electronic portable devices, it is possible to instantiate devices like laptops, mp3 players, monitoring sensors, etc. in order to get automatically the particular power management system, PMS, for each one. This process is illustrated by Figure 1.

Figure 1. MDA approach used for generating power management systems for portable devices

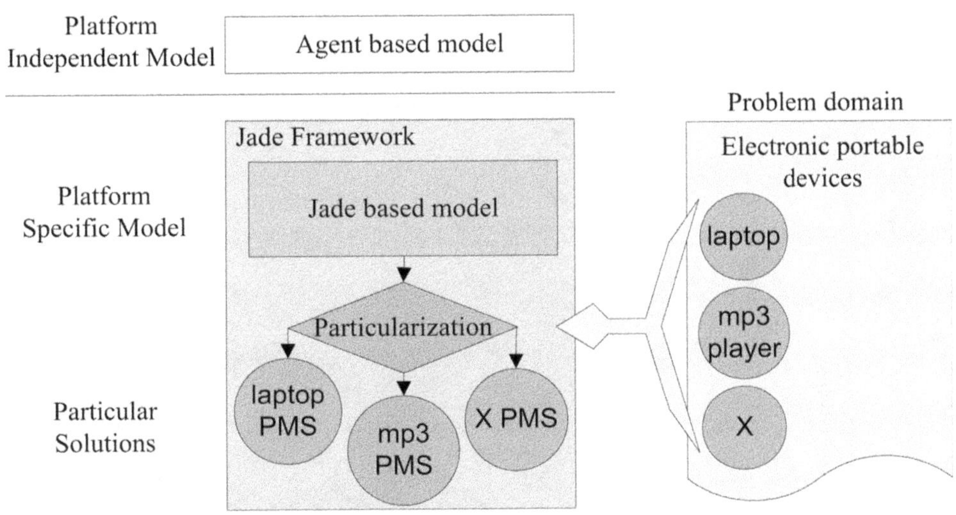

Portable Electric System Characterization

Components found in portable devices are analysed in order to classify them using the criterion of the power flow. There are available three kind of power flow: input, output and input-output. Components whose power flow are input, represent those that require electricity in order to develop their services or functionalities, they are called power loads. Components whose power flow are output, represent those that generate electricity, they are called power supplies. Finally, input-output components represent those that are able to consume or generate electricity, at different times or at the same time. These kinds of components are able to transport, store or transform electricity, and they are called power transporters. Moreover, components can be organised by the role played in the electric system. There are power generators and power consumers. According to this criterion, power loads belong to a consumer's role; power supplies belong to a generators role; and power transports belong to both, generators and consumers. This components characterization mixing power flow and roles is shown in Figure 2.

According to our aim, we have made some conceptual simplifications: considering electrical systems like a set of heterogeneous power sources, transmission lines, batteries storage and consumption centres; and not considering power transformations and transmissions losses due to short distances.

Agent-Based Model

The agent-based model represents the platform independent model defined in MDA paradigm. It represents all the components characterized in portable devices and some extra agents related to develop auxiliary tasks. The model has a set of agents and interactions among them.

The agent based model is composed by the agents set (A) which is composed by two kinds of subsets A={EA, NEA}. Energy Agents (EA) represent physical components which handle energy. Non-Energy Agents (NEA) are auxiliary agents that perform tasks to keep the multi-agent system in working order.

EA subset is composed by three kinds of agents EA={SA, TA, CA}. Source Agents (SA) represent power supplier components that provide electricity to the whole system; Transport Agents (TA) model power transporter components like batteries or transmission lines; and Consumption Agents (CA) represent power loads components. Each kind of agent is defined by a set of attributes that allow the particularization of a real power supply, battery, transmission lines and consumption centres. SA represent the role of energy producer and CA

Figure 2. Components characterization according to power flow and roles criteria

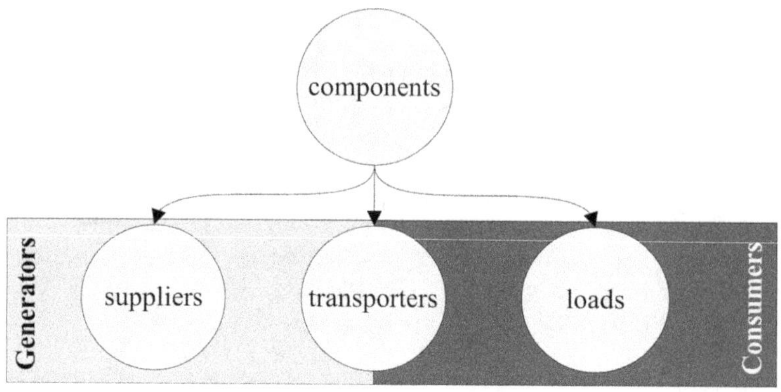

describe the role of energy consumer whereas TA contain both of them, because a transmission component can be considered like an energy producer or consumer.

SA are described with the tuple <volt, maxPow, genCost, GEN(t), envImp>, where volt represents the output voltage; maxPow shows the maximum power that the source is able to supply; genCost represents the economic cost of 1kWh; GEN(t) is a function which determines at every instant the power supplied; and envImp represents the environmental impact of generating 1Wh of electrical power.

CA is defined with the tuple <volt, maxPow, CON(t), OM>, where volt represents the input voltage; maxPow shows the maximum power that the component is able to demand; CON(t) is a function which determines at every instant the power required; and OM={OM1,…,OMn} represents a set of different Operating Modes (OM). Each one is composed with the tuple <crit, powCon>, where crit is a flag which shows whether the operating mode is critical or can be disabled for reducing power consumption; and powCon shows the power consumption required by this operating mode.

We denote TA with the tuple <volt, maxPow, genCost, envImp, chargeRate, criticLevel, GEN(t), CON(t)>, where volt represents the reference voltage; maxCap shows the maximum power able to store or transport; genCost represents the economic cost of supplying 1Wh; envImp represents the environmental impact of supplying 1Wh of electrical power; chargeRate shows the maximum charge rate that the battery is able to accept; and criticLevel represents the minimum percentage of the battery capacity which is starting to send request to reduce consumption to CA; GEN(t) and CON(t) are functions that determine the power supplied and consumed respectively at every moment.

NEA subset is composed by several kinds of agents. Interface Agent (IA) interacts with external user. It allows user to consult or modify the multi-agent system. Management System Agent (MSA) represents the middleware between IA and whole system. It provides security tasks enabling the communication among the external user and EA. Directory Agent (DA) stores an agent list of the system. It registers services offered by each EA and it allows finding out the services that other agents are offering. Monitoring agent (SIMA) saves the state of each EA in every simulation step. Agent Generator of Agents (AGA) is useful for replicate agents in order to keep on system integrity and security. Figure 3 shows every kind of agent, relationships among them, and the allowed interactions.

The agent-based model topology matches with network topology (Macal & North, 2010). The topology determines interactions among agents. Interactions show the kind of agents that are allowed to request actions to other agents. EA is able to interact with MSA to send or receive data. MSA interacts with every kind of agent because represents the middleware between IA and the whole system. Each EA is able to interact with MSA for send or receive data. DA allows to any agent searching for a particular service offered by other agent. SIMA develops two main tasks, sending a new simulation step flag for EA and saving each EA state at the end of every simulation step for monitoring the whole system. Interactions among EA are described as follows. CA is able to communicate with TA to request power. TA interact with CA to request a reduction of consumption, and with SA to request power. SA interact to TA to request reduction of consumption. The power flow among EA is as follows. SA can be electrically connected to CA through TA, and several TA can be connected among them. Connecting several TA allow designing schemes composed with several batteries, where the big one is able to recharge the smaller. There are two main interactions among SA, TA and CA in order to planning the power distributions. Power request interaction allows to consumer's roles, TA or CA, to request power for a given time period to a

Figure 3. Agent's typology defined in our agent based model

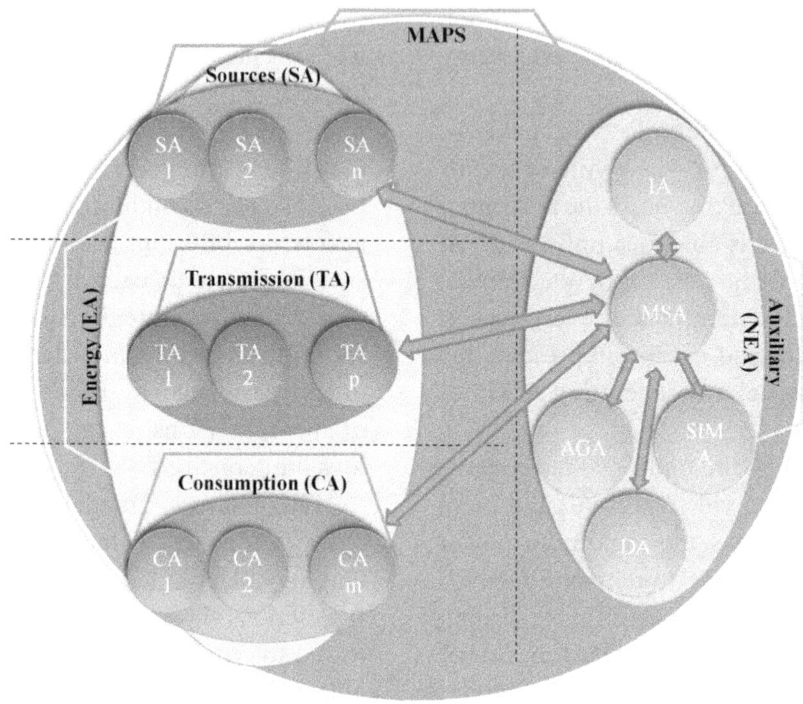

generator's roles TA or SA. Power reduction interaction is developed by generator's roles, SA or TA, in order to ask for power consumption reduction to consumer's roles, TA or CA. The power reduction is achieved turning any non-critical service off. Power reduction interaction is triggered by two situations: when a generator role has not got enough power for all the power requests; when a power request is higher than the average power requested at the previous simulations steps.

JADE-Based Model

JADE multi-agent software development framework has been chosen because it is a mature technology that is one of the most popular in scientific research. It complies with Foundation for Intelligent Physical Agents (FIPA) specification, it has a continuous development and it is fully implemented in Java Runtime Environment. The platform is deployed with a main-container that contains two extra agents: DF agent that allows looking up specific agent or services offered by any agent and AMS is the agent in charge of the platform. The platform can launch some agent's containers, so a platform can be composed by a main-container and several containers that can be on the same host or in remote host. A distributed platform can be deployed using different agent containers installed on several remotes hosts. A single container provides better communication performance but it has a negative impact on computational performance. Our specific model for portable devices defines a JADE platform composed by the main-container. The context of portable devices are characterized by small number of agents, so for this context is better to improve the inter-agent communication performance. The JADE based model defines several agent kinds available to create inside the platform, the behaviour that each agent is able to carry on, and the parameters data types, which defines the particular features of each agent. The agents available on the platform are defined in agent based model: SA, TA, CA, IA, MSA, AGA, DA, SIMA, and the compulsory agents introduced by JADE: AMS and DF, but MSA and AMS carry on similar functions, so ASM overwrites MSA, as the same way DF overwrites DA. Finally, the agents available are: SA, TA, CA, IA, ASM, AGA, DF and SIMA. The behaviour and interactions among these agents are described as follows. Interactions among agents for requesting data or services are implemented using a FIPA Request protocol (Foundation for Intelligent Physical Agents, 2002c) and interactions to propose actions are implemented according FIPA Propose protocol (Foundation for Intelligent Physical Agents, 2002b). The power request and power reduction interactions are implemented using an iterative FIPA Contract-net protocol (Foundation for Intelligent Physical Agents, 2002a) for messages exchange. Parameter's data types used for representing the data are real and boolean values.

Power Management as a result of Agents' Interactions

The double aim of power management strategies is based on distributing power from sources to consumers improving the use of renewable power sources and reducing power consumption when there is not enough power for the whole system. Reaching these goals is possible through the emergent behaviour of the whole system like a consequence of multiple interactions among EA.

Each agent is able to request the DA in order to know which agents are linked to him and which ones are able to provide a particular service, like power supply. The management process starts when an agent with the role of consumer, CA or TA (consumer), needs power so it triggers the contract-net power request interaction protocol. This power request is sent to the linked agents that are able to supply power, SA or TA (generator), which are agents with role of generator. There are two strategies to determine how much power is requested to each generator. The first one is to request all the power required to each generator linked, and whether a request is accepted then refuse others. The other strategy is to request in the three firsts iterations a proportional quantity of power to the number of available generators, after that, if the agent still needs power, then it requests for the total power needed. Each generator who has received a power request answers with the available power that is able to provide during a defined period of time. When consumer receives all the answers of the generators, it initiates a decision process to determine which generator are the best candidates to reach the goals. Taking decision process is modelled by function best, see Equation 1, which use attributes of each linked generator for ordering the candidates from best to worst. This function tries to give priority to generator, which have the lowest value of envImp attribute, considering maxPow and genCost. The function linkedProd returns a tuple of linked generators from the consumer given as parameter.

$$\langle A0,\ldots,An\rangle = best(linkedProd(X)) \,/\, X \in CA \vee TA,\ Ai \in SA \vee TA$$

Next, the consumer chooses the bests generators and requests a proportional or total quantity of required power. This action is iterated until the consumer does not need more power or all generators reject the power request. The following action is to send a finishing message to the linked generators to inform that the power request interaction is finished. When a generator receives the finishing message from all the linked consumers then it makes a decision to request for reduce consumption to the linked consumers or to send a no action message to indicate that this interactions has finished. If the generator decides to reduce power consumption, then it initiates a contract-net interaction protocol sending a reduce power request to all the connected consumers. The power to reduce can be determined by two ways. Requesting the total power to reduce to all the consumers linked or request for a proportional power reduction to the number of consumers linked. If a reduction request is received by a consumer then it check the power required to perform each functionality and it turn non-critical functionalities off until the power reduced is enough to satisfy the reduction request. In the case, that is not able to reduce power consumption then it replies with a refuse message. When the reduce consumption request is finished, the generator sends a no action message to consumers. Next, producers and consumers send their status to SIMA. It receives the status of all EA, it replies to indicate that the management can be scheduled for the next period of time.

There are some strategies or policies for determine the power requested to each generator: total or proportional, as the same way for calculating the power requested for reducing consumption. Focused on power request protocol, the use of total policy allows to get the total power required using a single negotiation protocol, but if the generator requested has not enough power to satisfy the total power requested, a second negotiation protocol has to be scheduled to get the remaining power from other generators. This policy is suitable for networks or devices, which are equipped with few power sources, which are powerful enough to provide the whole power requested. The power distribution solution obtained using this policy has a tendency to overload some transmission lines and to underuse other lines, so it is possible to get higher Joule losses using this policy. On the other hand, the proportional policy allows getting a proportional power required from all the available generators using a single negotiation protocol. Therefore, this policy is suitable for networks or devices, which are made up of a lot of small-medium capacity generators. Moreover, solutions obtained using a proportional power request policy allows to balance the power required over all the linked transmission lines, so the power losses due to Joule effect can be decreased because power losses depend on the square of the current flowing by the line. Although each policy can be more appropriate for a particular context, simulation step allows simulating each policy to make sure the best selection for each particular context, because selecting the best strategy depends of the particular features of the network or device to be managed.

LAPTOP CASE STUDY

The agent-based model of portable device is instantiated in a real laptop in order to evaluate the power autonomy using the power management system automatically obtained from the model. The power management system is conceived as a distributed system, particularly as a multi-agent system whose interactions determines the power distribution and the power strategies to reduce power consumption in order to increase autonomy in mobile context. The most relevant laptop's components have been taking into account for making simulations results lighter. The laptop power supplies are a grid connection (*GRID*) and a photovoltaic charger (*PHO*). This power sources are not always available, grid connection due to mobility constraints and photovoltaic charger because it depends on direct sun radiation.

The laptop is equipped with a chemical battery (*BAT*) that can be charged by grid connection and photovoltaic charger. The components that require power to carry on their functionalities are: processor (*CPU*), display (*DIS*), hard disk (*HD*), two Universal Serial Bus ports (*USB1*), (*USB2*) and the other components (*OTH*) whose power consumption cannot be managed are taking into account together. The grid connection and photovoltaic charger can charge the battery. The power requester components are connected to the battery and to the grid connection, so they are able to get power from these two components. The laptop is represented in a mobile context where is not possible to use a grid connection, the battery's state of charge is fully charged and the photovoltaic charger can be used to charge

Figure 4. Agent based model from laptop's components and available connections among them.

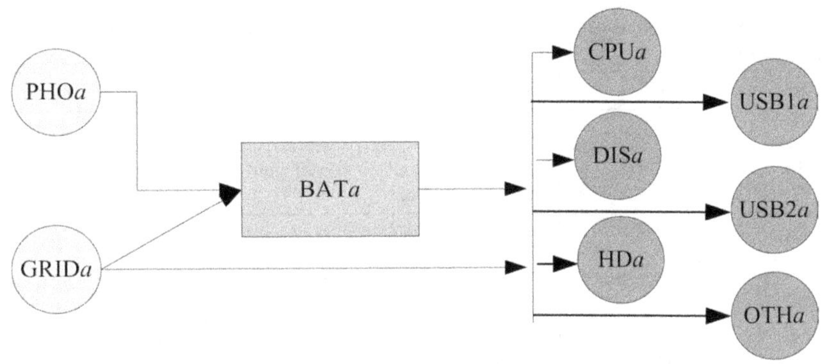

the battery if there is direct sun radiation. The components features are as follow: grid connection presents a voltage of 230V and the electric current is too much bigger than laptop requirements. Photovoltaic charger is able to provide a maximum power of 25Wh at 12VDC when the sun radiation is optimal. The battery capacity is 5.8Ah at 10.8VDC and the charge rate is about 25% of power. The *CPU* maximum power required is 25Wh at full performance although is able to reduce its consumption using DVFS techniques when the computational requirements are medium 15Wh, low 8Wh and sleep mode 0.5Wh. The *DIS* power consumption can be decreased reducing the brightness level. It is discretized in 5 levels: At maximum brightness 4.8Wh, at 80% 4.14Wh, at 60% 3.48Wh, at 40% 2.82Wh, at 20% 2.16Wh. The hard disk, *HD*, presents the maximum power consumption, 5Wh, when the engine starts to spin, when it is on idle requires 0.6Wh and in R/W operations it needs 1.9Wh, when it is seeking a position, it consumes 2.1Wh and when is on sleep mode it requires 0.13Wh. The *USB1* and *USB2* according to USB specification each USB requires the maximum power of 2.5Wh, although it can decrease power consumption at 2Wh, 1.5Wh, 1Wh and 0.5Wh, although if any USB device is connected the power consumption is 0Wh. It is assumed that the others components, *OTH*, present a fixed power consumption of 4Wh, in order to make simulation result easier to understand.

Agent-Based Model Instantiated for Laptop

Each laptop's component is represented by an instance of each agent kind. The components *GRID* and *PHO* are modeled using SA and respectively are called GRID*a* and PHO*a*. The battery *BAT* is represented by TA, and is called BAT*a*. The power consumer's components *CPU, DIS, HD, USB1, USB2,* and *OTH* are modeled like a CA, and respectively are called CPU*a*, DIS*a*, HD*a*, USB1*a*, USB2*a* and OTH*a*. Figure 4 shows the agents defined to represent each laptop's component and the available connections that allow power flow from power supplies to consumption components.

The relevant attributes of SA and TA are: maxPow, which represents the maximum power that is able to generate, and envImp, which represents the environmental impact of power generation. CA relevant attributes are: maxPow, which shows the maximum power that are able to request, and each OM parameter represents each function that is able to provide and it is defined like critical/non-critical and the power required to carry on. The non-critical operating modes can be disabled when a power reduction request is received. Table 1 summarizes the information related to the values assigned to the model parameters of each agent kind. The operative modes of USB1*a* and USB2*a* components represent the power consump-

Table 1. Parameters values for each agent based model to represent each laptop's component

Component id	Agent kind	Agent id	Parameters values
PHO	SA	PHO*a*	volt = 12v, maxPow=25Wh, genCost=0.1 €/kWh, envImp=0.2
GRID		GRID*a*	volt = 230v, maxPow=∞, genCost=0.14 €/kWh, envImp =1
BAT	TA	BAT*a*	volt = 10.8v maxPow=63Wh, genCost=0.2 €/kWh, envImp=0.5, chargeRate=25%, criticLevel=60%
CPU	CA	CPU*a*	maxPow=25Wh, OM1= <critic,8Wh>, OM2= <non-critic,7Wh>, OM3= <non-critic,10Wh>
DIS		DIS*a*	maxPow=4.8Wh, OM1=<critic,2.16Wh>, OM2=<non-critic,0.66Wh>, OM3=<non-critic,0.66Wh>, OM4=<non-critic,0.66Wh>, OM5= <non-critic,0.66Wh>
HD		HD*a*	maxPow=5Wh, OM1=<critic,0.6Wh>, OM2=<non-critic,1.3Wh>, OM3= <non-critic,0.2Wh>
USB1		USB1*a*	maxPow=2.5Wh, OM1=<critic,0.5Wh>, OM2=<non-critic,0.5Wh>, OM3=<non-critic,0.5Wh>, OM4=<non-critic,0.5Wh>, OM5= <non-critic,0.5Wh>
USB2		USB2*a*	maxPow=2.5Wh, OM1=<critic,0.5Wh>, OM2=<non-critic,0.5Wh>, OM3=<non-critic,0.5Wh>, OM4=<non-critic,0.5Wh>, OM5= <non-critic,0.5Wh>
OTH		OTH*a*	maxPow=4Wh, OM1= <critic,4Wh>

tion of the USB devices when they are connected to the laptop's USB ports, so when they are not connected, the power consumption is zero.

Power Management Results

The agents defined and parameterized have been implemented in JADE development framework in order to evaluate the power management behaviour. The power supply and demand is simulated during a period of twelve hours. Each simulation step represents one hour, so the power management interactions are triggered at each hour.

The simulation context represents a user who is travelling while he is using the laptop and there is no electrical grid connection available until 15:00 to 17:00, although there is direct sun radiation during the time period simulated, so photovoltaic panel can be used to charge the battery. Power generation is modeled according to (Faisal, 2008) where the main parameters considered are: solar irradiation and ambient temperature. The solar irradiance and ambient temperature depend on geographical location and weather, so for this experimentation the geographical location selected has a great solar irradiance, with maximum values of 900W/m2 at 13:00 and minimum value of 100W/m2 from 18:00 to 20:00. The ambient temperature presents its maximum 25°C at 14:00 and minimum 15°C at 8:00. Figure 5a shows the power generation capacity of grid connection, photovoltaic charger and battery maximum charge rate. Battery maximum charge rate is constant during the simulation time, for this context is taken into account the 25% of maximum capacity, so it is established at 15.75Wh or 1450mAh. This maximum charge rate is exceeded from 12:00 to 15:00, so BAT*a* avoids charging the battery during this time for do not overcharge it. Figure 5b illustrates the power consumption for each laptop's component taken into account for the power management system. The workload for each component has been designed randomly. The most representative component is the CPU, which has a maximum consumption of 25Wh at high workloads and minimum of 8Wh at low workloads. It is the most relevant component due to the minimum power consumption is almost double of the maximum power required by any other component.

Figure 5. (a) Power generated by photovoltaic panel, grid connection and battery maximum charge rate at each hour. (b) Power required by each laptop's components simulated for the power management system.

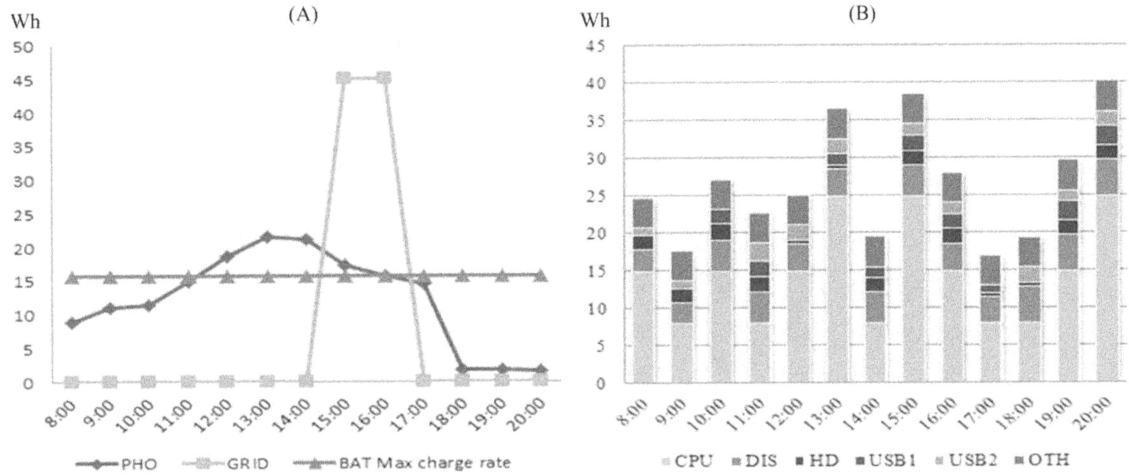

The power management system establishes the power distribution planning from power supplies to batteries or to laptop components according to the simulated generation and consumption patterns described in Figure 5. Simulations has been made using three alternatives strategies for determine the power distribution. The first one disables power reduction request from BAT*a* to laptop's components and following enables it. The difference among second and third is the event which triggers the power reduction requests. The second one triggers power reduction interaction when BAT*a* state of charge drops below 60% and the last one throws power reduction interactions when the total power requested is higher than the average of the power supplied previously. Figure 6a represents the first strategy, which illustrates the power distribution planning disabling power reduction interactions among BAT*a* to CPU*a*, DIS*a*, HD*a*, USB1*a*, USB2*a* and OTH*a*. This strategy represents the solution with lower time autonomy, because the BAT*a* provides all the power requested by laptop's components. This solution is the base case for discussing following strategies. The power request sent from laptop's components to BAT*a* at each simulation step are satisfied until de battery gets discharged, meanwhile BAT*a* sends power request to PHO*a* and GRID*a* in order to be charged. Figure 6-A shows the total power required by each component, the real power supplied by BAT*a* to the components and the battery state of charge evolution. BAT*a* is not able to supply power to components at 12:00, 13:00, 16:00, 19:00 and 20:00, so the laptop is turned off, meanwhile BAT*a* is been charged by PHO*a*.

The second strategy enables power reduction requests if battery state of charge is lower than the particular BAT*a* criticLevel defined at 60%. If any CA make an agreement with BAT*a* for supplying power at each simulation step and BAT*a* state of charge is lower than 60%, then it sends a broadcast power reduction request to ask CA for turn non-critical operating modes off. Results of simulating this strategy are shown by Figure 6b. Laptop's components power consumption is constant from 9:00 until 19:00 because components reduce their consumption at each simulation during this time. BAT*a* state of charge is constant from 9:00 to 18:00 because PHO*a* is charging BAT*a* at similar rate as the discharge rate. The power supplied by PHO*a* from 18:00 to 20:00 is between 1 and 2Wh so battery is discharged

Figure 6. (a) Power distribution planning with power reduction interactions disabled. (b) Power distribution planning with power reduction interaction triggered by battery critic level. (c) Power distribution planning with power reduction interaction triggered when the power request is above the average. (d) Power average supplied by BAT previously at each simulation step in the third strategy.

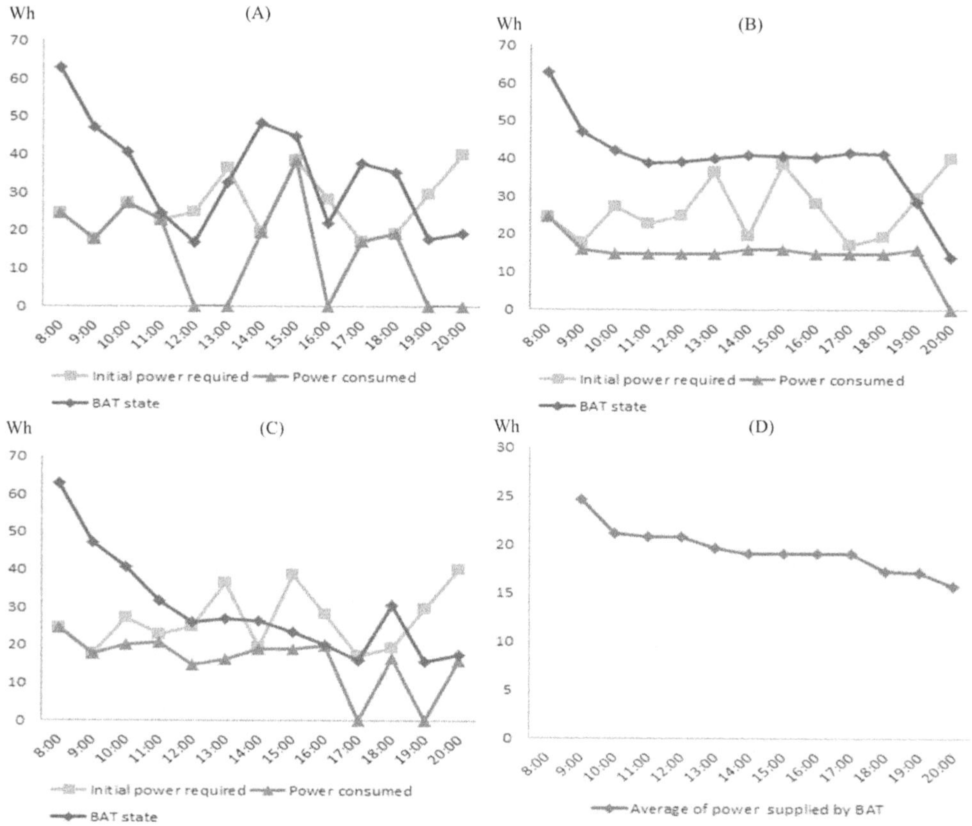

quickly. Laptop's components are not able to get power at 20:00 because BAT*a* is almost discharged.

Figure 6c illustrates the third strategy that triggers power reduction request when the total power requested is higher than the average power supplied during the previous simulation steps. In this strategy is relevant to recalculate the average of the power supplied at the beginning of each simulation step for determine if power reduction request must be sent. The evolution of average of power supplied is shown in Figure 6d. Average consumption is calculated from second hour because at the first one there is no data available. It is slightly decreasing at each simulation step due to the power reduction requests are higher at followings simulation steps, because power reduction is proportional to the difference among the average and the current power requested. This approach sends lower power reduction request, so it increases the number of hours that the laptop is working compared to the first strategy but it decreases the operative hours compared to the second strategy. BAT*a* has not enough power for supplying laptop components at 17:00 and 19:00.

Table 2 summarizes the discussion among the power management strategies simulation results, The first approach satisfies all the laptop's components power requirements although minimizes the number of hours that the laptop is working. The second strategy maximizes the

Table 2. Power management strategies comparative according to the number of working hours, user satisfaction, and number of component's functionalities turned off

Strategy	Laptop working hours	Number of components functionalities turned off	Laptop user satisfaction
No reduction request	Low	None	High
Reduction request at reaching 60% of battery state of charge	High	High	Low
Reduction request exceeding the average consumption	Intermediate	Intermediate	Intermediate

laptop working hours, achieving constant power consumption. This approach is the most aggressive because it minimizes the functionalities developed by laptop's components during ten of eleven hours that laptop is operative, so it would be negative for laptop's user satisfaction. The third approach gets power distribution planning between the first and second strategies, because it turns laptop's functionalities off slightly for getting better autonomy than first one and worse than the second one. In this approach, the laptop's user satisfaction is lower than first one and higher than the second one. This strategy has a dynamic behaviour because the average of power supplied is being updated continuously. This fact allows agents to learn user habits for getting an optimal balance between laptop working hours and user satisfaction.

FUTURE RESEARCH DIRECTIONS

Post-PC era is coming with many electronic portable devices, which are increasingly sophisticated and powerful. Increasing battery life of portable devices should be achieved not only improving on power managements strategies but researching in complementary topics together: more efficient renewable power supplies, larger batteries capacity and smaller size, reducing component's power consumptions, better power management strategies and user awareness of power usage. The last one requires user involvement so it would be perhaps hard to come by. Active load management systems can be an appropriate approach to obtain better power managements strategies according to user's satisfaction criteria to avoid them to manage power consumption of each component.

From our point view, the future trends of electronic portable devices are supposed to be equipped with better size-capacity ratio batteries, complementary renewable power sources which exploit sun, wind, body in motion, etc. and wide-ranging smart power management strategies for complex, dynamic, and heterogeneous environments.

CONCLUSION

A model-driven architecture approach is presented, which allows defining a general power management systems solution based on multi-agent systems for portable devices. The general solution is based on agent modeling. Each agent model is parameterized allowing automatically getting the power management multi-agent system for a particular device giving values to each agent's parameters. Power management strategies are based on giving priority to renewable power sources and reducing power consumption of components that requires power when there is not enough power. A laptop equipped with photovoltaic charger is introduced for simulating power demand and supply. The agent-based model is particularized for the laptop getting the multi-agent system for laptop power management system. Three power management strategies have been simulated for discussing the advantages and disadvantages of each one. The first power distribution planning is obtained disabling power reduction requests. Following approach has been obtained triggering power reduction requests when the battery state of charge is lower than 60%. The last power distribution planning triggers the power reduction requests when the current power request is higher than the average of previous request. Simulations results presents that the first approach satisfies all the laptop's components power requirements although minimizes the number of hours that the laptop is working. Second strategy maximizes the laptop working hours, achieving constant power consumption although it minimizes the functionalities developed by laptop's components. Third approach sends lower power reduction request than second one. It increases the number of hours that the laptop is working compared to the first strategy but it decreases the operative hours compared to the second. The last strategy presents better battery life and functionalities developed ratio, because it increases the battery life reducing laptop's components power consumption slightly turning few functionalities off. Moreover, the third strategy behaviour is dynamic because the average of power requested is being updated continuously, so it can learn user habits. Choosing the best strategy depends on the particular device and user satisfaction criterion, so our approach allows to automatically generates solutions based on different strategies to discuss the suitability of each approach.

REFERENCES

Abbasian, A., Hatami, S., Afzali-Kusha, A., & Pedram, M. (2008). Wavelet-based dynamic power management for nonstationary service requests. *ACM Transactions on Design Automation of Electronic Systems*, *13*(1), 1–41. doi:10.1145/1297666.1297679

Abras, S., Ploix, S., Pesty, S., Jacomino, M., Cetto, J. A., & Ferrier, J. (2008). A multi-agent home automation system for power management. *Informatics in Control Automation and Robotics*, *15*, 59–68. doi:10.1007/978-3-540-79142-3_6

Aydinalp, M., Ugursal, V. I., & Fung, A. S. (2003). Modelling of residential energy consumption at the national level. *International Journal of Energy Research*, *27*(4), 441–453. doi:10.1002/er.887

Aydinalp-Koksal, M., & Ugursal, V. I. (2008). Comparison of neural network, conditional demand analysis, and engineering approaches for modeling end-use energy consumption in the residential sector. *Applied Energy*, *85*(4), 271–296. doi:10.1016/j.apenergy.2006.09.012

Bellifemine, F., Poggi, A., & Rimassa, G. (2001). Developing multi-agent systems with a FIPA-compliant agent framework. *Software, Practice & Experience*, *31*(2), 103–128. doi:10.1002/1097-024X(200102)31:2<103::AID-SPE358>3.0.CO;2-O

Bevrani, H., Hiyama, T., & Mitani, Y. (2008). Power system dynamic stability and voltage regulation enhancement using an optimal gain vector. *Control Engineering Practice*, *16*(9), 1109–1119. doi:10.1016/j.conengprac.2008.01.001

Bouquain, D., Blunier, B., & Miraoui, A. (2008). A hybrid fuel cell/battery wheelchair: Modeling, simulation and experimentation. In *Proceedings of the Vehicle Power and Propulsion Conference, 2008*. IEEE Press.

Brundtland, G. H. (1987). *Our common future: Brundtland commission (WCED) report*. World Commission on Environment and Development.

Carrasco, J. M., Franquelo, L. G., Bialasiewicz, J. T., Galvan, E., Guisado, R. C. P., & Prats, M. A. M. (2006). Power-electronic systems for the grid integration of renewable energy sources: A survey. *IEEE Transactions on Industrial Electronics*, *53*(4), 1002–1016. doi:10.1109/TIE.2006.878356

Cerovsky, Z., & Mindl, P. (2005). *Regenerative braking by electric hybrid vehicles using super capacitor and power splitting generator.* Paper presented at the Power Electronics and Applications. London, UK.

Chang, C., Tran, V. H., Wang, J., Fuh, Y.-K., & Lin, L. (2010). Direct-write piezoelectric polymeric nanogenerator with high energy conversion efficiency. *Nano Letters*, *10*(2), 726–731. doi:10.1021/nl9040719

Faisal, A. M. (2008). *Microgrid modelling and online management*. Espoo, Finland: Helsinki University of Technology.

Farhangi, H. (2009). The path of the smart grid. *IEEE Power and Energy Magazine*, *8*(1), 18–28. doi:10.1109/MPE.2009.934876

Foundation for Intelligent Physical Agents. (2002a). *FIPA foundation for intelligent physical agents contract net interaction protocol specification*. FIPA.

Foundation for Intelligent Physical Agents. (2002b). *FIPA foundation for intelligent physical agents propose protocol specification*. FIPA.

Foundation for Intelligent Physical Agents. (2002c). *FIPA foundation for intelligent physical agents request protocol specification*. FIPA.

Gilberto, C., & Margaret, M. (2005). *Power prediction for intel XScale processors using performance monitoring unit events*. Paper presented at the Proceedings of the 2005 International Symposium on Low Power Electronics and Design. New York, NY.

Goodland, R., & Daly, H. (1996). Environmental sustainability: Universal and non-negotiable. *Ecological Applications*, *6*(4), 1002–1017. doi:10.2307/2269583

Irani, S., Singh, G., Shukla, S. K., & Gupta, R. K. (2005). An overview of the competitive and adversarial approaches to designing dynamic power management strategies. *IEEE Transactions on Very Large Scale Integration (VLSI) Systems*, *13*(12), 1349–1361.

Jian, W., Longyun, K., Binggang, C., & Liuchen, C. (2005). *Reliability study of a distributed generation system based on renewable energy.* Paper presented at the Electrical and Computer Engineering, 2005. Toronto, Canada.

Kihwan, C., Soma, R., & Pedram, M. (2005). Fine-grained dynamic voltage and frequency scaling for precise energy and performance tradeoff based on the ratio of off-chip access to on-chip computation times. *IEEE Transactions on Computer-Aided Design of Integrated Circuits and Systems*, *24*(1), 18–28. doi:10.1109/TCAD.2004.839485

Lagorse, J., Paire, D., & Miraoui, A. (2010). A multi-agent system for energy management of distributed power sources. *Renewable Energy*, *35*(1), 174–182. doi:10.1016/j.renene.2009.02.029

Li, J., Poulton, G., & James, G. (2007). Agent-based distributed energy management. *Advances in Artificial Intelligence*, *4830*, 569–578.

Lucia, U., & Gervino, G. (2006). Thermoeconomic analysis of an irreversible Stirling heat pump cycle. *The European Physical Journal B - Condensed Matter and Complex Systems*, *50*(1), 367-369.

Macal, C. M., & North, M. J. (2010). Tutorial on agent-based modelling and simulation. *Journal of Simulation*, *4*(3), 151–162. doi:10.1057/jos.2010.3

McArthur, S. D. J., Davidson, E. M., Catterson, V. M., Dimeas, A. L., Hatziargyriou, N. D., & Ponci, F. (2007). Multi-agent systems for power engineering applications part I: Concepts, approaches, and technical challenges. *IEEE Transactions on Power Systems*, *22*(4), 1743–1752. doi:10.1109/TPWRS.2007.908471

Mellor, S. J., Scott, K., Uhl, A., & Weise, D. (2004). *MDA distilled: Principles of model-driven architecture*. Boston, MA: Addison Wesley.

Miguelez, E. L., Rodriguez, L. R., Roman, T. G. S., Cerezo, F. M. E., Fernandez, M. I. N., & Lafarga, R. C. (2004). A practical approach to solve power system constraints with application to the Spanish electricity market. *IEEE Transactions on Power Systems, 19*(4), 2029–2037. doi:10.1109/TPWRS.2004.831704

Ogasa, M. (2008). Energy saving and environmental measures in railway technologies: Example with hybrid electric railway vehicles. *IEEJ Transactions on Electrical and Electronic Engineering, 3*(1), 15–20. doi:10.1002/tee.20227

Olfati-Saber, R., Fax, J. A., & Murray, R. M. (2007). Consensus and cooperation in networked multi-agent systems. *Proceedings of the IEEE, 95*, 215–233. doi:10.1109/JPROC.2006.887293

Paradiso, J. A., & Starner, T. (2005). Energy scavenging for mobile and wireless electronics. *IEEE Pervasive Computing / IEEE Computer Society and IEEE Communications Society, 4*(1), 18–27. doi:10.1109/MPRV.2005.9

Pathak, J., Li, Y., Honavar, V., & McCalley, J. (2007). A service-oriented architecture for electric power transmission system asset management In *Proceedings of Service-Oriented Computing ICSOC 2006* (*Vol. 4652*, pp. 26–37). Springer. doi:10.1007/978-3-540-75492-3_3

Ray, L. E., Lever, J. H., Streeter, A. D., & Price, A. D. (2007). Design and power management of a solar-powered "cool robot" for polar instrument networks. *Journal of Field Robotics, 24*(7), 581–599. doi:10.1002/rob.20163

Ren, Z., Krogh, B. H., & Marculescu, R. (2005). Hierarchical adaptive dynamic power management. *IEEE Transactions on Computers, 54*(4), 409–420. doi:10.1109/TC.2005.66

Starner, T., & Paradiso, J. (2004). Human generated power for mobile electronics In Piguet, C. (Ed.), *Low Power Electronics Design*. Boca Raton, FL: CRC Press.

Tsuchiya, H. (2008). Innovative renewable energy solutions for hydrogen vehicles. *International Journal of Energy Research, 32*(5), 427–435. doi:10.1002/er.1374

Xiaoling, H., & Hodgson, J. W. (2002a). Modeling and simulation for hybrid electric vehicles. *IEEE Transactions on Intelligent Transportation Systems, 3*(4), 235–243. doi:10.1109/TITS.2002.807781

Xiaoling, H., & Hodgson, J. W. (2002b). Modeling and simulation for hybrid electric vehicles. *IEEE Transactions on Intelligent Transportation Systems, 3*(4), 244–251.

Xu, M., Hu, Z., Wu, J., & Zhou, Y. (2008). A hybrid society model for simulating residential electricity consumption. *International Journal of Electrical Power & Energy Systems, 30*(10), 569–574. doi:10.1016/j.ijepes.2008.08.005

Zeng, J., Liu, J. F., Ngan, H. W., & Wu, J. (2009). *A multi-agent solution to energy management of distributed hybrid renewable energy generated system.* Paper presented at the Advances in Power System Control, Operation and Management (APSCOM 2009). New York, NY.

ADDITIONAL READING

Barroso, L. A., & Holzle, U. (2007). The case for energy-proportional computing. *Computer, 40*, 33–37. doi:10.1109/MC.2007.443

Bauer, B., Muller, J., Odell, J., Ciancarini, P., & Wooldridge, M. (2001). Agent UML: A formalism for specifying multiagent software systems In *Agent-Oriented Software Engineering*. Berlin, Germany: Springer. doi:10.1007/3-540-44564-1_6

Bellifemine, F., Caire, G., Trucco, T., Rimassa, G., & Mungenast, R. (2010). *JADE administrator's guide*. JADE.

Bellifemine, F., Rimassa, G., & Greenwood, D. (2007). *Developing multi-agent systems with JADE*. New York, NY: John Wiley and Sons. doi:10.1002/9780470058411

Benini, L., Bogliolo, A., & De Micheli, G. (2000). A survey of design techniques for system-level dynamic power management. *IEEE Transactions on Very Large Scale Integration (VLSI) Systems*, *8*, 299–316.

Bordini, R. H., Dastani, M., & Seghrouchni, E. F. A. (Eds.). (2009). *Multi-agent programming: Languages, tools and applications*. Berlin, Germany: Springer.

Brooks, R. A. (1991). Intelligence without representation. *Artificial Intelligence*, *47*, 139–159. doi:10.1016/0004-3702(91)90053-M

Brundtland, G. H., & Khalid, M. (1987). *Our common future. Brundtland Commission (WCED) report*. World Commission on Environment and Development.

Carrascosa, C., Bajo, J., Julian, V., Corchado, J. M., & Botti, V. (2008). Hybrid multi-agent architecture as a real-time problem-solving model. *Expert Systems with Applications*, *34*, 2–17. doi:10.1016/j.eswa.2006.08.031

Dincer, I., & Rosen, M. A. (2007). Exergy, environment and sustainable development In *Exergy: Energy, Environment and Sustainable Development*. London, UK: Elsevier.

Dushyanth, N., Austin, D., & Antony, R. (2008). Write off-loading: Practical power management for enterprise storage. *Transaction Storage*, *4*, 1–23.

FIPA. (2002). *Wepage*. Retrieved from http://www.fipa.org/repository/standardspecs.html.

Hall, P. J., & Bain, E. J. (2008). Energy-storage technologies and electricity generation. *Energy Policy*, *36*, 4352–4355. doi:10.1016/j.enpol.2008.09.037

Jenkins, N., Strbac, G., & Ekanayake, J. (2010). *Distributed generation*. New York, NY: Institution of Engineering & Technology.

Jennings, N. R., Sycara, K., & Wooldridge, M. (1998). A roadmap of agent research and development. *International Journal of Autonomous Agents and Multi-Agent Systems*, *1*, 7–38. doi:10.1023/A:1010090405266

Jurasovic, K., Jezic, G., & Kusek, M. A. (2006). Performance analysis of multi-agent systems. *International Transactions on Systems Science and Applications*, *1*(4).

Kansal, A., Hsu, J., Zahedi, S., & Srivastava, M. B. (2007). Power management in energy harvesting sensor networks. *ACM Transactions on Embedded Computer Systems*, *6*, 32. doi:10.1145/1274858.1274870

McArthur, S. D. J., Davidson, E. M., Catterson, V. M., Dimeas, A. L., Hatziargyriou, N. D., Ponci, F., & Funabashi, T. (2007). Multi-agent systems for power engineering applications part II: Technologies, standards, and tools for building multi-agent systems. *IEEE Transactions on Power Systems*, *22*, 1753–1759. doi:10.1109/TPWRS.2007.908472

Patterson, M. G. (1996). What is energy efficiency? Concepts, indicators and methodological issues. *Energy Policy*, *24*, 377–390. doi:10.1016/0301-4215(96)00017-1

Poslad, S. (2009). *Ubiquitous computing: Smart devices, environments and interactions*. New York, NY: John Wiley & Sons, Ltd.

Priya, S., & Inman, D. J. (Eds.). (2009). *Energy harvesting technologies*. Berlin, Germany: Springer. doi:10.1007/978-0-387-76464-1

Suzanne, R., Mehul, A. S., Parthasarathy, R., & Christos, K. (2007). JouleSort: A balanced energy-efficiency benchmark. In *Proceedings of the 2007 ACM SIGMOD International Conference on Management of Data*. Beijing, China: ACM Press.

Tran, Q. N. N., Low, G., & Williams, M. A. (2005). A preliminary comparative feature analysis of multi-agent systems development methodologies. *Lecture Notes in Computer Science*. Retrieved from http://www.angelfire.com/theforce/numi-tran/AOIS04.pdf.

Wang, M., Shi, Z., Jiao, W., & Lukose, D. (2009). Dynamic interaction protocol load in multi-agent system collaboration In *Multi-Agent Systems for Society*. Berlin, Germany: Springer. doi:10.1007/978-3-642-03339-1_9

Zambonelli, F., Jennings, N. R., & Wooldridge, M. (2003). Developing multiagent systems: The Gaia methodology. *ACM Transactions on Software Engineering and Methodology*, *12*, 317–370. doi:10.1145/958961.958963

KEY TERMS AND DEFINITIONS

Agent: It is a computer-based system, located in some environment, which is able to perceive the environment status, to determine the appropriate actions to achieve their goals and to perform actions, which changes the environment.

Distributed Generation: It is also called embedded generation and it is used in power grid contexts to express that the power is generated by many small-scale energy sources, which are located near loads.

Framework: It is a software solution that abstracts hardware, offering a common interface for the software entities that can be integrated on it, enabling communication mechanisms among them.

Interaction Protocol: It is a model of sequence of messages exchange among an initiator entity, which starts the interaction protocol, and the receiver entity, which receives the requesting or proposing action.

Load Management: It is a power management strategy, which reduces the loads power consumption in order to meet power generated with power demand.

Multi-Agent System: It is a system designed and implemented using interacting agents that are able to request actions among themselves according to their own goals. The global goal of the system is achieved from interactions of all agents.

Portable Device: It is also called wearable device and it represents a subset of electronic devices, which can be worn by people, animals, or other devices and usually uses energy storage elements like batteries.

Power Management: It is a feature for modifying the power state of a particular electrical device in order to achieve some goals like: reduce power consumption, keep electrical security constraints, or achieve power distribution planning, etc.

Chapter 13
Essential Technologies for Location-Aware Mobile, Online Map Prefetching, and Caching

Wen-Chen Hu
University of North Dakota, USA

Naima Kaabouch
University of North Dakota, USA

Lei Chen
Sam Houston State University, USA

Ming Yang
Southern Polytechnic State University, USA

ABSTRACT

Map navigation is one of the most popular applications used by mobile users. At the same time, it is also one of the time- and resource-consuming applications. Various methods such as most-recently used and first-in, first-out algorithms are used to reduce the map transmission time and delay. One of the popular methods is online mobile map prefetching and caching. However, the mobility and location features of mobile users are usually left out by these methods. Caching and prefetching maps based on a mobile user's location would greatly reduce the transmission time and hence the battery power consumption. For example, if a user is visiting a town, prefetching the maps of nearby interesting stores and caching the maps of the visited, neighboring landmarks would help the user's visitation experience and save the transmission time. Online mobile map prefetching or caching is useful, but is not widely employed because it involves several different subjects and developers usually are not familiar with all of them. This chapter intends to relieve the problem by introducing essential technologies for online mobile map prefetching and caching so more developers can start working on this kind of project or research. It consists of four themes: (1) green handheld computing, (2) location-based services and programming, (3) map tile system, and (4) location-aware map prefetching and caching methods. A summary is given at the end of this chapter.

DOI: 10.4018/978-1-4666-1839-8.ch013

INTRODUCTION

The number of smartphones shipped worldwide has surpassed the number of PCs and servers shipped in 2011 and the gap between them is expected to keep bigger. The emerging smartphones have created many kinds of applications that are not possible or inconvenient for PCs and servers, even notebooks. One of the best-seller applications is Location-Based Services (LBS) according to the following market research:

- Fleishman Hillard (2012, February 6) reports 80% of smartphone owners have location-based services and half of them use services that provide offers, promotions, and sales based on their current locations.
- The most convenient mobile shopping experience is price comparison and product research according to JiWire (2011, October 14).
- The number of location-based services users was increased from 12.3 million in 2009 to 33.2 million in 2010 (170% increase) in the US based on SNL Kagan (Cohen, 2011, January 20).

Map navigation is a critical function for most location-based services and is one of the most popular applications used by mobile users.

Despite the bright future of smartphones, they are not without problems, especially the short battery life. To save the transmission time and reduce the delay, mobile map navigation caches or prefetches maps by using various methods such as most-recently used and first-in, first-out algorithms. However, the mobility and location features of mobile users are usually left out by these methods. Caching and pre-fetching maps based on a mobile user's locations would greatly reduce the transmission time and hence the battery power consumption. For example, if a user is visiting a town, prefetching the maps of nearby interesting stores and caching the maps of the visited, neighboring landmarks would enhance the user's visitation experience and save the transmission time. On the other hand, though map-related mobile applications are popular, the details of their design and implementation are still not known to most developers. This chapter tries to relieve the problem by explaining the map-related application design and implementation, especially map prefetching and caching, from a developer perspective. It studies the four themes of map-related application design and implementation: (1) green handheld computing, (2) location-based services and programming, (3) map tile system, and (4) location-aware map prefetching and caching methods.

This chapter studies the essential technologies for mobile, online map prefetching and caching. The rest of this chapter is organized as follows. Section 2 introduces green handheld computing, which is the foundation of map prefetching and caching. Location-based services and programming are explained in Section 3, where a simple Android application development is given. Map prefetching and caching are closely related to map structure, which is studied in Section 4. Various methods of map prefetching and caching are given in Section 5. The final section summarizes this study.

GREEN HANDHELD COMPUTING

Green computing is the practice of designing, manufacturing, using, and disposing of computers, servers, and associated subsystems for reducing the use of hazardous materials, maximizing energy efficiency, and promoting the recyclability or biodegradability of products. This section discusses the green computing using mobile handheld devices.

Green Mobile Computing

According to Handheld Computing Research (2012), 1,775 million cell phones and 486 million smartphones were sold in 2011 worldwide. One of the reasons of the huge number of cell phones sold is companies encourage customers to upgrade their phones constantly so they can profit from it. It would greatly hurt the environment if used phones are simply thrown away. There are two ways to mitigate the problem:

- *Device Recycling*: A cell phone consists of tens or even hundreds of components and some of the parts such as Printed Circuit Board (PCB), Liquid Crystal Display (LCD), and batteries may be poisonous. Recycling is to process used materials into new products in order to reduce or prevent materials or energy wasted. An example of device recycling is the memory chip in a device could be used to replace a small-size chip in another device.
- *Device reusing*: Instead of recycling the old phones, they could be used by other people or for other purposes. Reusing is to use a product more than once for the same function or a different function. For example, plain cell phones may not meet people's needs, but people in under developed areas may be just happy to own a phone, even a plain cell phone.

Most cell phones use Lithium Ion batteries. This kind of batteries can only last for few hours of operations before needing to be re-charged. Extending its time of operations not only is an urgent need for users, but also is good for the environment because the battery would fail after hundreds of times of recharging. One way to extend the battery life is to use efficient algorithms, one of the methods of green mobile computing. Many methods are used to improve the algorithm efficiency; three of them are

- *Picking the right algorithms*: To accomplish a task, it is more important to pick the right algorithms than to use a more powerful device or acquire more memory. For example, to sort a long list of numbers, upgrading the device or adding more memory will not improve the efficiency significantly, but if the algorithm is switched from selection sort to quick sort, the time improvement would be obvious.
- *Scheduling*: Some mobile operating systems such as Android allow multithreading programming. Scheduling is to distribute the system resources such as CPU time and memory to processes effectively in order to make the maximum out of the resources. Examples of scheduling algorithms include First-Come First-Serve (FCFS), round robin, and shortest job first. Good scheduling algorithms could save time and space for devices.
- *Prefetching or caching*: Instead of waiting for the requests, the system is able to prefetch or cache the data based on previous experience or predictions. For example, if a user is visiting a town, prefetching the maps of nearby interesting stores and caching the maps of the visited, neighboring landmarks would help the user's visitation and save the transmission time.

There are many other methods such as resource reallocation and optimization can also be used for mobile green computing.

Mobile Handheld Devices

Handheld computing is the use of handheld devices like smart cellular phones to perform wireless, mobile, handheld operations such as browsing the mobile Web and finding the nearest gas stations. It is an essential part of location-based services. This section discusses two handheld computing subjects: mobile handheld devices and client-side

Figure 1. System structure of mobile handheld devices

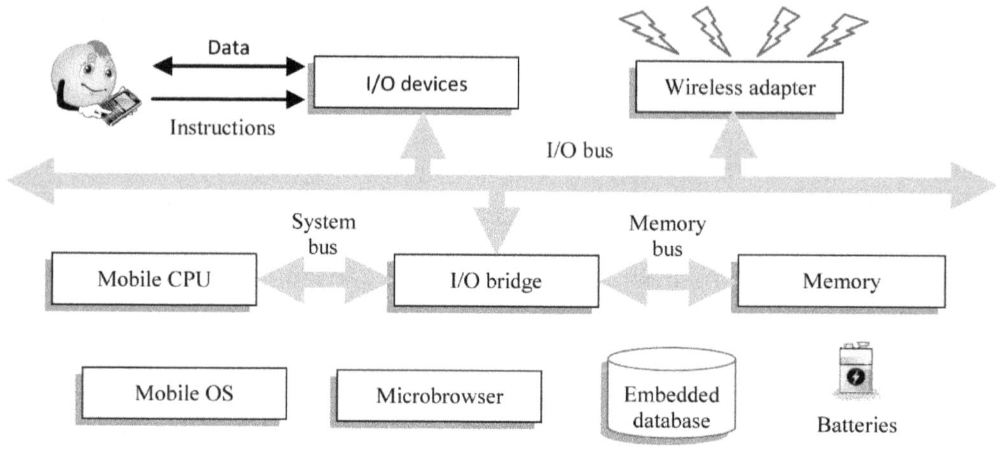

handheld computing. Mobile users interact with mobile commerce applications by using small wireless Internet-enabled devices, which come with several aliases such as handhelds, palms, PDAs, pocket PCs, and smart phones. To avoid any ambiguity, a general term, mobile handheld devices, is used in this article. Mobile handheld devices are small general-purpose, programmable, battery-powered computers, but they are different from desktop PCs or notebooks due to the following special features: (1) limited network bandwidth, (2) small screen/body size, and (3) mobility.

Short battery life and limited memory, processing power, and functionality are additional features, but these problems are gradually being solved as the technologies improve and new methods are constantly being introduced. The limited network bandwidth prevents the display of most multimedia on a microbrowser. Though the Wi-Fi and 3G networks go some way toward addressing this problem, the wireless bandwidth is always far below the bandwidth of wired networks. The small screen/body size restricts most handheld devices to using a stylus for input. Figure 1 shows a typical system structure for handheld devices, which includes the following six major components (Hu, Wiggen, & Yang, 2005):

- *Mobile operating systems*: Simply adapting desktop operating systems for handheld devices has proved to be futile. A mobile operating system needs a completely new architecture and different features to provide adequate services for handheld devices. A generalized mobile operating system structure can be visualized as a six-layer stack: (1) applications, (2) GUI, (3) API framework, (4) multimedia, communication infrastructure, and security, (5) computer kernel, power management, and real-time kernel, and (6) hardware controller.
- *Mobile central processing units*: Handheld devices are becoming more sophisticated and efficient every day and mobile users are demanding more functionality from their devices. To achieve this advanced functionality, in addition to the obvious feature, low cost, today's mobile processors must have the following features: (1) high performance, (2) low power consumption, (3) multimedia capability, and (4) real-time capability. The cores and architectures designed by Cambridge-based ARM Holdings Ltd. have begun to dominate the mobile CPU market.
- *Microbrowsers*: Microbrowsers are miniaturized versions of desktop browsers

such as Netscape Navigator and Microsoft Internet Explorer. They provide graphical user interfaces that allow mobile users to interact with mobile commerce applications. Microbrowsers usually use one of the following four approaches to return results to the mobile user: (1) wireless language direct access, (2) HTML direct access, (3) HTML to wireless language conversion, and (4) error.

- *Input/output components*: Various I/O devices have been adopted by mobile handheld devices. The only major output device is the screen, but there are several popular input devices, among them: (1) keyboards and (2) touch screens/writing areas that need a stylus.*Memory*: Three types of memory are usually employed by handheld devices: (1) RAM, (2) ROM, and (3) flash memory. Hard disks, which provide much more storage capacity, are likely to be adopted by handheld devices in the near future.
- *Batteries*: At present, rechargeable Lithium Ion batteries are the most common batteries used by handheld devices. However, the life of this kind of battery is short and the technology will not significantly improve unless and until manufacturers begin to switch to fuel cells, which may not happen for at least several years.

Synchronization connects handheld devices to desktop computers, notebooks, or peripherals to transfer or synchronize data. Without needing serial cables, many handheld devices now use either an Infrared (IR) port or Bluetooth technology to send information to other devices.

Client-Side Handheld Computing

Client-side handheld computing is the programming for handheld devices and it does not need the supports from server-side programs. Typical applications created by it include (1) address books, (2) video games, (3) note pads, and (4) to-do list. Various environments/operating systems/languages are available for client-side handheld programming. Table 1 lists the top-five mobile operating systems.

Table 1. Top-5 mobile operating systems and their features

Mobile OS	Android	BlackBerry OS	iOS	Symbian	Windows Phone
Creator	Open Handheld Alliance	RIM	Apple	Nokia	Microsoft
2011 Market Share (472.7 million)	1st (46.4%)	4th (12.1%)	2nd (18.9%)	3rd (18.7%)	6th (2.1%) (5th: Samsung's Bada)
Development Languages	Java	Java	Objective C/C++	C++, Java, others	Visual C++
Kernel Type	Linux	Unix	Hybrid	Microkernel	Windows CE 6/7
IDEs, Libraries, Frameworks	Android SDK; ADT plug-in for Eclipse	BlackBerry JDE	iPhone SDK		Windows Phone SDK (works with Visual Studio)
Source Model	Open	Closed	Closed (open for the core)	Closed	Closed
Initial Release	2008	1999	2007	2001	2010
Latest Version as of August 2011	3	7	4	Anna	7
Mobile Application Store	Android Market	App World	App Store	Ovi Store	Windows Marketplace for Mobile

Figure 2. A generalized client-side handheld computing development cycle

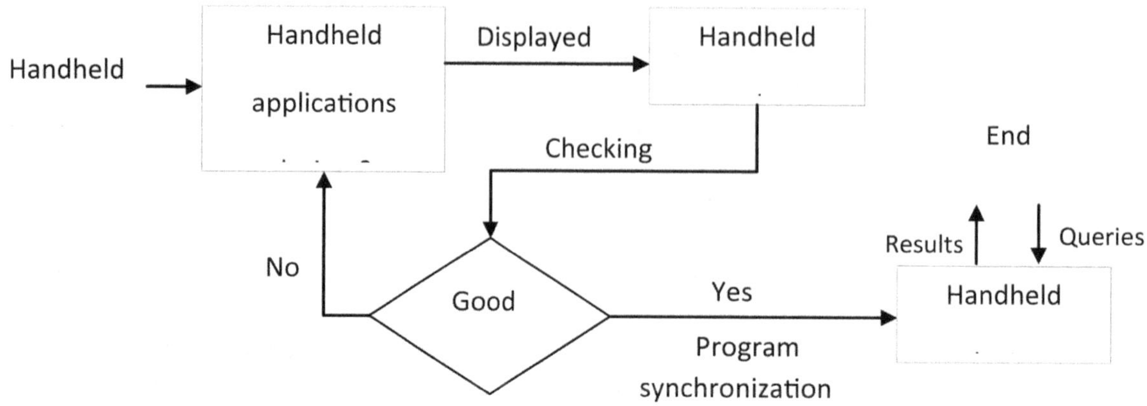

They apply different approaches to accomplishing the development of mobile applications. Figure 2 shows a generic development cycle applied by them. Handheld emulators instead of the handhelds themselves are used for the development because of the convenience reason.

LOCATION-BASED SERVICES AND PROGRAMMING

This section introduces location-based services by giving a generic LBS architecture and an LBS example. Additionally, it gives an example of mobile application programming.

LBS Architecture and Example

A location-based service is a service based on the geographical position of a mobile handheld device (Kupper, 2005; Kolodziej & Hjelm, 2006). Two of the LBS examples are (1) finding a nearby ethnic restaurants and (ii) locating a nearby store with the best price of a product. A system structure of location-based services, shown in Figure 3, includes five major components (Steiniger, Neun, & Edwardes, 2006):

a. Mobile handheld devices, which are small computers that can be held in one hand. For most cases, they are smartphones.

Figure 3. A system structure of generic location-based services

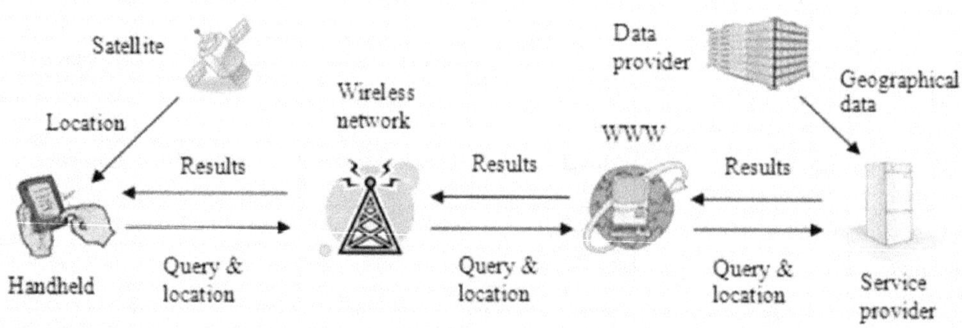

Figure 4. (a) A direction using a straight line or an actual route and (b) a location marker

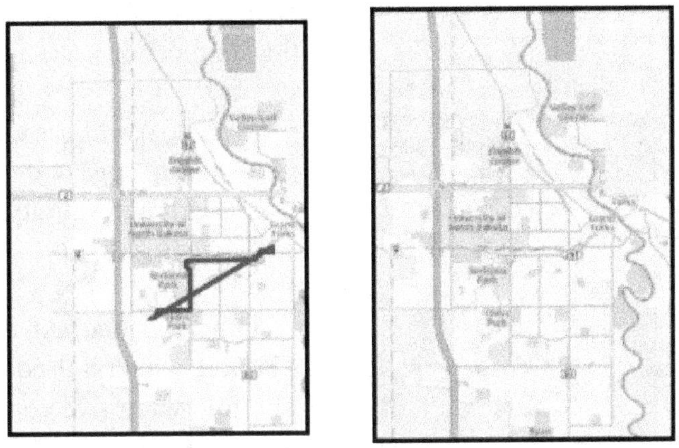

(a) (b)

b. Positioning system, which is a navigation satellite system that provides location and time information to anyone with a receiver.
c. Mobile and wireless networks, which relay the query and location information from devices to service providers and send the results from the providers to devices.
d. Service providers, which provide the location-based services.
e. Geographical data providers, which are databases storing a huge amount of geographical data such as information about restaurants and gas stations.

An LBS example of finding nearby gas stations is given step by step as follows:

1. The smartphone (a) includes an application of finding interesting locations locally.
2. A mobile user submits a query of locating nearby gas stations along with the location information from a positioning system (b) to the application program.
3. The application program calls the server-side programs (d) with the location information via mobile or wireless networks (c).
4. The programs at the servers try to locate nearby gas stations using the databases (e), which stores gas-station data.
5. The result information of nearby gas stations is sent back to the smartphone.

Popular LBS include mapping and navigation, search and information, social networking, entertainment, and tracking. Figure 4 gives two screenshots of typical LBS from Yahoo! Maps which show a direction and a location marker. A nice introduction of LBS technologies and standards is given by Wang, Min, and Yi (2008).

DEVELOPMENT OF A SIMPLE ANDROID APPLICATION

Android (2012) is a software stack for mobile devices such as smartphones and tablet PCs. It was developed by the Open Handset Alliance, a consortium of 80 hardware, software, and telecommunication companies led by Google devoted to advancing open standards for mobile devices. Google purchased the initial developer of the software, Android Inc., in 2005. Android includes an operating system, middleware, and

Figure 5. A screenshot of a Google Map with a location marker

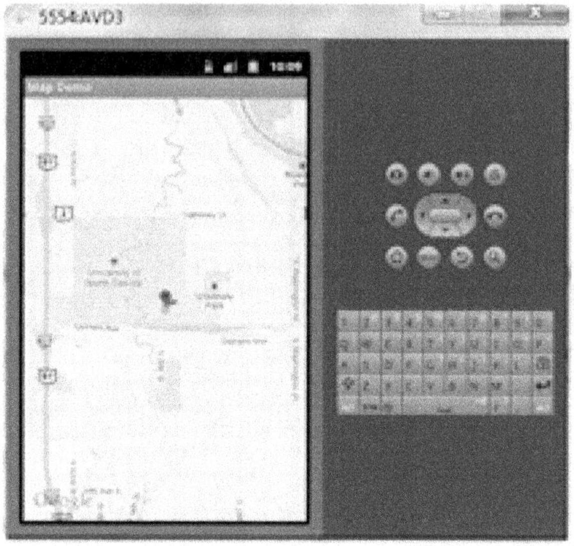

key applications. The Android SDK (Software Development Kit) provides the tools and APIs necessary for developing applications on the Android platform using the Java programming language. Google released most of the Android code under the Apache License, a free software license. Android is tightly integrated with Google Maps, which provides a powerful tool for location-based services. Figure 5 shows a screenshot of a Google map with a location marker.

This application is based on the default template when you create an Android project. It has two different screens and is able to move one screen to another screen. Figure 6 shows the result screenshots. Two activities/pages are included in this application:

- HelloWorldActivity.java, which is the startup page and
- NextPageActivity.java, which is the other page.

Other than the text, each of the pages contains a button, which is used to display the other page. The following five components of this app will be detailed next:

- Android manifest file (AndroidManifest.xml),
- Layout XML code (main.xml and next.xml),
- Strings (strings.xml),
- The R file (R.java), and
- Java source code (HelloWorldActivity.java and NextPageActivity.java).

Eclipse (2012) is a software development environment for multiple programming languages including Ada, C, C++, COBOL, Java, Perl, PHP, Python, R, Ruby, etc. It comprises an Integrated Development Environment (IDE) and an extensible plug-in system. Android Development Tools (ADT) is a plugin for the Eclipse IDE, which gives developers an integrated environment for building Android applications. Figure 7 shows a screenshot of the Eclipse IDE with Android plugin, which

Figure 6. Two screenshots of the Android application

provides the functionality of a well-established IDE, along with Android-specific features that are bundled with ADT. It allows developers to perform various tasks such as (1) setting up Android projects, (2) creating an application user interface, (3) adding components based on the Android Framework API (Application Programming Interface), (4) debugging applications using the Android SDK (Software Development Kit) tools, and (5) exporting .apk files in order to post your application.

Android Manifest File

The main missions of the manifest file are to explain what the application consists of, what its major building blocks are, what permissions it requires, etc. It must declare all activities, services, broadcast receivers, and content provider of the application. For example, if the application wants to access the Internet, it must define in this file that it would like to use the related permission, as shown in Box 1.

Layout XML Code

The user interface for activities is defined via layouts. The layout defines the UI elements, their properties and their arrangement. A layout can be defined via Java code or via XML:

- *Java code*: You typically use Java code to generate the layout if you do not know the content until runtime; e.g., if your layout depends on content which you read from the Internet. An example of this is the next three statements whose screenshot is shown in Figure 8:

```
TextView tv = new TextView(this);
tv.setText("Hello, Android");
setContentView(tv);
```

Figure 7. A screenshot of the Eclipse IDE with Android plugin

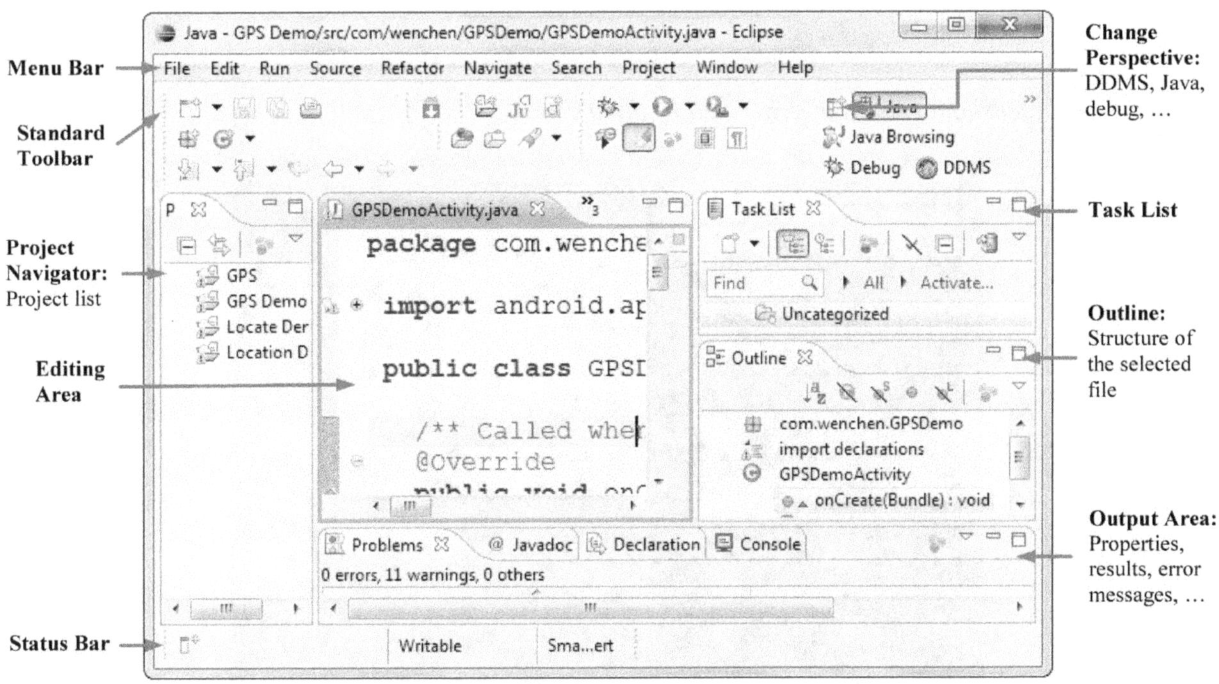

Box 1.

```
<?xml version="1.0" encoding="utf-8"?>
<manifest xmlns:android="http://schemas.android.com/apk/res/android"
   package="com.example.HelloWorld"
   android:versionCode="1"
   android:versionName="1.0">
  <uses-sdk android:minSdkVersion="8" />
  <application android:icon="@drawable/icon"
          android:label="@string/app_name">
   <activity android:name=".HelloWorldActivity"
          android:label="@string/app_name">
     <intent-filter>
       <action android:name="android.intent.action.MAIN" />
        <category android:name="android.intent.category.LAUNCHER" />
     </intent-filter>
   </activity>
   <activity android:name=".NextPageActivity"
          android:label="Next Page" />
  </application>
</manifest>
```

- *XML*: XML based layouts are defined via a resource file in the folder /res/layout. This file specifies the view groups, views, their relationship, and their attributes for a specific layout. Defining layouts via XML is usually the preferred way as this separates the programming logic from the layout definition. For example, the following two layouts are used in this application:
    - Home page at res/layout/main.xml and
    - Next page at res/layout/next.xml

The application loads the screen specified in main.xml by the following Java command in the HelloWorldActivity.java:

```
setContentView(R.layout.main);
```

Figure 8. A screenshot of a layout generated by Java code

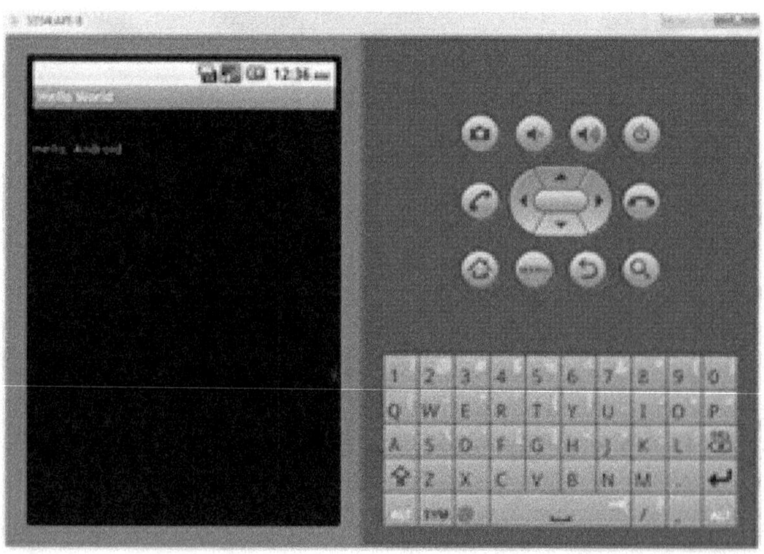

Box 2.

```
<?xml version=”1.0” encoding=”utf-8”?>
<LinearLayout xmlns:android=”http://schemas.android.com/apk/res/android”
  android:orientation=”vertical”
  android:layout_width=”fill_parent”
  android:layout_height=”fill_parent”
  android:gravity=”center”
  android:background=”@android:color/white”>
  <TextView
    android:layout_width=”fill_parent”
    android:layout_height=”wrap_content”
    android:gravity=”center”
    android:text=”@string/hello”
    android:textSize=”25sp”
    android:textStyle=”bold”
    android:textColor=”@android:color/black”
    />
  <Button android:id=”@+id/next”
    android:layout_width=”wrap_content”
    android:layout_height=”wrap_content”
    android:textSize=”15sp”
    android:text=”@string/button”
    />
</LinearLayout>
```

The following code in Box 2 shows the contents of the file main.xml, one of the two screens of this application:

Strings and the R File

It is an XML file containing all text, such as button names, labels, and default text, used by the application. The purpose of using several XML files in an Android application is each XML file having its specific tasks. For example, a layout XML file is for user interface and a strings XML file is for their textual content. The code in Box 3 shows the file strings.xml used in this application:

The attribute value @string/hello, where string is the resource type and hello is the resource name, represents the string “Hello, World\n!,” where the symbol ‘\n’ is the character newline, stored in the file:

```
/res/values/strings.xml
```

You can use this syntax in an XML resource any place where a value is expected that you provide in a resource.

The directory gen in an Android project contains generated values. R.java is a generated class containing references to resources of the res folder in the project. These resources can be values, menus, layouts, or icons/pictures/animations. It is an automatically generated file so it should not

Box 3.

```
<?xml version=”1.0” encoding=”utf-8”?>
<resources>
  <string name=”hello”>Hello, World!\n</string>
  <string name=”button”>Next Page</string>
  <string name=”app_name”>Hello World</string>
</resources>
```

be modified. The code below in Box 4 shows the file R.java for this app:

The Android system provides methods to access the resources specified in the R file. Assume that two string variables hello and app_name are defined in the file res/values/strings.xml. They could be accessed anywhere in the Java code by using the following syntax:

```
[<package_name>.]R.<resource_
type>.<resource_name>
```

An example of the string accesses is R.string.hello where string is the resource type and hello is the resource name. The reference is "final static int" instead of String. There are many Android APIs that can access your resources when you provide a resource ID in this format. For example, the method getString(R.string.hello) returns a string "Hello, World!\n" with the reference ID R.string.hello.

Java Source Code

The Java code is what drives everything. It is this code that ultimately is converted to Dalvik executable and runs the application. This application includes two screens/activities/classes:

- HelloWorldActivity.java, which is the main page, and
- NextPageActivity.java, which is the other page.

These two screens are similar. Each of the screens displays text and a button. The button is used to activate the other activity/class, which displays the other screen. Notice that the class is based on the Activity class. An Activity is a single application entity that is used to perform actions. The onCreate method is called by the Android system when your Activity starts—it is where you should perform all initialization and UI setup. An

Box 4.

```
/* AUTO-GENERATED FILE. DO NOT MODIFY.
 *
 * This class was automatically generated by the
 * aapt tool from the resource data it found. It
 * should not be modified by hand.
 */
package com.example.HelloWorld;

public final class R {
  public static final class attr {
  }
  public static final class drawable {
    public static final int icon=0x7f020000;
  }
  public static final class id {
    public static final int home=0x7f050001;
    public static final int next=0x7f050000;
  }
  public static final class layout {
    public static final int main=0x7f030000;
    public static final int next=0x7f030001;
  }
  public static final class string {
    public static final int app_name=0x7f040002;
    public static final int button=0x7f040001;
    public static final int hello=0x7f040000;
  }
}
```

activity is not required to have a user interface, but usually does. The code below in Box 5 shows the file HelloWorldActivity.java:

DDMS (Dalvik Debug Monitor Server)

The ability of receiving locations is the fundamental requirement for all location-based applications. Android allows its emulator to receive a location (including longitude and latitude) from the DDMS (Dalvik Debug Monitor Server). The DDMS can be activated from the standard toolbar of Eclipse or by selecting the following options of Eclipse:

a. Select the Window menu.
b. Select the Open Perspective option.
c. Select the DDMS option.
d. Fill in the longitude and latitude values and click the Send button:

Figure 9 shows an application receiving and displaying the location information sent from the DDMS.

Embedded Database SQLite

A location-based service does not require an embedded database in the handheld device. A server-side geographical database provides the necessary information for LBS. For example, GeoNames (2012) is a geographical database available and accessible through various Web services. However, they also charge users for their services. An embeddable database is a convenient tool for small LBS projects. This section discusses the embedded database SQLite that comes with Android packages. SQLite (2012a) is an open source embeddable relational database management system, which is embedded within an application process without the overhead associated with a client-server configuration. Embedded databases are lightweight because they require little memory

Box 5.

```
package com.example.HelloWorld;

import android.app.Activity;
import android.content.Intent;
import android.os.Bundle;
import android.view.View;
import android.widget.Button;

public class HelloWorldActivity extends Activity {
 /** Called when the activity is first created. */
 @Override
 public void onCreate(Bundle savedInstanceState) {
   super.onCreate(savedInstanceState);
   setContentView(R.layout.main);

   final Button button = (Button) findViewById(R.id.next);
   button.setOnClickListener(new View.OnClickListener() {
     public void onClick(View v) {
       /** Here i calls a new screen. **/
       Intent i = new Intent(
         HelloWorldActivity.this, NextPageActivity.class);
       startActivity(i);
     }
   });
 }
}
```

Figure 9. (a) A screenshot of the DDMS and (b) the application receiving and displaying the location information sent from the DDMS

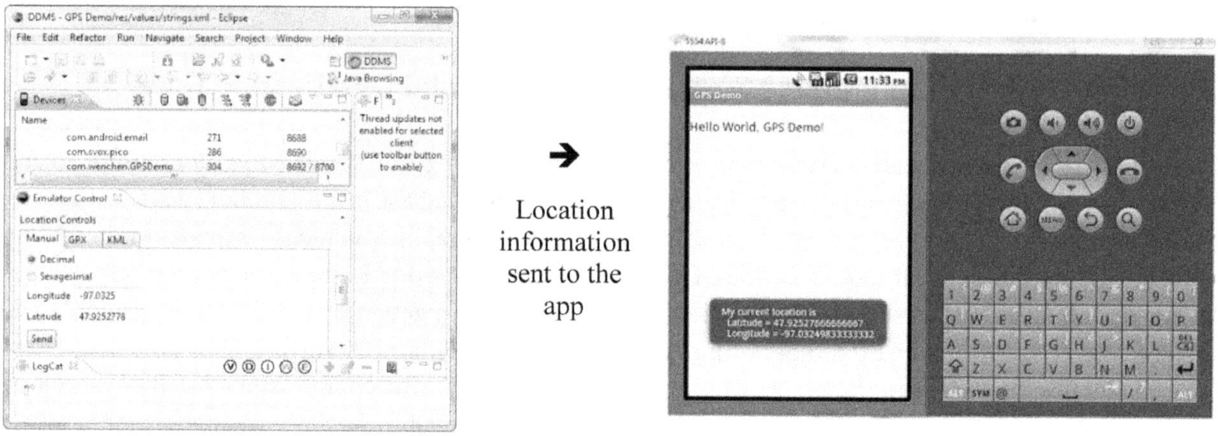

(a) (b)

during run time and are written in compact code. SQLite includes the following major features: (1) supporting most of the SQL-92 standard, (2) running on all major operating systems, (3) having support for the major computer languages, and (4) including multitasked read operations and sequential writes. The SQLite structure (2012b) consists of four components: (1) SQL Compiler, (2) Core, (3) Backend, and (4) accessories as shown in Figure 10. It can store data up to 2 TB with each database saved in a single disk using a B^+ tree data structure. The SQL statements are compiled into assembly code executed on the SQLite virtual machine, Virtual Database Engine (VDBE).

SQLite database is bundled with Android SDK (Software Development Kit). When the Android SDK is downloaded and installed, the SQLite database is ready to be used. Since no extra database setup and administration are needed, developers only have to define the SQL statements for defining and managing the database. SQLite supports standard relational database features like SQL syntax, transactions, and prepared statements

Figure 10. SQLite internal structure

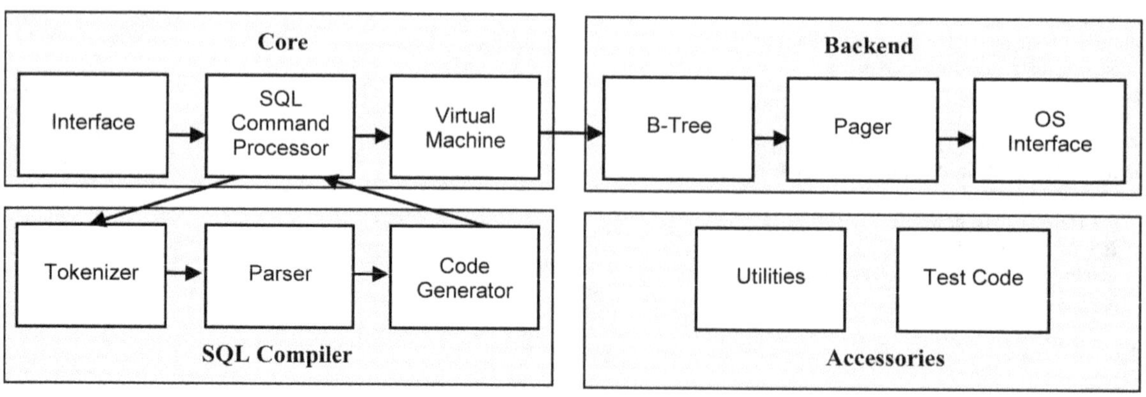

and the following data types: (1) NULL, a NULL value, (2) INTEGER, a signed integer, (3) REAL, a floating point value, (4) TEXT, a text string, and (5) BLOB, a binary large object. Compared to server-side databases, SQLite requires only little memory at runtime (about 250 KB). SQLite database is private to the Android application which creates it. In order to share data with other applications, the applications can use Content Provider, which manages access to a structured set of data. Content providers are the standard interface that connects data in one process with code running in another process. To query a content provider, a query string uses the following URI syntax:

```
<standard_
prefix>://<authority>/<data_
path>/<id>
```

For example, to retrieve all the personal information stored by the AddressBook application, the URI would look like this:

```
content://AddressBook/people
```

MAP TILE SYSTEM

A map is usually rendered from a number of map tiles for convenience and speed. A map tile system includes many tiles of different zoom levels. By using map tiles, a map can be panned and zoomed easily and quickly. For panning, some of the map tiles instead of the whole map are replaced. For zooming, different zoom-level tiles can be retrieved more efficiently because lengthy tile searches can be avoided if map tiles are used.

MAP REPRESENTATION

Several methods are used to find and display the correct maps. For example, Figure 11 shows a map using pixel coordinates (Microsoft, 2012). The pixel at the upper-left corner of the map has pixel coordinates (0, 0) and the pixel at the lower-right corner of the map has pixel coordinates (2047, 2047). One of the advantages of pixel coordinates is that the specific locations can be pinpointed quickly. At the same time, using an image to represent a world map of a specific zoom level is not realistic because the image size would be huge. Therefore, a map usually consists of a number of map tiles. For example, Figure 11 shows a world map consisting of 64 map tiles from $(0,0)_t$ to $(7,7)_t$ and each tile contains $(2048/8)\times(2048/8)=256\times256$ pixels. Given a pair of pixel $(x,y)_p$ coordinates, the tile coordinates $(a,b)_t$ containing that pixel is given by:

$$a = \lfloor x/256 \rfloor \text{ and } b = \lfloor y/256 \rfloor$$

On the other hand, most locations use latitude and longitude. Examples of how to convert latitude and longitude to tile numbers can be found from the CloudMade Web site (CloudMade, 2012).

PANNING AND ZOOMING USING MAP TILES

Traditional map panning and zooming is complicated and slow because they need an algorithm to find the correct map locations. Map panning and zooming are made easy by map tiles. Figure 12 shows an example of map tiles of three zoom levels. For panning and zooming, the following methods are used:

- *Zooming*: Zooming in the tile *x* will result in four tiles *x0*, *x1*, *x2*, and *x3* and zooming out will do the reverse. For example, zooming in the tile 32 in Figure 12 will bring in the tiles 320, 321, 322, and 323. On the other hand, zooming out the tiles 320, 321, 322, and 323 will result in the tile 32.

Figure 11. A world map consists of 64 map tiles

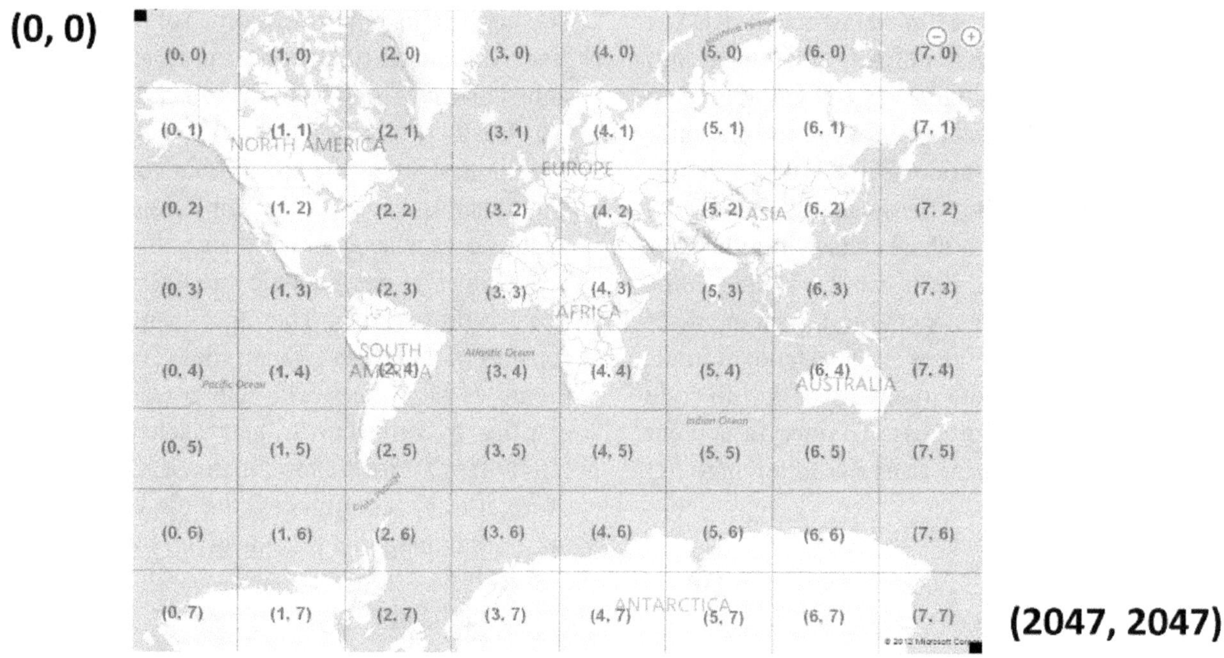

Figure 12. Map tiles of three zoom levels

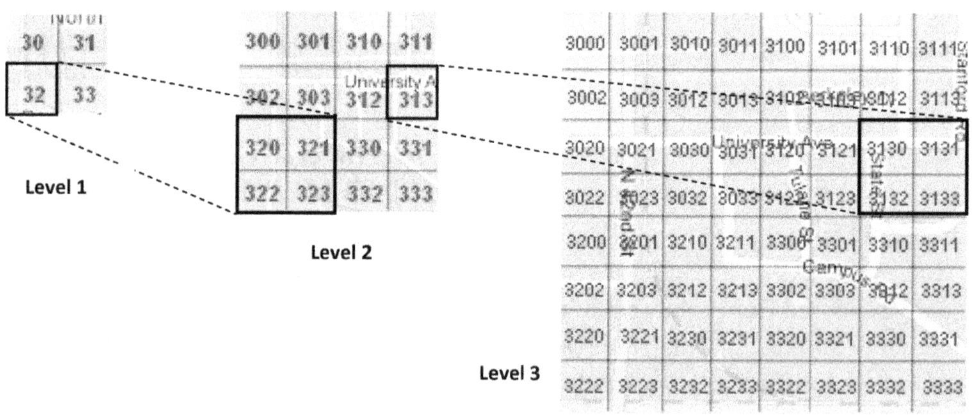

- *Panning*: There is no need to replace the whole maps; instead, only certain tiles are replaced. For example, panning the map including the tiles 320, 321, 322, 323 right may drop the tiles 320 and 322 and bring the tiles 330 and 332 in.

LOCATION-AWARE MAP PREFETCHING AND CACHING METHODS

This chapter studies map prefetching and caching based on human behavior recognition, which is to discover activity patterns from human activities. One of the human activity studies is travel route trajectory, the path that a human follows through space as a function of time and one of human behavior recognition studies. Human travel route trajectories can be used to facilitate map prefetching and caching.

Related Research of Human Travel Route Trajectories

Instead of waiting for the requests of downloading specific map tiles, map tiles are prefetched and cached based on human travel route trajectories. This section discusses several important location-based articles involving human travel sequences as follows.

Learning travel recommendations from user-generated GPS traces (Zheng & Xie, 2011): GPS-enabled devices allow people to record their location histories with GPS traces. The traces usually imply human behaviors and preferences related to travel. This research proposes a method to mine interesting locations and travel sequences. Three steps are employed:

- Model multiple individuals' location histories with a Tree-Based Hierarchical Graph (TBHG).
- Based on the TBHG, a HITS (Hypertext Induced Topic Search)-based inference model is constructed.
- Mine the travel sequences among locations considering the interests of these locations and users' travel experiences.

Finally, a large GPS dataset, collected by 107 users over a period of one year in the real world, is used to evaluate their system. The result shows their HITS-based inference model outperformed baseline approaches like rank-by-count and rank-by-frequency.

Smart itinerary recommendation based on user-generated GPS trajectories (Yoon, Zheng, Xie, & Woo, 2010): Previous routes can be used to recommend future travel patterns. This research proposes itinerary recommendation based on multiple user-generated GPS trajectories. Users only need to provide a start point, an end point, and travel duration to receive an itinerary recommendation. Liu (2009) presents a route recommendation system which guides the user through a series of locations. Their system uses the methods of sequential pattern mining to extract popular routes from a set of stored routes from previous users. It then recommends routes by matching the user's current route with the extracted routes.

Trajectory pattern mining (Giannotti, Nanni, Pedreschi, & Pinelli, 2007): The GPS is leading to the collection of large spatio-temporal datasets and to the opportunity of discovering usable knowledge about movement behavior. This chapter develops an extension of the sequential pattern-mining paradigm that analyzes the trajectories of moving objects. It introduces *trajectory patterns* as concise descriptions of frequent behaviors, in terms of both space and time. In this setting, they provide a general formal statement of the novel mining problem and then study several different instantiations of different complexity. The various approaches are then empirically evaluated over real data and synthetic benchmarks, comparing their strengths and weaknesses.

Location awareness through trajectory prediction (Liu & Karimi, 2006): Most existing location management methods record current location information periodically and passively. On the other hand, active strategies predict user movement through trajectories and locations. Trajectory prediction provides richer location and context information and facilitates the means for adapting to future locations. This chapter presents two models for trajectory prediction, namely *probability-based model* and *learning-based model*, both of which use long-term and short-term historical trajectories for prediction. The major difference between these two models is that the learning-based model characterizes trajectory features and adopts machine learning algorithms for prediction while the probability-based model adopts probabilistic information for prediction.

Discovering popular routes from trajectories (Chen, Shen, & Zhou, 2011): This research investigates the problem of discovering the *Most Popular Route* (MPR) between two locations. Travelers who are asking directions or planning a trip in an unfamiliar area is beneficial by this research, as past traveling experiences can show how people choose routes between locations. To achieve this goal, the authors develop a *Coherence Expanding* algorithm to retrieve a transfer network from raw trajectories, for indicating all the possible movements between locations. The Absorbing Markov Chain model is then applied to derive a reasonable *transfer probability* for each transfer node in the network, which is subsequently used as the popularity indicator in the search phase. Finally, they propose a *Maximum Probability Product* algorithm to discover the MPR from a transfer network based on the popularity indicators in a breadth-first manner.

Some related location-based research can be found from the articles (Duckham, Winter, & Robinson, 2010; Cao, Cong, & Jensen, 2010; Shang, Ding, Yuan, Xie, Zheng, & Kalnis, 2012; Gid'ofalvi & Pedersen, 2006).

SUMMARY

Map-related applications are one of the most popular mobile applications. Most of them require the support of GPS (Global Positioning Systems), which is a satellite navigation system providing location and time information. Smartphone users can subscribe a GPS service with a map. Some of the map-related applications are

- *Locations*: Pinpoint interesting locations such as the current locations, shopping malls, or friends.
- *Routes and distances*: Find various routes and distances between two locations.
- *Place recommendations*: Recommend interesting places such as movie theaters with specific cinemas, certain ethnic restaurants, and stores having products with the lowest prices.
- *Resource tracking*: For example, find the nearby taxis, locate the nearest truck with needed goods, and schedule fleets.

Map display and download consume a great deal of time and power from devices. Using map prefetching and caching could greatly save time and power of devices and therefore extend their length of operation time. Map prefetching and caching involve several subjects and developers are normally not familiar with all of them. This chapter studies four themes closely related to map prefetching and caching:

- *Green handheld computing*: This section introduces green computing using mobile handheld devices. It includes three topics: (1) green mobile computing, (2) mobile handheld devices, and (3) client-side handheld computing.
- *Location-based services and programming*: Map prefetching and caching are closely related to location-based services and programming. This section introduces location-based services and gives a detailed example of Android programming including using DDMS to send a mock location to an Android emulator. Location-based services require the support of databases and the embedded database SQLite is also introduced.
- *Map tile system*: A map is usually rendered from a number of map tiles for convenience and speed. A map tile system includes many tiles of different zoom levels. By using map tiles, a map can be panned and zoomed easily and quickly.
- *Map prefetching and caching methods*: Human travel route trajectories can be used to facilitate map prefetching and caching. Instead of waiting for the requests of

downloading specific map tiles, map tiles are prefetched and cached based on human travel route trajectories. This section discusses various methods of using human travel route trajectories for map prefetching and caching.

The four themes are the essential technologies for map-related application development, especially for mobile map prefetching and caching. Developers should be able to develop map-related applications smoothly after reading this chapter.

REFERENCES

Android. (2012). *Android developers*. Retrieved March 6, 2012, from http://developer.android.com/.

Cao, X., Cong, G., & Jensen, C. S. (2010). Mining significant semantic locations from GPS data. In *Proceedings of 36th International Conference on Very Large Data Bases*, (pp. 1009-1020). Singapore: IEEE.

Chen, Z., Shen, H. T., & Zhou, X. (2011). Discovering popular routes from trajectories. In *Proceedings of 2011 IEEE 27th International Conference on Data Engineering (ICDE)*. Hannover, Germany: IEEE Press.

CloudMade. (2012). *The CloudMade developer zone*. Retrieved March 21, 2012, from http://developers.cloudmade.com/.

Cohen, H. (2011). *Mobile marketing: 56 must have facts*. Retrieved December 9, 2011, from http://heidicohen.com/mobile-marketing-must-have-facts/.

Duckham, M., Winter, S., & Robinson, M. (2010). Including landmarks in routing instructions. *Journal of Location Based Services*, *4*(1), 28–52. doi:10.1080/17489721003785602

Eclipse. (2012) *About the eclipse foundation*. Retrieved March 12, 2012, from http://www.eclipse.org/org/.

Fleishman Hillard. (2012). *2012 digital influence study*. Retrieved from March 5, 2012, from http://www.factbrowser.com/facts/4671/.

Gartner. (2009a). *Gartner says consumer location-based services market will more than double in 2009*. Retrieved June 12, 2010, from http://www.gartner.com/it/page.jsp?id=1059812.

Gartner. (2009b). *Gartner identifies the top 10 consumer mobile applications for 2012*. Retrieved June 09, 2011, from http://www.gartner.com/it/page.jsp?id=1230413.

GeoNames. (2012). *About GeoNames*. Retrieved February 17, 2012, from http://www.geonames.org/about.html.

Giannotti, F., Nanni, M., Pedreschi, D., & Pinelli, F. (2007). Trajectory pattern mining. In *Proceedings of the 13th ACM SIGKDD International Conference on Knowledge Discovery and Data Mining*, (pp. 330-339). San Jose, CA: ACM Press.

Gid'ofalvi, G., & Pedersen, T. B. (2006). Mining long, sharable patterns in trajectories of moving objects. In *Proceedings of the Third Workshop on Spatio-Temporal Database Management (STDBM)*. Seoul, Korea: STDBM.

Handheld Computing Research. (2012). *Worldwide: Computer and device sales*. Retrieved March 2, 2012, from http://people.cs.und.edu/~wenchen/handheldresearch/facts/all.html.

Hu, W.-C., Wiggen, T., & Yang, H.-J. (2005). Mobile commerce systems and payment methods. In *Proceedings of the 2005 Information Resources Management Association International Conference (IRMA 2005)*, (pp. 769-771). San Diego, CA: IRMA.

JiWire. (2011). *JiWire mobile audience insights report Q2 2011*. Retrieved from February 18, 2012, from http://www.factbrowser.com/facts/3196/.

Kolodziej, K., & Hjelm, J. (2006). *Local positioning systems: LBS applications and services*. Boca Raton, FL: CRC Taylor & Francis.

Kupper, A. (2005). *Location-based services: Fundamentals and operation*. New York, NY: Wiley. doi:10.1002/0470092335

Liu, D. M. C. (2009). Recommend touring routes to travelers according to their sequential wandering behaviours. In *Proceedings of the 10th International Symposium on Pervasive Systems, Algorithms, and Networks (ISPAN 2009)*, (pp. 350-355). ISPAN.

Liu, X., & Karimi, H. A. (2006). Location awareness through trajectory prediction. *Computers, Environment and Urban Systems, 30*(6), 741–756. doi:10.1016/j.compenvurbsys.2006.02.007

Microsoft. (2012). *Bing maps tile system*. Retrieved December 17, 2011, from http://msdn.microsoft.com/en-us/library/bb259689.aspx.

OpenStreetMap. (2012). *CloudMade developers zone*. Retrieved January 12, 2012, from http://developers.cloudmade.com/projects.

Shang, S., Ding, R., Yuan, B., Xie, K., Zheng, K., & Kalnis, P. (2012). Users oriented trajectory search for trip recommendation. In *Proceedings of the 15th International Conference on Extending Database Technology (EDBT 2012)*. Berlin, Germany: EDBT.

SQLite. (2012a). *About SQLite*. Retrieved from March 5, 2012, from http://www.sqlite.org/about.html.

SQLite. (2012b). *The architecture of SQLite*. Retrieved January 12, 2012, from http://www.sqlite.org/arch.html.

Steiniger, S., Neun, M., & Edwardes, A. (2006). *Foundations of location-based services*. Retrieved January 13, 2012, from http://www.geo.unizh.ch/publications/cartouche/lbs_lecturenotes_steinigeretal2006.pdf.

Wang, S., Min, J., & Yi, B. K. (2008). Location based services for mobiles: technologies and standards. In *Proceedings of the IEEE International Conference on Communication (ICC)*. Beijing, China: IEEE Press.

Yoon, H., Zheng, Y., Xie, X., & Woo, W. (2010). Smart itinerary recommendation based on user-generated GPS trajectories. *Lecture Notes in Computer Science, 6406*, 19–34. doi:10.1007/978-3-642-16355-5_5

Zheng, Y., & Xie, X. (2011). Learning travel recommendations from user-generated GPS traces. *ACM Transactions on Intelligent Systems and Technology, 2*(1).

ADDITIONAL READING

Chen, Y., Jiang, K., Zheng, Y., Li, C., & Yu, N. (2009). Trajectory simplification method for location-based social networking services. In *Proceedings of the 2009 International Workshop on Location Based Social Networks (LBSN 2009)*, (pp. 33-40). LBSN.

Duffy, T. J. (2012). *Programming with mobile applications*. Boston, MA: Course Technology.

Hoisington, C. (2012). *Android boot camp for developers using Java*. Boston, MA: Course Technology.

Huang, L., Li, Q., & Yue, Y. (2010). Activity identification from GPS trajectories using spatial temporal POIs' attractiveness. In *Proceedings of the 2nd ACM SIGSPATIAL International Workshop on Location Based Social Networks (LBSN 2010)*, (pp. 27-30). San Jose, CA: ACM Press.

Liu, Q. (2009). *Web latency reduction with prefetching*, Doctoral Dissertation. Ontario, Canada: University of Western Ontario.

Lorenzo, D., Pinelli, G., Pereira, F., Biderman, F. C., Ratti, A., & Lee, C. (2009). An effective intelligent driving agent: Driver's trajectory and activities prediction. In *Proceedings of the IEEE 70th Vehicular Technology Conference (VTC 2009)*, (pp. 1-4). IEEE Press.

Yu, X. (2006). *Learning significant user locations with GPS and GSM.* Retrieved May 2, 2011, from http://www.google.com/url?sa=t&source=web&cd=7&sqi=2&ved=0CEMQFjAG&url=http% 3A%2F%2Fdspace.mit.edu%2Fbitstream%2Fhandle%2F1721.1%2F42121%2F227037360.pdf%3Fsequence%3D1&rct=j&q=pdf%3A%20ieee%20location%20travel%20sequence%20GPS&ei=v-IKTY3xJcuinwe-luSFDg&usg=AFQjCNFU5kJkt-Q07p4zb8428kg9eatngQ&cad=rja.

Yue, Y., Zhuang, Y., Li, Q., & Mao, Q. (2009). Mining time-dependent attractive areas and movement patterns from taxi trajectory data. In *Proceedings of the 17th International Conference on Geoinformatics*, (pp. 1-6). Fairfax, VA: Geoinformatics.

Zheng, Y., Liu, L., Wang, L., & Xie, X. (2008). Learning transportation mode from raw GPS data for geographic applications on the Web. In *Proceedings of the 17th International World Wide Web Conference (WWW 2008)*, (pp. 247-254). Beijing, China: WWW.

Zheng, Y., Zhang, L., Xie, X., & Ma, W.-Y. (2009). Mining interesting locations and travel sequences from GPS trajectories. In *Proceedings of the 18th International World Wide Web Conference (WWW 2009)*. Madrid, Spain: WWW.

Zheng, Y., Zhang, L., Xie, X., & Ma, W.-Y. (2009). Mining correlation between locations using human location history. In *Proceedings of the 17th ACM SIGSPATIAL International Conference on Advances in Geographic Information Systems*, (pp. 472-475). Seattle, WA: ACM Press.

KEY TERMS AND DEFINITIONS

Green Computing: It is computing is the practice of designing, manufacturing, using, and disposing of computers, servers, and associated subsystems for reducing the use of hazardous materials, maximizing energy efficiency, and promoting the recyclability or biodegradability of products.

GPS (Global Positioning System): It is a satellite navigation system, which provides location and time information. Smartphone users can subscribe a GPS service coming with maps.

Human Behavior Recognition: It is to discover activity patterns from human activities such as travel routes. The discovered patterns can be applied to many areas such as travel route recommendations.

Human Travel Route Trajectories: Each of the trajectories is the path that a human follows through space as a function of time.

Location-Based Services (LBS): A location-based service is a service based on the geographical position of a mobile handheld device. One of the LBS examples is to find a nearby ethnic restaurant by using a smartphone.

Map Tiles: A map is usually rendered from a number of map tiles, which in turn consist of more detailed tiles. By using map tiles, a map can be panned and zoomed more easily and quickly. For panning, some of the map tiles instead of the whole map are replaced. For zooming, more detailed or undetailed tiles can be retrieved more efficiently because lengthy tile searches can be avoided if map tiles are used.

Section 4

Computer Vision and Image Processing

Chapter 14
Green Video Compression for Portable and Low-Power Applications

Serdar Burak Solak
McGill University, Canada

Fabrice Labeau
McGill University, Canada

ABSTRACT

This chapter presents recent advances in the implementation of video compression that allow for energy efficiency and management. Algorithmic complexity reduction and complexity management are key to the implementation of video compression in a green computing environment. The authors concentrate on low-power applications (such as smartphones or autonomous cameras), in which the ability to manage algorithmic complexity (and thus energy consumption) to match battery conditions and the priority of other tasks is one of the key enablers for reduced energy consumption. First, a study of the state-of-the-art video coding standard H.264/AVC and an analysis of its encoder complexity will be presented. Secondly, low complexity H.264/AVC encoder implementations will be explored in two categories as: Low Complexity Motion Estimation Algorithms and Low Complexity Mode Decision Algorithms. Later, a discussion of Complexity Scalable Encoding Algorithms that can adaptively adjust their computational complexities will follow. During the discussions, the authors will also introduce a novel framework for managing the complexity of an H.264/AVC encoder in a processor or power constrained environment as well as a complexity reduction tool. The chapter will conclude with a discussion about the future of sustainable green computing in video compression, followed by summary and concluding remarks.

INTRODUCTION

Digital video technology has been recently finding its way in many portable and wireless applications. Advances in video compression and wireless communications have made it possible to transmit video streams over wireless networks. It is now common to record and watch video on portable devices such as cell phones, and wireless surveillance camera networks are being deployed.

Although video de-compression is a rather easy task, video compression remains a complex and power-hungry endeavor. In order to reduce the

DOI: 10.4018/978-1-4666-1839-8.ch014

energy footprint of such ubiquitous applications (and hence their carbon footprint), it is necessary to harness and manage the complexity involved in the video compression process.

This chapter aims to describe algorithmic ways to control and manage the complexity of digital video encoding, in order to allow integration of video compression in an overall green computing design, which can help prolong battery life and/or allow running more applications concurrently on low-power processors with limited computational capabilities.

Over the years, video compression algorithms have improved in their compression efficiency. H.264/AVC is the state-of-the-art international video coding standard developed by the *Joint Video Team* (JVT) (Wiegand, Sullivan, Bjontegaard, & Luthra, 2003). JVT is a joint workforce established by two study groups, ITU-T Video Coding Experts Group (VCEG) and ISO/IEC Moving Pictures Experts Group (MPEG), with the vision of finalizing the new video coding standards. H.264/AVC adopts highly advanced and flexible encoding techniques to achieve high compression efficiency which significantly surpass previous standards. The high performance of the new standard renders it the prevalent compression algorithm in many applications and platforms. However, the encoding techniques employed in the standard come with a price of increased computational complexity. Although designs that are capable of performing such complex functions are realizable with the improved computational power of the IC chips, the increased computational complexity poses a great problem particularly for the battery operated portable devices with limited processing capabilities such as mobile phones and PDAs. In fact, it has been shown that the processing of even a QCIF (176×144) video for communication over a wireless network consumes 2/3 of the total power which exceeds by far the transmission in terms of power consumption (Agrawal, Chen, Kishore, Ramanathan, & Sivalingam, 1998).

An essential green computing objective is to improve energy efficiency while maintain a good quality of service. In the case of video compression, this involves hunting down the causes of complexity (and hence energy inefficiencies) in a compression algorithm, and finding ways to avoid them, but beyond this, it also means being able to maintain service (i.e. video compression), in cases where the battery life of the device might prevent the algorithm from running at full performance, which means that a complexity-adaptive scheme is necessary. Therefore, the design of computational complexity control algorithms for video coding has become a highly active research field. A complexity control structure should comprise of two distinct components, a low complexity component, and a complexity scalable component that can adapt the computational complexity of the system and operate under variable resource constraints. In the recent years, different techniques have been developed which are either low complexity or complexity scalable but only a few that joins the two notions together.

The purpose of this chapter is four-fold. We will first describe the computational complexity incurred by individual components of the H.264/AVC encoding process. Then, we will investigate the existing energy efficient encoding algorithms developed to operate at low and variable complexity levels. Subsequently, we will introduce a novel framework for managing adaptively the complexity of an H.264/AVC encoder. Lastly, we will consider the sustainability of green computing in the future directions of video compression.

The following section constitutes a short review of video compression principles as they are applied in a recent standard (namely H.264 AVC), to allow readers not familiar with the video compression domain to fruitfully read further sections. It is mainly inspired by review papers on the subject (such as Wiegand, Sullivan, Bjontegaard, & Luthra, 2003), and is by no means an exhaustive review of the topic.

A SHORT REVIEW OF VIDEO COMPRESSION

Video technology has significantly evolved in the past two decades. With digital video growing more popular, the immense volume of data present in video renders its storage and transmission very expensive as bandwidth is an invaluable commodity in today's world. However, video contains copious amounts of redundancy, which can be grouped into two types:

- Temporal redundancy
- Spatial redundancy

A video is in fact a group of consecutively captured pictures (frames). Each frame exhibits spatial redundancy to a certain extent. Within a frame, there are typically some highly detailed regions but also some homogeneous regions representing similar content. In addition to spatial redundancy, temporally consecutive frames tend to be highly correlated and usually minor changes occur between two pictures (unless there is a scene change). Video compression algorithms operate on the premise of removing both the temporal and the spatial redundancies while maintaining an acceptable level of visual quality; and compress a video sequence into a bitstream for condensed storage and efficient transmission (Richardson, 2003).

Since the early 1990s, the major video coding standards have evolved around the same hybrid encoding structure, which incorporates a block-based predictive coding stage that removes the temporal redundancies and a transform-domain quantization stage that removes the spatial redundancies. This structure is referred to as the hybrid structure, due to its temporal/spatial duality. Similar to its predecessors, H.264/AVC is also based on the hybrid video encoding framework and conforms to the hybrid encoder structure. The block diagram of an H.264/AVC encoder is illustrated in Figure 1. As seen in the figure, the encoder consists of two distinct data flow paths. The forward data path is the encoding process of a macroblock MB), a basic unit of video which comprises an area of 16×16 pixels. A prediction of the macroblock is constructed either using an INTER

Figure 1. Simplified block diagram of a typical H.264 encoder

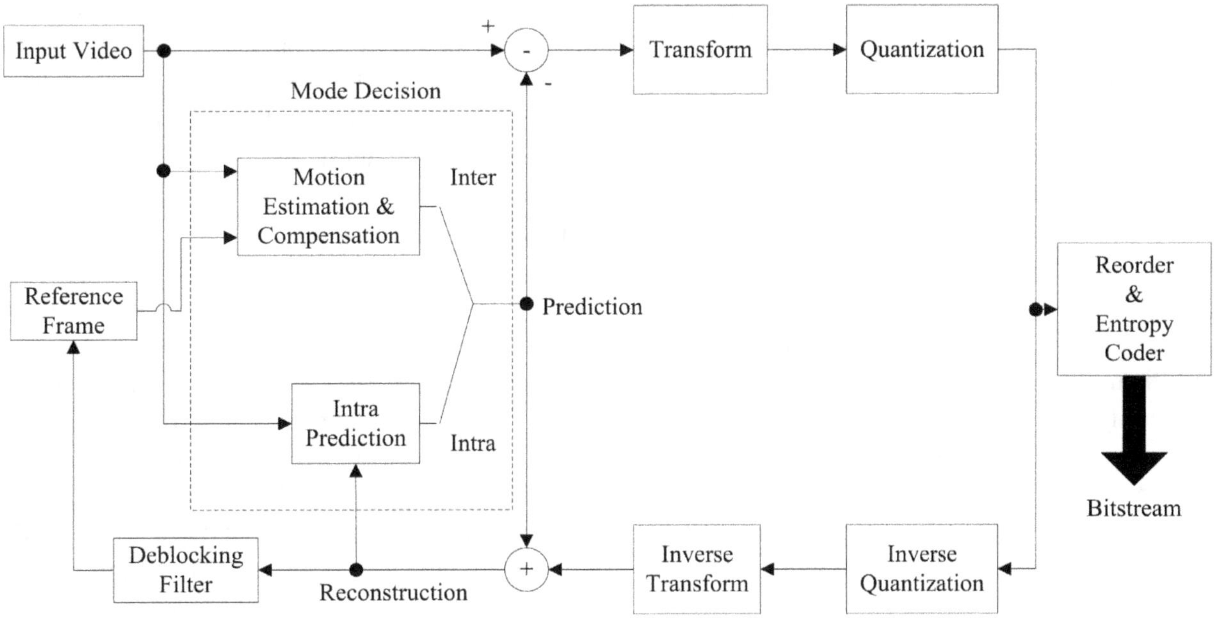

prediction through motion estimation and motion compensation or using an INTRA prediction, as detailed below. The decision of INTER or INTRA prediction is identified by the *Mode Decision* block. The *Motion Estimation and Compensation* block and the *Residue Subtraction* are responsible for removing temporal redundancies. The *INTRA Prediction* block together with the *Transform* and the *Quantization* blocks are responsible for removing spatial redundancies. The *Reorder and Entropy Coder* block is responsible for removing any remaining statistical redundancies in the data. Each of these blocks will be explained thoroughly in the following subsections. The backward data path is the reconstruction process of the encoded data. After a frame is encoded, it is also decoded and then reconstructed again by the encoder. These reconstructed frames are stored and used in the prediction of the proceeding frames as references. In hybrid encoders, the backward data path is crucial as it guarantees that both the encoder and the decoder use the same reconstructed frame for prediction.

Mode Decision

The encoder processes each input frame in units of macroblocks and constructs a prediction for each macroblock based on previously coded data. The previously coded data can come either from the current frame or from other frames. A macroblock that is predicted from one or more MBs in other frames is referred to as an INTER predicted MB. A macroblock that is predicted from one or more macroblocks in the current frame is referred to as an INTRA predicted MB. H.264/AVC standard offers more INTER and INTRA prediction methods than any previous standard. A 'coding mode' or simply a 'mode' is a synonym for a prediction method. The abundance of modes enables the encoder to construct a more accurate prediction, which yields significantly improved compression efficiency compared to previous standards.

INTRA Prediction

If a frame contains new information, e.g. a new scene that was not present in the previous frame, the temporal correlation becomes minimal. In such scenarios predicting from the spatially neighboring pixels instead of the temporally neighboring pixels is more desirable because the probability of having a spatially correlated region as opposed to a temporally correlated one is higher. Hence, the encoder chooses INTRA prediction over INTER prediction.

H.264/AVC encoder can form one single INTRA prediction for the entire MB using one of the four INTRA 16×16 coding modes (I_16×16). Alternatively, the encoder can form a separate INTRA prediction for each 4×4 partitions of the MB using one of nine INTRA 4×4 coding modes (I_4×4) (Wiegand, Sullivan, Bjontegaard, & Luthra, 2003). The latter is more suitable for coding of regions with significant detail and the former is more suitable for coding of smoother and more homogeneous regions with less detail. Depending on which of the thirteen coding modes is chosen, the INTRA prediction is constructed by either averaging the samples of the previously encoded and decoded adjacent MBs that are positioned on the left, on the top and on the top-right, or by copying the samples of one of these three adjacent MBs. Therefore, the standard requires these three respectively positioned MBs to be encoded prior to encoding a macroblock.

INTER Prediction

Video coding techniques achieve most of the compression by exploiting temporal redundancies and constructing INTER-frame predictions, i.e. predicting the current frame from previously encoded frames. H.264/AVC offers a vast number of coding options for INTER prediction. Each macroblock can be predicted from two lists of already encoded frames (Wiegand, Sullivan, Bjontegaard, & Luthra, 2003). These lists can contain temporally preceding frames or temporally proceeding frames, as long as these frames have already been coded; this is possible specifically for non-real time applications in which the coding order of frames may differ from their display order. Depending on the reference list used, INTER macroblocks are categorized in two groups. A P-macroblock (P*redicted*) can use only references from the first list whereas a B-macroblock (B*i-predicted*) can use references from both lists (Wiegand, Sullivan, Bjontegaard, & Luthra, 2003). In INTER prediction, the prediction of a MB is constructed upon searching for the best matching block of pixels in the reference frame and compensating for the motion between the MB and the matching block. Similar to INTRA prediction, H.264/AVC provides different INTER coding modes where each coding mode corresponds to a specific partitioning of the macroblock for motion search (estimation) and compensation. The available MB partitions in H.264/AVC are one 16 × 16 block, two 16 × 8 blocks, two 8 × 16 blocks, and four 8 × 8 blocks, i.e. P_8 × 8 mode. If P_8 × 8 is selected, the corresponding 8 × 8 blocks can be further divided into two 8×4 blocks, two 4×8 blocks or four 4×4 blocks (Wiegand, Sullivan, Bjontegaard, & Luthra, 2003).

The plethora of INTER coding modes (each with different MB partition sizes) supported by H.264/AVC complicates the macroblock partition decision process.

In H.264/AVC encoder implementations, a Lagrangian Rate-Distortion-Optimized (RDO) method is commonly used to select between the different INTER modes (Ortega & Ramchandran, 1998). Per this method, for each coding mode, the encoder searches an area in the reference frame to find a best matching block to each partition. It compares an $M \times N$ block with the $M \times N$ blocks in the reference region and picks the one that minimizes an optimized Lagrangian cost function of distortion and rate (Sullivan & Wiegand, 1998). Distortion is a quantification of the difference between the current $M \times N$ block and the candidate region, and rate is the number of bits needed to represent the motion vector, which holds the offset between the current block and the candidate region. The equation is as follows (Sullivan & Wiegand, 1998):

$$J_{i,\text{motion}}\left(MV\right) = D_{i,\text{motion}}\left(MV\right) + \lambda_{\text{motion}} R_{i,\text{motion}}\left(MV\right) \quad (1)$$

where $J_{i,\text{motion}}\left(MV\right)$ is the Lagrangian motion cost for block number i with candidate motion vector MV. $D_{i,\text{motion}}\left(MV\right)$ is either the Sum of Absolute Differences (SAD) or Sum of Squared Differences (SSD) (depending on the configuration options) between the current block and the candidate prediction block and $R_{i,\text{motion}}\left(MV\right)$ is the bitrate representing motion information and is calculated through a look-up table. λ_{motion} is the Lagrangian multiplier and a function of the Quantization Parameter (QP). QP is a scaling index which controls the quantization step size and adjusts the visual quality. The Lagrangian multiplier is computed as follows (Sullivan & Wiegand, 1998):

$$\lambda_{\text{motion}} = \sqrt{0.85 \times 2^{(QP-12)/3}} \quad (2)$$

For instance, for P_16×16 mode the MB is not partitioned into smaller blocks; therefore, the 16 × 16 block in the reference frame that minimizes $J_{i,\text{motion}}\left(MV\right)$ and the corresponding motion vector representing the offset are used to construct the prediction. The search for the best matching block is referred to as 'motion estimation' (Richardson, 2003), and consists in finding for a given block i motion vector that corresponds to:

$$\arg\min_{MV} J_{i,\text{motion}}\left(MV\right). \quad (3)$$

For P_16×8 mode the MB is partitioned into two 16 × 8 blocks hence motion estimation is performed for each block separately. Once the best matching blocks for a MB are found, the motion between the current blocks and the reference blocks are compensated to form the prediction, which is subtracted from the original MB and the residual, is obtained. This process is referred to as 'motion compensation' (Richardson, 2003).

H.264 supports sub-pixel motion estimation and compensation, which improves the prediction accuracy. The reference block is actually constructed by creating new pixels through interpolation inside a block of real pixels. Half-pixel and quarter-pixel accuracies are supported by the standard. For instance, in quarter-pixel accuracy, motion estimation happens in three stages. In the first stage, the best matching block is found on the integer sample pixels. In the second stage, the half-pixel positions neighboring the best match are searched for improvement. If a better matching block is found through half-pixels, then in the third stage, the quarter-pixels are searched for further improvement.

In addition to the motion-compensated modes, H.264 offers another mode known as the SKIP mode (Wiegand, Sullivan, Bjontegaard, & Luthra, 2003). For this coding mode, neither any motion estimation is done nor is any data transmitted. Simply a motion vector is predicted by averaging the motion vectors of the adjacent MBs and the vector points to the actual reconstruction. This mode is used to predict large areas with no change or constant motion such as the background.

Rate Distortion Optimized (RDO) Mode Decision

Typically, an H.264 encoder offers a total of 21 coding modes for a macroblock (Wiegand, Sullivan, Bjontegaard, & Luthra, 2003):

- Nine INTRA 4 × 4 modes
- Four INTRA 16 × 16 modes
- Seven INTER modes with different partition sizes
- One SKIP mode

The encoder is expected to choose one of these coding modes for each macroblock such that for a given bit-rate constraint the visual quality is maximized. To accomplish this task, the encoder needs a reliable mode decision method. Similar to the INTER mode selection process; H.264/AVC encoder implementations typically employ a so-called *Rate-Distortion-Optimized Mode Decision* process (Ortega & Ramchandran, 1998). For each coding mode, the process forms a prediction of the MB and computes the residual which is then frequency transformed and quantized. Then, the MB is decoded and reconstructed: The quantized transform coefficients are inverse quantized and inverse transformed to obtain the residual block, which is then added to the prediction block to reconstruct the MB. Using the reconstruction and the transform coefficients, the following Lagrangian cost function is computed (Sullivan & Wiegand, 1998):

$$J_{i,\text{mode}}(M) = D_{i,\text{mode}}(M) + \lambda_{\text{mode}} R_{i,\text{mode}}(M) \tag{4}$$

The above equation is similar to the one used in motion estimation yet there are several important differences. $D_{i,\text{mode}}(M)$ is the distortion representing the difference between the original MB i and its reconstruction (typically SSD), if mode M is used. $R_{i,\text{mode}}(M)$ is the bit-rate calculated from the output of the entropy coder which represents the total number of bits required to hold the residual MB and the accompanying motion information in the bitstream. The Lagrangian multiplier λ_{mode} is again a function of the QP, yet different:

$$\lambda_{\text{mode}} = 0.85 \cdot 2^{(QP-12)/3} \tag{5}$$

The Lagrangian optimization used in RDO mode decision provides a method of minimizing distortion subject to a rate constraint which coincides well with the goal of a video compression technique. Since the process is repeated for every possible coding mode, it guarantees that the optimal mode will be picked, as the mode corresponding to:

$$\arg\min_{M} J_{i,\text{mode}}\left(M\right). \tag{6}$$

COMPLEXITY MANAGEMENT OF H.264/AVC ENCODER

As explained in the previous section, video compression algorithms are comprised of a series of highly complex encoding tools including motion estimation/compensation, transform/quantization, and reorder/entropy coder. Prior to H.264/AVC, the video coding standards had more or less an equal complexity distribution over their employed coding tools (Richardson & Zhao, 2002). However, due to the increased number of supported coding modes in H.264/AVC, motion estimation, and mode selection processes are more computationally intensive in comparison to transform/quantization and reorder/entropy coder. Together the processes may consume up to 90 to 95 percent of the total encoding time. Hence, complexity management research for H.264/AVC has concentrated on developing:

- Low complexity motion estimation algorithms
- Low complexity mode decision algorithms
- Complexity scalable encoding algorithms

Complexity Analysis of H.264/AVC Video Encoder

The increased number of available coding modes in H.264/AVC demands a more discriminating mode decision operation to be performed by the encoder such that all the coding modes are carefully analyzed to guarantee the optimal decision. The H.264/AVC encoder implementations typically utilize a Rate-Distortion-Optimized (RDO) mode decision that ensures the optimal choice of mode for each macroblock. For each coding mode, the RDO mode decision carries out a number of operations, which examine both the rate and the distortion effects of coding the macroblock with the mode, and it calculates a joint rate-distortion penalty. From this point on, we will refer to this critical RD analysis of a mode as a "mode trial." Each mode trial requires a number of complex operations to be performed; including motion estimation/compensation for INTER modes, transform/quantization and reorder/entropy coding. Although the overall complexity of the RDO mode decision is large, the contributions of different mode trials to this complexity are not equal. To categorize the complexity of each mode, we profile the mode decision process. The time spent evaluating each mode and the outcome of the mode decision process for each macroblock are recorded for a total of fifteen test video sequences and twenty quantization parameter values varied between 8 to 48 with a step size of 2.

In Table 1, the average processing times for each mode as well as the number of macroblocks coded with each mode are presented. The RDO mode decision process is used. It is seen that the INTER modes are the major complexity contributors, consuming together over 95 percent of the total time spent in the mode decision process. The highly complex nature of INTER modes is due to the exhaustive motion estimation operation performed for these modes. INTRA modes together consume just under three percent of the total time and SKIP mode's consumption is negligible with less than one percent. Similarly, more than 60 percent of the macroblocks are coded in one of the seven INTER modes and more than 35 percent of the macroblocks are coded in SKIP mode. INTRA modes are preferred for only two

Table 1. Percentages of computational time spent evaluating each mode and number of macroblocks coded in each mode

<table>
<tr><th>Modes</th><th>Mode Exploration Time (%)</th><th>Number of MBs Coded (%)</th></tr>
<tr><td>SKIP</td><td>0.12</td><td>35.64</td></tr>
<tr><td>INTRA</td><td>2.79</td><td>1.41</td></tr>
<tr><td>INTER 16×16</td><td>10.60</td><td>13.84</td></tr>
<tr><td>INTER 16×8</td><td>11.00</td><td>5.65</td></tr>
<tr><td>INTER 8×16</td><td>11.83</td><td>6.81</td></tr>
<tr><td>INTER 8×8</td><td>12.69</td><td rowspan="4">36.56</td></tr>
<tr><td>INTER 8×4</td><td>14.69</td></tr>
<tr><td>INTER 4×8</td><td>15.97</td></tr>
<tr><td>INTER 4×4</td><td>19.82</td></tr>
</table>

percent of the macroblocks. The prevalence of SKIP and INTER modes over INTRA modes is due to the high temporal correlation between subsequent video frames.

In fact, the negligible complexity of the SKIP mode trial is in striking contrast to its occurrence. For the 35 percent of the macroblocks that are eventually coded in SKIP mode, the RDO mode decision has to evaluate the remaining INTRA and INTER modes to conclude that the SKIP mode is in fact the optimal mode. Our conclusion from this observation is that the encoder complexity could be significantly reduced if the SKIP coded macroblocks or, even better, the approximate final RD costs of the macroblocks were known prior to mode decision. The potential benefits of early detection of SKIP coded macroblocks have also been noted by other researches in the past (Kannangara, et al., 2006; Jeon & Lee, 2003). In the pursuit of this inference, we have designed a novel RD cost prediction method which predicts the RD cost of a macroblock by exploiting the readily available information from the previous frame prior to any mode trials.

Low Complexity Motion Estimation Algorithms

For motion estimation, the *Full Search Algorithm* (FSA) is the most straightforward search method for block matching. It examines all the pixel positions and finds the motion vector pointing to the best matching block. Although the optimal solution to motion estimation is guaranteed, FSA can be computationally intensive as the search range increases. For instance, if the search range for block matching in the reference picture is a square with w pixels in size for each direction and with a step size of one pixel, there exist a total of $(2w+1)^2$ possible motion vectors. As this operation is repeated for every block in a sequence, the total number of motion search operations is very large. In order to relieve the encoder of the computational burden of FSA, many fast motion search algorithms have been proposed to decrease the computational complexity of the motion estimation operation by reducing the number of candidate motion vectors and eliminating the motion vectors that are likely to yield a large residue. Earlier examples of such algorithms are the two-dimensional logarithmic search (Ghan-

Table 2. Comparison of fast motion search algorithms with the full search algorithm

Motion Search Algorithm	Number of candidate motion vectors	
	Minimum	Maximum
Full Search *(with search window size of 7)*	255	255
Two-dimensional logarithmic *(with search step size of 2)*	13	26
Three-step search *(with search window size of 7)*	25	25

bari, 1990) and the three-step search (Koga, Iinuma, Hirano, Iijima, & Ishiguro, 1981).

Two-dimensional logarithmic search is one of the first fast motion estimation algorithms. For a block with coordinates (x, y), the search starts at the zero-displacement location and includes the following four pixel locations: S pixels north, S pixels south, S pixels east, and S pixels west of (x, y), where S is the search step size. For each iteration, the four neighboring locations are examined as well as the central location; and if the central pixel yields the least residue, then the search step, S, is halved for the next iteration; otherwise the search step size remains the same. When the search step is reduced to one, this operation is repeated for the last time and the vectors indicating the offset between the original macroblock and the best matching block are stored as the motion vectors. In three-step search, four extra pixel locations are examined additional to the five locations examined in the two-dimensional logarithmic search. The four pixels mark the corners of a square with a side of $2S$ pixels and with its center as the zero-displacement pixel (x, y). The initial search step S depends on the motion estimation search window size. For instance, if the search window size is $2^N - 1$, then the initial search step size is set to 2^{N-1}; and after each iteration, the search step is halved until it is equal to one.

Both algorithms reduce the complexity of motion estimation by reducing the number of candidate motion vectors (MVs) examined, i.e. the number of times equation (1) needs to be evaluated. In Table 2, the maximum and minimum number of candidate MVs required by the algorithms are compared with the full search algorithm. The number of candidates for three-step search is constant while the number varies for two-dimensional logarithmic search. This is due to the fact that for the latter, the search step size is halved after an iteration if and only if the central pixel location yields the smallest residue. Whereas for the former, the search step size is halved regardless. However, it is evident that either algorithm significantly reduces the computations required by the full search algorithm.

Fast motion estimation algorithms have been developed specifically for H.264/AVC. In fact, *The Hybrid Unsymmetrical-Cross Multi-Hexagon-Grid Search* (UMHexagonS) (Chen, Zhou, & He, 2002) is adopted in a H.264/AVC reference test model software (Lim, Sullivan, & Wiegand, 2004) developed by JVT. UMHexagonS exploits previous motion information such as the motion vectors of the neighboring blocks, motion vectors for the INTER coding modes with larger partitions than the current coding mode and the motion vectors of the co-located block in the previous frame for current coding mode. Using such information, UMHexagonS predicts an initial minimum cost integer-pixel motion vector and a sub-pixel motion vector. An integer-pixel motion search is performed around the integer-pixel motion vector. Different search shapes (unsymmetrical cross, local square full search, and a hexagon based search) are used in order to prevent stalling in a local minimum. After the integer-pixel motion search, a sub-pixel motion search is performed around the sub-pixel motion vector. Results show that UMHexagonS saves from 41 to 67 percent of the total encoding time compared with FSA while achieving the same rate-distortion performance. Yi, Zhang, Ling, and Shang (2005) extend UMHexagonS by using a simpler motion vector prediction method (excluding the motion information from the co-located block in the previous frame) and by replacing the local square full search with a hexagonal search. The authors claim further 40 to 60 percent encoding time savings.

Low Complexity Mode Decision Algorithms

In addition to the low complexity motion estimation algorithms, there has been a great amount of research work directed into developing fast macroblock mode decision algorithms. Such algorithms aim to alleviate the computationally intensive mode decision process. H.264/AVC encoders in particular need to perform a computationally expensive mode selection operation because the new standard supports a larger number of macroblock coding modes than any previous standard.

Fast mode decision algorithms for both INTRA mode decision (Wu, et al., 2003) and INTER mode decision (Lim, et al., 2003) have been proposed by the same authors and have been incorporated in the H.264/AVC reference test model software (Lim, Sullivan, & Wiegand, 2004). Both methods exploit the local characteristics of the current block. Specifically, using the local edge directional information obtained from the edge direction histogram along the INTRA prediction directions, only the more probable INTRA modes are favored. Results presented by the authors indicate average coding time savings between 20 and 30 percent while the impacts on rate and distortion are less than two percent increase and 0.1dB decrease in Peak Signal to Noise Ratio (PSNR) respectively (Wu, et al., 2003). The fast INTER mode decision algorithm works similar to its INTRA counterpart. It uses the same edge detection technique to recognize the homogeneous regions and the difference between the current block and the co-located block in the reference frame to recognize the stationary regions. Based on these analysis, the algorithm zeros in on the more probable INTER or motion estimation, predicting block sizes. For instance, if a region is detected to be both homogeneous and stationary, then the P 16×16 mode is chosen without any further mode evaluation. From the experimental results, the fast INTER mode decision is shown to reduce the encoding time by 30 percent on average while the impacts on rate and distortion are again less than two percent increase and 0.1dB decrease respectively (Lim, et al., 2003). Jeon and Lee (2003) propose a selective INTRA mode decision algorithm where they put forward the hypothesis that an INTRA prediction mode is likely to be chosen for a block if the block is more correlated with its previously encoded spatial neighbors than with its previously encoded temporal neighbors. Following this hypothesis, their proposed method restricts the INTRA prediction mode evaluation to only the blocks conforming to the hypothesis. The results indicate on average 30 percent encoding time saving with minimal loss in rate-distortion performance.

A common approach taken to reduce mode decision time is to utilize an early SKIP mode detection algorithm, i.e. a mode trials termination algorithm. An early SKIP mode detection algorithm aims to detect the macroblocks that are likely to be coded in the SKIP mode and to avoid the evaluation of other modes. Since the evaluation of SKIP mode is computationally simple, it would only be wise not to evaluate the INTER or the INTRA prediction modes for such MBs. This is possible only if these MBs could be identified a priori. Jeon and Lee propose an early SKIP detection where a macroblock is coded in the SKIP mode only if the P_16×16 mode yields both a set of all zero transform coefficients and a motion vector pair that is equal to the predicted motion vector (Jeon & Lee, 2003). Their algorithm is also adopted in the H.264/AVC reference test model software (Lim, Sullivan, & Wiegand,

2004). Its performance depends on the motion characteristics of the input sequence and on the quantization parameter. The saved encoding time is greater for the slow moving sequences and for large quantization parameters. The computational savings are less than 50 percent with negligible impact on rate-distortion performance.

Rate-Distortion (RD) cost estimation is another tool that is used to define benchmarks during the mode decision process. For instance, Kannangara et al. (2006) developed a RD cost estimation method for a macroblock using the distortion and the rate values of the co-located block in the reference frame. This estimated RD cost is used in their proposed early SKIP detection algorithm such that the MBs for which the estimated RD cost is less than the SKIP mode RD cost, are coded in SKIP mode without any additional mode evaluation. The method achieves on average 60 percent encoding time savings surpassing the early SKIP detection method by Jeon and Lee (2003). However, the rate distortion performance is significantly inferior as the rate-distortion-optimized mode decision is disabled. Liao, Yang, and Sun propose another RD cost estimation for both INTER and INTRA prediction modes by exploiting the statistical information from the transform coefficients. However the encoding time savings are only after a residual block is integer transformed; hence, computational time spent in neither the motion estimation nor the mode decision processes is reduced.

More recently, we have introduced a new RD cost estimation algorithm which we used to predict the macroblocks that are likely to be SKIP coded, forming a new early SKIP termination (EST) algorithm (Solak & Labeau, 2010). The final Rate-Distortion (RD) costs are estimated through the distortion and the rate values of the corresponding co-located macroblock in the previous frame. The distortion and the rate values are scaled by two distinct prediction coefficients α_d and α_r which are functions of the Quantization Parameter (QP). The prediction function and the prediction coefficients are respectively formulated below:

$$\hat{J}_i^n = \alpha_d \cdot D_i^{n-1} + \alpha_r \cdot \lambda \cdot R_i^{n-1} \tag{7}$$

$$\alpha_d = -0.0001 \cdot QP^2 + 0.0069 \cdot QP + 0.9809 \tag{8}$$

$$\alpha_r = -0.0002 \cdot QP^2 - 0.0025 \cdot QP + 1.0164, \tag{9}$$

where $\hat{J}_i^n$ is the predicted RD cost for macroblock X_i^n in frame n, D_i^{n-1} and R_i^{n-1} are the distortion and the rate values of the co-located macroblock X_i^{n-1} in frame $n-1$. The scaling factors for distortion and rate are α_d and α_r respectively. For a macroblock, the EST algorithm first evaluates the SKIP mode and compares the predicted RD cost with the RD cost of SKIP mode. If SKIP has a cost less than or equal to the predicted cost, then the mode decision process is terminated and the macroblock is skipped.

Table 3 records the percentages of the encoding time saved by our EST algorithm and by the early SKIP termination method proposed by Jeon and Lee. It must be noted that in this case execution time has been chosen as a metric of performance; it is clear that these improvement depend somewhat on the particular implementaation and processor architecture used. Power consumption figures would depend even more on the partiuclar implementation, and would therefore render these results less general. Our analysis of early SKIP termination algorithms with different sequences highlights several critical points. While such algorithms achieve significantly high complexity savings for slower sequences, they fall short of performing at the same level for faster sequences as the number of skipped macroblocks in such sequences is insufficient for the algorithm to perform at a desired level. Motivated by providing higher computational savings for faster

Table 3. Comparison of the saved encoding time percentages by the EST algorithm proposed by Solak and Labeau (2010) and by the JM EST proposed by Jeon and Lee (2003), for different sequences and different values of the quantization parameter QP

Sequence	Algorithm	QP=16	QP=20	QP=24	QP=28	QP=32
Akiyo	EST	65.42	69.67	74.39	78.57	82.62
	JM EST	29.76	45.34	33.78	39.29	40.61
Silent	EST	45.24	52.50	57.59	62.60	66.88
	JM EST	8.50	22.77	33.78	39.29	40.61
Foreman	EST	10.76	19.44	27.59	33.99	40.74
	JM EST	-0.12	1.91	3.51	9.22	15.58
Football	EST	2.99	5.71	10.60	17.73	26.26
	JM EST	-0.02	-0.13	0.30	1.22	4.67

sequences, we extended the early termination criterion to the remaining INTRA and INTER modes. This extension forms the basis of new complexity reduction tool called the Early Mode Termination (EMT) algorithm. Early Mode Termination codes a macroblock with the first mode that yields a rate-distortion cost less than or equal to its predicted cost.

Unlike in EST, the termination criterion is applied to the INTER and INTRA modes as well as the SKIP mode in Early Mode Termination; therefore, the mode evaluation order in its entirety affects the encoder complexity and should be carefully chosen. Our goal was to determine a single order with ascending computational complexity that could be used for any sequence. At this point, referring to Table 1, we derive a mode mapping function, $NextModeToTest()$, as presented in Table 4. This function dictates the next mode to be evaluated based on the previously evaluated mode and follows an ascending computational complexity. With the mode evaluation order finalized, the steps of the Early Mode Termination algorithm are as follows:

1. For each MB, using equation (7), predict the RD cost prior to any mode decision step.
2. Set mode M to START
3. Evaluate mode M such that $M = NextModeToTest(M)$.
4. If the RD cost of mode M is less than or equal to the predicted RD cost, terminate the RDO mode decision process and code the MB in mode M.
5. If not, go to step 3.

The motivation behind developing EMT was to provide a better complexity reduction tool for faster sequences; therefore, four of the faster sequences are used to present the performance of the algorithm: *Foreman, Stefan, Football, and NBA*. The encoding time savings of EMT

Table 4. The definition of the mode mapping function $NextModeToTest()$

$NextModeToTest()$								
Input Mode	Skip/Intra	P16x16	P16x8	P8x16	P8x8	P8x4	P4x8	P4x4
Output Mode	P16x16	P16x8	P8x16	P8x8	P8x4	P4x8	P4x4	N/A

Table 5. Comparison of the saved encoding time by EST (Solak & Labeau, 2010) and by EMT algorithms as a percentage of the RDO encoding time, for different sequences and different values of the quantization parameter QP

Sequence	Algorithm	QP=16	QP=20	QP=24	QP=28	QP=32
Foreman	EST	10.76	19.44	27.59	33.99	40.74
	EMT	41.89	46.13	49.39	54.07	57.69
Stefan	EST	5.95	9.64	14.36	21.05	30.64
	EMT	34.72	36.64	38.69	41.88	48.40
Football	EST	2.99	5.71	10.60	17.73	26.26
	EMT	33.25	35.00	38.47	43.63	52.67
NBA	EST	0.03	1.00	2.41	4.32	7.99
	EMT	27.76	25.93	24.99	24.39	27.25

will be compared with EST. A standard full rate-distortion-optimization (RDO) will also be included in the results as a golden standard (i.e. the one achieving the best possible PSNR). Table 5 compares the encoding time savings by EMT and by EST as a percentage of the full RDO encoding time. Clearly, EMT achieves higher complexity savings. In fact, the improvement over EST is more evident for smaller QPs where EST barely saves any encoding time and EMT saves at least 25 percent of the encoding time. Rate-distortion performance of EMT is compared with EST and RDO mode decision for the four sequences as shown in Figure 2. Although the difference between the RD curves of EMT and RDO is slightly more discernible than the difference between EST and RDO, it is still negligible.

Complexity Scalable Encoding Algorithms

Low complexity encoding methods that we have discussed so far attempt to reduce the encoding time and complexity with minimal impact on

Figure 2. RD curves of the proposed early mode termination (EMT) algorithm compared with early SKIP termination (EST) and full rate-distortion optimization (RDO) algorithms. Most curves are indistinguishable at this scale for a given sequence.

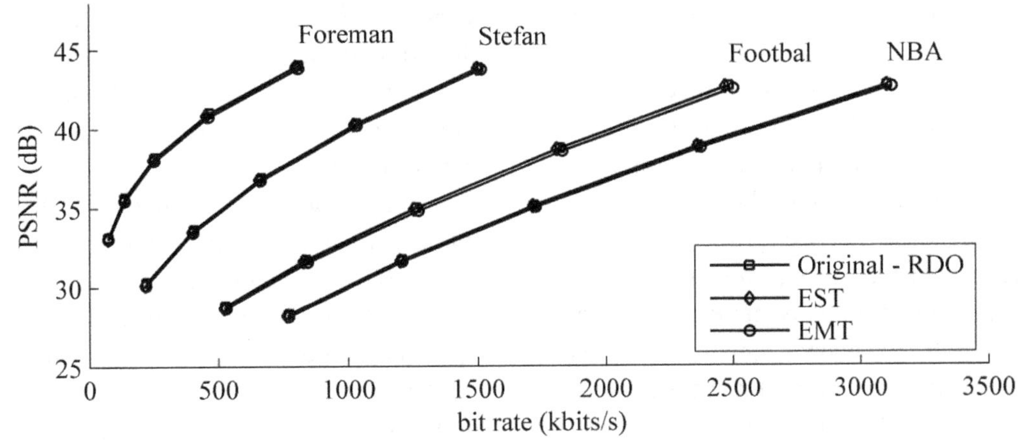

the RD performance of the output. These methods replace some of the encoding components with faster and complexity optimized designs and they generally utilize information from the previously encoded data to expedite the overall encoding process. However, devices with limited processing power such as mobile devices require more than reduced complexity. These devices are often subject to volatile resource constraints and low complexity algorithms provide a reduced yet rigid complexity; thus falling short of optimizing the complexity as per these changing constraints. The third category of encoding schemes, namely the complexity scalable encoding algorithms, provides the desired flexibility in the encoder complexity to meet such constraints.

Complexity scalability is a means of regulating the amount of processing power used by the encoder according to the pressing constraints on the host platform or the encoder itself, while still producing a satisfactory outcome. A complexity scalable encoder is one that offers a variety of operational modes with the encoder complexity descending from maximum to minimum as the achieved quality also degrades gracefully. The idea behind a complexity scalable encoder is to translate the high-level constraints (i.e. at a frame level or more usually at a sequence level) into a systematic allocation of the present processing power down to individual processing units in a video. Most complexity scalable frameworks prefer a frame level allocation of the resources where frames are dropped if their encoding time exceeds some threshold. Such implementations are popular as using a high-level processing unit simplifies the translation of the constraints that are already defined at high level; however, frame dropping results in low and choppy perceptual quality that disturbs a viewer gravely.

Kannangara, Richardson, and Miller (2008) proposed a novel complexity management technique for H.264/AVC encoding which comprises of two control levels: a frame-level control and a per-frame level control. The frame-level control algorithm initially calculates a target coding time for the next video frame based on the overall coding delay and on the target frame rate. If the coding delay hinders the target frame rate, the frame is dropped. If the frame is kept, the decision power is passed to the per-frame level control algorithm, which manages the computational complexity within the frame. It uses their previous work on early SKIP detection (Kannangara, et al., 2006) and varies the threshold that is used to identify the skipped macroblocks. The variable block skipping guarantees that the frame is encoded within the target encoding time. The advantage of this method over the other complexity scalable methods is its per-frame complexity control. Complexity scalable methods generally control the complexity only at a frame level by dropping frames, which in turn leads to unwanted choppy perceptual video quality at the decoder. However, the disadvantage of their method is the fact that the rate-distortion-optimized mode decision is disabled in order to enable the employed early SKIP detection.

Several complexity scalable algorithms have been proposed that extend the low complexity motion estimation algorithms discussed earlier (Braspenning, Haan, & Hentschel, 2002; Gallant, Cote, & Kossentini, 1999; Lengwehasatit & Ortega, 1998). These methods introduce control parameters that vary the search termination criteria of their fast motion estimation algorithms; thus adjusting the computational complexity at a very fine level. However, with these algorithms, the extent to which the complexity of the encoder can be reduced to is restricted by the ratio of the motion estimation complexity to the total encoding complexity.

Fonseca and de Queiroz (2009) propose a complexity scalable framework for the H.264/AVC prediction module which employs only a subset of the prediction modules, namely the most "popular" modes. The popularity ranking of the modes is done by sampling macroblocks of the previous frames. From each frame, a set of macroblocks are randomly chosen as the test domain for which all the prediction modes are evaluated. The set of dominant modes are the modes that have been the most frequently used in coding the MBs in the test domain. The complexity of the algorithm is controlled by removing modes from the dominant modes set as the complexity budget is exhausted. For the remaining macroblocks, only the dominant modes from the previous frame's test domain are evaluated. The authors state that choosing the number of macroblocks in the test domain as one tenth of the number of macroblocks in a whole frame yields a good trade-off between RD performance and complexity. The results indicate that the algorithm can save 50 percent in complexity for 0.1 dB loss in PSNR and for 5 percent loss in bit-rate. The degradation in RD performance is due to the errors in finding the accurate dominant modes.

An H.264/AVC encoder processes video frames in units of Macroblocks (MBs); therefore, implementing complexity scalability on a MB level would be an ideal solution. However, macroblocks within a frame bear dependencies to some of their neighbors due to the fact that INTRA coding relies on the neighbors of a macroblock to form a prediction. Careful consideration must be shown to adhere to these dependencies if a MB level scalable encoding framework is to be implemented. The Wave-Front (WF) macroblock scheduling technique is a way of processing macroblocks which achieves this scalability without disturbing the interdependencies. Macroblocks in a frame are grouped together in wave-fronts such that the MBs in a wave-front are independent of each other and can be encoded simultaneously given that the previous wave-fronts are fully encoded. Figure 3 illustrates the wave-front grouping of macroblocks. Each box symbolizes a macroblock and the numbers inside the boxes indicate the wave-fronts the MBs belong to.

Figure 3. Illustration of the wave-front concept. MBs in wave-front #11 are highlighted; they can be coded independently.

1	2	3	4	5	6	7	8	9	10	11
3	4	5	6	7	8	9	10	11	12	13
5	6	7	8	9	10	11	12	13	14	15
7	8	9	10	11	12	13	14	15	16	17
9	10	11	12	13	14	15	16	17	18	19
11	12	13	14	15	16	17	18	19	20	21
13	14	15	16	17	18	19	20	21	22	23
15	16	17	18	19	20	21	22	23	24	25
17	18	19	20	21	22	23	24	25	26	27

Tan, Lee, Thai, and Rahardja (2009) have recently proposed a macroblock level complexity scalable H.264/AVC encoding framework which adopts the so-called 'wave front macroblock scheduling' to provide MB-level scalability. Each wave front is incrementally allocated some complexity through a predefined complexity budget. This budget is controlled by a single parameter, which scales the complexity of the encoder by limiting the number of mode decision operations performed per wave-front. Since the macroblocks in a wave front can be encoded simultaneously, the encoding of a wave front is handled in multiple stages. Initially, the encoder computes the Rate-Distortion (RD) cost of SKIP and INTRA modes and saves the minimum of the two for each macroblock in the wave-front. After the initial computation, at every stage, the RD cost of an INTER mode is computed for the macroblock with the largest RD cost from the previous stage. This procedure is repeated until the predefined complexity budget for the wave front is exhausted. The motivation behind the practice of spending the complexity budget on macroblocks with high RD costs from the previous stage is to give priority to macroblocks with worse RD performance in order to achieve a homogenous quality over a frame, as the RD cost of a macroblock is the indication of its coding performance. However, this method falsely assumes that the RD costs of macroblocks are uniform.

Figure 4. Histogram of the macroblock RD cost values for Foreman sequence with a QP of 24

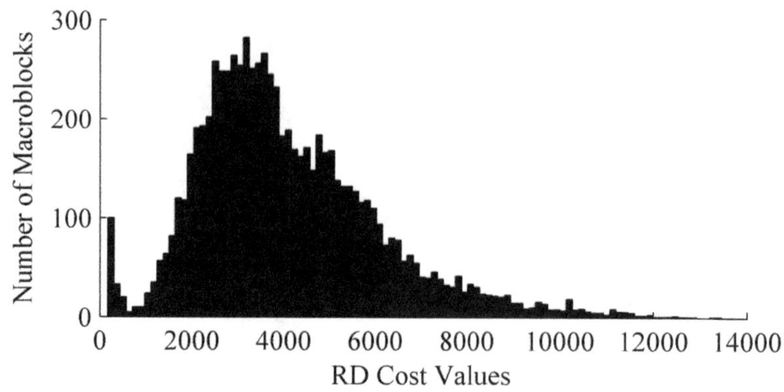

In fact, Figure 4 plots the histogram of the final RD costs of macroblocks for Foreman sequence encoded with a QP of 24 using full RDO mode decision. We observe a wide range of RD cost values extending from a little over 0 to as large as 14000. For instance, let us assume that there exists two macroblocks in a wave front that is presently processed and that we know their final RD costs and their coding modes which are 1400 and SKIP and 100 and P_16 × 8. Let us also assume that we have tried SKIP and INTRA modes for both MBs where the best RD costs are 1400 and 1000 respectively. Based on the algorithm, the encoder will exhaust its complexity budget for the first MB with the RD cost of 1400 and ignore the second MB with the RD cost of 1000, not realizing that the first MB although with a larger RD cost has found its optimal RD cost and that the second MB with the smaller RD cost actually needs more mode trials and RD refinement. This is of course an extreme case constructed to illustrate a point; nevertheless, it is plausible. Therefore, the distribution of the complexity budget over the macroblocks can be improved.

Another drawback of the encoding frame of Tan, Lee, Thai and Rahardja is that they compute the RD costs for four INTER modes, P_8×8, P_8×4, P_4×8, and P_4×4, at a single stage. This assumes that this quartet of INTER modes has the same collective complexity as the remaining INTER modes have individually. From Table 1 it is obvious that this is not correct, and in fact, these four INTER modes add up to almost 60 percent of the whole mode decision complexity. This assumption undermines the modeling of the encoder complexity.

We have recently introduced a complexity scalable encoding framework, which uses the wavefront scheduling of macroblocks (Solak & Labeau, 2010). In this study, we did away with grouping of P_8×8, P_8×4, P_4×8, and P_4×4 modes in one stage and instead, evaluated each INTER mode at a separate stage. This practice allowed us to obtain a more accurate modeling of the encoder complexity; and in a complexity scalable encoder, it is desirable to have a linear relation between the encoding time (i.e. the encoder complexity) and the complexity control parameter(s) of the encoder. Such a relation ensures that the effect of a complexity change on the encoding performance is predictable and maintainable. A linear relation signifies the preciseness of the complexity control of the encoder and the fine controllability of the complexity modeling, which is achieved by our algorithm.

Further, the proposed Complexity Scalable Encoding Framework (CSEF) gives priority to macroblocks whose RD performance from the previous stage is furthest from its optimal RD performance. In other words, we used a method to estimate the optimal RD costs of every macroblock in a wave-front such that before the next stage, we can measure how far the RD costs of macroblocks are from their optimal values: we incorporated the RD cost prediction algorithm that we have employed in our Early SKIP Termination (EST) and Early Mode Termination (EMT) algorithms in our complexity scalable encoding framework. The steps of the encoding framework can be summarized as follows: for each wave-front WF_j:

1. Reset the complexity counter to zero.
2. For each MB in the wave-front:
    a. Compute the predicted RD cost $\hat{J}_i$ by Equation(7).
    b. Evaluate SKIP and INTRA modes.
    c. Store the mode with the minimum RD cost and its RD cost as the current best mode and as the current best RD cost $\hat{J}_i^{\text{best}}$.
3. Pick the MB with $\max\left(\hat{J}_i^{\text{best}} - \hat{J}_i\right)$ as the current MB.
4. Increase the complexity counter by 1.
5. Evaluate the mode given by the output of the function $NextModeToTest()$ whose input is the current best mode of the MB in progress.
6. Update the current best mode with the newly tried one if the newly tried mode has a smaller RD cost.
7. Repeat steps from 3 to 7 until the counter is equal to the complexity budget $N_{op}(WF_j, \beta)$ for wave-front WF_j

β is the single control parameter that sets the maximum number of RD operation allowed for a wave-front, i.e. the complexity budget $N_{op}(WF_j, \beta)$. The budget is set by the following equation: $N_{op}(WF_j, \beta) = 7 \cdot N_{MB}(WF_j) \cdot \beta$ where $N_{MB}(WF_j)$ is the number of macroblocks in wave-front WF_j. β forms a link between the available processing resources and the processing resources made available to a wave-front by the encoder and may be varied within the range of 0 to 1.0 per wave-front, depending on the imposed resource constraints. Setting β to 1.0 amounts to allowing the encoder to explore all the mode trials within a wave front, as all seven INTER modes can be evaluated for each MB in this case.

Step 3 of the algorithm gives priority to the macroblocks whose current RD cost has the maximum difference with its predicted RD cost. This practice takes into consideration the fact that the RD costs of different macroblocks do not need to be equal or even similar and guarantees that a scenario similar to the one mentioned above never occurs. As a result, we expect the new resource allocation technique to improve the RD performance, particularly when β is small (less than 0.5). A small β value limits the allotted number of RD operations per wave front and the new framework should do a smarter job of distributing the RD operations over the macroblocks.

Therefore, we will compare the RD performances of the complexity scalable framework with and without the improved resource allocation. In Tables 6 and 7, the average PSNR and bit-rate gains with the new resource allocation method for sequences *Akiyo, Silent, Foreman,* and *Football* are presented. We observe considerable gains in both PSNR and bit-rate for different complexity levels and QPs. In fact, as anticipated, the gain is especially significant for β values less than 0.5 because with such small values, the benefits of a smarter resource distribution are greater.

The next set of tests will demonstrate the effect of two complexity reduction tools that we have discussed previously, Early SKIP Termination (EST) and Early Mode Termination (EMT) algorithms, on the complexity scalable framework. We anticipate EST to not only further reduce the encoding complexity that has already been limited by the control parameter β but also to eliminate the possible SKIP macroblocks from the mode decision process so that the complexity budget will be spent on other macroblocks. In

Table 6. PSNR improvements (in dB) with improved resource allocation, as compared to Tan et al. (2009), over different QPs and different β values. The values are averaged over four sequences: Football, Foreman, Silent, and Akiyo.

QP \ β	0.1	0.2	0.3	0.4	0.5	0.6	0.7	0.8	0.9
32	0.62	0.71	0.74	0.63	0.49	0.32	0.14	0.09	0.00
28	0.55	0.68	0.69	0.62	0.51	0.31	0.17	0.11	0.02
24	0.51	0.69	0.67	0.64	0.51	0.30	0.17	0.12	0.00
20	0.46	0.63	0.63	0.59	0.48	0.32	0.20	0.11	0.01
16	0.40	0.52	0.53	0.51	0.43	0.29	0.19	0.12	0.01

Table 7. Bit-rate improvements (in %) with improved resource allocation, as compared to Tan et al. (2009), over different QPs and different β values. The values are averaged over four sequences: Football, Foreman, Silent, and Akiyo.

QP \ β	0.1	0.2	0.3	0.4	0.5	0.6	0.7	0.8	0.9
32	38.16	41.38	38.15	34.03	25.19	17.19	9.73	6.48	0.78
28	39.08	42.50	39.40	35.18	25.96	16.60	9.61	6.40	0.62
24	37.49	40.63	38.27	33.82	25.48	15.46	7.92	4.36	0.06
20	35.03	39.23	36.58	32.64	25.26	13.99	6.60	3.63	-0.11
16	32.12	37.29	34.49	30.98	23.76	13.27	6.43	3.18	-0.23

scalable encoding with EST, when a macroblock that has an RD cost less than its predicted RD cost for an INTRA or an INTER mode but not for SKIP, the encoder does not take any special action and keeps evaluating the remaining modes as long as the RD operations budget allows it. However with EMT, the encoder will take such situations into consideration and will terminate the mode decision for macroblocks at the first mode satisfying the early termination condition. In other words, the encoder will stop trying different modes for a macroblock when it finds a mode that yields a RD cost that is less than its predicted RD cost.

During experiments, we have observed that for some macroblocks the predicted RD costs are slightly higher than their optimal values. Therefore, while encoding such macroblocks, the SKIP mode RD costs tend to be smaller than the predicted RD costs causing the early termination to kick in. In scalable encoding, we will observe the same trend to follow and regardless of the value of the control parameter β, both the EST and the EMT will cause an increase in the number of skipped macroblocks. In fact, we will see the number saturates for increasing β values. In return, this effect causes a slight saturation in the RD performances and a significant saturation in the encoding time which jeopardizes the linear relation between encoding time and β. In order to recapture the linearity, we replace the calculation of complexity budget:

$$N_{op}(WF_i, \beta) = 7 \cdot N_{MB}(WF_i) \cdot \beta \qquad (10)$$

with

$$N_{op}(WF_i, \beta) = 4 \cdot N_{MB}(WF_i) \cdot \beta \qquad (11)$$

by changing the coefficient of 7 to 4. The reasoning behind the choice of 4 is that 4 gives a good cut-off point for which the preceding points do not cause a significant saturation in either the RD performance or the encoding time.

Figure 5 plots the encoding time against varying β values of Football sequence for CSEF, CSEF-EST and CSEF-EMT. CSEF is our proposed Complexity Scalable Encoding Framework with improved resource allocation method. CSEF-EST and CSEF-EMT are the cases when the EST and the EMT are turned on respectively. In the full budget scenarios, for both sequences, we observe the encoding time saturating as β increases in CSEF-EST and CSEF-EMT encoders. In the latter case, the saturation is quicker due to the stabilization of the number of macroblocks coded in INTER modes with large partition sizes, particularly INTER 16×16. The same effect carries on to the RD curves. We can clearly see the RD points cluster together as β increases in Figure 6 for Football sequence when encoded with full complexity budget. When the RD operations budget is reduced, in Figure 5b, we see that the linearity between the encoding time and β is preserved to some extent. Similarly, the RD points are more clearly separated from one another as demonstrated in Figure 7.

The analysis of the experimental findings leads to several conclusions. The Complexity Scalable Encoding Framework (CSEF) has the ability to scale the encoder complexity up and down linearly through its control parameter β. As the encoder complexity varies, the CSEF successfully adjusts the PSNR and bit-rate of the compressed bitstream. Including Early SKIP Termination (EST) and Early Mode Termination (EMT) in CSEF, boost the encoding speed, further reducing the encoder complexity, with a slight penalty in the RD performance. At the same time, since

Figure 5. Total encoding time in seconds vs. β curves for Football sequence with different scalable encoders: the complexity scalable encoding framework (CSEF), CSEF-EST, and CSEF-EST. (a) Full complexity budget (equation 10), (b) reduced complexity budget (equation 11).

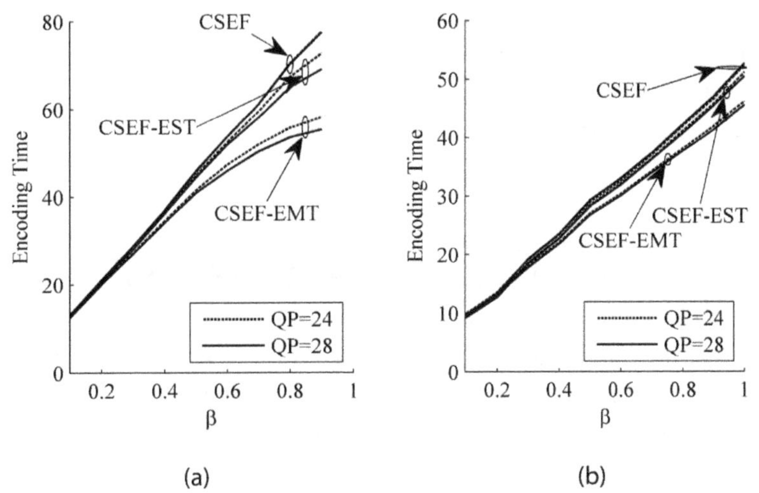

Figure 6. RD curves of football sequence with varying β points for full budget (equation 10) for CSEF, CSEF-EST, and CSEF-EMT encoders, (a) QP of 24, (b) QP of 28.

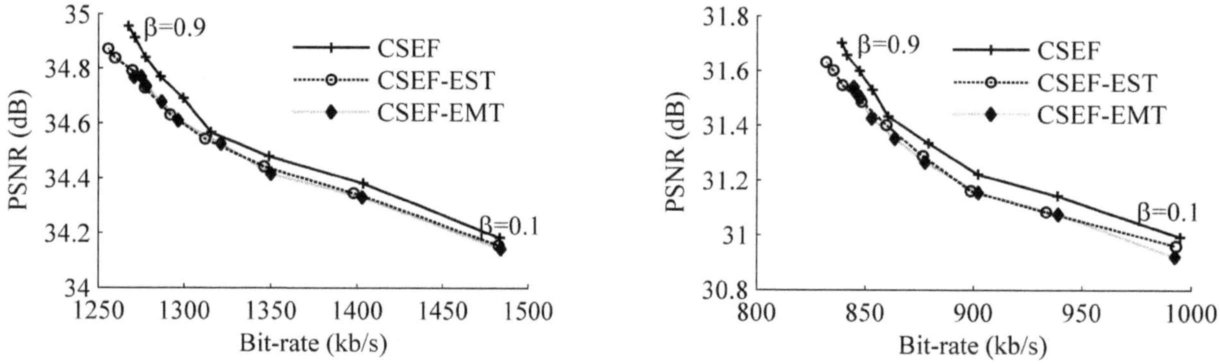

(a) (b)

Figure 7. RD curves of football sequence with varying β points for reduced budget (equation 11) for CSEF, CSEF-EST, and CSEF-EMT encoders, (a) QP of 24, (b) QP of 28.

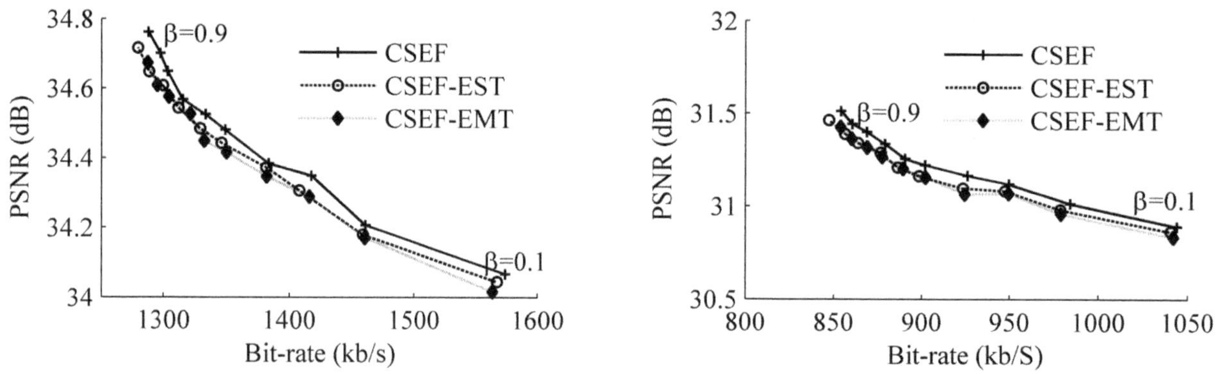

(a) (b)

the complexity reduction rate is proportional to the β value (i.e. the greater the β value, the greater the complexity reduction rate), the linearity of the encoder complexity control is compromised. However, for scenarios where the RD operations budget is reduced or for the RD operations budget is full with a reduced range of, yet finer tuned β (e.g. 0.1 to 0.5 with steps of 0.05), the linearity is retained and the complexity is reduced. Therefore, in practice, CSEF can work with full budget (though high complexity) scenarios; but, in low complexity scenarios, in addition to its ability reduce the encoder complexity, it has the option to activate the EST or the EMT in order to reduce the complexity to greater extents.

FUTURE RESEARCH DIRECTIONS

Similar to previous ISO/IEC MPEG and ITU-T VCEG standards, H.264/AVC has been based on a structure that joins a motion compensated coding stage and a transform domain quantization stage.

Such a hybrid design implies a complex encoding structure that performs an extensive analysis on the input video data to capture and remove as much redundancy as possible both in temporal and spatial domain (Dufaux, Gao, Tubaro, & Vetro, 2010). H.264/AVC has been equipped with various tools to facilitate this analysis such as the variable block-size motion compensation with quarter-pixel precision, multiple reference frames, improved spatial INTRA prediction, etc. The complex encoder and lightweight decoder nature of H.264/AVC is applicable to be used in architectures with a single powerful transmitter and multiple receivers. However, with the developments in wireless communications, application such as wireless video surveillance and low power sensor networks, requiring mobile users with limited processing capabilities are emerging. In such scenarios high encoder complexity is unwanted and instead lightweight encoding with high compression efficiency is preferred (Dufaux, Gao, Tubaro, & Vetro, 2010). Therefore there is a growing need for power efficient and low complexity video encoding paradigms in the industry.

In January 2010, a final call for proposals has been issued for a new video coding standard, The High Efficiency Video Coding, namely the H.265. The initial evaluation of the proposed algorithms indicates a doubled compression efficiency (half the bit-rate at a similar quality level) compared with H.264/AVC. Some of the new features include increased number of block sizes, advanced motion vector prediction, up to 34 intra prediction directions and transform block size from 4 × 4 to 32 × 32 (Richardson, 2010). These improvement, of course, comes at the expense of significantly higher computational complexity (estimated at three times more) and brings even greater challenges for the designers.

CONCLUSION

This chapter addresses the problem of computational complexity in H.264/AVC video encoders. In the scenarios of real-time multimedia and mobile video, where the host devices are typically battery powered and limited in processing power, the encoder computational complexity becomes a key constraint. The achieved video quality and bit-rate of a coded sequence depend on the encoder complexity. The distribution of the complexity over different encoding tools is examined and the profiling test results showed that the mode decision process, which is carried out for each and every processing unit of a video, i.e. a macroblock, to obtain its coding mode, is the most computationally intensive operation in H.264/AVC encoding. We have discussed the literature in three categories:

- Low complexity motion estimation algorithms
- Low complexity mode decision algorithms
- Complexity scalable encoding algorithms

The algorithms in the first two categories aim to reduce the computational complexity of motion estimation and mode decision processes respectively. While some of the algorithms achieve significant computational savings, they lack the scalability notion, which is essential for sustaining the encoding process under volatile resource constraints. The algorithms of the third category seek to solve this problem. We also proposed a novel complexity reduction algorithm, the Early Mode Termination. The algorithm estimates the rate-distortion cost of a macroblock through a prediction function and codes the macroblock with the first mode with a cost less than or equal to the prediction, thus saving computation for further mode evaluations. Results demonstrate savings of 33%–88% dependent on the source sequence. Secondly, an improved resource distribution technique is introduced as an extension to our recently proposed complexity scalable encoding framework. With the new technique, results demonstrate striking improvements in coding efficiency, as in some cases, the PSNR gain and the bit-rate savings are greater than 1dB and 60 percent, respectively.

REFERENCES

Agrawal, P., Chen, J. C., Kishore, S., Ramanathan, P., & Sivalingam, K. (1998). Battery power sensistive video processing in wireless networks. In *Proceedings of IEEE PIMRC 1998*, (vol. 1), (p. 116). Boston, MA: IEEE Press.

Braspenning, R., de Haan, G., & Hentschel, C. (2002). Complexity scalable motion estimation. In *Proceedings of VCIP 2002*, (vol. 4671), (pp. 442–453). San Jose, CA: VCIP.

Chen, Z., Zhou, P., & He, Y. (2002). Fast integer and fractional pel motion estimation for JVT. In *Proceedings of JVT of ISO/IEC MPEG and ITU-T VCEG 6th Meeting*. Awaji Island, Japan: JVT.

da Fonseca, T. A., & de Queiroz, R. L. (2009). Complexity-constrained H.264 HD video coding through mode ranking. In *Proceedings of PCS 2009*. Chicago, Illinois: PCS.

Dufaux, F., Gao, W., Tubaro, S., & Vetro, A. (2010). Distributed video coding: Trends and perspectives. *EURASIP Journal on Image and Video Processing*. Retrieved from http://www.merl.com/reports/docs/TR2010-026.pdf.

Gallant, M., Cote, G., & Kossentini, F. (1999). An efficient computation-constrained blockbased motion estimation algorithm for low bit rate video coding. *IEEE Transactions on Image Processing*, *8*(12), 1816–1822. doi:10.1109/83.806627

Ghanbari, M. (1990). The cross-search algorithm for motion estimation. *IEEE Transactions on Communications*, *38*(1), 950–953. doi:10.1109/26.57512

Jeon, B., & Lee, J. (2003). Fast mode decision for H.264. In *Proceedings of JVT of ISO/IEC MPEG and ITU-T VCEG 10th Meeting*. Waikoloa, HI: JVT.

Kannangara, C. S., Richardson, I. E., & Miller, A. J. (2008). Computational complexity management of a real-time H.264/AVC encoder. *IEEE Transactions on Circuits and Systems for Video Technology*, *18*(9), 1191–1200. doi:10.1109/TCSVT.2008.928881

Kannangara, C. S., Richardson, I. E. G., Bystrom, M., Solear, J. R., Zhao, Y., Maclennan, A., & Cooney, R. (2006). Low complexity skip prediction for H.264 through Lagrangian cost estimation. *IEEE Transactions on Circuits and Systems for Video Technology*, *16*(2), 202–208. doi:10.1109/TCSVT.2005.859026

Koga, T., Iinuma, K., Hirano, A., Iijima, Y., & Ishiguro, T. (1981). Motion compensated interframe coding for video conferencing. In *Proceedings of the National Telecommunication Conference 1981*. (pp. G5.3.1-G5.3.5). New Orleans, LA: NTC.

Lengwehasatit, K., & Ortega, A. (1998). A novel computationally scalable algorithm for motion estimation. In *Proceedings of VCIP, 1998*, 68–79.

Lim, K. P., Sullivan, G., & Wiegand, T. (2004). Joint model reference encoding methods and decoding concealment methods. In *Proceedings of JVT of ISO/IEC MPEG and ITU-T VCEG*. Munich, Germany: JVT.

Lim, K. P., Wu, S., Wu, D. J., Rahardja, S., Limn, X., Pan, F., & Li, Z. G. (2003). Fast inter mode selection. In *Proceedings of JVT of ISO/IEC MPEG and ITU-T VCEG 9th Meeting*. San Diego, CA: JVT.

Malvar, H. S., Hallapuro, A., & Karzewicz, M. (2003). Low-complexity transform and quantization in H.264/AVC. *IEEE Transactions on Circuits and Systems for Video Technology*, *13*(7), 598–603. doi:10.1109/TCSVT.2003.814964

Ortega, A., & Ramchandran, K. (1998). Rate distortion methods for image and video compression. *IEEE Signal Processing Magazine*, *15*(6), 23–50. doi:10.1109/79.733495

Rao, K. R., & Yip, P. (1990). *Discrete cosine transform: Algorithms, advantages, applications*. San Diego, CA: Academic Press.

Richardson, I. E. G. (2003). *H.264 and MPEG-4 video compression: Video coding for next-generation multimedia*. New York, NY: John Wiley & Sons Ltd. doi:10.1002/0470869615

Richardson, I. E. G. (2010). *High efficiency video coding / HEVC / H.265: Beyond H.264*. Retrieved from http://www.vcodex.com/h265.html.

Richardson, I. E. G., & Zhao, Y. (2002). Adaptive management of video encoder complexity. *Real-Time Imaging*, *8*(4), 291–301. doi:10.1006/rtim.2001.0275

Solak, S. B., & Labeau, F. (2010). Complexity scalable encoding for power-aware application. In *Proceedings of IEEE IGCC 2010*. Chicago, IL: IEEE Press.

Suhring, K. (2011). *JM reference software*. Retrieved from http://iphome.hhi.de/suehring/tml/.

Sullivan, G. J., & Wiegand, T. (1998). Rate distortion optimization for video compression. *IEEE Signal Processing Magazine*, *15*(6), 74–90. doi:10.1109/79.733497

Tan, Y. H., Lee, W. S., Tham, J. Y., & Rahardja, S. (2009). Complexity rate distortion optimization for real-time H.264/AVC encoding. In *Proceedings of the IEEE ICCCN 2009*. San Francisco, CA: IEEE Press.

Tourapis, A. M., Cheong, H. Y., & Topiwala, P. (2005). Fast ME in the JM reference software. In *Proceedings of the JVT of ISO/IEC MPEG and ITU-T VCEG, 16th Meeting*. Poznan, Poland: JVT.

Wedi, T. (2003). Motion compensation in H.264/AVC. *IEEE Transactions on Circuits and Systems for Video Technology*, *13*(7), 577–586. doi:10.1109/TCSVT.2003.815171

Wiegand, T., Sullivan, G. J., Bjontegaard, G., & Luthra, A. (2003). Overview of the H.264/AVC video coding standard. *IEEE Transactions on Circuits and Systems for Video Technology*, *13*(7), 560–575. doi:10.1109/TCSVT.2003.815165

Wu, S., Lim, K. P., Wu, D. J., Rahardja, S., Limn, X., Pan, F., & Li, Z. G. (2003). Fast mode decision for intra prediction. In *Proceedings of the JVT of ISO/IEC MPEG and ITU-T VCEG 7th Meeting*. Pattaya II, Thailand: JVT.

Yi, X., Zhang, J., Ling, N., & Shang, W. (2005). Improved and simplified motion estimation for JM. In *Proceedings of JVT of ISO/IEC MPEG and ITU-T VCEG, 16th Meeting*. Poznan, Poland: JVT.

ADDITIONAL READING

Carrillo, P., Tao, P., & Kalva, H. (2010). Low complexity H.264 video encoder design using machine learning techniques. In *Proceedings of the Digest of Technical Papers International Conference on Consumer Electronics*, (pp. 461-462). Digest of Technical Papers.

Chien, M. C., Huang, J. Y., & Chang, P. C. (2009). Complexity control for H.264 video encoding over power-scalable embedded systems. In *Proceedings of the IEEE 13th International Symposium on Consumer Electronics, ISCE 2009*, (pp. 221-224). IEEE Press.

Colin, P. (2003). *RTP: Audio and video for the internet*. Boston, MA: Addison-Wesley Professional.

Eisenberg, Y., Luna, C. E., Pappas, T. N., Berry, R., & Katsaggelos, A. K. (2002). Joint source coding and transmission power management for energy efficient wireless video communications. *IEEE Transactions on Circuits and Systems for Video Technology*, *12*(6), 411–424. doi:10.1109/TCSVT.2002.800309

Haggag, M. N., El-Sharkawy, M., & Fahmy, G. (2010). Efficient fast multiplication-free integer transformation for the 2-D DCT H.265 standard. In *Proceedings of the 2010 17th IEEE International Conference on Image Processing (ICIP)*, (pp. 3769-3772). IEEE Press.

He, Z., Liang, Y., Chen, L., Ahmad, I., & Wu, D. (2005). Power-rate-distortion analysis for wireless video communication under energy constraints. *IEEE Transactions on Circuits and Systems for Video Technology*, *15*, 645–658. doi:10.1109/TCSVT.2005.846433

Hoyoung, L., Bongsoo, J., Jooyoung, J., & Byeungwoo, J. (2009). Computational complexity scalable scheme for power-aware H.264/AVC encoding. In *Proceedings of IEEE International Workshop on Multimedia Signal Processing*, (pp. 1-6). IEEE Press.

Huijbers, E. A. M., Ozcelebi, T., & Bril, R. J. (2011). Complexity scalable motion estimation control for H.264/AVC. In *Proceedings of IEEE International Conference on Consumer Electronics (ICCE)*, (pp. 49-50). IEEE Press.

Kannur, A. K., & Baoxin, L. (2009). Power-aware content-adaptive H.264 video encoding. In *Proceedings of the IEEE International Conference on Acoustics, Speech and Signal Processing, 2009, ICASSP 2009*, (pp. 925-928). IEEE Press.

Kim, W., You, J., & Jeong, J. (2010). Complexity control strategy for real-time H.264/AVC encoder. *IEEE Transactions on Consumer Electronics*, *56*(2), 1137–1143. doi:10.1109/TCE.2010.5506050

Li, D., Sun, Y., & Feng, Z. (2005). Joint power allocation and rate control for real-time video transmission over wireless system. In *Proceedings of IEEE GLOBECOM 2005*, (vol. 4), (p. 5). IEEE Press.

Li, S., Yan, L., Feng, W., Shipeng, L., & Wen, G. (2009). Complexity-constrained H.264 video encoding. *IEEE Transactions on Circuits and Systems for Video Technology*, *19*(4), 477–490. doi:10.1109/TCSVT.2009.2014017

Lian, S. J., Chien, S. Y., Lin, C. P., Tseng, P. C., & Chen, L. G. (2007). Power-aware multimedia: Concepts and design perspectives. *IEEE Circuits and Systems Magazine*, *7*(2), 26–34. doi:10.1109/MCAS.2007.4299440

Liang, Y., & Ahmad, I. (2006). Adaptive techniques for simultaneous optimization of visual quality and battery power in video encoding sensors. In *Proceedings of International Conference on Image Processing*, (pp. 2477-2480). Atlanta, GA: IEEE.

Liang, Y., Ahmad, I., & Luo, J. (2005). Joint power and distortion control in video coding. *SPIE Video Communications and Image Processing*, *5685*, 885–895.

Liu, L., Cohen, R., Sun, H., Vetro, A., & Zhuang, X. (2010). New techniques for next generation video coding. In *Proceedings of the 2010 IEEE International Symposium on Broadband Multimedia Systems and Broadcasting (BMSB)*, (pp. 1-6). IEEE Press.

Merkle, P. (2007). Efficient prediction structures for multiview video coding. *IEEE Transactions on Circuits and Systems for Video Technology*, *17*(11), 1461–1473. doi:10.1109/TCSVT.2007.903665

Mohammed, G. (1999). *Video coding: An introduction to standard codecs*. Inspec/IEE.

Muhammed, A. M., Canagarajah, C. N., & Bull, D. R. (2002). *Video coding for mobile communications: Efficiency, complexity and resilience*. San Diego, CA: Academic Press.

Peter, S. (2000). *Video compression demystified*. New York, NY: McGraw-Hill Professional Publishing.

Pu, W., Lu, Y., & Wu, F. (2006). Joint power-distortion optimization on devices with MPEG-4 AVC/H.264 codec. In *Proceedings of ICC 2006*, (pp. 441-446). Istanbul, Turkey: ICC.

Shafique, M., Zatt, B., Bampi, S., & Henkel, J. (2010). Power-aware complexity-scalable multiview video coding for mobile devices. In *Proceedings of the Picture Coding Symposium (PCS)*, (pp. 350-353). Picture Coding Symposium.

Tan, Y. H., Lee, W. S., Tham, J. Y., & Rahardja, S. (2009). Joint source coding-transmission power control for H.264 video delivery. In *Proceedings of the IEEE Region 10 Conference TENCON*, (pp. 1-6). IEEE Press.

Tsai, T. H., Pan, Y. N., & Hou, Z. H. (2010). A green computation H.264/AVC encoder with reconfigurable motion estimation. In *Proceedings of the 2010 International Conference on Green Circuits and Systems (ICGCS)*, (pp. 539-542). ICGCS.

Wes, S. (2008). *Video over IP: IPTV, internet video, H.264, P2P, web TV, and streaming: A complete guide to understanding the technology* (2nd ed). New York, NY: Focal Press.

Yu, Y. J., Hsiu, P. C., Pang, A. C., & Lai, C. P. (2010). Power-aware scalable video multicast in 4G wireless systems. In *Proceedings of the 2010 IEEE Global Telecommunications Conference GLOBECOM 2010*, (pp. 1-6). IEEE Press.

KEY TERMS AND DEFINITIONS

Video Compression: Video compression is the process of representing uncompressed/raw video data in a more compact form with fewer bits for efficient transmission or storage. Compression algorithms reduce the data size by recognizing and exploiting the redundancies present in data.

Complexity Scalable Encoding: Complexity scalability is a means of regulating the amount of processing power used by the encoder according to the pressing constraints on the host platform or the encoder itself, while still producing a satisfactory outcome. A complexity scalable encoder is one that offers a variety of operational modes with the encoder complexity descending from maximum to minimum as the achieved quality also degrades gracefully.

Low Complexity Motion Estimation: Low complexity motion estimation is the process of finding the block with the most similar content in the previously coded frame(s) for a particular block in the current frame fast and in fewer operations.

Low Complexity Mode Decision: Low complexity mode decision is the process of finding the coding mode that yields the least residual for a particular block in the current frame fast and in fewer operations.

Power-Aware: A power-aware design is a power-efficient and also power-adaptable one that is capable of operating under variable power constraints.

Temporal Redundancy: Temporal redundancy is the existing similarites between frames apart in time.

Spatial Redundancy: Spatial redundancy is the existing similarites within a single frame.

Chapter 15
Display Energy Management based on Eye Tracking

Vasily G. Moshnyaga
Fukuoka University, Japan

ABSTRACT

With the explosive use of personal computers or PCs, reducing computer energy consumption is paramount for sustainability. The display is the largest energy consumer in a personal computer. Current display energy management technologies ignore the attention of the PC user and therefore may either switch the display off when the user looks at the screen or lose energy by keeping the display on when nobody looks at it. This chapter discusses a new display energy management technology and outlines its implementation in a personal computer system. Unlike existing technologies, which "sense" a PC user through keyboard and/or mouse or the other sensors, this technology "watches" the user through a single camera or CMOS vision sensor. The technology tracks the user's eyes, keeping display active only if the user looks at its screen. Otherwise, it dims the display down or even switches it off to save energy. The authors implemented the technology in software and hardware and present the results of their experimental evaluation.

1. INTRODUCTION

As personal computers become more intertwined with everyday life of people, minimizing their energy consumption becomes very important. According to SMART 2020 report (Climate Group, 2008), the number of PCs globally is expected to increase from 592 million in 2002 to 2 billion in 2014 and 4 billion in 2020. The vast majority of these PCs are expected to be portable devices, which rely on rechargeable batteries. Unfortunately, the battery capacities are increasing at a much slower pace than the overall power consumption growth of such PCs. Therefore, it is critical to develop energy-aware design methodologies to keep the energy consumption of portable PCs in line with the battery capacity increase.

DOI: 10.4018/978-1-4666-1839-8.ch015

Portability is by no means the sole force behind the push for energy reduction. There exists a strong pressure for producers of desktop computers to decrease power consumption. For desktops or servers, high power consumption raises temperature, deteriorates performance and reliability. Besides, due to the large grows of PCs, their share in the total ICS CO2 emission footprint will rise by 2.7 times in comparison to 2002 and reach 57% in 2020 (Climate Group, 2008). According to the International Energy Agency, the energy used by computers will not only double by 2022 but increase threefold by 2030 (IEA Press, 2009). From the global warming perspective, the improvement of energy efficiency of computers, reducing their energy consumption during lifetime has become an urgent issue for global sustainable development.

The primary goal of the emerging discipline of IT sustainability is to enable firms to develop and use computing resources more efficiently while maintaining or increasing overall performance (Harmon, 2010). The first wave of these efforts is commonly identified as "green computing" where the emphasis has been primarily minimizing power usage for datacenters and technical equipments, such as computers, projectors, etc.

In a typical personal computing system, display accounts 30% ~ 50% of the total energy (Mahesri, 2005; Robertson, 2002). For example, 19" Sony SDM-S93 LCD monitor (1280 x 1024pixels) burns in active mode 50W or almost 38% of the total desktop system power (130W). With large popularity of video and gaming applications, LCD makers are being called on to cut power consumption while providing better images. Rapid utilization of multiple displays - each consuming tens of watts—throughout homes, offices, and buildings increases cost and environmental impact of energy consumption significantly. Although most computer displays support power management, new robust methods are needed for evolving display usage scenarios.

This chapter discusses a new camera-based approach to display power management and outlines its implementation in personal computer system. Unlike existing display power management technology, which "senses" its user through keyboard and/or mouse, our technology "watches" the user through a single camera. The technology tracks the user's eyes keeping the display active only if the user looks at the screen. Otherwise, it dims the display down or even switches it off to save energy. We implemented the technology in hardware and present the results of its experimental evaluation.

The chapter is organized as follows. In the next section, we survey related research. Section 3 describes the proposed camera-based display energy management approach. Section 4 presents the design and outlines its implementation features. Section 5 reports on experimental evaluation. Section 6 summarizes our findings and outlines work for the future.

2. RELATED RESEARCH

The first widely used adaptive display power management technology was introduced almost two decades ago with adoption of VESA's Display Power Management Signaling Specification. The technology specifies one or more power states (e.g. standby, sustain, and off) that are intermediate between *on* and *off,* turning the display to low power state after a specified period of inactivity on mouse and/or keyboard. Over the years, many additions and revisions to this technology have occurred, including Advanced Configuration and Power Interface (ACPI, 2004), developed by HP, Intel, Microsoft, Phoenix, and Toshiba. The OS-based ACPI links the display brightness as well as inactivity intervals to application1.

One problem with ACPI is that it strongly depends on inactivity intervals, either set as default (see Table 1) or by the user. From one hand, if the inactivity intervals are improperly short, e.g. 1 or 2 minutes, the ACPI can be quite troublesome by shutting the display off when it must be on. From another hand, if the inactivity intervals are set to

Table 1. Default APM setting for Sony Vaio PCG-Z1V/P monitor

Mode	APM off	Word Text	DVD	Music	Camera
Brightness Level	9	7	9	5	9
Inactivity interval	never	10	never	10	10

be long, the ACPI efficiency decreases. Because modifying the intervals requires system setting, most users never adjust the power management of their PCs for fear that it will impede performance (Fujitsu-Siemens, 2004). Those who do the adjustment, usually assign long intervals. HP inspected 183,000 monitors worldwide and found that almost a third was not set to take advantage of the energy saving features. Just enabling these features after 20 minutes of inactivity can save up to 381 kWh for a monitor per year (Hewlett-Packard, 2006). Evidently, to prevent such a problem the PC energy management must employ more efficient user monitoring.

Another approach to save the display energy consumption is to scale down or dim the backlight luminance. This approach is based on observation that transmissive and transflective color TFT LCD panels (Sharp, 2002) do not illuminate itself but filter a backlight, the primary source of display energy dissipation. Because simply dimming the backlight degrades the display visibility, Choi et al. (2002) proposed to maintain brightness or contrast of the LCD panel when the backlight is dimmed down. To reduce the average energy demands of the backlight, Gatti et al. (2002) suggested the backlight auto regulation scheme. Cheng and Pedram (2004) showed that a Concurrent Brightness and Contrast Scaling (CBCS) technique further enhances image fidelity with a dim backlight, and thus saves an extra power. Chang et al. (2004) introduced a dynamic luminance scaling or DLS technique that dimmed the backlight while allowing more light to pass through the screen panel to compensate for the loss of brightness in the original image. Shim et al. (2004) combined the DLS technique with dynamic contrast enhancement and applied it for transflective TFT LCD panels. Pasricha et al. (2004) presented an adaptive middleware-based technique to optimize backlight power when playing streaming video. Iranli et al. (2006) presented HVS-aware dynamic backlight scaling in TFT LCD. A modification of the LCD panel to permit zoned backlighting has been discussed in Flinn (1999).

There is also a variety of techniques for automated adjustment of brightness in high dynamic range panoramic images (e.g. Pattanai, 2000; Tumblin, 1999). These techniques dynamically brighten or darken image regions depending on the scene content and average local luminance.

Despite differences, the proposed brightness and/or contrast adjustment techniques have one feature in common. Namely, they work independently on the viewer attention. The techniques are able to lower the display energy in active mode especially when showing images or video. But if a display delivers text or idles for some time, the APM remains the only energy savior. If the relation between the inactivity interval (*T*) and the time of keyboard (*K*) and/or mouse (*M*) typing is as in Figure 1a, the display will be active

Figure 1. An illustrative example

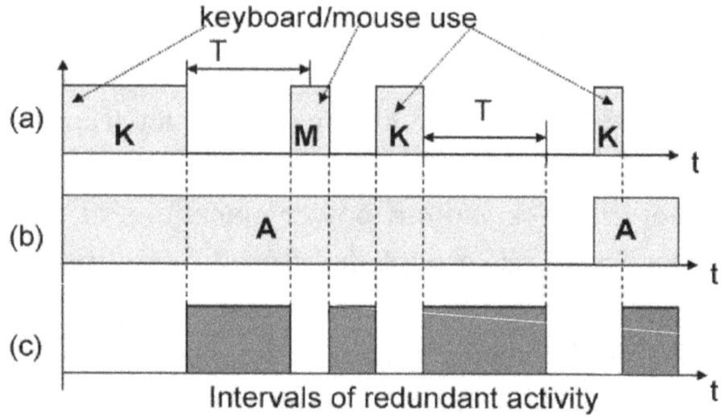

(*A*) almost all the time (see Figure 1b). Surely, if nobody looks at the screen in between key presses, activating the display is unnecessary for all the intervals shown in Figure 1c.

Several methods also have been proposed to improve the PC user presence detection. Extending the touch-pad function for user presence detection (Park, 1999) and placing thermal sensors around the display (Dai, 2003) are some of them. Despite differences, all these methods have one drawback in common. Namely, they ignore the viewer attention. Paradoxically, while the display is needed only for our eyes, none of existing approaches, up to our knowledge, takes them into account. Neither ACPI nor temperature sensors nor the advanced touchpad screeners can distinguish whether the user looks at screen or not. As a result, they may either switch the display off inappropriately (i.e. when the user looks at screen without pressing a key) or keep the display active when nobody looks at it.

We proposed to apply user's eye tracking for PC energy management. Unlike existing methods, our technology "watches" the user through a single camera or CMOS vision sensor. More precisely, it tracks the user's eyes to detect whether he or she looks on the screen or not, and based on the tracking results control the display brightness and power consumption. Up to the date eye tracking has been applied for applications, such as human-computer interface, security systems, health care, drowsy-driver detection, assistive technologies, ubiquitous computing, etc. Surveys on eye-gaze tracking technology and applications can be found in (Duchowski, 2002; Morimoto, 2004; Hansen, 2010). Our technology is a new application of eye tracking, namely to the field of energy management.

In order to make the technology effective, the eye-detection and tracking system must meet several requirements:

1. Initialization simplicity (i.e. minimal personal calibration);
2. Be non-intrusive (i.e. not attached to the user's head);
3. Support unrestricted head movements;
4. Operate in real time;
5. Be robust (i.e. have high detection accuracy);
6. Low energy consumption.

The latter requirement is especially important for battery operated personal computers because it keeps the PC battery lifetime and weight reasonable. Majority of currently available eye tracking systems, unfortunately, do not satisfy above requirements making a gaze-based power management a real challenge. For instance, some systems need calibration procedures; the others require users to wear head-mounted gears (a hat or helmet or glasses on which camera is mounted) or restrict head positions within a quite narrow area. Those systems which meet the requirements are unfortunately very energy consuming (Baluja, 1994; Ji, 2004; Kawato, 2005; Ohno, 2003; Park, 2005; Theocharides, 2004).

Computational complexity is one of the main factors attributing to power. Although it is easy for a human to detect and track eyes, performing it on hardware requires complex algorithms and many energy-dissipating computations. For example, the system (Theocharides, 2004) initially scales an input 300 x 300 image into five images of 240 x 240, 180 x 180, 120 x 120, 60 x 60, and 20 x 20 pixels, respectively, and then processes them in parallel for histogram equalization, Neural Network (NN) classification, image rotation, brightness and contrast adjustment, NN-based face detection, etc. To perform over 163 Million multiply-accumulate operations per image frame, the system employs a wide network of processors and memories, consuming over 5W of power. Lowering the number of computations of eye-gaze-tracking is imperative for energy saving.

In our previous work (Moshnyaga, 2005, 2009; Yamamoto, 2009) we have outlined the idea and a straightforward software-based implementation. In this chapter, we summarize the technology, present a low complexity eye-tracking algorithm and hardware implementation, which ensure real-time accurate eye-tracking by a single camera with less than 1 Watt power overhead.

3. THE PROPOSED DISPLAY ENERGY MANAGEMENT TECHNOLOGY1

3.1. An Overview

Our technology is based on the following assumptions:

1. The PC is equipped with a color video camera (or CMOS vision sensor). The camera is located at the top of display. When the user looks at display it faces the camera frontally.
2. The display has a number of backlight intensity levels with the highest level corresponding to the largest power consumption and the lowest level to the smallest power, respectively.
3. The highest level of backlight intensity is enabled either initially or whenever the user looks at the screen.

Figure 2 outlines the power management flowchart. Here C counts the number of consecutive image frames in which the user's gaze is off the screen. If no human face is detected in the current video frame, the display is turned off. Otherwise, the method tracks the user's eye-gaze. If the gaze has been off the screen for more than N consecutive frames, the current backlight luminance is dimmed down to the next level. Any on-screen gaze reactivates the initial backlight luminance by moving the display onto power up mode. However, if no on-screen gaze has been detected for more than N frames and the backlight luminance has already been lowest, the display enters the standby mode. Returning back from either standby or off modes is done by pushing the ON button. Below we describe the method in details.

3.2. User Presence Detection

The goal of this task is to determine from the camera (or CMOS sensor) readings whether or not the user is currently present in front of display. To detect the user's presence in front of the display, we first localize the face search by applying background subtraction and skin-color segmentation to the RGB representation of input image. The skin is defined by the following criteria (Douxchamps, 2008): $0.55<R<0.85$, $1.15<R/G<1.19$, $1.15<R/B<1.5$ and $0.6<(R+G+B)<1.8$.

To accelerate the face-area extraction, we use two filters. The first one limits the size of the head in reasonable range. The second one verifies that the face contains a minimum of 25% of skin colored pixels. Thus, if the total number of pixels in the derived face area exceeds the threshold, the user is assumed present.

Figure 2. Flowchart of the proposed approach

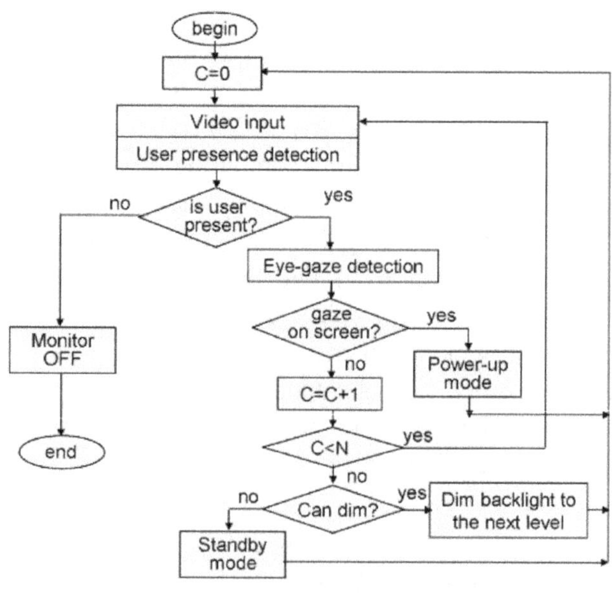

3.3. Eye-Gaze Detection

The eye-gaze detector implements the algorithm proposed by Kawato (2005), which scans the Six Segment Rectangular filter (SSR) over the integral representation (Viola, 2001) input image. to define the Between The Eyes (BTE) point of human face (see Figure 3) and then searches the regions 1 and 3 from the left and right side of the BTE point to locate the eyes. The algorithm does not depend on illumination, face occlusion, eye closure. It is more stable, robust, and less complex than the other eye-tracking formulations. However, it is still very computationally demanding. In a quest to locate all faces (without restriction on face size, motion and rotation) the algorithm repeats scan six times over the whole image frame performing over 28 Million operations, such as addition or subtraction per frame (640 x 480 pixels frame size). Though such a full search might be necessary in some applications, it seems redundant when tracking eyes of the PC user.

In our eye-tracking application, we can assume that:

1. The target object is a single PC user. The user sits in front of PC at a relatively close distance (50-70cm).
2. The user's motion is slow relatively to the frame rate.
3. The background is stable and constant.

Based on these assumptions, we apply the following optimizations to reduce eye-tracking complexity:

a. Single SSR filter scan;
b. Fixed SSR filter size;
c. Three pixel displacement of the SSR filter during the scan;
d. Low frame processing rate (5-10 fps);

As our study shows (Yamamoto, 2009), such enhancements reduce computational complexity considerably without affecting the quality of the results.

Figure 4 shows the modified algorithm. For the first frame or any frame in which the search for BTE candidate was unsuccessful, we search the image area reduced by background and skin-color extraction; otherwise, the search is restricted to a small area (S) of ±8 pixels around the BTE point. For the chosen area, the algorithm first computes the integral image and then scans it by the Six Segment Region filter (SSR) to select the BTE candidate. If the BTE candidate is found, the system uses the BTE as a starting point to locate eyes. If eyes have been detected, the user is assumed to be looking at screen. If no face candidate has been found for N consecutive frames, the user is considered not present in front of the display.

The BTE search is organized as follows. The green component of the search area is transformed to integral image representation (Viola, 2001) and scanned by the Six-Segment Rectangular (SSR)

Figure 3. An illustration of the SSR filter

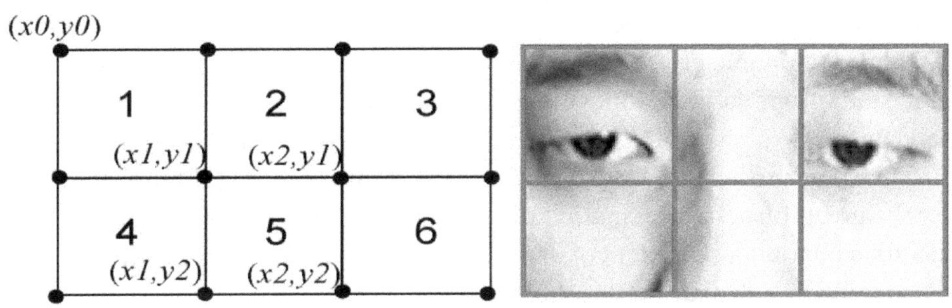

Figure 4. The modified eye-tracking algorithm

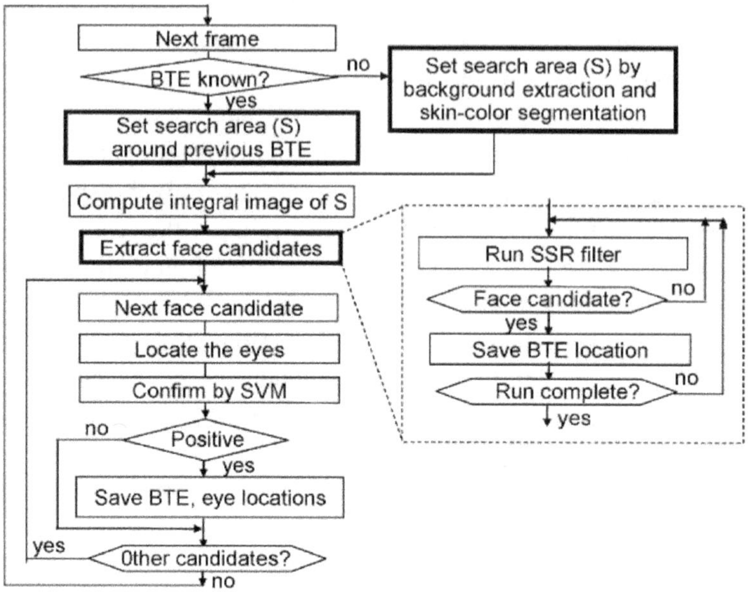

filter (Figure 3). At each location, the SSR filter compares the integral sums of the segments as follows: At each location, the SSR filter compares the integral sums of the segments as follows:

Sum< *Sum* (2) & *Sum* (1) < *Sum* (4) (1) (1)

Sum (3) < *Sum*& *Sum* (3) < *Sum* (6) (2) (2)

If the above criteria (1) are satisfied, the SSR is considered to be a candidate for the BTE pattern (i.e. face candidate) and two local minimum (i.e. dark) points each are extracted from the regions 1 and 3 of the SSR for left and right eyes, respectively.

In the search for eyes, we ignore 2 pixels at the boarder of the regions to avoid effects of eyebrows, hair and beard. In addition, because the eyebrows have almost the same grey level as the eyes, the search starts from the lowest positions of regions 1 and 3.

The eye localization procedure is organized as a scan over the green plane representation of regions 1 and 3 for a continuous segment of dark pixels (i.e. whose value is lower than the threshold k). As in Kawato (2000), we assume that eyes are located if the distance between the located eyes (*D*) and the angle (*A*) at the center point of the BTE area 2 (see Figure 5, left) satisfy the following:

$$30 < D < 42 \ \& \ 115° < A < 180° \quad (3)$$

If both eyes are detected, the user's gaze is considered to be on screen. Otherwise, it is assumed to be off the screen. The eye positions found in the current frame are then used to reduce complexity of processing the successive frames. Since the face/eye motion of computer user is slow in practical circumstances, the search in the next frame is limited to a small region, which spans by 8 pixels in vertical and horizontal direction around the eye points of the current frame.

Figure 5 (right) demonstrates the search area reduction by our algorithm: dashed line shows the area defined by background extraction; dotted line depicts the area obtained by skin-color segmentation; the plain (dark line) shows the area around the previous BTE point; white crosses show the computed locations of eyes.

Figure 5. An illustration of the eye detection heuristics (left) and the search area reduction (right)

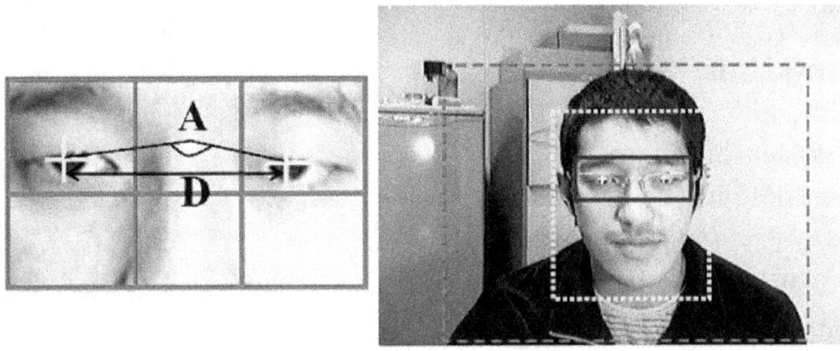

If for any frame, the user gaze is detected to be on screen, the backlight luminance is set-up to the highest level by resetting the backlight voltage to the initial value. If the gaze has been detected off the screen for N consecutive frames, the backlight voltage is reduced ($V=V-\Delta V$) to the next level to dim the backlight luminance down. If the lowest level of the backlight voltage supply has been reached and the user still is not found in the camera readings, the system switches the display off to save energy

4. IMPLEMENTATION

4.1. System Overview

Figure 6 outlines the block-diagram of the proposed computer display power management system. The user tracking unit receives an RGB color image and outputs two logic signals, u1,u0. If the user is detected in the image, the signal u0 is set to 1; otherwise it is 0. The zero value of u0 enforces the voltage converter to shrink the backlight supply voltage to 0 Volts, dimming the display off. If the eye-gaze detector determines that the user looks at screen, it sets u1=1. When both u0 and u1 are 1, the display operates as usual. If the user's gaze has been off the screen for more than *N* consecutive frames, u1 becomes 0. If u0=1 and u1=0, the input voltage (Vb) of the high-voltage inverter is decreased by ΔV. This voltage drop lowers backlight luminance and so shrinks the power consumption of the display. Any on-screen gaze in this low power mode reactivates the initial backlight luminance and moves the display onto normal mode. However, if no on-screen gaze has been detected for *N* consecutive frames and the backlight luminance has already reached the lowest level, the display is turned off. Returning

Figure 6. System overview

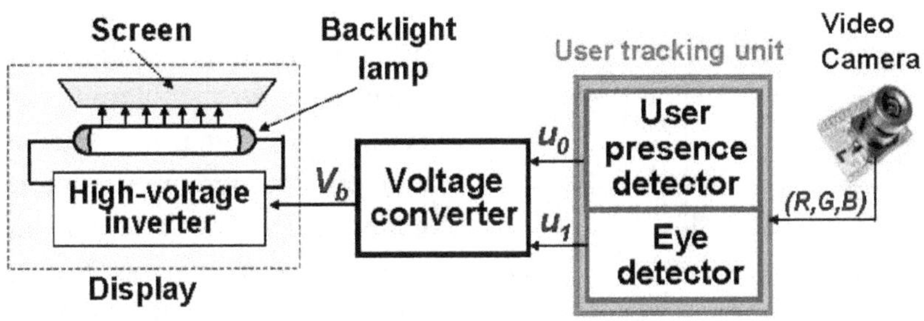

back from the OFF mode requires pushing the ON button.

4.2. User-Tracking Unit

We implemented the system in hardware. Figure 7 illustrates the hardware design of the user tracking unit. The design was synthesized in Verilog HDL using Synopsis Design Compiler and realized on a single FPGA (Xilinx XC3S250E) board. The board is connected to VGA CMOS camera, through parallel I/O interface.

Table 2 summarizes parameters of the FPGA design. The user-tracking unit operates at 48 MHz frequency using 3.3 V external voltage and provides user presence detection at 20 fps rate. Due to capacity limitations of the on-chip SRAM memory, input images were 160 x 120 pixels in size. The SSR filter was 30 x 20 pixels in size. The total power consumption of the gaze detector design was 150mW, which is 35 times less than software implementation of the user presence detector on desktop PC (Pentium4@2.53GHz) (Moshnyaga, 2005).

4.3. Display Voltage Convertor

Conventional TFT displays utilize the Pulse Width Modulation (PWM) scheme to link brightness of the Cold Cathode Fluorescent Lamps (CCFL)

Table 2. FPGA design parameters

Parameter	Value
System clock frequency	48MHz
External voltage	3.0V
Internal voltage	1.2V
System gate count	250000
Logic cell count	18508
Memory size	216Kb
Frame size (pixels)	160x120
Detection rate	20fps
Power	150mW

Figure 7. The user-tracking hardware

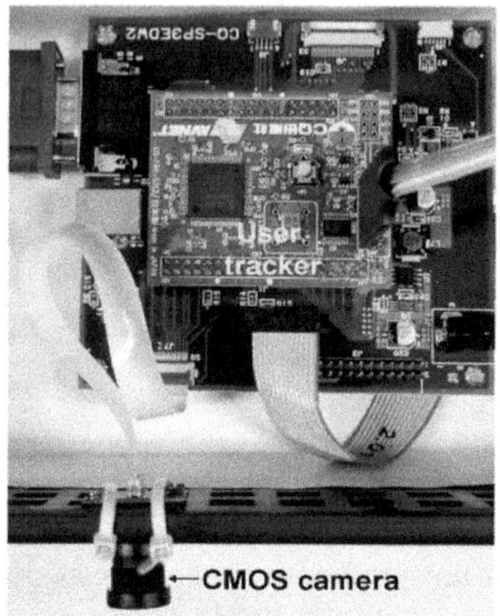

to the input voltage, Vb, of the CCFL AC-DC inverter. Usually, the display brightness is the highest when Vb=0 and the lowest when Vb=5V. To change the brightness dynamically, we modified the display voltage convertor circuit by adding control circuitry, shown in Figure 8. Table 3 outlines parameters of the display voltage control and display the power consumption observed at the corresponding levels of input voltage, Vb. As it clearly shows, changing the brightness leads to large power savings of display.

Figure 9 outlines block-diagram of the modified voltage controller. The inputs (u0,u1) are set by the eye-gaze detector (see Figure 6). When u0=0, the relay disconnects the 12V power supply to the display switching it OFF. Turning the display ON automatically sets u0 to 1, which enforces +12V voltage supply to display. When u0=1, the display power consumption is controlled by the output (Vb) of the voltage controller. Namely, when u1=1, the microcontroller (PIC12F629) nulls outputs L1, L2 setting Vb to 0V. When u1=0, the microcontroller increments its state with each new rise of signal *clk*, thus setting the voltage converter to increment its output Vb by approximately 0.7V. The state transition diagram of the controller is shown in Figure 10. Here "off" and "on" denote states of the ON/OFF switch, respec-

Figure 8. The photograph of the modified display circuit

Table 3. Display voltage control parameters

u1 u0	Vb (V)	Brightness	Power (W)
1 1	0	bright	34.03
1 0	1.2	Level 1	27.43
1 0	2.4	Level 2	22.21
1 0	3.6	Level 3	17.54
1 0	4.8	Level 4	14.62
0 -	-	OFF	0.01

tively. Any signal "u0=0" or "off" sends the display OFF independently of the display power state and the level of screen brightness. When neither of these signals is set, the screen remains ON, while its brightness is controlled by the value of signal u1.

5. EXPERIMENTAL EVALUATION

To evaluate accuracy of the gaze detector, we ran four different tests each of each conducted by different users. The users were free to look at the camera/display, read from the materials on the table, type text, wear eyeglasses, move gesticulate or even leave the PC whenever wanted. Figure 9 illustrates the detection results on 4 images. The + marks depict positions where the system assumes the eyes to be. As we see, even though the lighting conditions of faces vary, the results are correct. Ordinary pairs of glasses (see Figure 11) have no bad effect on the performance for frontal faces. In some face orientations, however, the frame of pair of glasses can hide a part of eye ball, causing the system to lose the eye, or sometimes it takes eyebrow or hair as an eye and tracks it in the following frames. Figure 12 exemplifies two

Figure 9. The modified voltage converter

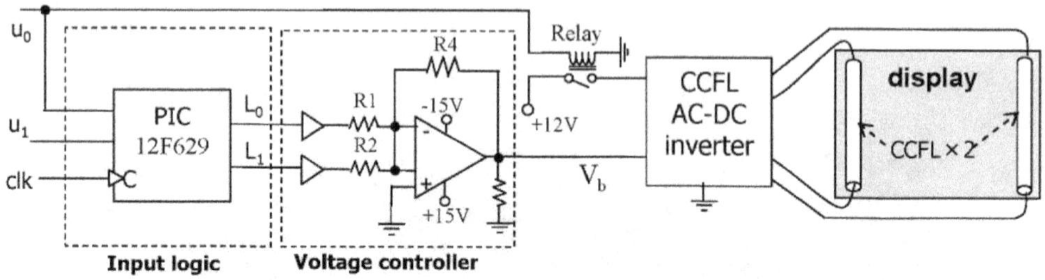

Figure 10. The state transition diagram of the voltage converter

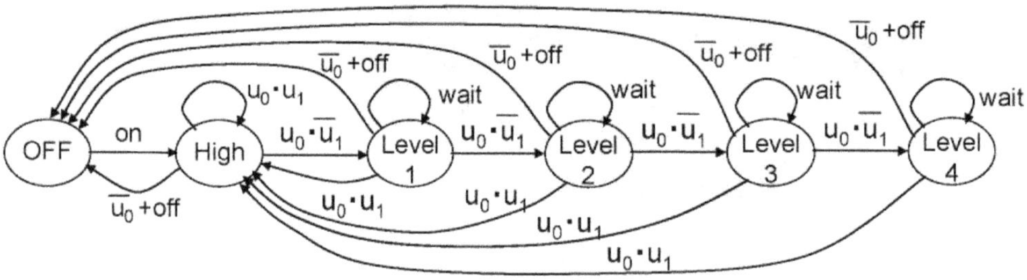

image frames in which eye positions were incorrectly found.

Table 4 summarizes the results. Here, the second column depicts the total number of frames considered in the test; columns marked by 'True' and 'False' reflect the number of true and false detections, respectively; the *Accuracy* column shows the ratio of true decisions to the total number of all decisions made. As the tests showed, the eye detection and tracking accuracy of proposed hardware is quite high (88% on average).

Next, we estimated the energy efficiency of the proposed system by measuring the total power consumption taken from the wall by the system itself and the 17" IO-DATA TFT LCD display (35W on peak, eight different levels of screen brightness) controlled by the system.

Figure 12 profiles the results measured per frame on a 100sec (2000frames) long test. In the test, the user was present in front of the display (frames 1-299, 819-1491, 1823-2001); moved a little from the display but still present in the camera view (frames 1300 to 1491); and stepped away from the PC disappearing from the camera (frames 300-818, 1492-1822). The system was set to step down from the current power level if the eye-gaze off the screen was continuously detected for more than 15 frames (i.e. almost 1 sec). The ACPI line shows the power consumption level ensured by the conventional ACPI power management.

We observe that our technology is very effective. It changes the display power accordingly to the user behavior; dimming the display when the user gaze is off the screen and illuminating the screen (by elevated power) when the user looks on it. Changing the brightness from one power level to another in our system takes only 20ms, which is unobservable for the user. Figure 10 shows the brightness of the screenshots and the corresponding power consumption level (see the figures which are displayed on the down-right corner of the pictures; the second row from the bottom shows the power).

The total power overhead of the system is 960mW. Even though the system takes a little more power than ACPI (see horizontal line in

Figure 11. Examples of correct eye-detection

Figure 12. Examples of incorrect eye-detection

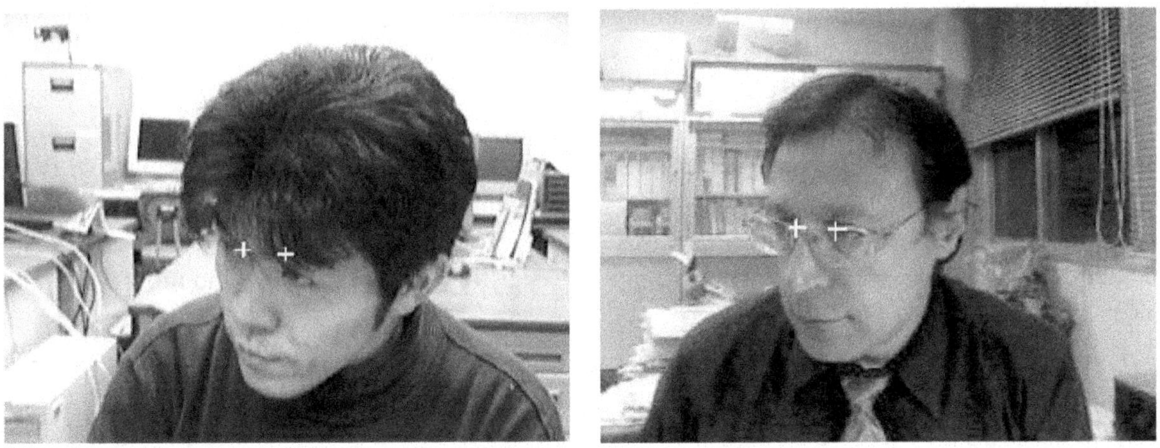

Table 4. Results of evaluation on test sequences

Test	Frames	True	False	Accuracy (%)
1	151	133	18	88
2	240	214	26	89
3	100	90	10	90
4	180	152	28	84
Average	167	147	20	88

Figure 12) in active mode on this short test, it saves 36% of the total energy consumed by the display. In environments when users frequently detract their attention from the screen or leave computers unattended (e.g. school, university, office) the energy savings could be significant (see Figure 13).

CONCLUSION

In this chapter, we presented a new technology for display power management by camera-based user monitoring and outlined its hardware implementation. Unlike existing technologies, our method takes tracks eyes of the user keeping the display active only when it needed. Experiments showed that the technology provides user monitoring in real-time with 88% accuracy while consuming only 1W of power. However, these power figures can be reduced even further should custom design of both chip and board performed. Such board can also accommodate a CMOS image sensor in order to reduce power consuming data transfers. There are many applications, which require eye tracking hardware; so we are currently working on smart vision sensor design.

The research presented here is a work in progress and the list of things to improve it is long. In the current study, we restricted ourselves to a simple case of a singular user. However, when talking about the user-gaze monitoring in general, some critical issues arise. For instance, how to handle several users? The main PC user might not look at screen while the others do. Concerning this point, we believe that a feasible solution is to keep the display active while there is someone looking at the screen. Monitoring the users located far away from the camera also is far from trivial. Besides, some users can use PC in low illumination conditions, wearing sunglasses or hats, which may affect the quality of eye tracking. Though in some cases the use of IR-camera may help, the correct real-time face monitoring with low-energy overhead might be quite difficult. We are currently investigating the issue as well as the influence of camera positioning, user gender/race etc.

Finally, due to the shortage of energy resources the cost of energy will eventually increase in the near future. We hope that making energy-management technologies more intelligent and user-centric will help us consume less energy; make our living less expensive and more enjoyable.

Figure 13. Power consumption vs. image frame for the tested display

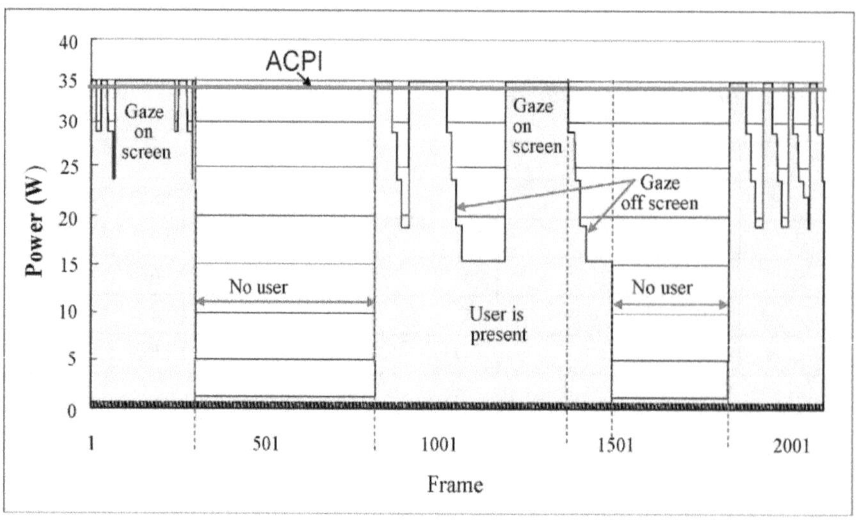

ACKNOWLEDGMENT

The work was sponsored by The Ministry of Education, Culture, Sports, Science, and Technology of Japan under Regional Innovation Cluster Program (Global Type, 2nd Stage) and Grant-in-Aid for Scientific Research (C) No.21500063.

REFERENCES

ACPI. (2004). *Advanced configuration and power interface specification, rev. 3.0*. Retrieved from http://www.acpi.info/spec.htm.

Baluja, S., & Pomerlau, D. (1994). *Non intrusivegaze tracking using artificial neural networks*. Technical report CMU-CS-94-102. Pittsburgh, PA: Carnegie Mellon University.

Chang, N., Choi, I., & Shim, H. (2004). DLS: Dynamic backlight luminance scaling of liquid crystal display. *IEEE Transactions on VLSI Systems*, *12*(8), 837–846. doi:10.1109/TVLSI.2004.831472

Cheng, W.-C., & Pedram, M. (2004). Power minimization in a backlight TFT-LCD display by concurrent brightness and contrast scaling. *Design Automation and Test in Europe*, *1*, 16–20.

Choi, L., Shim, H., & Chang, N. (2002). Low power color TFT LCD display for hand-held embedded systems. In *Proceedings of the International Symposium on Low-Power Electronics and Design*, (pp. 112-117). IEEE.

Climate Group. (2008). *SMART2020: Enabling the low carbon economy in the information age*. Retrieved from http://www.smart2020.org.

Dai, X., & Raychandran, K. (2003). *Computer screen power management through detection of user presence*. US Patent 6650322. Washington, DC: US Patent Office.

Douxchamps, D., & Campbell, N. (2008). Robust real time face tracking for the analysis of human behavior. *Lecture Notes in Computer Science*, *4892*, 1–10. doi:10.1007/978-3-540-78155-4_1

Duchowski, A. T. (2002). A breadth-first survey of eye tracking applications. *Behavior Research Methods, Instruments, and Computers*. Retrieved from http://andrewd.ces.clemson.edu/research/vislab/docs/BET107cr.pdf.

Flinn, J., & Satyanarayanan, S. (1999). Energy-aware adaptation for mobile applications. In *Proceedings of the Symposium on Operating Systems Principles*, (pp. 48-63). Operating Systems Principles.

Fujitsu-Siemens. (2006). *Energy savings with personal computers*. Retrieved from http://www.fujitsu-siemens.nl/aboutus/sor/energy_saving/prof_desk_prod.html.

Gatti, F., Acquaviva, A., Benini, L., & Ricco, B. (2002). Low-power control techniques for TFT LCD displays. In *Proceedings of the International Conference on Compilers, Architecture and Synthesis for Embedded Systems*, (pp. 218-224). IEEE.

Hansen, D. W., & Ji, Q. (2010). In the eye of the beholder: A survey of models for eyes and gaze. *IEEE Transactions on Pattern Analysis and Machine Intelligence*, *32*(3), 478–500. doi:10.1109/TPAMI.2009.30

Harmon, R., Demirkan, H., Auseklis, N., & Reinoso, M. (2010). From green computing to sustainable IT: Developing a sustainable service orientation. In *Proceedings of the 43rd Hawaii International Conference on System Sciences*, (pp. 1-10). IEEE.

Hewlett-Packard Co. (2006). *Global citizenship report*. Retrieved from www.hp.com/hpinfo/globalcitizenship/gcreport/pdf/hp2006gcreport_lowres.pdf.

IEA. (2009). *Gadgets and gigawatts -- Policies for energy efficient electronics*. IEA Press.

Iranli, A., Lee, W., & Pedram, M. (2006). HVS-aware dynamic backlight scaling in TFT-LCDs. *IEEE Transactions on Very Large Scale Integration Systems*, *14*(10), 1103–1110. doi:10.1109/TVLSI.2006.884151

Ji, Q., & Zhu, Z. (2004). Eye and gaze tracking for interactive graphic display. *Journal of Machine Vision and Applications*, *15*(3), 139–148.

Kawato, S., & Ohya, J. (2000). Two-step approach for real-time eye-tracking with a new filtering technique. In *Proceedings of the IEEE International Conference on Systems, Man and Cybernetics*, (pp. 1366-1371). IEEE Press.

Kawato, S., Tetsutani, N., & Osaka, K. (2005). Scale-adaptive face detection and tracking in real time with SSR filters and support vector machine. *IEICE Transactions on Information & Systems (Norwalk, Conn.)*, *88-D*(12), 2857–2863.

Mahesri, A., & Vardhan, V. (2005). Power consumption breakdown on a modern laptop. *Lecture Notes in Computer Science*, *3471*, 165–180. doi:10.1007/11574859_12

Morimoto, C., & Mimica, M. R. M. (2004). Eye gaze tracking techniques for interactive applications. *Computer Vision and Image Understanding*, *98*(1), 4–24. doi:10.1016/j.cviu.2004.07.010

Moshnyaga, V. G., Hashimoto, K., Suetsugu, T., & Higashi, S. (2009). A hardware implementation of the user-centric display energy management. *Lecture Notes in Computer Science*, *5953*, 56–65. doi:10.1007/978-3-642-11802-9_10

Moshnyaga, V. G., & Morikawa, E. (2005). LCD display energy reduction by user monitoring. In *Proceedings of the IEEE International Conference on Computer Design*, (pp. 94-97). IEEE Press.

Nordman, B., Piette, M.-A., & Kinney, K. (1996). *Measured energy savings and performance of power-managed personal computers and monitors*. LBL-38057. Retrieved from http://eetd.lbl.gov/EA/Reports/38057/.

Nordman, B., Piette, M.-A., Kinney, K., & Webber, C. (1997). *User guide to power management for PCs and monitors*. LBNL-39466. Retrieved from http://eetd.lbl.gov/EA/Reports/39466.

Ohno, T., Mukawa, N., & Kawato, S. (2003). Just blink your eyes: A head-free gaze tracking system. In *Proceedings of the Computer-Human Interface*, (pp. 950-951). Computer-Human Interface.

Park, R., & Kim, J. (2005). Real-time facial and eye gaze tracking system. *IEICE Transaction on Information & Systems (Norwalk, Conn.)*, *88-D*(6), 1231–1238.

Park, W. I. (1999). *Power saving in a portable computer*. EU Patent EP0949557. Geneva, Switzerland: European Union.

Pasricha, S., Luthra, M., Mohapatra, S., Dutt, N., & Venkatasubramanian, N. (2004). Dynamic backlight adaptation for low-power handheld devices. *IEEE Design and Test Magazine*, *21*(5), 398–405. doi:10.1109/MDT.2004.57

Pattanai, K. S. N., Tumblin, J. E., Yee, H., & Greenberg, D. P. (2000). Time dependent visual adaptation for realistic image display, in *Proceedings of SIGGRAPH,* (pp. 47-54). SIGGRAPH.

Robertson, J., Homan, G. K., Mahajan, A., Nordman, B., Webber, C. A., & Brown, R. E. … Koomey, J. G. (2002). *Energy use and power levels in new monitors and personal computers*. LBNL-48581. Berkeley, CA: UC Berkeley.

Sharp Microelectronics of the Americas. (2002). *Display modes: Transmissive/reflective/transflective*. Retrieved from http://www.sharpsma.com/sma/Products/displays/AppRefGuide/DisplayModes.htm.

Shim, H., Chang, N., & Pedram, M. (2004). A backlight power management framework for the battery-operated multi-media systems. *IEEE Design and Test Magazine*, *21*(5), 388–396. doi:10.1109/MDT.2004.43

Theocharides, T., Link, G., Vijakrishnan, N., Irwin, M. J., & Wolf, W. (2004). Embedded hardware face detection. In *Proceedings of the 17th IEEE International Conference on VLSI Design*, (pp. 33-138). IEEE Press.

Tumblin, J. E., Hodgins, J. K., & Guenter, B. K. (1999). Two methods for display of high contrast images. *ACM Transactions on Graphics*, *18*(1), 56–94. doi:10.1145/300776.300783

VESA. (2011). *Display power management signaling specification (DPMS)*. Retrieved from http://vesa.org/vbe3.pdf.

Viola, P., & Jones, M. (2001). Rapid object detection using a boosted cascade of simple features. In *Proceedings of the IEEE Conference on Computer Vision and Pattern Recognition*. IEEE Press.

Yamamoto, S., & Moshnyaga, V. G. (2009). Algorithm optimizations for low-complexity eye tracking. In *Proceedings of the IEEE International Conference on Systems, Man and Cybernetics*, (pp. 18-22). IEEE Press.

ENDNOTE

1. Apple PCs employ power management technology that is distinct from but similar to ACPI (Nordman, 1996, 1997).

Chapter 16
Energy-Aware Intelligence in Smart Spaces:
A Case Study using Computer Vision and Machine Learning for User-Behavior Analysis

Sangho Park
University of North Dakota, USA

Henry Kautz
University of Rochester, USA

ABSTRACT

The improvement of energy efficiency in our society has become an urgent issue for sustainability under global warming. The authors present research issues on sensor-based smart environments for energy-aware intelligence, and showcase a study of algorithms for monitoring human activities that provides the context awareness to the smart environments. In order to build energy-aware environments, it is desirable to embed intelligence into the environment itself so that the environment can interpret human behavior in order to adjust itself to human activities occurring in the environment. This is achievable by integrating the environment and the intelligent computing facilities. The computing facility embedded in the environment is equipped with intelligent algorithms that can monitor salient features indicative of the events and learn and recognize changes in the environment. Recent developments in sensor-based intelligent systems can provide suitable algorithms and facilities for building such energy-aware smart environments. The authors present a framework for monitoring human activities in daily living toward the energy-aware intelligence that can detect and learn inhabitants' behavior patterns in the smart environment.

DOI: 10.4018/978-1-4666-1839-8.ch016

INTRODUCTION

The improvement of energy efficiency in our society has become an urgent issue for sustainability in the current era of global warming. The energy efficiency is closely related to the design of infrastructure such as buildings, transportation systems, Information Technology (IT) devices, etc. The energy efficiency also depends on how people use such infrastructure. There is a conflict, however, between the need for energy efficiency in human-inhabited environments and the inhabitants' desire for comfort. For example, reducing the cost of air conditioning or heating for a building would result in the increased humidity in summer or the chilly indoor temperature in winter, respectively. Therefore, it is imperative to achieve the balance between the cost and the comfort. For example, in case of room temperature, the typical practice of controlling the room temperatures in a building is to have a building maintenance technician adjust the lowest and highest levels of the thermometer setting at each room for each season. In case of room illumination, human involvement of such kind can be minimized by installing an infrared motion sensor to detect the inhabitants' motion to recognize their existence and controlling the on/off switches of the lighting equipments. In these examples, however, the control of actuators for heating, air conditioning, and lighting, is fulfilled in a brute-force manner. That is, once set, the actuators will work consistently in the same way no matter what situation the environment would be in. For example, the temperature of a classroom with 50 students at class will be the same as the temperature of the classroom without anybody during off-class period on weekend. It is of course possible to manage the classroom temperature differently on weekdays and weekends by incorporating the day's information into the control system. This would require more sophisticated control systems and extra manual interfaces. The systems that require extra human involvement, however, are not scalable. How much manpower would be needed to manage several hundred classrooms on a university campus? What happens to the system if the technician's maintenance is delayed? What if the pattern of human activity changes unexpectedly (such as accidents, multiple users, and guests, etc.)? We need more intelligent systems for energy-aware environments that can automatically adjust themselves according to the activities of the inhabitants in the environment. One of the core research issues in energy-aware environment is the development of the computer vision and machine learning algorithms to understand human activity in the environment.

In this chapter, we present a case study of computer vision and machine learning that recognizes human activity in sensor-augmented smart spaces for energy-aware intelligence. Figure 1 shows the schematic diagram of the proposed system framework.

Smart environment or *smart space* is "a small world where different kinds of smart device are continuously working to make inhabitants' lives more comfortable" (Cook & Das, 2004). The terms *environment* and *space* are interchangeably used in this context. Mark Weiser who coined the term 'ubiquitous computing' says that smart environment is "a physical world that is richly and invisibly interwoven with sensors, actuators, displays, and computational elements, embedded seamlessly in the everyday objects of our lives, and connected through a continuous network"(Cook & Das, 2004). With the richness of sensor types and their numbers embedded in the environment, smart environment is closely related to "ubiquitous computing."

From the perspective of energy-aware intelligence, the smart environment is regarded as an intelligent system that detects certain changes in the environment, autonomously learns patterns of inhabitants' activities, and adjusts the environment itself to optimize resource usage such as heating, cooling, and lighting, etc., while providing the inhabitants with best convenience in an autonomous manner. In this sense, the smart environment is

Figure 1. Schematic diagram of the proposed system framework

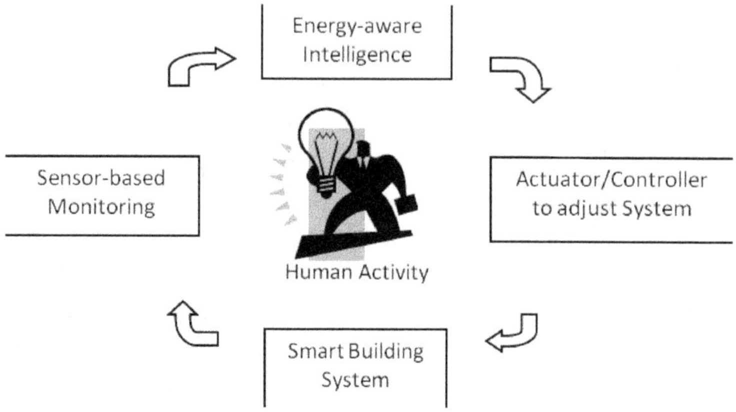

closely related to "automatic control" of various actuators, but in a more sophisticated and intelligent way than traditional control systems.

Research on smart environment involves many subareas, each with diverse issues and challenges on its own, but in this chapter, we focus on the energy-aware *intelligent system*: i.e., the computer system equipped with sensors and information processing algorithms. More specifically, we focus on computer vision and machine learning algorithms applicable to smart spaces that can recognize inhabitants' activities in an autonomous manner. The machine-learning algorithm of the system can be trained to learn the usual patterns of the inhabitants' several daily activities and events such as walking around, watching TV, preparing meal, arranging utensil, etc. With the gained knowledge of the human inhabitants' usual patterns, the algorithm can classify newly detected events that occur in the environment, and provide intelligent interface for the inhabitants or the facility managers. The algorithm of the system for automatic recognition of human activities could also be used to control the various actuators to adjust the environmental settings such as illumination levels, temperature, air circulation, door locking, etc., and provide comfortable and energy-aware environments to the inhabitants.

BACKGROUND

The Information Technologies (IT) to solve energy and environmental problems are called *green computing* or *green IT*, and categorized into two types: 'Green of IT' and 'Green by IT' (Cheng, et al., 2011). The goal with 'Green of IT' is to reduce the energy consumed by IT equipments (e.g., computer systems, consumer products, network devices), which constitutes 2% of the total global energy consumption. In contrast, the goal with 'Green by IT' is to apply IT methods to the problem of energy reduction in non-IT areas, which accounts for 98% of the total global energy consumption. OECD (Organization for Economic Co-Operation and Development) published a report on smart sensor networks for green growth in the era of global warming (OECD, 2009). The report focuses on 'Green by IT' and surveys the promising applications of smart sensor technology to cope with environmental challenges and reduce CO_2 emissions. The OECD report also emphasizes that the success of such efforts not only depends on sensor technology but also on a proper use of these systems by humans. Facility management examples include space heating, water heating, cooling, lighting, cooking, and electrical appliances, etc. Development of smart buildings using Green by IT is one of the promising areas to reduce

CO_2 emissions. The environmental impact of smart buildings is closely related to the optimal facility management where sensor technology contributes to energy savings (OECD, 2009).

Regardless of approaches to building smart homes, 'context awareness' is the most salient feature of such an intelligent environment (Roy, 2006). 'Context awareness' is the idea of intelligent automation that computers can both sense and react based on their environment. Ontology and semantic Web technology have been used for representing such context awareness in smart home research (Wang, 2004; SWRL, 2011), but the design / engineering of the proper ontology are not trivial (Protégé, 2011); it requires significant efforts and costs (Cheng, 2011). It is because the ontology-based approach is mainly rule-based, and the ontology design involves the careful manual crafting of proper rules. Different personal preferences make it even harder for such developed ontology to cope with personal differences in reality. Cheng (2011) suggests the incorporation of the machine learning techniques as a future study. Machine learning is promising in that it can eliminate manual engineering to design the rules and enable the system to learn the proper rules autonomously. Our study presented in this chapter belongs to such approaches to use probabilistic machine learning methods to implement energy-aware intelligence.

Building a generic sensor-based intelligence system is indeed its early stage (Roy, et al., 2006). Roy et al. proves that it is a NP-hard problem to achieve the optimal location prediction of multiple inhabitants. Therefore, most of the sensor-based intelligent systems developed so far aim at a single target for object recognition (Rekimoto, 2000) or 3D scene reconstruction (Boukraa, 2002). Rekimoto et al. use a camera system to recognize object category by scanning the 2D barcode attached to the object. In (Boukraa, 2002) users use RFID (Radio-Frequency Identification) to identify object types and retrieve database and use vision sensors to match the 3D model to the imaged object. In Hontani (2003), the users apply a sophisticated illumination to recover 3D model of objects. In McDaniel (2008), the authors integrate RFID and vision sensors for object perception for blind people. One of the major limitations of such systems is that the systems assume the cooperative user who handles the sensor device in the predefined proper way so that the system works. It is not the goal of such systems that human inhabitants are moving unobtrusively without the awareness of the existence of the sensors and actuators/controllers.

We argue that in order to build energy-aware intelligence it is important to monitor natural behavior patterns of human inhabitants in an unobtrusive manner. It is crucial not to require the inhabitants to wear special sensors or manipulate devices in specific ways in a cooperative manner for the monitoring of the naturalistic human activities in daily living in smart environments. It is reality that people are often non-cooperative or unfamiliar with sensors or controllers even though they are willing to save energy usage (Logica Survey, 2007).

In order to build an energy-aware smart environment, it is desirable to make the environment itself aware of what is happening and adjust itself according to the recognized events. This is achievable by integrating the environment with intelligent computing facilities. The computing facility is equipped with intelligent algorithms that can detect observable features salient to particular events, and learn and recognize changes in the environment. Recent development in sensor-based intelligent systems can provide suitable algorithms for building such an energy-aware smart environment.

Before we proceed, it would be informative to take a short review of historical improvement in smart environments in a broad sense in conjunction with automatic control and ubiquitous computing. The early stage of controlling the human-inhabited environments was achieved by manual operations of the equipments for heating the space, lighting the rooms, locking the doors, etc. A traditional house with a fireplace and several dead bolts for main doors is a good example. The equipments for controlling the environment in this stage were typically used with the option of manual activation and suspension (i.e., manual on/off mechanism). Manual tools for controlling the environment were then replaced by mechanical or electrical actuators. The mechanical or electrical actuators in this stage were optionally controlled by a clock with preset on/off times with manual configuration of the parameter values. Electric washer with mechanical knob switches is a good example. Electro-mechanical actuators were then integrated with negative-gain feedback loop mechanism for *automatic control*. The preset parameter is then replaced by dynamic parameter settings adjusted by environmental sensors. A good example is a modern building equipped with wireless remote controls for air conditioning, infrared motion detectors for lighting and door opening, pressure sensors for adjusting beds, etc. Advanced smart controllers accessible via the Internet are commercially available these days (X10, 2011; Wattsup, 2011) for monitoring the power usage of various devices and switching the power of such devices remotely or programmatically. These dynamic sensors and controls enable certain degree of dynamic self-adaptation of the human-inhabited environment. However, most of such systems still require the user involvement for the activation and suspension of the system and the configuration of the operational ranges of the system.

The development of embedded systems and ubiquitous computing in recent decades opened a new level of innovation for controlling the environments: that is, *intelligence for smart environments*. From the perspective of information systems, the smart environment is characterized by *computation*; the core innovation resides in computational algorithms that can integrate various sensor data, machine learning capability for recognizing specific patterns of events, probabilistic inference, and autonomous decision making for the self regulation of the system. In a nutshell, the intelligent system for smart environments needs to extract and learn dynamic contexts of a wide range of events and quickly transform the signal-level data to higher-level semantic information for the interpretation of the events and for the decision making to adjust itself to human inhabitants. Computer-based recognition of human activities in daily living has gained increasing interest from engineers and scientists as a promising technology for enabling the next-generation smart environments. Understanding human activities can be approached from different levels of detail; for example, the body transition across a room at a coarse level, vs. the hand motion that manipulates objects at a detailed level (Aggarwal & Park, 2004). Therefore, different levels of representation are used for different application purposes.

Previous approaches to recognizing human activities in daily living have adopted various sensing modalities to reliably detect salient features and to ease the task of recognizing human activities. In general, different sensors provide different information characteristics. Mozer (2005) built an 'adaptive house' that minimizes user involvement in controlling resources such as heating, air conditioning, and lighting based on the inhabitant's occupancy and motion patterns detected by low-cost sensors including motion sensors. In Mozer's system, the decision making for regulating the resources was based on reinforcement learning using an artificial neural network. The main goal of Mozer's 'adaptive house' is to minimize resource usage and maximize user comfort by learning the inhabitant's life style from sensor data. He also suggested machine-learning approaches for classifying the inhabitants' activities as a future study.

Another example of using many low-cost sensors for smart home is MIT's PlaceLab project (Intille, et al., 2005). The various objects in the smart home at the PlaceLab such as furniture, kitchen items, toilet items, etc., are attached by hundreds of sensors including wired reed switches, water flow sensors, current sensors, temperature sensors, etc. Inhabitants' activity patterns are logged by recording the on / off signals of the sensors when the user manipulates the items. Most of the objects in the PlaceLab project are fixed (i.e., non-disposable) items. However, if fine-detailed activities such as cooking are of interest, then disposable items should be recognized as well. Patterson et al. used wearable sensors (RFID readers and accelerometers) on the glove form factor to detect various disposable objects identifiable with RFID tags attached on them (Patterson, 2005). Various activities in daily living involve different series of object manipulations. For example, cooking involves the manipulation of special utensils in specific order, while making a phone call involves a very different set of objects (i.e., a phone and a yellow book or a personal address book.) The glove-attached RFID reader identifies the manipulated objects and registers the sequential order of manipulated objects. A dynamic Bayesian network was used in Patterson (2005) to learn activity patterns and recognize new data. This system assumes cooperative users willing to wear the glove and requires the attachment of RFID tags on each object. The application domain was indoor activities in a smart home environment.

In office environment, however, it is cumbersome and distractive to wear the RFID-reader glove during the work; unobtrusive sensing modalities are needed. Oliver et al. used audio-visual signals captured by many cameras and microphones deployed across an office environment to detect human activities (Oliver, 2002). Their system can recognize typical activities occurring in an office such as entering / exiting the office, making a phone call, using a computer keyboard, etc. A layered Hidden Markov Model was used to train and recognize activity patterns. A broader survey on audio-visual information fusion for human-computer interface and intelligent environments is in Shivappa et al. (2010).

If the complexity of the environmental configuration is not so high, much simpler systems can be used for monitoring a specific category of activities. A single-camera vision system was used by Abowd (2007) to develop an intelligent healthcare system for monitoring autistic children. Mihailidis (2007) also used a single camera system to develop an intelligent healthcare system that can interact with Alzheimer patients for their independent living in a smart home. This system can be trained to learn hand-washing patterns of people and provide voice instruction to help Alzheimer patients follow the correct sequence of normal hand washing. Activities in daily living may involve various kinds of object. For example, cooking activities requires many heterogeneous objects including food ingredients, utensils, silverware and tools. Wu et al. (2007) developed a sophisticated intelligent system that combined vision and RFID. The system was able to learn various object types based on appearance. RFID tag information was used for labeling object categories instead of manual labeling. The system is, however, limited by a single camera view, which entails view dependency of the processing. In real-world situation, objects frequently occlude one other depending on camera viewpoints. Object occlusion is still a challenging and open problem in vision-based systems.

THE OVERVIEW OF THE PROPOSED SYSTEM

In this chapter, we present a multi-sensor based activity recognition system that uses multiple cameras and RFID sensors integrated with machine learning based pattern classification algorithms. The multiple cameras resolve object occlusion and the RFID sensors automatically labels object categories. Our related previous work is in (Park, 2008a, 2008b). Figure 2 shows the snapshot of our prototype smart home testbed and the multi-camera configuration. The smart home includes

Figure 2. Multi-camera configuration in a prototype smart home. Two wide field-of-view (FOV) cameras (1 and 2) monitor the living room area, while two narrow-FOV cameras (3 and 4) monitor the kitchen counter top. Cameras 1-2 and 3-4 form the approximately orthogonal viewing axes, respectively. The picture on the right shows the view from camera 1.

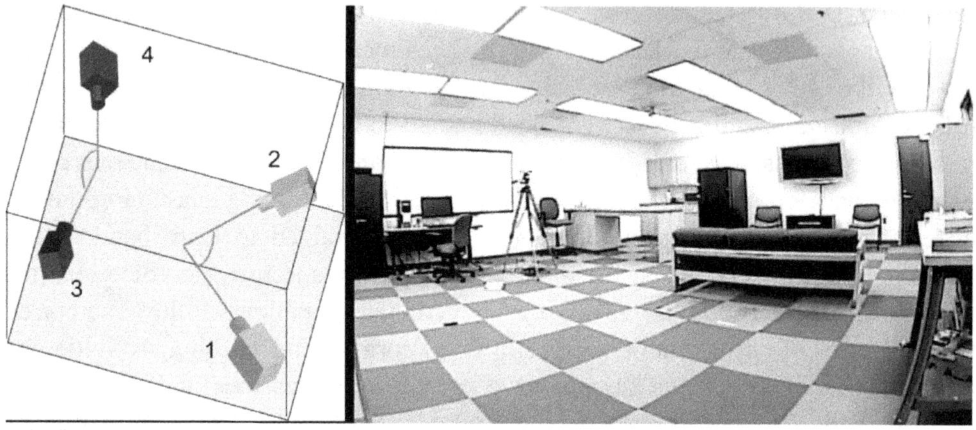

a living-room space with a couch and a wall-mounted TV as well as a large open area, and a kitchen space with a counter top, a cupboard, a microwave oven, and a refrigerator.

The synchronized cameras enable us to capture multi-scale, multi-view videos of human activities, enhance visual recognition under the occlusions, and monitor both large- and small-scale motions of object manipulation. The camera system covers various areas of the smart home environment in different viewing resolutions from different viewpoints. We also use a short-range, bracelet form factor, RFID reader, *iBracelet* developed at Intel Research Seattle, to remotely transmit to the vision system's computer the time-stamped RFID readings of RFID tags attached on objects. The RFID tags are attached on various objects including furniture, appliances, and utensils around the smart home. RFID tags and reader(s) detect without false positives the nearby objects that are handled by the user.

The synergistic integration of vision and RFID provides us with many advantages such as robust segmentation of objects, view-independent tracking, and representation of objects and persons in 3D space, efficient handling of occlusion, efficient learning of temporal boundary between activities without human intervention, and the recognition of human activity at both a coarse and fine level.

Sensor-Based Modeling of Smart Spaces

Our vision system based on multiple cameras enables the view-independent recognition of human activities. As shown in Figures 2 and 3, four cameras are used to capture the scene from different perspectives. Two wide-FOV cameras (1 and 2) form the approximately orthogonal viewing axes to capture the overall space (views 1 and 2 in Figure 3), while the two narrow-FOV cameras (3 and 4) form the approximately orthogonal viewing axes to capture more details of certain focus zones such as the kitchen (views 3 and 4 in Figure 3.) The four synchronized views are overlaid with a virtual grid to compute scene statistics such as pixel counts in each grid cell.

Human activities in daily living may involve multiple objects moving simultaneously, which can create challenges for a vision system: for example, changing backgrounds and object occlusion. We adopt a dynamic background model using K-means clustering (Kim & Davis, 2005). The background model is updated with a memo-

Figure 3. Rectified images 1 to 4 from the activity recognition experiment in which a person performs the "prepare utensil" activity. The virtual grid is overlaid to compute scene statistics.

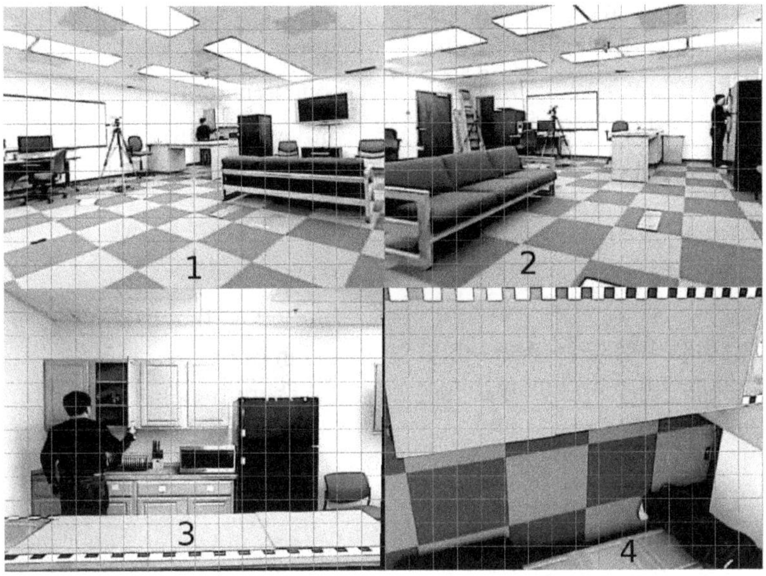

Figure 4. Binary foreground maps corresponding to the multi-view images in Figure 3. The juxtaposed whole image in this figure forms the 'super foreground map' Γ^t *captured at time t. Black areas represent effective foreground regions (i.e., inverted for visualization only).*

ry decay factor to adapt to the changes in the background, and foreground-background segmentation is achieved at each pixel. Any vision-based recognition of inhabitant's activities may incur the concern of privacy. Silhouette representation of the foreground regions may obscure object appearances to preserve privacy (Figure 4), but it also loses the useful information about object identities. RFID reading provides a smart way for identifying the object types even from the obscure silhouette. That is, the onset and offset of a specific RFID label stream may provide a useful clue for the onset/offset of a certain activity that typically manipulates the corresponding objects.

Recognition Framework Based on Vision and Machine-Learning Algorithms

Figures 3 and 4 show the process of representing scene statistics. We denote the *m*-th camera image and its foreground map at time *t* as I_m^t and F_m^t, respectively, ($m \in \{1,2,3,4\}$, See Figure 3). A *super image* Υ^t and its associated *super foreground map* Γ^t at time *t* are obtained by juxtaposing the individual images $I_1^t,\ldots, I_4^t$ and $F_1^t,\ldots, F_4^t$, respectively. Therefore if I_m^t is of size $W \times H$ pixels, the *super image* Υ^t and the *super foreground map* Γ^t are of size $2W \times 2H$. (In our implementation, image width $W = 640$ pixels and image height $H = 480$ pixels).

A virtual grid overlays the *super foreground map* Γ^t (Figure 4) for decimation as follows. Each of the grid cells with cell size of $S \times S$ pixels ($S = 40$ pixels) counts the number of foreground pixels (in Figure 4) within its cell boundary and divides the number with the cell area as follows.

$$\delta_{i,j}^t = \frac{\sum \#\ foreground\ pixels}{S^2}$$

Therefore, $\delta_{i,j}^t$ forms the value of a *super pixel* representing the ratio of foreground occupancy in the cell, ranging [0, 1]:

$$\delta_{i,j}^t \in [0,1] \in R$$

In our current implementation, the original $2W \times 2H$ (= 1280 × 960)$ super foreground map Γ^t is converted to $\frac{2W}{S} \times \frac{2H}{S} (= 32 \times 24)$ *occupancy map* O^t which contains super pixel values $\delta_{i,j}^t$, $i \in \{1,\ldots,32\}, j \in \{1,\ldots,24\}$. We concatenate all the super pixels $\delta_{i,j}^t$ across O^t into a vector format, Δ^t:

$$\Delta^t = \left[\delta_{1,1}^t,\ldots,\delta_{32,24}^t\right]^T$$

where the dimensionality of Δ^t is 768 (= 32 × 24) in our experiments.

Therefore, the entire foreground silhouettes from the four simultaneous camera views at time *t* (e.g., Figure 4) are represented as a *single* point in the 768-dimensional vector space. Consecutive frames of a given video sequence are mapped as sample points in this vector space.

Our conjecture is that two sample points, Δ^{t_1} and Δ^{t_2} from nearby frames (*i.e.*, $|t_2 - t_1| \ll \tau$) involved in the same activity will be grouped together, whereas the sample points from distant portions of the video sequence (*i.e.*, $|t_2 - t_1| \gg \tau$) involved in different activities will get separated in this vector space. The proper threshold value τ can be learned from training data.

Clouds of sample points in the vector space can be grouped into clusters by clustering algorithms such as K-means clustering or Expectation-Maximization (EM) based Gaussian mixture modeling (Duda, et al., 2010). Note that each element of vector Δ^t was normalized into the range of [0, 1]. We adopt the Linde-Buzo-Gray Vector Quantization algorithm (Sayood, 1996), an efficient variant of K-means clustering, to cluster the high-dimensional vector space representation of the video data into *M* clusters. In this approach, the enormous size of raw data containing the long-term video sequence that involves multiple sequential activities is therefore succinctly represented as the transition of cluster labels across the *M* clusters. The patterns of the cluster label transitions among different activities show characteristic differences per activity as well as certain degree of noises within each activity due to personal differences and/or sensor noises, making a stochastic model appropriate for the analysis. Therefore, instead of relying on manual engineering to specify the pattern characteristics, we adopt the Hidden Markov Models (HMMs) (Rabiner, 1989) to learn and classify the activity patterns.

Hidden Markov Models have been successfully used for speech recognition for decades, and then adopted to computer vision research. We use a bank of multiple Hidden Markov Models to train and recognize the activities depicted in Table 1. We addressed that video sequence of activities was converted to succinct patterns of transition among the cluster labels. The specific transition patterns for the human activities depicted in Table 1 should be trained by proper machine learning algorithms; which amount to the training of the HMMs for the activities. We use the standard Baum-Welch and Viterbi algorithms (Rabiner, 1989) to train the individual HMM for each activity categories. We experimentally determined the HMM model parameters, including the number of hidden node states and the number of observation node states. The number of observation node states for RFID was fixed to the number of RFID tags used in the experiments.

As for generating the ground-truth annotations of the data, however, the proper parsing of the temporal sequence of feature streams for activity recognition is still an open research question. Traditional approaches for generating ground-truth annotation of training data are based on cumbersome manual segmentation of feature stream by user discretion. Such approaches are not effective and error-prone for natural activities that may vary in duration.

We use the RFID sensing for segmenting and labeling the training data of the human activity. The RFID reader has the detection range of about 10-15 centimeters. As the person's hand wearing the RFID reader approaches an RFID tagged object, the bracelet detects the tag and wirelessly transmits the time-stamped ID information to the PC-based activity recognition system. In our current configuration, the ID transmission is repeated every second until the person's hand leaves the object. The RFID sensing in our system provides an efficient way for learning the temporal segmentation of salient motions in video by providing the indices of the corresponding frame numbers.

The combination of vision and RFID was proposed by Wu *et al.* (2007) to train object appearance models from high-quality color image frames. In their system, the RFID labels were used only to infer object use. A single detailed-view camera was used in their system. Using raw images from the color camera, however, may raise the issue of privacy in practical situations. Therefore, we choose to use the foreground silhouette (i.e., binary foreground map) or the motion map (i.e., frame-differenced binary image map) instead of using the multi-view raw frames to represent human body and objects while blocking the identity.

Activity Recognition Experiment

Activities in daily living occur in certain contexts. Such contexts may include short-range history of preceding activities, as well as long-range information such as an individual's health conditions, the time of day, the time since the last occurrence of a regularly repeated activity (*e.g.*, toileting), etc. Activities in such contexts may be observed at a coarse level, such as moving across multiple rooms during a day, or at a fine level, such as detailed cooking behavior in a kitchen. Our goal is to encompass both levels of the activity analyses by incorporating the coarse-level and fine-level features simultaneously extracted from the multiple camera views. In the current experiments, we consider the six activity patterns in Figure 5.

Note that each of the six gross-level activities in Figure 5 is composed of a series of fine-level *unit actions*. Also note that activity classes 1 and 2 in Figure 5 occur in the wide living-room space in our smart home testbed, and are monitored by

Figure 5. Activity class descriptions

Enter / exit scene Walk	Take/put remote control Sin on couch Turn on/off TV Watch TV	Open/close cupboard Open/close drawer Bring utensil (bowl, cup) Bring flatware (spoon, knife)
Open/close cupboard Open/close fridge Pour cereal/milk in bowl Eat cereal with spoon	Open/close refrigerator Bring water jar Pour water in a cup Drink water from cup	Open/close cupboard Open/close drawer Return utensil Return flatware

Figure 6. Example frames for activity classes 1 - 3 in Table 1. 1: 'walk around,' 2: 'sit and watch TV,' and 3: 'prepare utensil,' from top to bottom row, respectively. The example frames show raw multi-view input frame, background subtraction, and frame differenced motion map, from left to right column, respectively.

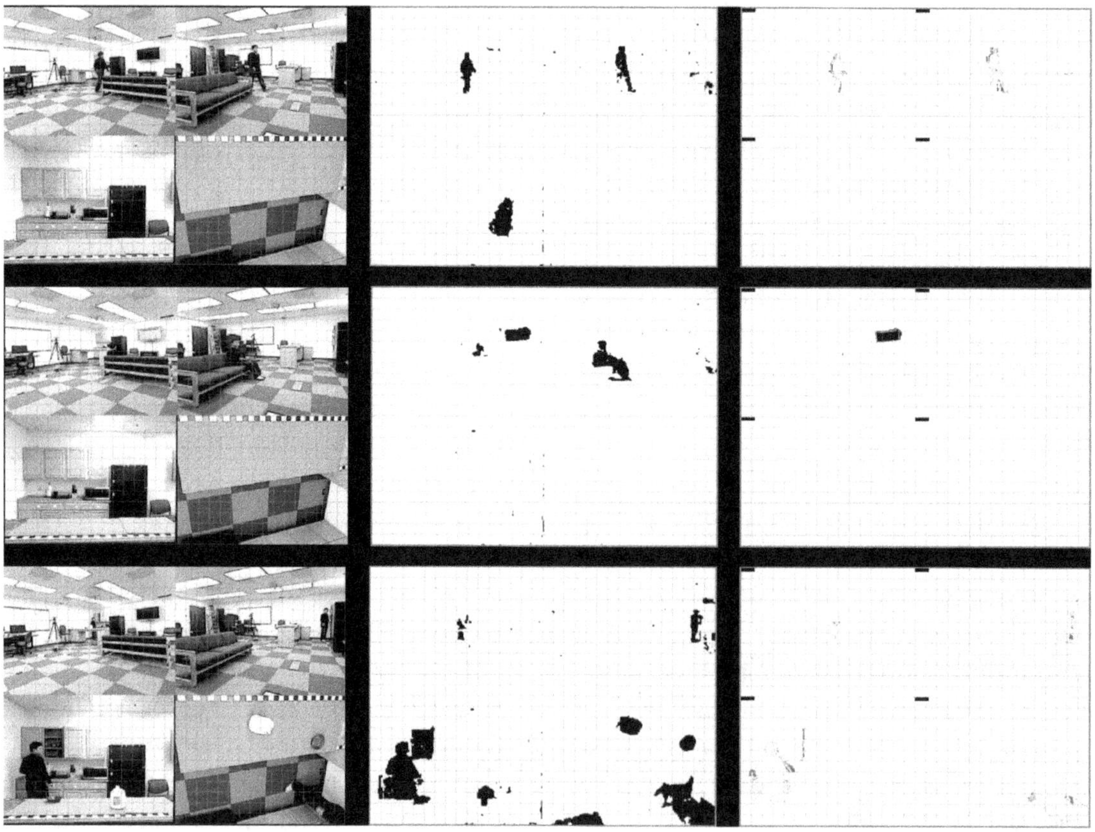

the two wide FOV cameras covering the entire space. (We call the activity classes 1 and 2 *wide activity*.) Activities 3 through 6 in Figure 5 occur in the confined kitchen room space and are monitored by the two narrow FOV cameras monitoring the kitchen area. (We call the activity classes 3 though 6 *narrow activity*.)

Figures 6 and 7 shows example snapshot frames captured by the multiple cameras for the activity categories depicted in Table 1. Each row

Figure 7. Example frames for activity classes 4 - 6 in Table 1. 4: 'prepare cereal,' 5: 'drink water,' and 6: 'store utensil,' from top to bottom row, respectively. The example frames show raw multi-view input frame, background subtraction, and frame differenced motion map, from left to right column, respectively. Notice that some activities may look very similar at certain moments during the sequences (e.g., activity classes 4 and 5).

in Figures 6 and 7 shows different activity classes, while each column in Figures 6 and 7 shows different image maps: i.e., raw frame, foreground silhouette map, and frame-differenced motion map, respectively. The wide activity classes (*i.e.*, wide activities 1 and 2 in Figure 5) have very different characteristics from the narrow activity classes (*i.e.*, narrow activities 3 through 6) in terms of interaction styles with different objects in the environments and available evidence from salient imaging features. Different environments such as living room versus kitchen involve different objects such as couch/TV *vs.* utensil/food, respectively. In such structured environments, human behavior is constrained by the existing setup of objects. This constraint makes it easy to capture reliable visual evidence from stationary cameras. Furthermore, multi-view cameras can capture rich evidence from different perspectives.

We use the foreground map obtained by background subtraction as the *scene statistics* feature, *S*, that describe scene changes from the learned background scene. We use the motion map obtained by frame differencing as the *object statistics* feature, *J*, which represents the local motion of objects in the foreground map. The RFID signals are used as the *reference evidence, R,* of object touch, which is regarded as the baseline performance in our experiments. The *S, J* and *R* represent different sensing modalities about the scene. We conducted the comparative evaluation of the performances achieved by our recognition algorithms using the S, J, and R, respectively.

Performance Evaluation

In order to verify our framework, we conducted experiments for monitoring activities in daily living performed in our prototype smart home environment (Figure 2).

Figure 5 depicts the six activity classes of interest occurring in the smart home testbed as shown in Figure 2. K persons (K = 5) participated twice in the experiments in separate sessions to conduct the activities from 1 through 6 in a sequential manner, which defines an epoch. In total, 5 epochs (i.e., repetitions) per each of the 6 activity classes per each of the 10 trials (i.e., 5 persons × 2 sessions) were collected. Participants were free to choose different sub-sequences of the fine-level actions in each of the 6 coarse-level activity classes. The persons wore the RFID reader on their right wrists, which detected the nearby objects' RFID labels in a sequential manner.

Figure 8 shows the confusion matrices of activity recognition using RFID, scene statistics, and object statistics, respectively. The rows and columns of the table denote the true and the recognized classes, respectively. The leave-one-out cross validation scheme uses four epochs for training and the remaining one epoch for testing in a round-robin manner. The lower right cell of each sub-table shows the average of the recall rates in the corresponding sub-table.

The comparison of the sub-tables in Figure 8 shows that RFID-based recognition accuracy is higher for kitchen activities (activity classes 3 - 6 in Figure 5), while vision-based recognition accuracy is higher for living room activities (activity classes 1 and 2.) The cells of low accuracy are complementary between the two sensing modalities as follows: living-room activities (e.g., Walk around) are better recognized with evidence from vision-based scene statistics, while cooking activities (e.g., Prepare cereal and Drink water) are better recognized with the evidence from the sequence of RFID reading. This result is consistent with the environmental configuration of the testbed; the kitchen area contains many occluding objects with RFID tag and the living room area is a less-occluded open space with fewer RFID tags, which are sparsely distributed.

The most difficult class of activity in the recognition task is Drink Water (DW), as indicated by the low recall rates in Figure 8. This activity often gets confused with Prepare Cereal (PC), because the two activities are not well distinguishable from human silhouettes. Figure 8 also shows the significant increase of the recognition accuracy of the Drink water (DW) class from 50% to 72% by using the frame-differenced motion maps (e.g., in the last columns of Figures 6 and 7) as input observation. Unlike the background subtraction, the motion maps obtained by frame differencing effectively captures the moving parts (e.g., the arms) within the silhouette (e.g., the upper body).

Figure 8 shows that our multi-view vision system achieves reasonable performance rates (compared to the RFID-based system) with very coarse visual representation in terms of foreground silhouette or frame-differenced motion map as input features.

The experimental evaluation shows the following findings:

- Multi-perspective vision enhances the accuracy and robustness in the recognition of activities in daily living.
- Different input modalities (e.g., foreground silhouettes, frame differenced motion maps, and RFID readings) capture different aspects of activity.
- Data fusion by an integrated model is desirable.

The experimental results demonstrate that the proposed framework enables the smart space to acquire the 'context awareness' regarding the hu-

Figure 8. Performance evaluation for the recognition of the six activity classes using different features as input data: The input features are: (a) ***RFID*** *labels, (b) occupancy maps O from foreground silhouettes* ***(FoSi)****, and (c) occupancy maps O from frame-differenced* ***(FrDi)*** *motion maps, from top to bottom sub-table, respectively.*

RFID	WA	ST	PU	PC	DW	SU	*recall*
WA	0.70						*0.70*
ST		0.86					*0.86*
PU			0.88				*0.88*
PC				0.88			*0.88*
DW					0.78		*0.78*
SU						0.98	*0.98*
precision	*0.95*	*0.74*	*0.86*	*0.85*	*0.95*	*0.80*	*0.85*

FoSi	WA	ST	PU	PC	DW	SU	*recall*
WA	0.84						*0.84*
ST		0.96					*0.96*
PU			0.94				*0.94*
PC				0.68			*0.68*
DW					0.50		*0.50*
SU						0.90	*0.90*
precision	*0.95*	*0.87*	*0.87*	*0.58*	*0.63*	*0.94*	*0.80*

FrDi	WA	ST	PU	PC	DW	SU	*recall*
WA	0.92						*0.92*
ST		0.92					*0.92*
PU			0.94				*0.94*
PC				0.66			*0.66*
DW					0.72		*0.72*
SU						0.94	*0.94*
precision	*0.92*	*0.94*	*0.92*	*0.70*	*0.69*	*0.92*	*0.85*

man activities in daily living by means of computer vision and machine learning algorithms. The proposed framework solves the problem of labeling the human activity by using the RFID tags / readers for the training data and builds a multi-camera based unobtrusive system for monitoring the human activity. The context awareness acquired in the form of activity classification could trigger the actuators / controllers linked via network (e.g., X10, 2011; Wattsup, 2011) in order to automate and optimize the facility management targeting at space heating, water heating, cooling, lighting, ventilation, cooking, electrical appliances, etc.

FUTURE RESEARCH DIRECTIONS

The current system provides 80 – 85% overall accuracy in recognizing the categories of human activities in daily living depicted in Table 1. Since the stochastic nature of the uncertainty is inherent in almost all kinds of the probabilistic inference approaches, it is more practical to develop a smart environment system with humans in the loop than to aim at developing a fully automated system for smart environments. This notion leads us to consider various directions for future research. One of the future research directions is the study of the human-computer interface applicable to smart environments. In the human-in-the-loop system, it is important to provide the users with the on-time feedback in an adequate and non-cumbersome format. How to minimize human involvement and yet to provide useful information to the users is an open research issue. Another future research direction is related to actuator control. Some kind of fuzzy-logic is desirable to incorporate the uncertainty with the actuator control unit. Certain measure of safeguard is also needed to ensure the desired operation of the actuators. Yet another direction for future research is related to the development of efficient machine learning algorithms for training and inference.

CONCLUSION

This chapter presented a prototype system for the sensor-augmented smart environment as an energy-aware intelligent system. Among various subareas of the smart environment research, we focused on the energy-aware intelligence module (in Figure 1) implemented with computer vision and machine learning algorithms to train and recognize the human inhabitants' activity patterns in daily living.

Our approach uses a distributed multi-view vision system and RFID sensing system for view independence and robustness in acquiring the scene statistics, object statistics, and RFID labels.

RFID-based evidence better indicates cooking activities involving multiple utensils, while multi-view vision-based evidence better indicates living-room activities in large space. Overall, the RFID-based recognition accuracy is comparable to the multi-view based recognition accuracy that uses coarse silhouette mosaic or even coarser representation by frame-differenced images. The complementary nature of the two sensing modalities provides useful integration for better recognition of activities in the smart space. We have presented the performance evaluations for each of the sensing methods. The multi-view vision system achieves reliable performance in the recognition accuracy from very coarse silhouette mosaic and frame differenced motion maps; this looks promising for privacy preserving and user acceptance in real applications. The current system assumes that the situation involves a single person. Our future research will include the development of more robust algorithms for multi-object tracking and the investigation of more efficient machine-learning algorithms for diverse human activity scenarios. The empirical experiments with the new algorithms and the actuators/controllers integrated together will also be part of our future research.

REFERENCES

X10. (2011). *Website*. Retrieved from http://www.x10.com.

Abowd, G., & Mynatt, E. (2005). Designing for the human experience in smart environments In Cook, D., & Das, S. (Eds.), *Smart Environments: Technology, Protocols, and Applications* (pp. 153–174). New York, NY: Wiley. doi:10.1002/047168659X.ch7

Abowd, G. D., Rehg, J., Starner, T. E., & Arriaga, R. (2007). *Using information technologies to support detection of developmental delay*. Paper presented at the Assisted Cognition Workshop. Rochester, NY.

Aggarwal, J. K., & Park, S. (2004). Human motion: Modeling and recognition of actions and interactions. In *Proceedings of the 2nd International Symposium* on *3D Data Processing, Visualization and Transmission(3DPVT)*, (pp. 640-647). 3DPVT.

Boukraa, M., & Ando, S. (2002). A computer vision system for knowledge-based 3D scene analysis using radio-frequency tags. In *Proceedings of International Conference on Multimedia and Expo*, (vol 2), (pp. 245-248). Conference on Multimedia.

Cheong, Y., Kim, Y., Yoo, S., Lee, H., Lee, S., Chae, S., & Choi, H. (2011). An ontology-based reasoning approach towards energy-aware smart homes. In *Proceedings: IEEE Conference on Consumer Communications and Networking Conference (CCNC)*, (pp. 850 – 854). IEEE Press.

Cook, D., & Das, S. (Eds.). (2004). *Smart environments: Technology, protocols and applications*. Hoboken, NJ: John Wiley & Sons. doi:10.1002/047168659X

Duda, R., Hart, P., & Stork, E. (2001). *Pattern classification* (2nd ed.). New York, NY: Wiley.

Hontani, H., Baba, K., Kugimiya, T., Sato, K., & Nakagawa, M. (2003). Visual tracking system using an ID-tag and the network. In *Proceedings of SICE 2003 Annual Conference*, (vol 3), (pp. 2375-2380). SICE.

Intille, S. S., Larson, K., Beaudin, J., Munguia Tapia, E., Kaushik, P., Nawyn, J., & McLeish, T. J. (2005). A live-in laboratory for pervasive computing research (video) In *Proceedings of Pervasive 2005 Video Program. Pervasive Video Program*. The PlaceLab.

Kim, K., Chalidabhongse, T., Harwood, D., & Davis, L. (2005). Real-time foreground-background segmentation using codebook model. In *Proceedings of the Sixth Annual IEEE Conference on Real-Time Imaging*. IEEE Press.

Logica Survey. (2007). *Turning concern into action: Energy efficiency and the European consumer.* Logica.

McDaniel, T., Kahol, K., Villanueva, D., et al. (2008). Integration of RFID and computer vision for remote object perception for individuals who are blind. In *Proceedings of the 1st International Conference on Ambient Media and Systems*, (pp. 1-10). Ambient Media and Systems.

Mihailidis, A., Hoey, J., Boutilier, C., & Boger, J. (2007). *Towards the development of an intelligent home to support aging-in-place*. Paper presented at the Assisted Cognition Workshop. Rochester, NY.

Mozer, M. C. (2005). Lessons from an adaptive house In Cook, D., & Das, R. (Eds.), *Smart Environments: Technologies, Protocols, and Applications* (pp. 273–294). Hoboken, NJ: John Wiley & Sons. doi:10.1002/047168659X.ch12

Oliver, N., Horvitz, E., & Garg, A. (2002). Layered representations for human activity recognition. In *Proceedings of the IEEE International Conference on Multimodal Interfaces*, (pp. 3-8). IEEE Press.

Park, S., & Kautz, H. (2008a). Hierarchical recognition of activities in daily living using multi-scale, multi-perspective vision and RFID. In *Proceedings: IET International Conference on Intelligent Environments*. Seattle, WA: IET.

Park, S., & Kautz, H. (2008b). Privacy-preserving recognition of activities in daily living from multi-view silhouettes and RFID-based training. In *Proceedings of the AAAI Fall 2008 Symposium*. Washington, DC: AAAI.

Patterson, D., Fox, D., Kautz, H., & Philipose, M. (2005). Fine-grained activity recognition by aggregating abstract object usage. In *Proceedings: IEEE International Symposium on Wearable Computers*. IEEE Press.

Protégé. (2011). *The protégé ontology editor and knowledge acquisition system*. Retrieved from http://protege.stanford.edu/.

Rabiner, L. (1989). A tutorial on hidden Markov models and selected applications in speech recognition. *Proceedings of the IEEE*, *77*(2), 257–286. doi:10.1109/5.18626

Rekimoto, J., & Ayatsuka, Y. (2000). Designing augmented reality environments with visual tags In *Proceedings of Designing Augmented Reality Environments (DARE)*. *DARE*. CyberCode.

Report, O. E. C. D. (2009). *Smart sensor networks: Technologies and applications for green growth*. OECD.

Roy, N., Roy, A., & Das, S. (2006). Context-aware resource management in multi-inhabitant smart homes: A Nash h-learning based approach. In *Proceedings of the Fourth Annual IEEE International Conference on Pervasive Computing and Communications*. IEEE Press.

Sayood, K. (1996). *Introduction to data compression*. New York, NY: Morgan Kaufmann Publishers.

Shivappa, S., Mohan, M., Trivedi, M. M., & Rao, B. D. (2010). Audio-visual information fusion in human computer interfaces and intelligent environments: A survey. *Proceedings of the IEEE, 98*(10), 1692–1715. doi:10.1109/JPROC.2010.2057231

SWRL. (2011). *Website*. Retrieved from http://www.w3.org/Submission/SWRL/.

Takemura, K., Ohara, K., Ohba, K., Chong, N. Y., Hirai, S., & Tanie, K. (2004). Knowledge distributed tag-based vision system. In *Proceedings of the 1st International Workshop on Networked Sensing Systems*. Networked Sensing Systems.

Wang, X., Dong, J. S., Chin, C. Y., Hettiarachchi, S. R., & Zhang, D. (2004). Semantic space: An infrastructure for smart spaces. *IEEE Pervasive Computing / IEEE Computer Society and IEEE Communications Society, 3*(3), 32–39. doi:10.1109/MPRV.2004.1321026

Wattsup. (2011). *Website*. Retrieved from https://www.wattsupmeters.com.

Wu, J., Osuntogun, A., Choudhury, T., Philipose, M., & Rehg, J. M. (2007). A scalable approach to activity recognition based on object use. In *Proceedings of International Conference on Computer Vision*. IEEE.

KEY TERMS AND DEFINITIONS

Green Computing: A generic term for the Information Technologies (IT) used to solve energy and environmental problems.

Energy Awareness: The consciousness and notion about the importance of energy efficiency and power reduction, especially in the context of the current chapter, the practice and efforts to reduce the CO_2 emission-involved power which is critical to global warming.

Smart Space/Environment: A small world where different kinds of smart device are continuously working to make inhabitants' lives more comfortable.

Information System: Any combination of information technology and people's activities that support operations, management, and decision-making. It also frequently refers to the interaction between people, processes, data, and technology.

Hidden Markov Model: A stochastic pattern-recognition methodology to represent, learn, and recognize temporal patterns from data.

Machine Learning: A scientific discipline, as a branch of artificial intelligence, concerned with the design and development of algorithms that allow computers to evolve behaviors based on empirical data, such as from sensor data or databases.

RFID: Radio-frequency identification

Section 5

Education and Power Generators

Chapter 17
The Education Part of Green Computing in Higher Education and Beyond

Mary V. Connolly
Saint Mary's College, USA

ABSTRACT

In order for the population as a whole to consider green computing an essential part of environmental responsibility, the average citizen must be made aware of the issues and motivated to act. Often the users of technology are not technically trained; hence, information must be presented in language suitable for a lay person in this area. This chapter addresses current efforts being made to provide this education. Resources available from the federal government, state governments, non-profit groups, trade associations, and colleges and universities are discussed. In particular, in preparation for this chapter, the websites of all 50 states were surveyed for appropriate information. The entire life cycle of computing equipment is covered so that businesses and individual households are able to obtain the information needed to make environmentally sound technology decisions.

INTRODUCTION

As this book indicates, there is a great deal of good information now available about how to accomplish green computing. However, unless those who purchase, use, and dispose of computer technology know and act on this information, the environmental cost is not changed at all. While some of the purchasers and users of computer technology are well informed about the environmental consequences of their work, most of the general citizenry is not. However, it will take the efforts of a large portion of the citizenry to solve some of the environmental issues around information technology. This requires both education of the general public as well as motivation, all in a discipline which changes almost constantly. The federal government, state governments, non-profit agencies, trade associations, and colleges and universities are all working on the education and motivation issues. Many of these will be cited in the pages, which follow. These are all

DOI: 10.4018/978-1-4666-1839-8.ch017

places where someone who is not in a technical field can go for current information, information which is technical enough but understandable for the lay person.

GUIDELINES FOR PURCHASING COMPUTER TECHNOLOGY

Where does the need for education start? Presumably, the process should start when someone decides to purchase a computer or other related electronic device. It is fine to encourage such people to "buy green," but what does this really mean? Factors to consider are the use of fewer toxic constituents, use of recycled material in the new product, energy efficiency, ability to upgrade or disassemble easily and use of minimal packaging. The Green Electronics Council, a program of the non-profit corporation International Sustainable Development Foundation, manages the Electronic Product Environmental Assessment Tool (Green Electronics Council, 2009). Products registered in EPEAT must meet 23 required environmental performance criteria. Ratings are upgraded as additional criteria are met. The criteria cover such areas as the reduction and/or elimination of environmentally sensitive materials, materials selection, concerns addressing what happens when the item is no longer usable, ENERGY STAR specifications, end of life management, and packaging. The website provides a quick and easy way to get information about the rating of a particular product provided it is registered with EPEAT.

Often other sites will recommend this tool. For example, the State of Pennsylvania's Department of Environmental Protection encourages its citizens to evaluate, compare and select electronic items based on their environmental attributes. It directs citizens to EPEAT to do this. Ohio's Environmental Protection Agency also refers citizens to EPEAT as well as the federal government's Energy Star program (Energy Star, 2011). Recent research, however, adds a caveat for those who would quickly replace current equipment with more environmentally friendly equipment. Researchers from Arizona State University and Rochester Institute of Technology claim that up to 70% of the energy used in the lifetime of a laptop is consumed during the manufacturing stage. They suggest we would be better off designing devices that lasted longer and could more easily be upgraded, so that not so many would have to be manufactured (Deng, Babbitt, & Williams, 2011).

ENVIRONMENTALLY FRIENDLY USE OF COMPUTER TECHNOLOGY

Purchase of a computer which is more environmentally friendly certainly helps, but what happens after the purchase? Questions of proper maintenance, power conservation during use, and use in general should be considered. It is not hard to find a site on the Internet which encourages users of technology to power down when a device is not in use. However, the State of Maryland has done a particularly good job. There is a logo on the state's home page entitled "Smart, Green, and Growing." If one follows a link to information on lowering energy bills, there is good information on the savings achieved when a laptop not in use is unplugged (Maryland Energy Association, 2011). The Maryland Institute College of Art and the Maryland Energy Administration partnered to develop an energy-saving interactive mobile information station that demonstrates to users how to save money and reduce power consumption. The station is designed to education and inspire as it travels around the state. This mobile display helps to reach people who do not resort to the Internet for information, and it certainly attracts one's attention (Maryland Energy Association, 2010). New Jersey also makes it easy to find information. There is a "NJ Green" link on the home page, which quickly leads a reader to a site full of helpful ideas. Suggestions are made both for the home and for commercial settings, including such things as turning computers off when not in use, not using screen savers, and making use of PC power management settings (New Jersey, 2011). The State of Colorado maintains a website entitled "Recharge Colorado" which provides a comprehensive list of resources for both residential and business units, including helpful hints on saving energy while using electronics (Recharge Colorado, 2011). Delaware's Department of Technology and Information maintains a link to a site called "Green IT." This site has great motivating pictures encouraging readers to turn off the computer when not in use and to control the amount of printing (Go Green, 2010). Idaho provides similar tips through its Office of Energy Resources. Many of these sites will remind readers to power down ancillary devices as well as computers.

The federal government also has sites designed to help readers learn how to save energy. Federal sites do a good job of providing a bit more motivation by explaining just how much energy one might save. The Department of Energy has a clickable picture of a home office with suggestions and information about savings (Earth Day, 2011). Together with the Environmental Protection Agency, it has also developed another clickable picture which does much the same thing (EPA, 2011). The first site points out that stand-by power (power used when the device is not actually in use but is left on) can account for 5-10% of home energy use. Many of these sites recommend connecting all devices to a power strip which can be turned off, therefore eliminating the need to turn off several individual electronic devices. All of this is helpful, but consumers also need to think about their overall use of electronics. It is an instructive exercise to assess how much of an individual's computing time is necessary for the job, how much is necessary for personal business, and how much is pure recreation. None of this is necessarily bad (recreation is important!), but consumers need to be aware that all of it uses energy. A single Google search does not use that much power (Google estimates .0003kwh), but Google is not the only search engine, and worldwide the number of searches done per day is growing at an incredible rate.

REUSE RATHER THAN RECYCLE

Given the state of technology, most consumers will find that their electronic devices need to be upgraded or replaced frequently. The tendency is to replace, but there are steps a consumer can take before jumping to that conclusion. Often simply tuning-up a computer, removing unnecessary software, or doing some file maintenance can enhance the performance of the computer. Components can be updated (for example, memory added) which may add time to the life span of the device. Ohio's Environmental Protection Agency has a great title on one of its publications: "Electronic Equipment from Your Home…Don't Just Throw it Away" (Electronic Equipment, 2010). Virtually every state that offers advice on how to recycle a computer or other electronic device begins by suggesting that reuse is a better way to go.

How does someone figure out how to pass on a computer for reuse? Most states which have pages on their websites devoted to e-waste have information about items to consider when donating an older, but working computer. This should be considered if the computer is less than 5 years old and is in working condition. Schools and charities prefer equipment which has been checked by a national clearing house or upgraded by a refurbisher. Refurbishers will usually take care of clearing personal information from the machine, and they then pass on ready-to-use equipment to non-profits, schools and others at little or no cost. Donors should remember to include the operating system, ancillary devices and as much original documentation as possible. The federal Environmental Protection Agency is an excellent source of information. Major producers of PCs often participate in programs to foster reuse. For example, Dell works with the National Cristina Foundation, a clearinghouse that lets consumers donate to non-profits. Computers for Learning is another non-profit umbrella organization that serves as a clearinghouse for connecting unwanted computers to those in need. Dell runs a TechKnow program to get refurbished computers into schools. Philips has a partnership with Intel's Students Recycling Used Technology (the StRUT program) in which students do the reconditioning. The Electronics TakeBack Coalition provides a list of manufacturers that offer TakeBack programs (Electronics Takeback Coalition, 2011). Another good source of information is TechSoup. The Go Green link provides all kinds of helpful tips (Techsoup, 2009).

WHY RECYCLE?

However, if a computer is more than 5 years old reuse may be more difficult. In that case recycling may come into play. The first important point at this stage is that dumping the computer in a landfill is a terrible idea and, in many states, is illegal. Education of the populace on this point is critical. Citizens need to understand that there is a good deal of toxic material in electronic equipment. A processor and monitor can contain 4 to 8 pounds of lead. These items may also contain chromium, cadmium, mercury, beryllium, nickel, zinc, and brominated flame retardants. These can leach into the soil and water supplies and/or contaminate the air if dumped in a landfill or incinerated. This is bad in itself, but it becomes a major problem when the quantity of electronic waste being produced is considered. This is the fastest growing waste stream in the United States. According to the Consumer Electronics Association, Americans own approximately 24 electronic products per household (Consumer Electronics Association, 2008). The US EPA estimates that of 2.25 million tons of televisions, cell phones and computer products disposed of in 2007, 80% went to landfills. The quantity has only increased since then. All of this should be enough to convince consumers to handle their electronic waste in a more environmentally friendly way.

However, almost half the states have laws requiring that some or all of electronic waste must be recycled, often with stricter requirements for businesses than consumers. The federal government is still working on this. Currently the EPA regulates e-waste generated commercially under the Resource Conservation and Recovery Act (RCRA) and the Cathode-Ray Tube Rule. Household e-waste is not regulated at this level. On November 8, 2010 an executive order was issued requesting the EPA and the General Services Administration to join with the Council on Environmental Quality to co-chair an inter agency task force to prepare a national strategy for electronic stewardship. Part of the charge, which is due in 180 days from the time of issue, is to plan to build partnerships in the public and private sector for sustainable electronics management. Those states that do not have laws requiring recycling of electronic waste certainly do encourage it.

HOW STATES HANDLE RECYCLING

The problem of handling the recycling, whether dictated by law or not, is complex and currently is handled state by state. Consider first all the entities involved in recycling in addition to local, state and federal governments. Businesses and consumers certainly have a vested interest in how this is handled. Manufacturers and retailers may have to play significant roles. Those who actually do the work, namely collection centers and recycling centers, are certainly involved. Who, among all these entities, actually pays the cost of recycling? It may help to examine the policies of a few states to discover different models, both models in which recycling is voluntary and models in which it is required by law.

California led the way in 2003 with a law which requires recycling of any electronic device with a video display of 4 inches or larger measured diagonally (this include CRTs, laptops and monitors with LCD display, televisions with CRT or LCD screens and portable DVD players with LCD screens). California adopted the advanced recycling fee model. Consumers originally paid \$6 to \$10 at the point of sale to cover the cost of recycling at the end of the life of the device. The amount required is evaluated annually to maintain sufficient funding (and it never seems to decrease!). The retailer keeps 3% of the fee. The used device goes first to a collector and then to a recycler. The fee is used to cover program implementation through the Integrated Waste Management Board, permits, and inspections through the Department of Toxic Substance Control, payments to recyclers, and payments to collectors. Again these rates are set and reevaluated at regular intervals, but they are based on a per pound scale. The original bill included a provision that not more than 1% of the funds annually deposited in the Electronic Waste Recovery and Recycling Account could be used for establishing a public information program designed to educate the populous on why recycling of electronic devices is important and how to find opportunities to recycle (DTSC, 2004). Note that education is certainly an important part of implementing a program of this sort.

Most states with mandatory programs have not opted for the advanced recover fee model, but rather have looked at the producer responsibility model. This model requires the manufacturers to bear the responsibility for recycling electronics. They may choose to take back the devices themselves or to reimburse recyclers, either directly or through a state administered program. Oregon's program is typical (Oregon E-Cycles, 2010). Manufacturers either create their own take-back program or join a state sponsored program. All manufacturers must pay an annual registration fee based on their market share. The cost to the state for this program is not trivial—\$1.4 million in 2009. This covered payments to the contractor managing the program as well as costs for compliance and enforcement activities. The education part of all of this is left to the retailers, who are required to provide public education. Although seemingly the cost is borne by the manufacturers, it is quite likely that this is simply passed on to the consumer.

Maine was actually the first state to implement manufacturer responsibility for recycling at the end-of-life (E-Waste, 2011). Each manufacturer is responsible for paying the consolidators for the costs of handling and recycling their own computer monitor products plus a pro rata share of orphan products received by the consolidator through recycling. It is possible to have a producer responsibility model with less overhead for the state than the Oregon program. Maryland requires manufacturers to pay an initial \$10,000 fee and then \$500 annually if they implement their own take-back program or \$5,000 otherwise (Maryland Department of the Environment, 2007). The state maintains a list of those manufacturers who are registered and grants funds to counties and municipalities to operate e-waste programs. This program has considerably less overhead for the state once it is set up, but, again, it is quite likely that the manufacturers' fees are actually paid by the consumer.

It is important to note that there are costs involved in any recycling plan. Many states that have not adopted e-waste legislation have hesitated in part due to the costs involved in setting up a mandatory system. Recycling is a business, which must be profitable to those involved in running the business, which certainly depends on supply and demand. Some states also choose not to develop mandatory programs due to geographic isolation from reprocessing facilities and markets. States with lots of land and relatively small populations may face this isolation problem. Even when there is no mandatory program, however, most states will still encourage citizens to recycle electronics and will offer list of resources for doing this. Since some states require (or strongly encourage!) manufacturers to have take-back programs, these manufacturers may well make the programs available nationwide. For example, Dell offers customers a no strings attached take-back program. If you bought from Dell, the company will recycle that product from the customer's home at no charge. HP also has a take-back program. Some retailers also offer nationwide programs. Office Depot, Costco, and Best Buy are among these.

RECYCLING IN THE EUROPEAN UNION

The European Union began to cope with the recycling problem as early as did California. The WEEE (Waste Electrical and Electronic Equipment) and RoHS (Restriction of Hazardous Substances) directives have been in force since 2003. The WEEE directive was designed to increase recycling and/or reuse of electrical and electronic equipment; costs for handling the equipment used by private households were to be borne by the producers of the equipment. However, the directive has not had the desired effect, since only about 1/3 of such waste is being handled appropriately. As much as 54% of the e-waste has been going to potentially sub-standard treatment (see Table 1).

Hence the European Commission, in December, 2008, proposed to revise the directive. The goal is to increase the amount of e-waste that is handled appropriately and to reduce administrative burdens. The revisions are making their way through the European parliament and the Council of the European Union. This process should be complete by the end of 2011. The proposed directives (with possible amendments by the two bodies) are extensive and detailed. Normally member states are given 12-18 months to transpose EU legislation into national law. Member states may pass laws with more stringent restrictions if so desired. (International Trade Administration)

STANDARDS FOR RECYCLING FACILITIES

Once the decision is made to recycle an electronic device (whether voluntarily or mandated), the next question is where to take it. Most states have websites that list recycling centers and offer information on what the centers will take. Contact information is usually easy to obtain. Such lists usually include links to manufacturers and retailers with take-back programs. However, what happens next? There currently is no federal ban on shipping recycled electronic devices to other countries with lower environmental standards. Part of the executive order mentioned above which forms an interagency task force (with the EPA, General Services Administration and the Council on Environmental Quality) asks the task force to plan to reduce exports of used electronics to developing countries that lack the capacity to manage them.

Even when electronics are not shipped out of the country, how can a consumer know that the recycling will be done in an environmentally sound way? The key to this is established standards for use in third party certification of recyclers. The ANSI-ASQ National Accreditation Board (ANAB) offers accreditation for three recycling standards: R2, RIOS and e-Stewards. The EPA facilitated a multi-stakeholder group including

Table 1. The problem

Amount Being Recycled	Going to Landfills	Going to Sub-Standard Treatment
33%	13%	54%

state representatives, manufacturers, recyclers and trade groups to develop a document listing responsible recycling practices (referred to as R2) for use in certification programs (Responsible Recycling Practices, 2010). The practices cover the entire process, from the on-site work at the recycler to what happens at the end-of-life for the recycled equipment and materials (i.e. the downstream part of recycling). To meet the standards, a recycler should use a management system that covers environmental, safety and public health practices through the entire stream. Policies that promote reuse and material recovery for the end-of-life materials are required to meet the standards. Quite obviously, the standards require a recycler to comply with health and safety legal requirements and to reduce exposures and emissions during the recycling operation. Equipment and components that are reusable must be refurbished and properly tested. The standards pay lots of attention to what happens to the materials and components after they leave the recycler—no passing these on to other facilities that will not handle them appropriately.

The Recycling Industry Operating Standard (RIOS) provides a framework for an integrated management system designed to continuously improve environmental and health and safety performance (Recycling Industry Operation Standards, 2010). It was developed by the Institute of Scrap Recycling Industries (ISRI). Many appropriate tools are offered to facilities which hold RIOS membership. Among these are tools to keep environmental compliance current and decrease environmental risks and costs. The e-Stewards Standard for Responsible Recycling and Reuse of Electronic Equipment serves as the basis for e-Stewards certification (e-Stewards, 2010). It was developed by the Basel Action Network with input from the recycling industry, the environmental community, and occupational health and safety professionals and is meant to be an international standard. BAN is an organization focused on confronting the global environmental injustice and economic inefficiency of toxic trade. It works in opposition to exporting toxic waste from rich to poor countries. The e-Stewards standard covers data security, worker protection, on-site practices for toxic materials found in electronic waste, parameters for reuse and refurbishment, and accountability for the entire recycling chain for toxic materials (including export issues). Other recognized recycler certifications are available from the Industry Council for Electronic Equipment Recycling (ICER) and the International Association of Electronics Recyclers.

ARE RECYCLERS HELD TO STANDARDS?

Given that there are established standards, do states require that recycling facilities meet these or other standards? Many states do require that recyclers meet certain standards. For example, Maine's Department of Environmental Protection has developed guidelines, which provide a framework for environmentally sound management of electronic waste by recyclers (Guidelines, 2005). Consolidators may not transport electronic wastes to a recycling facility which has not provided a sworn certification that it has met the guidelines. The guidelines cover, among other requirements, having appropriate measures to safeguard environmental health and safety, due diligence in selecting facilities to which components and materials are sent, and proper management of all hazardous materials. Wisconsin's program, E-Cycle Wisconsin, also has requirements for recyclers and collectors to ensure that recycling is done in an environmentally sound way (E-Cycle Wisconsin, 2011). Either certification of recyclers or state standards are really critical to ensuring that electronic waste is handled well since often the recycling facility is not the first stop for electronic waste. If a business or household is using a take back program, the electronic waste may go first to the manufacturers, which then might send it to a collector. Finally, the collector will send it to a recycling facility.

However, in some states, household electronic waste will be taken to a recycling facility directly by the consumer. In those cases, how can a consumer find an acceptable facility and be assured that the facility follows appropriate standards? There are a number of questions which can be asked to determine if the electronic devices will be handled correctly. Among these are questions about how personal data will be destroyed, whether the management system is certified, whether the facility has had any environmental or safety violations in the last few years, what percentage of the materials collected are recycled versus being disposed of, and, if wastes are sent to other providers, are their practices appropriate (particularly if they export). The Telecommunications Industry Association (TIA) maintains a website which gives people a quick way to find reuse, recycling and donation programs in their areas (TIA, 2011). Also earth911 offers a similar service (earth911, 2011). When this author went to that site and entered the appropriate location, asking for recycling information for computers, a substantial list of possible places appeared, including retailers and non-profits which accept electronic waste. The local county facility was listed as well as the local ReStore for Habitat for Humanity, which sponsors take back programs on the last Wednesday of each month and guarantees that recycling is done with R2 certified recyclers and that hard drive protection is guaranteed. This last guarantee is an important one.

DATA PROTECTION

Data protection is part of the R2, RIOS and e-Stewards standards. This is critical since simply reformatting the hard drive or deleting files does not permanently erase the information stored there. While leaving files on a discarded hard drive is not really an environmental matter, it is a critical privacy matter. If consumers can not be guaranteed that their files will be deleted and made unavailable to anyone, they will be less likely to recycle their computing equipment. Consumers should actually take initial responsibility for this before taking the device to a manufacturer or a recycling center.

Although most recyclers will provide assurance that personal data is destroyed, it is better to be safe and take action before recycling a computer. This can be accomplished either by overwriting or by physical destruction of the hard drive. Operating systems installation CDs or DVDs often contain disk management tools that can be used to overwrite the data, which means that the previously stored data is replaced with a predetermined pattern of meaningless data. There is also commercial software available for this purpose. For example, Symantec's Wipe Info in Norton Utilities and System Works is a possibility for Windows PCs and Macs. Webroot Software Inc.'s Window Washer also works on Windows, and Jiiva Inc.'s SuperScrubber is available for Macs. The EPA's page "Do the PC Thing" has a long list of commercial software available for this purpose as well as references to shareware which can be used for the same purpose (EPA, 2007). Among the available shareware is Active@KillDisk, BCWipe, and DBAN "Darik's Boot and Nuke."

Businesses may face a larger problem when recycling electronic equipment since the amount of data to be dealt with is much larger. North Carolina has a direct link from its home page to Cyber Security awareness (North Carolina Office of Information Technology Services, 2010). A link from that page takes the reader to the Erasing and Disposal Guide. This is a non technical guide for business and office managers with excellent information on how to dispose of electronic storage media. The guide points out that there are many laws which require information to be protected and it is critical to prevent inadvertent disclosure of sensitive information.

CREATING AND SUPPORTING THE RECYCLING MARKET

As many states introduce requirements for recycling of electronic waste, the need for creating and supporting the recycling market becomes greater. In fact, states with low density populations may have a hard time requiring recycling simply because a significant portion of its citizens live at considerable distance from the necessary facilities. For example, The Department of Environmental Quality in the state of Montana maintains a website on recycling with information about electronics (Montana Department of Environmental Quality, 2010). It points out that due to Montana's distance from most recycling markets, it is not yet feasible to ban electronic materials from landfills until markets for these are established. Idaho's Department of Environmental Quality has a fine site on recycling on which it is mentioned that recycling in Idaho is limited by geographic isolation from appropriate facilities and markets (Idaho Department of Environmental Quality, 2011). Recycling facilities are businesses, and, when the cost of collection and transportation over long distances outweighs the value of the materials being recycled, it is hard to stay in business. Nevada's Division of Environmental Protection issued an extensive report in 2011 on electronic waste reuse and recycling (Nevada Electronic Waste, 2011). Nevada operates under a voluntary system for recycling of electronic waste. According to the study 95% of the state's population has access to appropriate services, either through local facilities, special collections or non-profit or manufacturer take back programs. The study recommended continuing and enhancing the current system, rather than making it mandatory. The study did recognize that recycling services are limited in the more rural part of the state, but it concluded that a mandatory program probably would not solve this problem and considerable costs would be involved, costs to be borne by the state and local governments. Many recyclers in Nevada are already working at close to capacity, and developing new collection programs can take from 3 to 5 years. The study also pointed out that, although the EPA is working on the problem, what a federal program might look like is unknown at this point.

DOES MANDATORY RECYCLING WORK?

Another logical question to ask is whether mandatory recycling of electronic waste really works, that is, is significantly more electronic waste kept out of landfills. Over the first 18 months of Maine's program (January 2006 – June 2007) almost 6 million pounds of electronic materials was diverted from landfills and incinerators (Recycling of Electronic Waste, 2008). Oregon's recycling programs began on January 1, 2009. The Oregon E-Cycles Program (which includes collection sites and recycling facilities) reported collecting and recycling 27% more in the second year of the program over the first year (measured in pounds taken in) (Oregon E-Cycles, 2011). In the first two years of operation, the manufacturer funded E-Cycle Washington program collected over 39,000 tons of electronic waste (Washington State Department of Ecology, 2010).

The E-Cycle Wisconsin 2010 report issued through the Department of Natural Resources Waste and Materials Management Program suggests that manufacturers are on board and that the public is paying attention and participating (E-Cycle Wisconsin, 2011). The Wisconsin electronics recycling law took effect on January 1, 2010 for manufacturers of certain electronics. They were required to register the brands they sell to households and schools, and recycle a number of pounds based on their sales. Collectors and recyclers also register. On September 1, 2010, the other part of the law took effect, forbidding certain kinds of electronic waste from being put in the trash or sent to landfills. The report shows that collection under these requirements began slowly but is picking up speed as more recycling options become available. However, the report also showed the need for careful and continuing education as these programs are put in place. The Department of Natural Resources conducted a survey of households in the fall of 2010 (after the September 1 deadline) to determine if residents were aware of the law and to learn how residents were handling their electronic waste. Over half had heard of the ban and about one-quarter had heard of the E-Cycle Wisconsin program. Not surprisingly, most residents indicated that they wanted free or low-cost recycling; nearly half said they were unwilling to pay anything. The survey pushed a bit on this last issue, asking if residents would change their minds if offered some kind of coupon to offset the fee or a guarantee of responsible recycling and safe data destruction. Only 12% were then still not willing to pay anything, even with these features. Almost a third indicated that they did not know where to take their electronics, and some indicated an inability to transport the electronics to a collection site.

The evidence indicates that mandatory programs do prevent much electronic waste from going to landfills, and often voluntary programs do also. It is also clear that there are real costs involved, particularly as new programs are introduced. Wisconsin's survey also highlights the needed for continuing education of the general population. Most states, whether their programs are voluntary or mandatory, do maintain good lists of places which will take your electronic waste, including non-profits, manufacturers, and retailers.

COLLEGE AND UNIVERSITY CONTRIBUTIONS

Education on all phases of green computing really should not wait until adulthood. While parents can train young children to adopt good habits (turning off the computer when not in use, controlling printing, among others), there is a unique educational opportunity in this area when these children grow up and go off to college. Often for the first time, these students are faced with working in a rather different environment. Many are living in residence halls and sharing all kinds of resources with others, including electronic resources. They encounter requirements not faced when they were living at home, such as acceptable use policies if they wish to connect to the campus network. As these students begin to think critically about issues, which extend beyond the concerns of their earlier years, often a core of students on a campus will be passionate about the environment. This core will push their peers to learn to think and act in ways, which respect rather than destroy the environment. In fact, often it is exactly these students who drive the change on campus. Students on a number of campuses have come up with clever and motivating ideas to encourage better environmental practices, and, in particular, green computing practices. The general population can learn a great deal from these projects. Some of these projects are described below and could well be used to education the general population.

At Stanford University, students working with the Office of Sustainability updated the "2010 Students Guide to Sustainable Living at Stanford" (Stanford Office of Sustainability, 2010). This guide covers all aspects of green computing which affect students, including purchasing and controlling power usage. Students are referred to EPEAT for information on purchasing. Special power strips which sense when a computer goes to sleep are recommended. There is an excellent chart listing the energy used by various computers and accessories. Printing is also covered. This guide is given out at orientation for new students—a great time to introduce students to green computing before patterns of work are set.

At Yale University, the Student Taskforce for Environmental Partnership (STEP) is a student-run organization whose goal is to help Yale become a sustainable community. They produced a PowerPoint presentation entitled "Recycling Made Easy" which thoroughly addresses e-waste (Yale Office of Sustainability, 2010). Columbia University lists at least a dozen student groups working on various aspects of sustainability, including e-waste issues.

Colleges and universities do not leave all the work to students. Virtually every university now has an office of sustainability, and some pay considerable attention to the students in the university community as well as to larger issues such as building green buildings. Both the example set by the university itself as well as the environmental educational opportunities offered to students have an important lasting effect on the students when they leave the university. It is these students who will then make the critical decisions which impact the environment no matter where they work. Harvard's Office for Sustainability sponsors a "Green Tip of the Month" which is available to anyone who wishes to subscribe (Harvard Office of Sustainability, 2011). This is a poster and email campaign, which offers monthly tips which can be implemented in the classroom, office, lab or home.

Pomona College's Sustainability Integration Office has put out the second edition of "The Little Green Book: A Guide to Sustainable Living at Pomona College" (Pomona College Sustainability Integration Office, 2010). It includes a great deal of information on electronic equipment. It encourages students to purchase Energy Star equipment, and it includes an extensive table showing the energy use of items such as desktop computers, laptops, monitors, printers, and wireless routers. It addresses power usage, urging students to eliminate phantom loads and to reduce brightness on the screen. Printer issues are also addressed, pointing out that it might be a better idea to use the campus printers rather than buy your own. Much of the advice in this guide and in Stanford's guide applies to anyone using a computer. In fact these are guides are much more complete and easy to understand than many of the state sites visited for this chapter.

Although many of the schools already listed above are large and well known, many smaller colleges are also concerned about educating the student body in green computing practices. In fact, in many colleges the students themselves are putting the effort into programs, which educate their peers and the campus community. Clarkson University in New York, a school of approximately 3300 students, has a student run organization called Sustainable Synergy, which has a focus on alternative energy and energy conservation. The organization offers a sustainability quiz for the campus community which is educational in nature. Under the "Take Action!" link there is a link to "Adopt Our Green Tips." The tips offer very specific information on how to control power usage on a computer. Over all the organization's site is full of good information (Clarkson Sustainability Initiatives, 2011).

Sweet Briar College in Virginia (748 students) has a link for green computing under training on the Academic Computing site. There is a good discussion there of how to enable power management as well as *why* one should do this. There is also appropriate information on buying computer equipment, with making sure that the products are Energy Star compliant at the top of the list (Green Computing, 2011).

Reed College in Oregon (1477 students) has a Green Computing Task Force. The mission of the task force is to identify and evaluate ways in which the College can make more energy efficient and environmentally sensitive use of its technology resources. The task force offers the expected suggestions on recycling, turning off equipment when not in use, and buying Energy Star compliant equipment. However, the task force does something unusual for a college group. It regularly performs due diligence to assure that recyclers and associated downstream providers meet environmental standards (Green Computing at Reed, 2011).

Table 2. Information

Life Cycle Phase	Information Sources and Requirements
Original purchase of equipment	Energy Star and EPEAT frequently recommended by states and colleges
Day-to-day use of equipment	US Department of Energy, many states, and many colleges offer comprehensive list of suggestions to control power usage
Reuse rather than recycle	States promote reuse on virtually any site on recycling
Recycling	About 50% of the states make this mandatory; all states encourage it
Who pays for recycling?	It depends on the state! Ultimately it is often the consumer, either directly or indirectly
Standards for recycling facilities	Standards exist, and many states require that standards be met
Does mandatory recycling work?	Yes, but it certainly does not produce 100% compliance!
Recycling markets	Difficult to establish state wide coverage in sparsely populated states

Grinnell College in Iowa (1600 students) believes it has a duty to be a leader in environmental stewardship, education, and policy. Additional student tips can be found under the Environmental Stewardship page. The viewer is directed to a "What Can You Do?" page with great tips for students on powering down computers when not in use, looking for the Energy Star label when purchasing equipment, and eliminating the phantom load produced when a device is turned off but still plugged in (Environmental Stewardship, 2007).

The above examples show that students do care about green computing issues and often are the campus leaders in initiatives to foster environmentally sound computing practices. Relatively small colleges are making contributions just as valuable as the larger, more widely recognized colleges and universities.

CONCLUSION

Table 2 summarizes the kind of information an average citizen can use to make environmentally sound decisions about technology.

The education part of green computing is critical. The work of colleges and universities as well as the appropriate state offices show that it is possible and can be done well if built in to any required or voluntary green computing program. Clearly, all entities will need to continue to work together to continue to keep information up to date and easily available.

REFERENCES

Basal Action Network. (2010). *e-Stewards*. Retrieved from http://e-stewards.org/certification-overview/program-details/.

California Department of Toxic Substances Control. (2004). *Electronic waste recycling act*. Retrieved from http://www.dtsc.ca.gov/HazardousWaste/EWaste/.

Clarkson University. (2011). *Clarkson sustainability initiatives*. Retrieved from http://www.clarkson.edu/green/.

Consumer Electronics Association. (2008). *Market research report: Trends in CE reuse, recycle and removal*. Washington, DC: Consumer Electronics Association.

Delaware. (2010). *Go green*. Retrieved from http://dti.deleware.gov/greenit/.

Deng, L., Babbitt, C., & Williams, E. (2011). Economic-balance hybrid LCA extended with uncertainty analysis: Case study of a laptop computer. *Journal of Cleaner Production, 19*(11), 1198–1206. doi:10.1016/j.jclepro.2011.03.004

Earth911. (2011). *Website*. Retrieved from http://earth911.com/.

Electronics TakeBack Coalition. (2010). *Take it back, make it green, recycle responsibly*. Retrieved from http://www.electronicstakeback.com.

EPA. (2007). *Do the "PC" thing: Donate computers*. Retrieved from http://www.epa.gov.wastes/partnerships/plugin/pdf/pcthing-con.pdf.

EPA. (2010). *Responsible recycling practices*. Retrieved from http://www.epa.gov/wastes/conserve/materials/ecycling/r2practices.htm.

EPA. (2011). *EnergyStar*. Retrieved from http://www.energystar.gov/.

EPA & US Department of Energy. (2011). *Save energy at home*. Retrieved from http://www.energystar.gov/index.cfm?c=products.pr_save_energy_at_home.

Governor's Energy Office. (2010). *Recharge Colorado*. Retrieved from http://rechargecolorado.com/.

Green Electronics Council. (2009). *Green electronics made easy*. Retrieved from http://www.greenelectronicscouncil.org/.

Grinnell College. (2007). *Environmental stewardship*. Retrieved from http://www.grinnell.edu/etal/green.

Harvard Office for Sustainability. (2011). *Green tip of the month*. Retrieved from http://green.harvard.edu/green-tip.

Idaho Department of Environmental Quality. (2011). *Waste management: Recycling in Idaho*. Retrieved from http://www.deq.idaho.gov/waste/recycling/recycling.cfm.

Institute of Scrap Recycling Industries, Inc. (2010). *Recycling industry operation standards*. Retrieved from http://www.isri.org/iMIS15_PROD/ISRI/_Program_and_Services/.

International Trade Administration. (2011). *Website*. Retrieved from http://export.gov/europeanunion/weeerohs/.

Maine Department of Environmental Protection. (2005). *Guidelines for the environmentally sound management of e-waste*. Retrieved from http://www.maine.gov/dep/rwm/recycle/tvcomputer-guidelines.htm.

Maine Department of Environmental Protection. (2008). *Recycling of electronic waste from households in Maine*. Retrieved from http://maine.gov/dep/rwm/ewaste/pdf/2008ewastereport.pdf.

Maine Department of Environmental Protection. (2011). *E-waste*. Retrieved from http://www.maine.gov/dep/rwm/ewaste/.

Maryland Department of the Environment. (2007). *Maryland's statewide electronics recycling program*. Retrieved from http://mde.maryland.gov/programs/Land/RecyclingandOperationsprogram/.

Maryland Energy Administration. (2010). *EmPower Maryland: You owe it to yourself*. Retrieved from http://energy.maryland.gov/events.html.

Maryland Energy Administration. (2011). *Residential energy saving tips: Learn 15 things that will empower you to save money*. Retrieved from http://energy.maryland.gov/Residential/index.html.

Montana Department of Environmental Quality. (2010). *Why recycle electronics*. Retrieved from http://deq.mt.gov/Recycle/Electronics/whyelectronics.mcpx.

Nevada Division of Environmental Protection. (2011). *Nevada electronic waste reuse & recycling study*. Retrieved from http://nevadarecycles.gov/study2010/index.html.

New Jersey. (2011). *Living green: What you can do*. Retrieved from http://www.nj.gov/nj/green/living/.

North Carolina Office of Information Technology Services. (2010). *Cyber security awareness*. Retrieved from http://www.esrmo.scio.nc.gov/CyberSecurity/.

Ohio Environmental Protection Agency. (2010). *Electronic equipment from your home ... don't just throw it away*. Retrieved from http://www.epa.ohio.gov/ocapp/p2/mercury_pbt/ewaste/index.aspx.

Oregon Department of Environmental Quality. (2010). *Oregon e-cycles*. Retrieved from http://www.deq.state.or.us/lq/ecycle/manufacturers/index.htm.

Oregon Department of Environmental Quality. (2011). *News release: Oregon e-cycles program show big gains in 2010*. Retrieved from http://www.oregon.gov/DEQ/.

Pomona College Sustainability Integration Office. (2010). *The little green book: A guide to sustainable living at Pomona College* (2nd ed). Retrieved from http://www.pomona.edu/adminstration/sustainability/resources/publications/littlegreenbook.pdf.

Reed College. (2011). *Computing and information services: Green computing at Reed*. Retrieved from http://www.reed.edu/cis/facilities/green_computing.html.

Stanford Office of Sustainability. (2010). *A student's guide to sustainable living at Stanford*. Retrieved from http://ssu.stanford.edu/sites/sustainable.stanford.edu/files/documents/Sustainable_at_Stanford_New.pdf.

Sweet Briar College. (2011). *Green computing*. Retrieved from http://www.sbc.edu/ac/green-computing.

TechSoup. (2009). *Reduce your environmental impact and save money*. Retrieved from http://www.techsoup.org/greentech/.

TIA. (2011). *E-cycling begins with you*. Retrieved from http://ecyclingcentral.com.

US Department of Energy. (2011). *Earth day 2011*. Retrieved from http://www.energy.gov/earthday/.

Washington State Department of Ecology. (2010). *2010 annual report*. Retrieved from http://www.ecy.wa.gov/programs/swfa/eproductrecycle/.

Wisconsin Department of Natural Resources. (2011a). *E-cycle Wisconsin*. Retrieved from http://dnr.wi.gov/org/aw/wm/ecycle/.

Wisconsin Department of Natural Resources. (2011b). *E-cycle Wisconsin 2010 report*. Retrieved from http://dnr.wi.gov/org/aw/wm/publications/anewpub/wa1529/pdf.

Yale Office of Sustainability. (2010). *Student taskforce for environmental partnership: Recycling made easy*. Retrieved from http://www.yalestep.com.

ADDITIONAL READING

Consumer Electronics Association. (2011). *Website*. Retrieved from http://www.ce.org.

EPA. (2011). *Website*. Retrieved from http://www.epa.gov.

Green Electronics Council. (2011). *Website*. Retrieved from http://www.greenelectronicscouncil.org.

Harvard Office for Sustainability. (2011). *Website*. Retrieved from http://green.harvard.edu.

International Trade Administration. (2011). *Website.* Retrieved from http://trade.gov.

Pomona College Sustainability. (2011). *Website.* Retrieved from http://www.pomona.edu/sustainability.

US Department of Energy. (2011). *Website.* Retrieved from http://energy.gov.

KEY TERMS AND DEFINITIONS

Acceptable Use Policy: Requirements issued by a school or business, which must be followed if a student or employee uses the organization's computer network(s).

Data Protection: When recycling, this refers to the destruction of personal data on a computer before the computer is passed on to another user or destroyed.

EPEAT: Electronic Product Environmental Assessment Tool, managed by the Green Electronics Council.

Mandatory Recycling: Recycling required by law.

Phantom Load: Power used when an electronic device is turned off but still plugged in.

Refurbisher: A person or company which prepares a computer to be passed on to another user, usually done when the computer is donated.

WEEE: Waste Electrical and Electronic Equipment directive (from the European Union).

Chapter 18
Study of Hybrid Sustainable Energy System Based on PEM Fuel Cells and Photovoltaic-Module Power Generator

Fatima Zohra Zerhouni
University of Sciences and Technology Mohamed Boudiaf (USTOMB), Algeria

M'hammed Houari Zerhouni
University of Sciences and Technology Mohamed Boudiaf (USTOMB), Algeria

Mansour Zegrar
University of Sciences and Technology Mohamed Boudiaf (USTOMB), Algeria

Amine Boudghene Stambouli
University of Sciences and Technology Mohamed Boudiaf (USTOMB), Algeria

ABSTRACT

The computer is the greatest innovation of the 20th century. It has changed our lives. It executes tasks with precision. There is no limit with what we can do with software. Computers are seductive. Companies and students cannot work without them. They help students to perform mathematical computations. It is very important that mathematical ideas are expressed in computer programs in order to have theoretical results and to verify them practically. Nowadays, the development of new and non-polluting energy producing and energy-storage systems is a great challenge for scientists. An alternative to the nuclear and fossil fuel power is renewable energy technologies. Due to ever-increasing energy consumption, rising public awareness of environmental protection, and steady progress in power deregulation, alternative (i.e., renewable and fuel cell based) distributed generation systems have attracted increased interest. There is an accelerating world demand for environmentally friendly power. Among the renewable energy sources, the Photovoltaic (PV) energy is the most promising candidate for research and development for large scale users.

DOI: 10.4018/978-1-4666-1839-8.ch018

Fuel cells have been receiving a lot of attention lately due to their potential of becoming a new energy source with a large range of applications. Fuel cells can be incorporated with other components to create high efficiency industrial power plants. Fuel cells permit clean and efficient energy production. The purpose of the work is to optimize the system's operation. The main reason to build described system is to supply stand-alone systems using renewable energy sources. Therefore, the power plant has to produce energy independent of any weather fluctuations. Integrating photovoltaic energy sources with fuel cells, as a storage device replacing the conventional lead-acid batteries, leads to a non-polluting reliable energy source. In this chapter, an energy system comprising different energy sources, namely PV and fuel cells, is proposed. Photovoltaic cells coupled with electrolytic devices can be used to produce hydrogen and oxygen in a sustainable manner. With the produced hydrogen from the electrolysis process, it is possible to generate electricity through fuel cells. Photovoltaic panels in particular can provide a good source of producing green electricity. It is autonomous, its operation does not pollute the atmosphere, and it is an inexhaustible and renewable source with great reliability. The simulation program developed also allows the exportation of different configurations. The experimental system described has permitted the validation of the proposed method.

INTRODUCTION

Computers are great resources to own. They are everywhere and society depends on them to manage our world. The computers have unlimited applications. Many of these applications are benefiting society directly or indirectly. There exist other functions of these computers, such as their use for running simulations and calculating large, complex data sets. In this present work, a hybrid system, particularly a photovoltaic fuel cell one, in Algeria, is studied. The fuel cell power supply is a very attractive option to be used with an intermittent power generation source like PV power because the fuel cell power system is characterized with many attractive features such as efficiency, fast load-response, modular production and fuel flexibility.

Photovoltaic (PV) has emerged in last decades since it has the previously mentioned advantages and less maintenance. Photovoltaic panels have received increasing attention over the last decade as one response to the challenges of global warming, increasing demand for energy, high fuel costs, and local pollution. Solar modules have the disadvantage of high initial installation costs. So, different methods have been proposed in the literature to increase the photovoltaic generator (SCA: Solar Cells Array) efficiency. The method, proposed here is based on selecting in real time, the SCA optimum configuration for a given load and given working conditions. This method is supposed to be particularly convenient for all direct coupling between a photovoltaic generator and a load system such as a pumping system. Experimental results strongly confirm the effectiveness of the proposed method. The system is an environmentally friendly solution since it tries maximising the use of a renewable energy source.

GENERAL SYSTEM DESCRIPTION

The alternative energy system that we propose to study in Algeria, more precisely in Oran city, is represented in Figure 1.

The overall system includes:

- Photovoltaic field containing photovoltaic polycrystalline silicon cells;
- Fuel cells stack;
- Gas storage equipment;
- Electrolyser;

Figure 1. Overall system

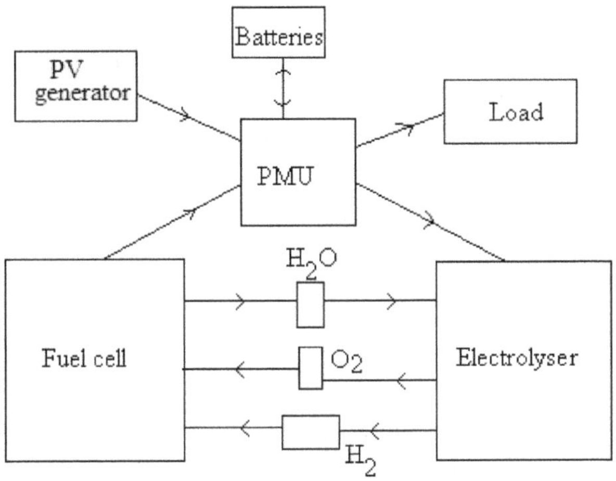

- Stand-by battery; and
- Power Management and Control device (PMU).

The load is fed by a photovoltaic field. Electricity excess supplies thus the electrolyser producing hydrogen and the oxygen. The gases are stored for a length or a short term. In the event of insufficient energy, the PV generator is adopted to feed the load. A battery is added to the system in the transitory moments where fast load's variations occur. The energy management (PMU) equipment is used to control the power between each component. A Proton Exchange Membrane (PEM) fuel cell is used to keep the system's reliability at the same level as for the conventional system while decreasing the environmental impact of the whole system. The PEM fuel cell consumes gases which are produced by an electrolyser to meet the user load demand when the PV generator energy is deficient, so that it works as an auxiliary generator (Busquet, et al., 2004). The experimental system used a microcontroller. It is robust when undergoing environment changes and load variations. The performance of the methods is verified through simulations and experiments. The experimental system used a microcontroller. It is robust to environment changes and load variations.

PHOTOVOLTAIC SOURCE

Introduction

The fact that oil is running out fast, alternatives should be adopted. Renewable energy is one of the most promising solution to the above problems (Boudghéne, 2011a). In particular, photovoltaic panels can provide a good source of producing *clean* electricity (Boudghéne, 2011b, 2011c). The most attractive features of photovoltaic panels are the non-existence of movable parts, the very slow degradation of the sealed photovoltaic cells, the flexibility in the association of modules, and the extreme simplicity of its use and maintenance. In addition, photovoltaic energy is a very relevant source. It is autonomous, its operation does not pollute the atmosphere, and it is an inexhaustible and renewable source with great reliability. To deal with these problems have been proposed several methods (Salamah, et al., 1990a; Applbaum, 1989; Zerhouni, 1996; Weidong, et al., 2007; Zerhouni, et al., 2008). Sun trackers have been utilized to face the PV system to the direction of the sun so that maximized power at any point in time is achieved. The Maximum Power Point Tracker (MPPT) method can increase the efficiency and the performance of the PVG (Photovoltaic Generator) (Enrique, et al., 2007; Zerhouni, et al., 2010; Mutoh, et al., 2007; Zerhouni, et al., 2005; Kwon, et al., 2006; Zerhouni, 2006a).

The methods vary in complexity, convergence speed, and cost (Zerhouni, et al., 2006b; Salameh, et al., 1990b; Benmessaoud, 2010).

The method presented in this chapter is extended to any load and based on reconfiguring on line the PVG.

Figure 2. Photovoltaic cell's equivalent model

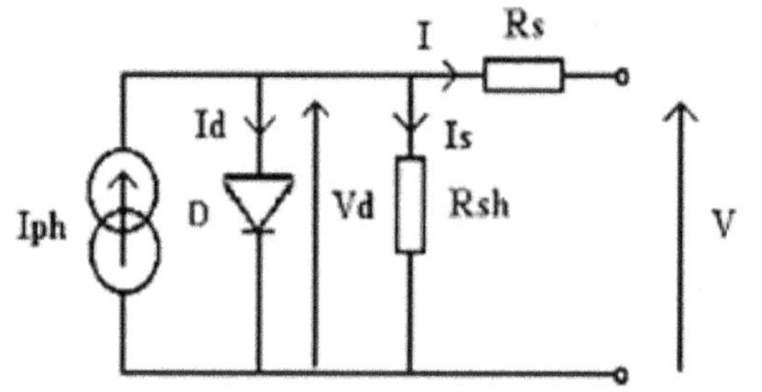

Fundamentals

The equivalent electrical circuit of a single cell is represented in Figure 2 (Busquet, 2003; Appelbaum, 1989; Zerhouni, 1996).

In Figure 2, we can find the following parameters:

- The photocurrent generator I_{ph} which is compared to the short circuit current I_{cc}. Its value is proportional to the solar radiation
- The diode D symbolizing the junction PN.
- The serial resistance Rs which takes into account contact's and connection's resistances.
- The parallel resistance Rsh representing the leakage current at the junction.
- The diode's voltage V_d
- The diode's forward current I_d

The simple equivalent circuit is sufficient for most applications. Kirchhoff's first law provides the current I–voltage V characteristics of the simple photovoltaic cell equivalent circuit. From Figure 1, we can drive equation 1 which gives the cell current I function of the cell voltage V.

$$I = I_{cc} - I_o \left[\exp \frac{(V + R_s I)}{A.U_T} - 1 \right] - \left(\frac{V + R_s I}{R_{sh}} \right) \quad (1)$$

where:

I_o is the saturation current, which varies with the temperature.

A: is the ideality factor. It depends on the semiconductor used and its nature (single- crystal, polycrystallin, amorphous...);

U_T: thermal potential.

$$U_T = \frac{K_B T}{q} \quad (2)$$

K_B: Boltzmann constant
T: junction temperature in Kelvin degrees.
q: electron charge.

If we neglect Rsh which presents an important value, equation (1) becomes:

$$I = I_{cc} - I_o \left[\exp \frac{(V + R_s I)}{A.U_T} - 1 \right] \quad (3)$$

The Characteristic of a Photovoltaic Module

The effective solar radiation comprises the spectral components such as wavelength of the solar spectrum, which the photovoltaic cells convert into electrical energy. A photovoltaic system is an uncontrolled power source whose behavior is quite different from conventional power supplies. Solar cells naturally exhibit a nonlinear characteristic, which vary with the solar irradiation and cell temperature. In the climatic conditions of solar radiation ($1000 W/m^2$ corresponding to Es=100%), temperature T=25°C, the characteristic of the current according to the voltage I=f(V) of a photovoltaic module, as given by equation (1) is illustrated by Figure 3.

From Figure 3, we can highlight three relevant parts (Zerhouni, 1996):

Figure 3. I-V characteristics of a photovoltaic module

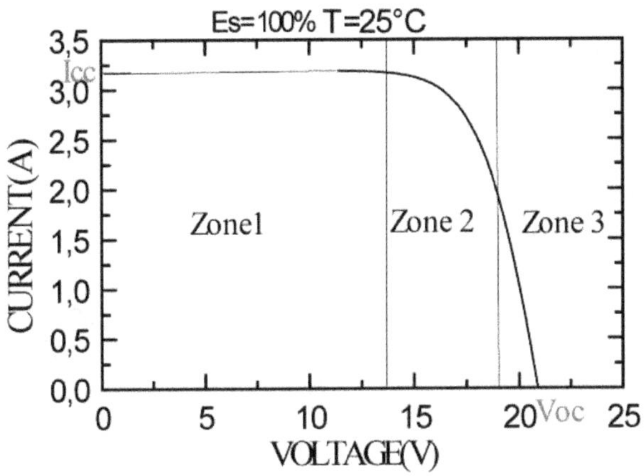

- Part 1 where the current remains almost constant for voltage variations. In this area, the photovoltaic module is comparable to a current generator.
- Part 2 which is characterized by a current fluctuation corresponding to an appreciably constant tension. In this area, the photovoltaic module is a voltage source.
- Finally, part 3 has the feature of being to be an intermediate zone between an operation of the current generator and the voltage generator. It is affirmed by an evolution, very significant of the output variables (I,V). This area includes a point M of co-ordinates $I_{opt,} V_{opt}$ delivering the optimum power P_{opt}.

The PVG electrical characteristics are essentially affected by the solar radiation and the temperature (Zerhouni, 1996, 2006c).

The Connection PVG-Load

It is the simplest connection without any system between the PVG and the load. This system is represented by the Figure 4, which directly associates the SCA to a load (Busquet, et al., 2004; Applbaum, 1989; Zerhouni, 1996).

Figure 4. The direct connection PVG-load

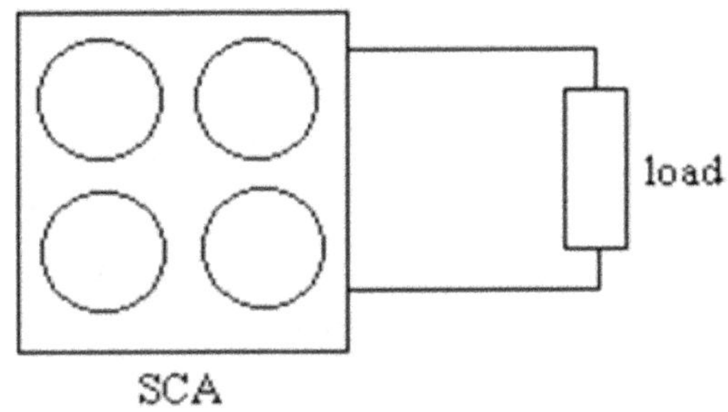

In a directly coupled system it is important to match the load to the photovoltaic generator otherwise the power loss resulting from this mismatch would lead to a lower efficiency of the overall system. To overcome this problem different methods have been proposed (Zerhouni, 2009).

PV arrays are built up with combined series/parallel combinations of PV solar modules (Salameh, et al., 1990a, 1990b). The number of modules required to build a SCA depends on the power required and also on the modules types used and the operating conditions (photovoltaic radiation, temperature, etc).

The method presented in this chapter is extended to any load and based on reconfiguring on line the PVG by changing the connections between different modules in order to minimise these losses. We determine the current Icom and the voltage Vcom corresponding to the commutation point against the photovoltaic radiation Es and/or the temperature T change. Consequently, the global system performance is improved.

Commutation Factors

We consider the case where a load is connected to a PVG. The commutation from a configuration to another can occur when the load change or when the climatic conditions change. In Figure 5, in the case of two configurations Np1xNs1 noted "parallel configuration" and Np2xNs2 noted "series configuration" we assume a constant load of characteristics P-V noted 'Z,' a constant tempera-

Figure 5. The commutation factors

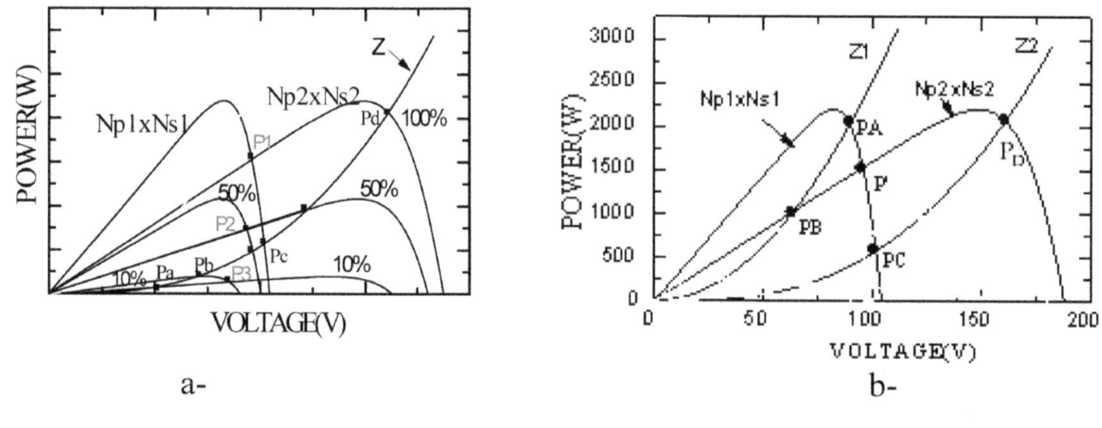

ture and a change of solar radiation (insulation) Es equals to 10% which corresponds to 100W/m2, 50% (500W/m2), 100% (1000W/m2). For example, at Es=10%, if the parallel configuration is adopted the operating power point is Pb, if the series one is adopted the operating power point is Pa, but Pb>Pa then it is necessary to adopt the configuration parallel because the power extracted by the load is the best. If the solar radiation varies, the operating point for example at Es=100% is Pc because the configuration is parallel. Pd is the power if the series one is adopted. In this case Pd>Pc then it is necessary to switch to the series one for a good exploitation of the PVG (Zerhouni, 1996, 2009).

The other case is for the constant climatic conditions, a load variation. As illustrated in Figure 5b, the load is noted Z1. If the parallel configuration is adopted the load Z1 has an operating point of power PA. PB is the power extracted by Z1 if the series configuration is adopted, or PA>PB so the configuration parallel is the best: it is adopted. A load variation suddenly occurs, the load becomes Z2, and the operating point is then Pc because the parallel configuration is maintained. The series configuration provide to the load PD with PD>Pc. Therefore, a switch from the parallel configuration to the series one is necessary to provide a best power to the load.

If NT is the total photovoltaic modules used:

$$N_T = N_s x N_p \tag{4}$$

With:

Np: number of branches in parallel;
Ns: number of modules in series in a branch;

The points P1, P2, P3 in Figure 5a and P' in Figure 5b are the intersect points of the PV characteristics for series and parallel configurations under the given climatic conditions. In these points for example P' these configurations have equal power. These points are the reference. So in Figure 5b as example, for any load with operating point situated at the right of P, configuration 2 (Np2xNs2) will be used. On the other hand, for any load with operating point situated at the left of P', configuration 1 (Np1xNs1) will be used.

In our work, NT=6 with the configurations: parallel Np1=2 Ns1=3 and Series Np=1 Ns=6. At T=25°C Es=100%, the I-V characteristic (current-voltage) is in series Np2=1 Ns2=6

$$I = 3.192 - 1.1295e^{-7}\left[\exp\frac{(V+2.4I)}{2.842e^{2}.U_T} - 1\right] \tag{5}$$

parallel Np1=2 Ns1=3

Figure 6. Different characteristics

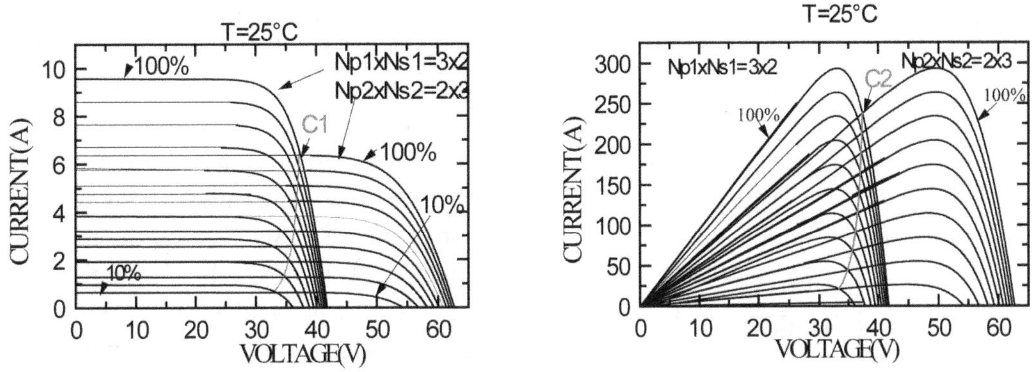

a- b-

$$I = 6.384 - 2.25e^{-7}\left[\exp\frac{(V+0.6I)}{1.421e^{2}.U_{T}} - 1\right] \quad (6)$$

The idea to improve the PVG operation, extended to any load is based on the following principle:

with a number of modules NT constant which satisfy the load demand, one sets either two or three configurations.

A simulation program developed calculates at each photovoltaic radiation the co-ordinates of the commutation points Pcom (Icom, Vcom) of the characteristics current-voltage I=f (V) or power-voltage P=f(V) of the various arrangements.

The points 1-5 of curve noted 'C1' on Figure 6a or P-V noted 'C2' on Figure 6b are calculated at each solar radiation Es and temperature T. In Figure 6 the solar radiation varies from 10% to 100% at T=25°C. They are independent of the constraint load but are a function of the climatic conditions and the selected configurations. They are the marks of commutation towards the suitable configuration.

The commutation mechanism from one configuration to the other, independently of the load constraint will use the commutation points in curve 'C1' or 'C2' for which the two configurations have an equality in power for the given conditions of temperature and the photovoltaic radiation. Each point on the curve C1 is defined by its current Icom and its voltage Vcom. It will be a matter of comparing the load voltage Vfct and/or current Ifct to Vcom and/or Icom and switch the PVG onto appropriate configuration (Zerhouni, 1996, 2009).

The Complexity of the Switches Number

To allow the commutation passage from one configuration to the other one, we use switches. By control the closing of some and the opening of others, one can obtain the desired configuration. In the practical case, it was observed that to allow only 2 changes of the system configuration is sufficient. The number of switches, for a specific number NT of photovoltaic modules follows especially the configurations considered. These last are determinated according to the particular needs for the equipment to supply and the modules total number. If two modules are adopted, there will be 2 configurations: NpxNs=1x2: the series configuration and the parallel one NpxNs=2x1. As shown in Figure 7, this choice leds us to the use of

Figure 7. Case of NT=2

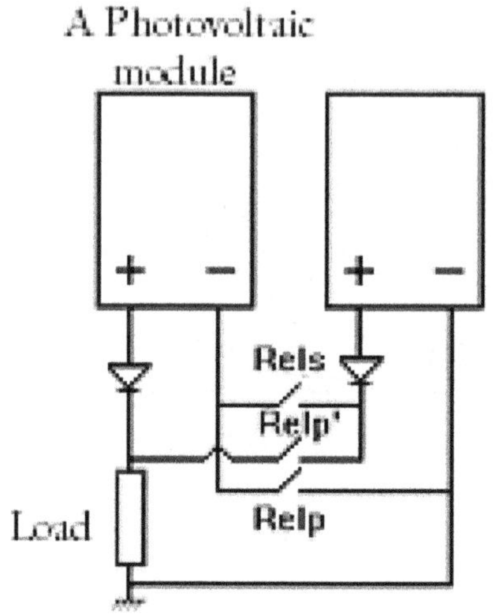

3 switches: Rels allows the series configuration. Relp, Relp' allow the parallel one.

As for NT=4, one can have 3 configurations (parallel NpxNs=4x1, mixed NpxNs=2x2, series NpxNs=1x4) as shown in Figure 8:

For this case, for the choice of:

- Parallel and series configurations, one needs 9 switches,
- Parallel and mixed configurations, one needs 6 switches,
- Series and mixed configurations, one needs 4 switches,
- Parallel, mixed, and series configurations, one needs 9 switches.

These results show that for the photovoltaic modules same total number, the switches number to be auctioned differs according to the selected configurations and not only to the total number of modules used.

Commutation with many configurations can be characterized by a switches significant number. For example, 9 switches for the three configurations of the 4 photovoltaic modules. If we tolerate a change between several configurations, we must make a compromise between the number of switches used and the power's gain obtained. In the case of our study, we will tolerate only the change of 2 configurations with NT=6.

To allow the opening or closing of the switches used in order to have the required commutation, one must have a reference mark which can be either the photovoltaic radiation (Zerhouni, 2009) or another reference. The photovoltaic radiation reference mark was already studied (Zerhouni, 2009). The drawback of this method is that we must change this reference when the load varies. It is not applicable on real time and for any load. The object of the next study is the our reference mark utilization which is applicable for any load and climatic conditions. This mark is the points on curve C1 in Figure 6a (Zerhouni, 2009).

The Variations of the Commutation Points

The commutation points (Icom-Vcom) vary with Es and T as it is illustrated in Figure 9. In Figure 9,

Figure 8. Case of Nt=4

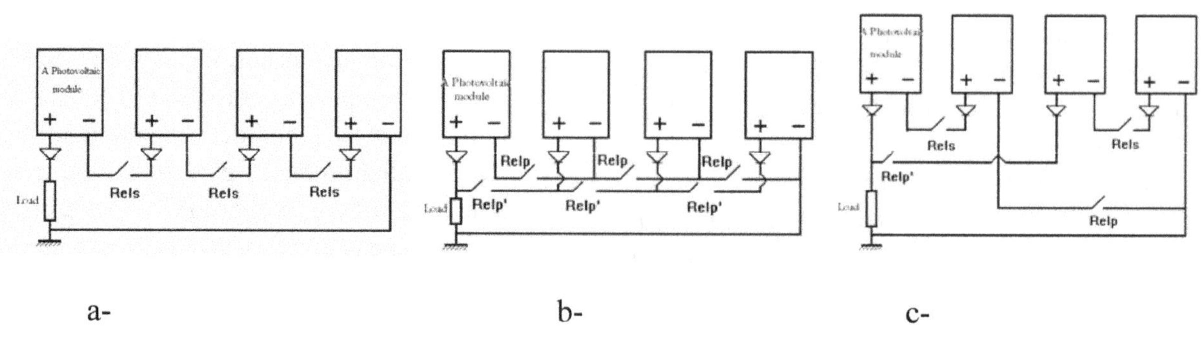

Figure 9. Variation of the commutation points

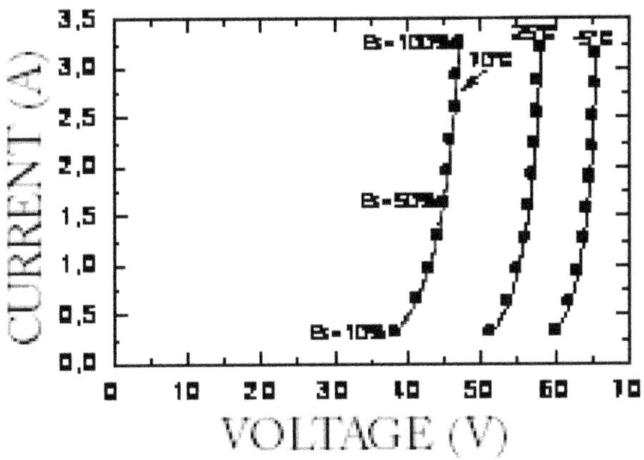

Figure 10. Icom variation against Es

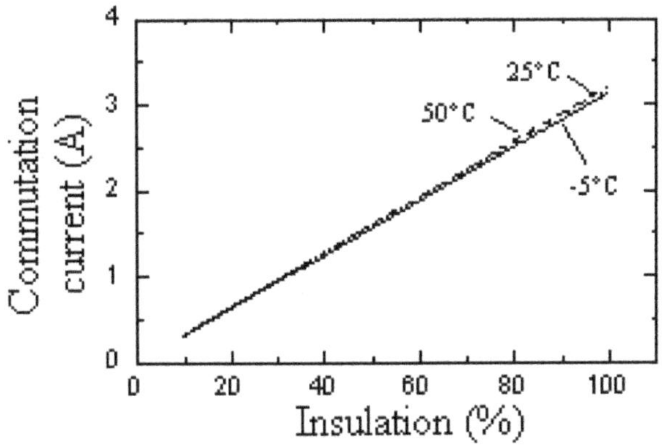

one represents these points variation for Es=10% to 100% at T=25°C and extreme temperatures of 70°C and -5°C. As observed, these points are not constant and varie if the climatic conditions change. The important deduction is that the current Icom remains constant in spite of the temperature changes but it is affected by the solar radiation variations. Vcom is affected by both Es and T variations. Computer simulations were made to investigate this method. A simulation program developed calculates at each photovoltaic radiation the commutation points co-ordinates Pcom (Icom, Vcom) of the current-voltage characteristics I=f(V) of various arrangements (Zerhouni, 2009, 2010).

The New Approach using the Commutation Current Icom

To concretize the idea of a configuration commutation from one towards the other without any influence of the load, our attention is particularly focused on the current Icom. The case considered 6 modules in NpxNs=2x3 named parallel configuration and NpxNs=1x6 named series one. The results can be applied to another number of modules. We are interested in the Icom variations according to climatic conditions. A simulation program has been developed. The Icom layout according to the photovoltaic radiation for a constant temperature of 25°C is represented in the Figure 10.

It is noticed from Figure 10 that Icom is linear according to the photovoltaic radiation. It can thus be written:

$$I_{com} = K_i.E_s \tag{7}$$

With Ki: constant which depends on the adopted configurations and the modules selected number.

For our example: Ki=0.032 A.m2/w.

Figure 10 also shows the simulation results of Icom versus T for the configurations chosen of the 6 modules. From Figure 9, it is clear that the commutation current varies linearly with the photovoltaic radiation and hardly affected by the temperature. The simulation program has been run for other different configurations and the results led to the same conclusion.

To work with Icom linearly proportional to the photovoltaic radiation and slightly depend on the temperature will simplify the practical realization according to the relation:

$$I_{com} = K_i.E_s, \forall T \tag{8}$$

Es is deduced from Icc which varies linearly with it.

Figure 11. The implementation flowchart follows using Icom

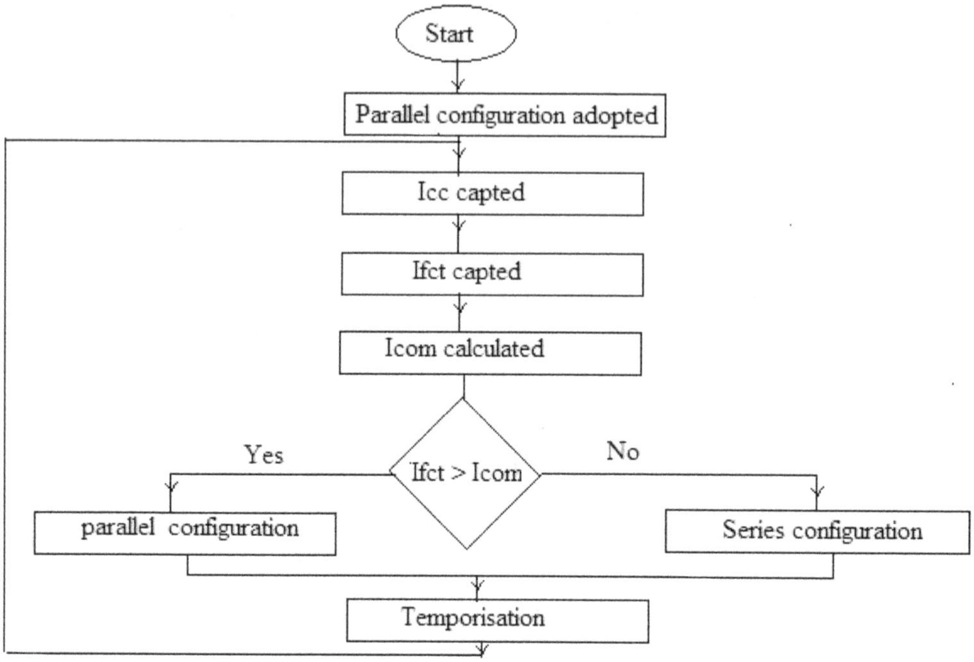

For a load supplied by a photovoltaic generator, one is interested in the operation current position Ifct compared to Icom. The control system has been implemented. A variable load has been used in order to cover different operating conditions. The practical realization has been based on the following flowchart in Figure 11.

The experimental characteristics for Es=24%, T=27°C are illustrated in the Figure 12. The Figure 12a is related to the commutation from parallel to series configuration. The commutation occurs at 'A.' The Figure 12b illustrates the commutation from series to parallel configuration, but the major drawback was summarized in the switch

Figure 12. Experimental results using Icom

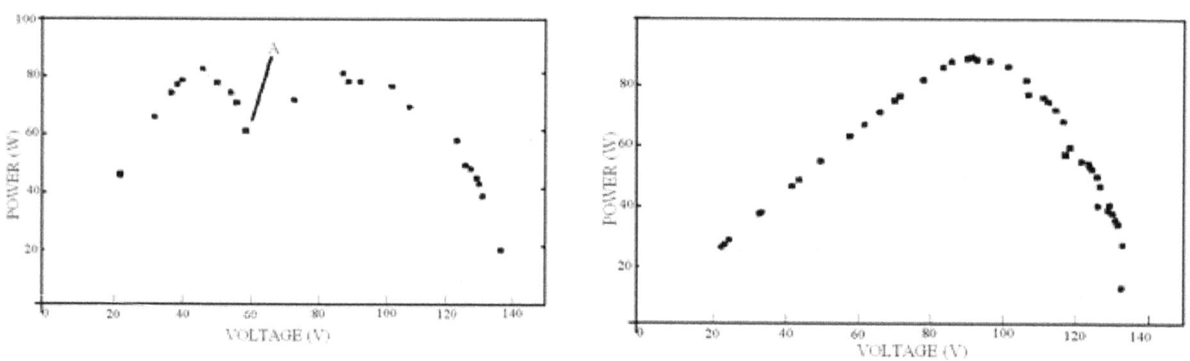

a-

b-from the series towards the parallel configuration never obtained. The commutation does not occur where it should be or later. Indeed, the difference between Ifct and Icom is very small in the flat region of the (I,V) characteristic of the PVG (current generator). Therefore, the commutation using Icom is not appropriate.

The New Approach using the Commutation Voltage Vcom

Instead of considering the currents, one is interested in the voltages. For a load supplied by photovoltaic way, one is interested in the operation voltage position Vfct compared to Vcom. Computer simulations were made to investigate this idea. The variation of Vcom as a function of the photovoltaic radiation for different temperatures has been determined and is illustrated in Figure 13a. As it can be seen the characteristic is non linear at low photovoltaic radiation level, so a simple linear procedure has been applied. As a consequence for a given temperature the commutation voltage Vcom could be written as follows:

$$V_{com} = a_i.E_s + b_i \tag{9}$$

Where, ai and bi are constants, which are derived from the linear procedure at 25°C. The Figure 13a also shows that the commutation voltage Vcom is significantly affected by the temperature T. As shown in Figure 13, as T increases Vcom diminishes linearly. To implement the reconfiguration method using the commutation voltage Vcom as reference, (9) should be corrected in order to take into account the linear variation of Vcom as a function of the temperature. After modification (9) becomes:

$$V_{com} = a_i.E_s + b_i - c_i.(25 - T) \tag{10}$$

Where ai and bi are as defined previously and ci is the slope of the curve Vcom=f(T) at 25°C.

The control system based on equation (10) has been implemented following the flowchart in Figure 14 and an example of the practical results is given in Figure 14. A variable load has been used in order to cover different operating conditions. One distributes the characteristic I=f(V), P=f(V) in three noted zones Z1, Z2, Z3, as the Figure 13a.

The calculation of the coefficients for the example of six modules led to the following values:

ES€[0,20%] ai=0.23V.m2/W bi= 49.19V ci=-0.288V/°C

Es€[20%,50%] ai= 0.09V.m2/W bi= 52.21V ci= -0.27V/°C

Figure 13. Vcom variation against climatic conditions

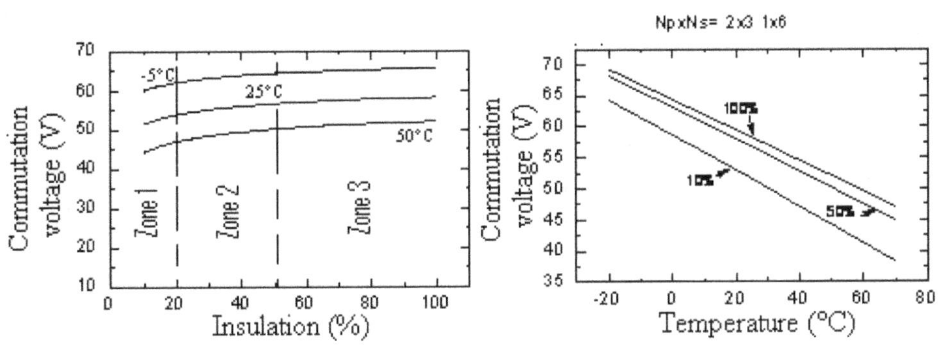

a- b-

Figure 14. The implementation flowchart using Vcom

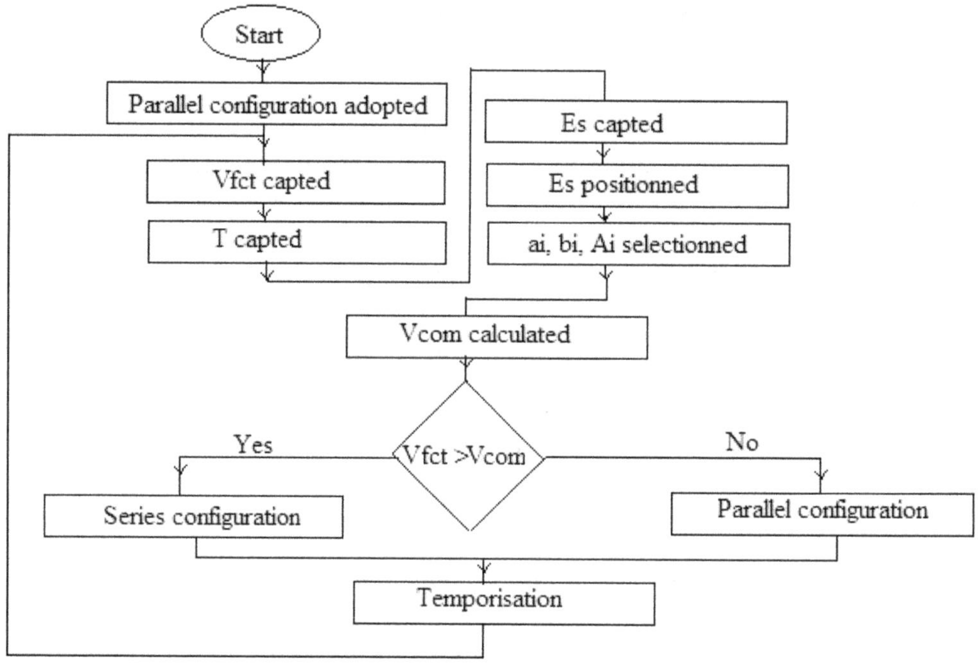

Es€[50%,100%] ai=0.03 V.m2/W bi= 55.21V ci=-0.253V/°C

The implementation follows the flowchart given in Figure 14.

The experimental switching characteristics for Es=24%, T=27°C are illustrated by the Figure 15. The Figure 15a is related to the commutation from parallel to series configuration. The commutation occurs at 'Pps.' The Figure 15b illustrates the commutation from series to parallel configuration. The commutation occurs at 'Psp.' But, the major handicap is that 'Pps' differs from 'Psp.' The zone of descent of the curve I - V (current - voltage) of the parallel configuration is similar to a generator of voltage that presents a stiff slope. The

Figure 15. Experimental results using Vcom

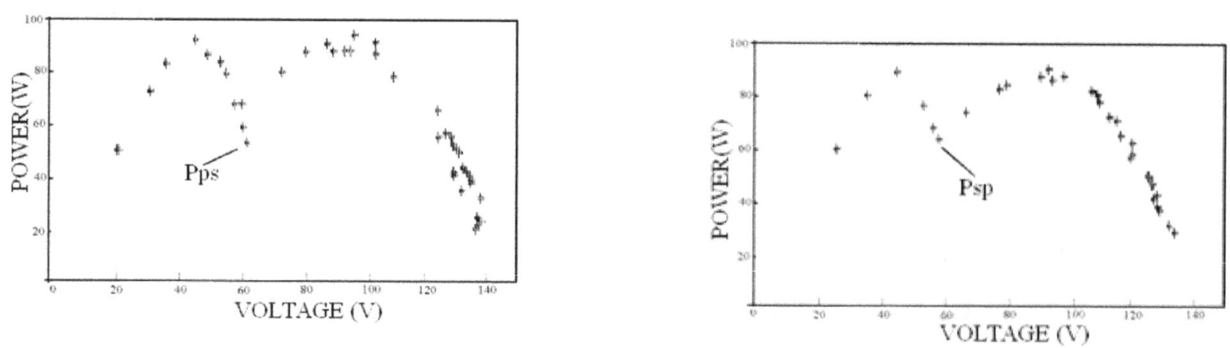

a-

b-difference in voltage between the voltages in this area is weak and the detection of the threshold of commutation is delicate. Thus, the commutation will be erroneous during the passage from the parallel configuration to the series one.

The New Approach using Icom and Vcom

In order to avoid the cited problems, the solution is to use both Icom and Vcom. An experimental hardware has been implemented in the laboratory as shown in Figure 16.

The panels are composed of 3 modules in series. To allow the commutation passage from one configuration to the other one, we use switches Rels, Relp, Relp'. By control the closing of some and the opening of others, one can obtain the desired configuration automatically, in real time. The control card is based on microcontroller. The control card permits the insulation Es and the temperature T measure, to determine then Vcom or Icom using equations 8 and 10. If the SCA is in the parallel configuration then the current flowing into the load is measured and compared to Icom. If it is higher, the SCA remains in its actual configuration otherwise it is reconfigured into the series configuration (Relp, Relp' opened and Rels closed). If however the SCA is in the series configuration, then the load's voltage is measured and compared to Vcom. If it is greater, then the SCA remains in its actual configuration otherwise it is switched onto the parallel configuration (Relp, Relp' closed and Rels opened). As a result, the SCA is always switched onto its best configuration for any load and any weather conditions, on line. A variable load has been used in order to cover a large currents and voltages range.

Figure 16. Experimental system synoptic

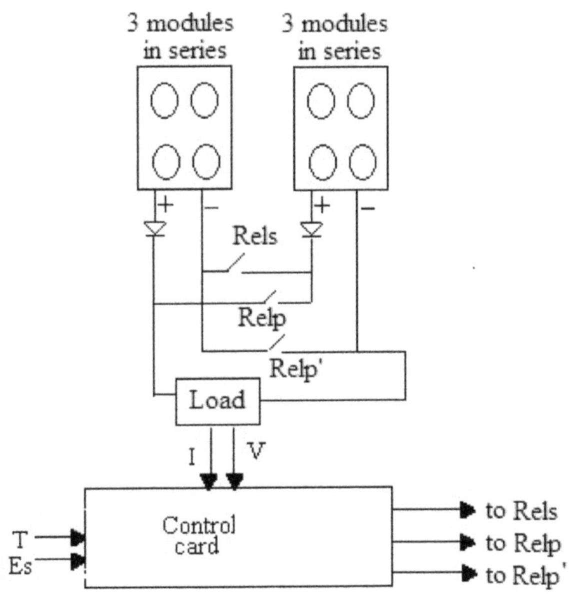

The flowchart in Figure 17 shows the principle of the method.

The experimental results for Es=24%, T=27°C are illustrated in Figure 18. Figure 18a shows the commutation at 'Pps' from the parallel to the series configuration. The Figure 18b shows the commutation at 'Psp' from the series to the parallel one. The commutation occurs precisely at the same point. This point coincides too with 'Pcom' that obtained by the intersection of the layout of the characteristic series alone and that parallel in Figure 19. There is no hysteresis phenomenon observed. Consequently, this idea is considered adequate. The PVG is always switched onto the best configuration for any load and any weather conditions.

The Uncertainty Analysis

Figure 19 illustrates the voltage commutation real curves and the approximate one for the chosen configurations (NpxNs=1x6 and 2x3) at solar radiation from 10% to 100%. Figure 20a is for T=25°C. Figure 20b is for T=-5°C. Figure 19c is for T=50°C. The curves noted 'b' in Figure 20 used the relationship (10). As seen in Figure 20, the curves are overlaped each other. The mean square error EQM calculus obtained is:

Figure 17. Implementation flowchart using Vcom and Icom

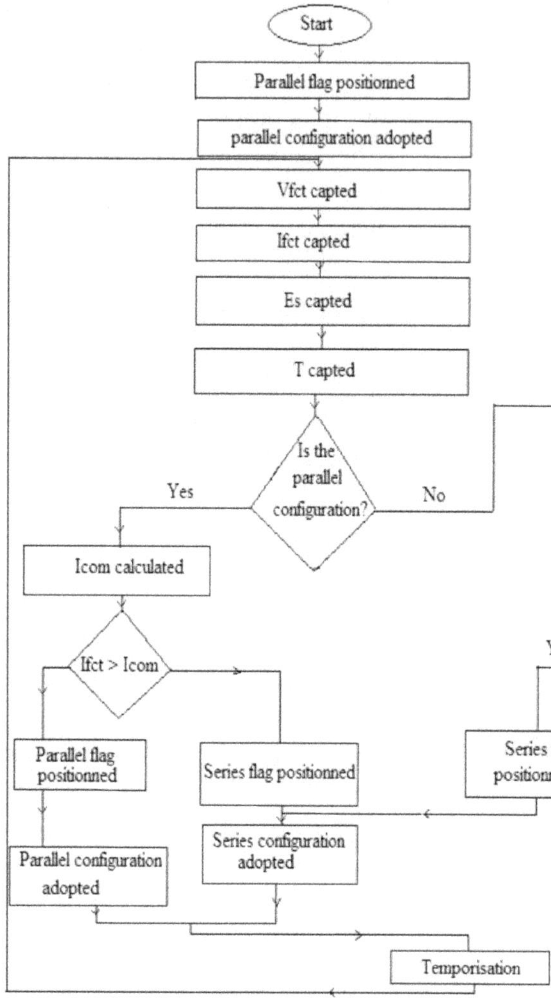

Figure 18. Experimental results using Icom and Vcom

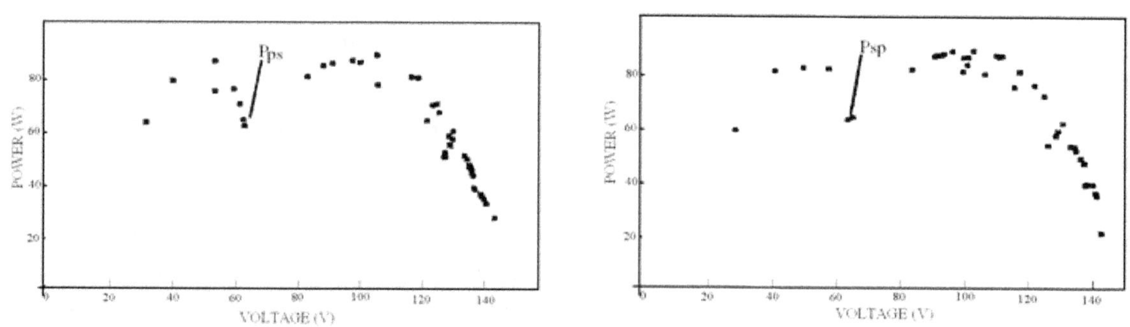

a- b-

Figure 19. Series and parallel experimental results

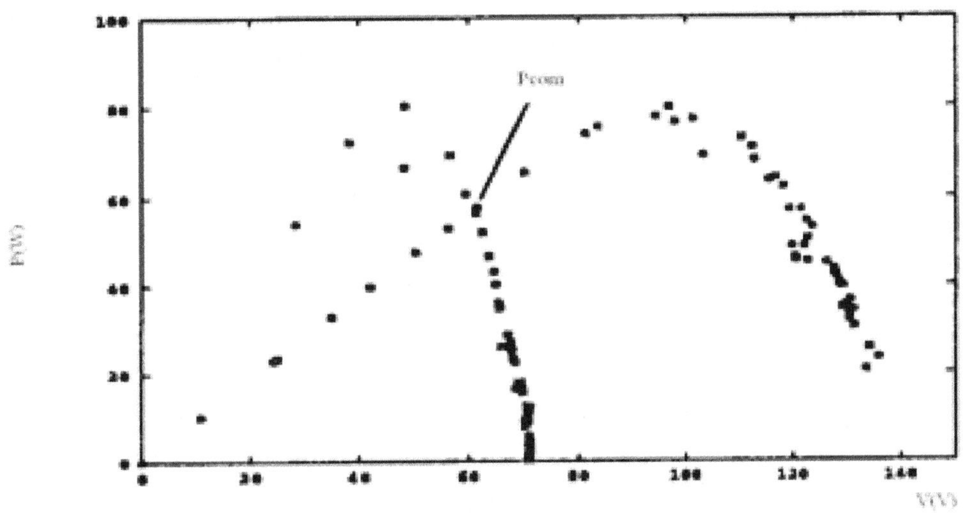

EQM=0.007 at 25°C.

EQM=0.027 at -5°C.

EQM=0.047 at 50°C.

Figure 21 represents the current real and approximate curves obtained for the adopted configurations at several solar radiations from 10% to 100% and temperatures of -5°C, 25°C and 50°C. In Figure 21, the approximate curve using equation (8) is noted "approximate curve." As seen in Figure 21, the curves are overlapped each other whatever is the temperature and these curves are closed to the approximate one.

The mean square error EQM calculus for this case is:

For -5°C, EQM=0.004;

for 50°C, EQM=0.002;

for 25°C, EQM=0.001.

Uncertainty analysis has been made on some other experimental works for different modules number and configurations and was assessed to be around 2.5% for the commutation voltage and 0.002% for the commutation current.

Conclusion

From this combined study, the use at the same time of Icom and Vcom is the best solution. Icom is a reference to allow the commutation from the parallel configuration towards the series one; Vcom is to allow the reverse commutation from series configuration towards the parallel one. This method is in real time, on line independently of the load and climatic conditions. This technique has been concretized in our laboratory. It was tested on a resistive load and a DC motor. The method of configuration change on line, in real time is a promising method since it guarantees the most power to the load. Thus, advantage of power is transmitted to the motor, more is the rise of the fluid through pump, and most important is the depth of water pumping reached. Another advantage of this configuration change is that a system motor-pump can sometimes begin its pumping earlier for such configuration contrary to another according to the current request. One can commutate towards another configuration if this last provides more power. To tolerate a configuration

Figure 20. Commutation voltage curves

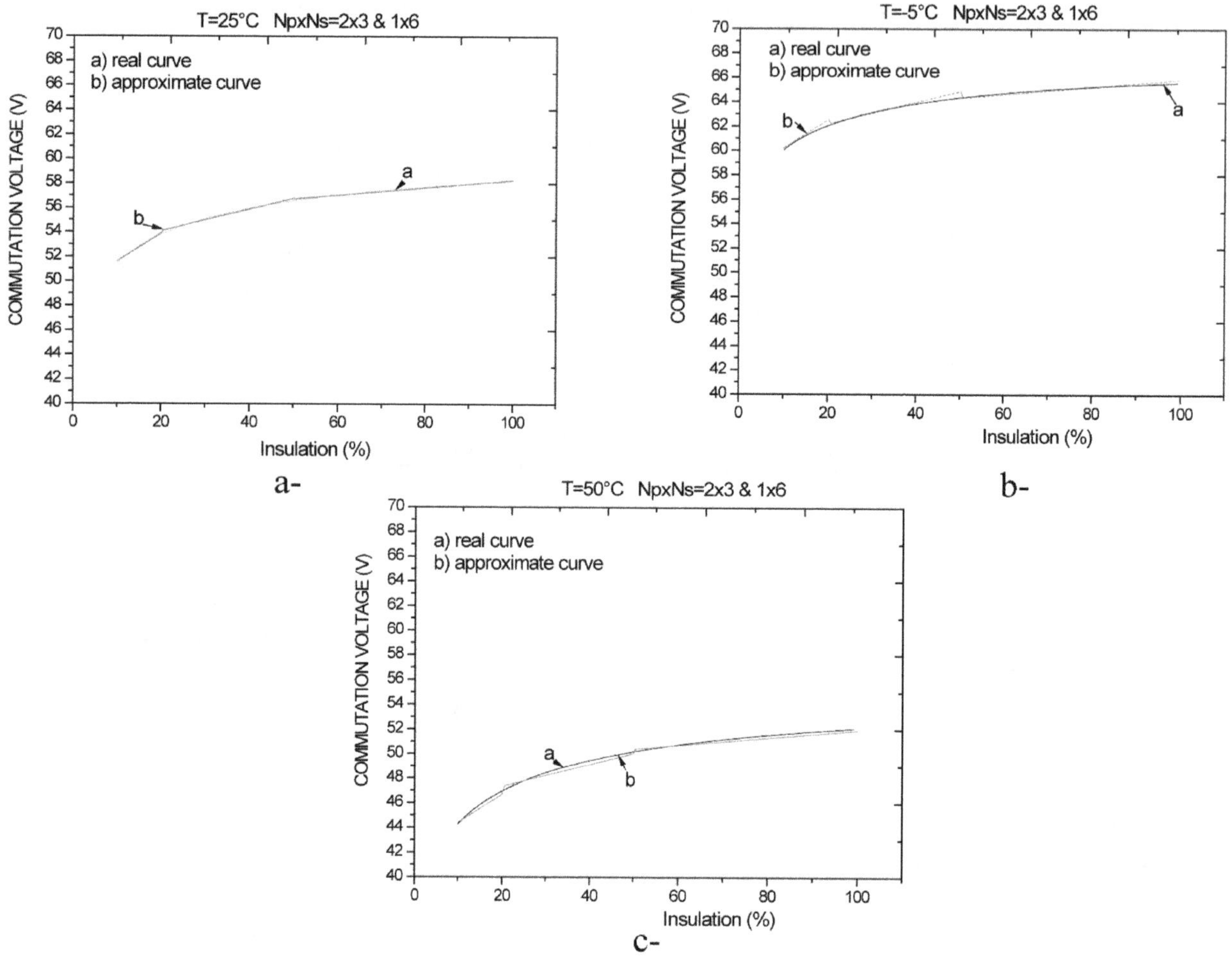

Figure 21. Commutation current curves

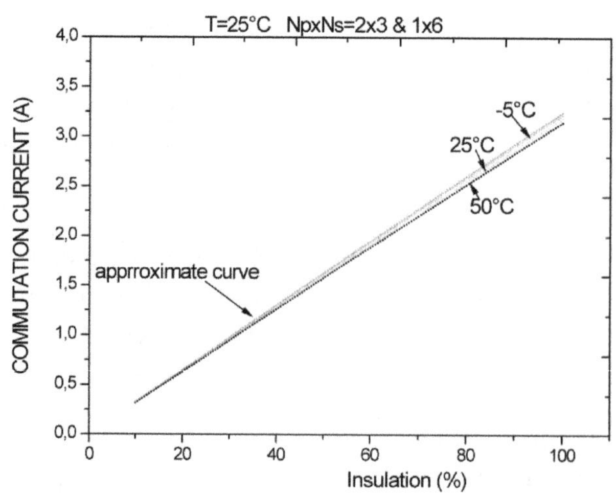

change will allow a PVG good exploitation and improve its operation. The technique based on the change of configurations is a simple technique, at moderate price.

FUEL CELLS

Fuel cells, although they have existed for more than 50 years, are just starting to capture the general public interest. Fuel cells promise clean and efficient operation along with high energy densities (Boudghéne, 2011a). While there are many different fuel cell types (Zerhouni, 2009; Boudghéne, et al., 2002a), the proton exchange membrane fuel cell stands out as one of the most promising for transportation and small stationary applications (Boudghéne, et al., 2002b).

The Proton Exchange Membrane or PEM fuel cell is a compact type of hydrogen fuel cell that has a relatively fast electrical response time and a low operating temperature and pressure (Busquet, 2003).

Polymer Electrolyte Membrane (PEM) fuel cells have received much attention, in scientific and industrial research (Boudghéne, 2011a). Proton Exchange Membrane (PEM) fuel cells take hydrogen and air and produce power and water. The mechanism is exactly opposite to that of electrolysis (Boudghéne, 2003).

$H_2 \Rightarrow 2H^+ + 2e^-$ Anodic half reaction

$\frac{1}{2}O_2 + 2H^+ + 2e^- \Rightarrow H_2O$ Cathodic half reaction

$$H_2 + \frac{1}{2}O_2 \Rightarrow H_2O \text{ The overall reaction} \quad (11)$$

Due to the increasing and highly promising use of fuel cells as an important electrical energy source, its modeling has been a major research issue, trying to find mathematical models to accurately predict how a cell will behave in a real system. The overall goal is the verification of fuel cell characteristics to provide a diagnostic tool to ensure the safe operation of the system. The model is based on the empirical electric model developed by NR Albrieux (Busquet, 2003). The used model is illustrated in Figure 22.

The characteristic equation is given by:

$$Vcell = E_{oc} + \frac{b}{Ln\left(\frac{J}{J_d \times e^2}\right)} + \left(\frac{b}{4 \times J_d} - \Delta\right) \times J \quad (12)$$

Vcell is the cell's potential, J the current density, E_{oc} is the cell's open circuit voltage, J_d and Δ are model's parameters. This equation takes into account the diffusion problems which exist for strong current densities, but the zone where the diffusion problems are dominating is not modelled by this model.

Characterization of the fuel cell polarization curves based on pressures, temperature allows analysis and simulation of the fuel cell power generation. The parameters Eoc, B, Jd, and Δ depend on the temperature T and on the partial oxygen pressure Po2. Each parameter is dissociated in three components according to the following equation:

$$\begin{bmatrix} E \\ Jd \\ b \\ \Delta \end{bmatrix} = \begin{bmatrix} E1 & E2 & E3 \\ Jd1 & Jd2 & Jd3 \\ b1 & b2 & b3 \\ \Delta1 & \Delta2 & \Delta3 \end{bmatrix} \times \begin{bmatrix} 1 \\ T \\ T \times \ln\left(P_{O2}\right) \end{bmatrix} \quad (13)$$

The pressure Po2 is maintained constant at 1.4 Bar (Figure 23a). For various temperatures, the

Figure 22. Fuel cells simulink model

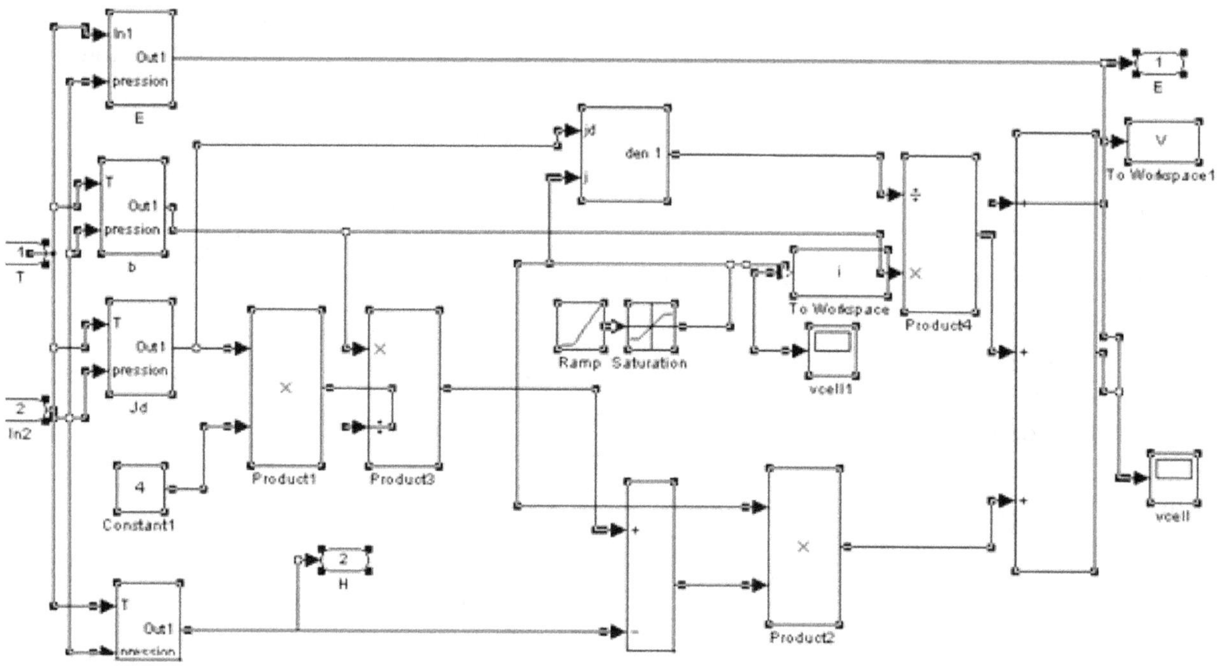

temperature improves the fuel cell performances as seen in the figure. In addition, the temperature is maintained constant at T=35°C with various pressures. The simulation results are described in the Figure 23b. The partial oxygen pressure increase increases the cell's voltage V for the case of the strong densities J. The temperature's influence is comparable to that of the partial oxygen pressure,. It improves the fuel cell's performances (Zerhouni, 2009; Busquet, 2003).

ELECTROLYSER

Electrolysis is the process of decomposition of a substance into its chemical components by an electric current. An electrolyser produces both hydrogen and oxygen gas for use in a fuel cell by the process of electrolysis. In the process of electrolysis, the atoms that comprise a water molecule, hydrogen and oxygen, are disassociated from each other. This is the only method of hydrogen production with no requirements for fossil fuels. In addition, with polymer membrane technology it is possible to have an electrolyser with a solid polymer membrane. In a solar hydrogen system, the electrolyser is the component, which changes electrical energy from the photovoltaic panels into hydrogen gas. Actually, if properly designed, the same PEM unit can be operated as an electrolyser to produce hydrogen, as well as hydrogen driven fuel cell to produce electricity. Hydrogen storage is chosen for its potential to make a clean energy option and low cost in construction.

The basic difference between the electrolyser and PEM fuel cell is that the hydrogen and oxygen reactions on the electrodes are opposite. Therefore, the electrolyser produces H2 at the cathode and O2 at the anode, while the fuel cell consumes H2 at the anode and O2 at the cathode. Also, the water's decomposition into H2 and O2 in the electrolyser is non-spontaneous and requires an electric power for the reaction to occur, while the opposite fuel cell reaction is spontaneous and produces an electric power. In electrolysis, an electric current is run through water and the water reacts to form hydrogen and oxygen.

Figure 23. Simulation results

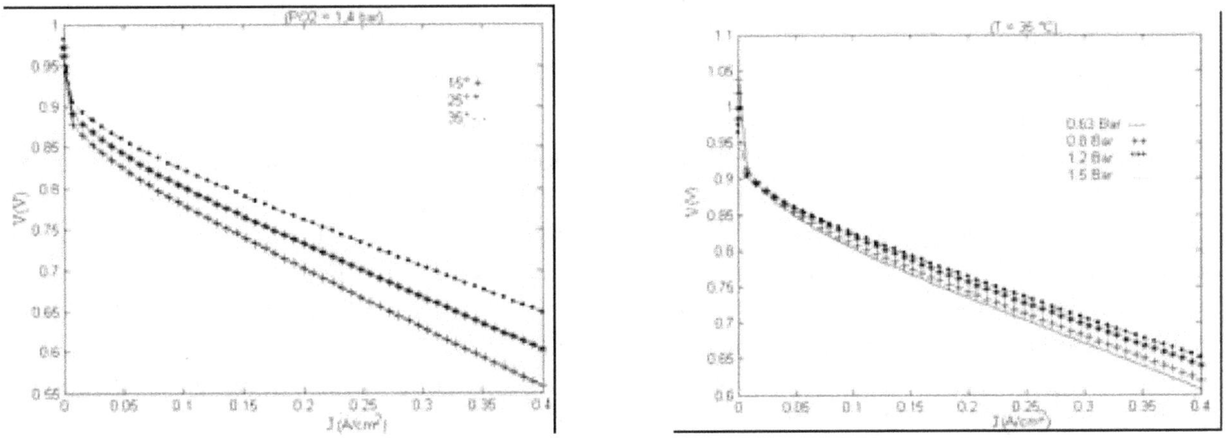

a- b-

The Simulink model developed is the same illustrated in Figure 22 with other constants values. The parameter pressure is kept constant ($Po_2 = 9.6$ Bar). For various temperatures (Figure 24a), it is noticed that the tension decreases when the temperature increases. The temperature's influence increases with the intensity. The electrolyser's performances are thus a temperature function. And when the temperature is maintained at a constant value of T=53°C, for example, with various pressures (Figure 24b), it is noticed that the pressure increase improves the electrolyser's performances (Busquet, 2003; Zerhouni, 2009).

CONCLUSION

Nowadays, the development of new and non-polluting energy producing and energy-storage systems is a great challenge for scientists. Hybrid power systems are based on renewable energy sources and especially on photovoltaic systems. Among the renewable energy sources, the Photovoltaic (PV) energy has been widely utilized in low power applications. The electricity generated from solar cells is one of the best options for the sustainable future energy requirements of the world. It is also the most promising candidate for

Figure 24. Simulation results

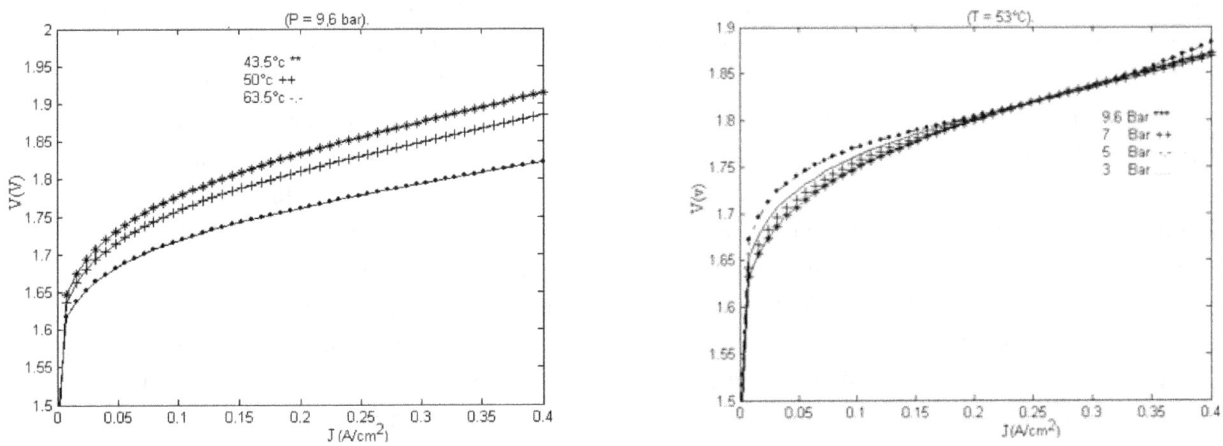

a- b-

research and development for large-scale users as the fabrication of low cost PV devices becomes a reality.

Fuel cells have been receiving a lot of attention lately due to their potential as becoming a new energy source with a large range of applications. Fuel cells can be incorporated with other components to create high efficiency industrial power plants. Fuel cells permit clean and efficient energy production for stationary and transportation applications. Considering economical and ecological aspects, low temperature fuel cells like Polymer Electrolyte Fuel Cells (PEMFC) seem to represent well-suited systems. Fuel cells can help in contributing in environment by maximizing the use of renewables in the energy production and using hydrogen produced with no pollutant emission. A simple and accurate fuel-cell model is studied for green power-system. The results of the model are used to predict the output voltage, and power of FC as a function of pressure and temperature. Circuit model parameters of many series-connected cells can be acquired using the results of a single stack simulation. Electrolysis, which is the simplest process to generate hydrogen, is powered by the DC voltage from the photovoltaic cells. To allow for simultaneous utilisation of emission-free renewable energy sources, the hydrogen shall be produced by implementation of electrolysis of water. The application of fuel cell technologies to advanced power generation systems implies the most significant advancement in energy conservation and environmental protection for the next years.

The purpose of the work is to optimize the system's operation. The main reason to build described system is to supply stand alone system using renewable energy sources. Therefore, the power plant has to produce energy independent of any weather fluctuations. In order to compensate for their cost, Photovoltaic (PV) systems need to be operated in a manner that ensures the optimum use of the available photovoltaic energy. In this chapter, the concentration is on photovoltaic systems. A simple and efficient method to optimise the SCA's power output has been presented. This method applies a dynamical electrical array reconfiguration strategy on the Photovoltaic (PV) generator, in order to improve its energy production. The illustrated method allows a continuous SCA reconfiguration in order to insure a good load matching. The principle of the proposed method consists of determining on line and in real time for a fixed number of photovoltaic modules, which configuration is the best for a given load under given working condition and the switching the SCA into that configuration. Unfortunately, the reference point is not fixed but varies with weather conditions (solar radiation, temperature, etc.). Positions are found according to the variation of the already mentioned conditions. A better way of calculating the current and voltage is therefore determined. The reconfiguration could occur during the insulation variation or when a load's change occurs. By rearranging connections between different modules, the matching between the load and PVG is improved. This enables the system to react to changes in loads and environmental conditions. The simulation program developed also allows to explore different configurations and similar results were obtained. The experimental system described has permitted to validate the proposed method. To avoid the untimely commutations if the operating point situates in the intersection zone of I - V(or P - V) of the two configurations, we have to tolerate a certain time between two succeed commutations: an hysteresis.

This method, prevent the decrease of the output power when environmental changes occured. The advantage of the method we presented appears for an engine suing a hydraulic pump for example. If advantage of power is transmitted to the engine, more will be the climbing of the fluid through the pump and more important will be the depth of water pumping reached. More, an engine pump system can sometimes begin its pumping earlier and more rapidly for such configuration to the other following the required current of beginning.

The technique of configuration change is simple, its price moderated.

REFERENCES

Appelbaum, J. (1989). The operation of loads powered by separate sources or by a common source of photovoltaic cells. *IEEE Transactions on Energy Conversion*, *4*(3), 351–357. doi:10.1109/60.43235

Benmessaoud, M. T., Zerhouni, F. Z., Zegrar, M., Boudghene Stambouli, A., & Tioursi, M. (2010). New approach modeling and a maximum power point tracker method for solar cells. *Computers & Mathematics with Applications (Oxford, England)*, *60*(44), 1124–1134. doi:10.1016/j.camwa.2010.04.001

Boudghéne Stambouli, A. (2003). *Electricity from photovoltaics and fuel cells: The expectations for an enviromental-friendly and sustainable source of energy*. Paper presented at the Environment 2010 International Symposium. Porto, Portugal.

Boudghéne Stambouli, A. (2011a). Fuel cells: The expectations for an environmental-friendly and sustainable source of energy. *Renewable & Sustainable Energy Reviews*, *15*(9), 4507–4520. doi:10.1016/j.rser.2011.07.100

Boudghéne Stambouli, A. (2011b). Promotion of renewable energies in Algeria: Strategies and perspectives. *Renewable & Sustainable Energy Reviews*, *15*(2), 1169–1181. doi:10.1016/j.rser.2010.11.017

Boudghéne Stambouli, A. (2011c). Algerian renewable energy assessment: The challenge of sustainability. *Energy Policy*, *39*(8), 4507–4519. doi:10.1016/j.enpol.2010.10.005

Boudghéne Stambouli, A., & Traversa, E. (2002a). Solide oxide fuel cells (SOFCs): A review of an environmentally clean and efficient source of energy. *Renewable & Sustainable Energy Reviews*, *6*, 433–455. doi:10.1016/S1364-0321(02)00014-X

Boudghéne Stambouli, A., & Traversa, E. (2002b). Fuel cells, an alternative to standard souces of energy. *Renewable & Sustainable Energy Reviews*, *6*(3), 295–304. doi:10.1016/S1364-0321(01)00015-6

Busquet, S. (2003). *Etude d'un système autonome de production d'énergie couplant un champ photovoltaïque, un électrolyseur et une pile à combustible: Réalisation d'un banc d'essai et modélisation*. Ph. D Thesis. Paris, France: Ecole des Mines de Paris.

Busquet, S., Hubert, C. E., Labbé, J., Mayer, D., & Metkemeijer, R. (2004). A new approach to empirical electrical modelling of a fuel cell, an electrolyser or a regenerative fuel cell. *Journal of Power Sources*, *134*(1), 41–48. doi:10.1016/j.jpowsour.2004.02.018

Enrique, M., Dura'n, E., Sidrach-de-Cardona, M., & Andu'jar, J. M. (2007). Theoretical assessment of the maximum power point tracking efficiency of photovoltaic facilities with different converter topologies. *Photovoltaic Energy*, *81*(3), 1–38.

Kwon, J.-M., Nam, K. H., & Kwon, B. H. (2006). Photovoltaic power conditioning system with line connection. *Transactions on Industrial Electronic*, *53*(4), 1048–1054. doi:10.1109/TIE.2006.878329

Mutoh, N., & Inoue, T. (2007). A control method to charge series-connected ultraelectric double-layer capacitors suitable for photovoltaic generation systems combining MT control method. *Transactions on Industrial Electronics*, *54*(1), 374–383. doi:10.1109/TIE.2006.885149

Salameh, Z. M., & Chaozi, L. (1990a). Optimum switching points for array reconfiguration controller. In *Proceedings of the 21st IEEE Photovoltaic Conference*, (vol 2), (pp. 971-976). IEEE Press.

Salameh, Z. M., Mulpur, A. K., & Dagher, F. (1990b). Two stage electrical array reconfiguration controller. *Photovoltaic Energy*, *44*, 51–55.

Weidong, X., Nathan, O., & William, D. G. (2007). Topology study of photovoltaic interface for maximum power point tracking. *Transactions on Industrial Electronics*, *54*(3), 1696–1704. doi:10.1109/TIE.2007.894732

Zerhouni, F. Z. (1996). *Adaptation optimale d'une charge à un générateur photovoltaïque*. Master Thesis. Oran, Algeria: University of Sciences and Technology Mohamed Boudiaf.

Zerhouni, F. Z. (2005). *Contribution dans l'amélioration du fonctionnement d'un système photovoltaïque*. Paper presented at the International Conference, ICEL05. Oran, Algeria.

Zerhouni, F. Z. (2006a). Etude des l'influence de différents agents sur les caractéristiques d'un générateur photovoltaïque. *Technological and Scientific Review*, *4*, 39–44.

Zerhouni, F. Z. (2009). *Développement et optimisation d'un générateur énergétique hybride propre à base de PV-PAC*. Doctoral Thesis. Oran, Algeria: University of Sciences and Technology Mohamed Boudiaf Oran.

Zerhouni, F. Z., Benmessaoud, M. T., & Boudghene Stambouli, A. (2006b). Développement de modèle de système à énergie renouvelable pour son utilisation en électronique de puissance et optimization. In *Proceedings of JNEA 2006*. Oran, Algeria: JNEA.

Zerhouni, F. Z., Zegrar, M., Benmessaoud, M. T., Boudghene Stambouli, A., & Midoun, A. (2008). A novel method to optimize photovoltaic generator operation. *International Journal of Energy Research*, *32*(15), 1444–1453. doi:10.1002/er.1461

Zerhouni, F. Z., Zegrar, M., Benmessaoud, M. T., Boudghene Stambouli, A., & Midoun, A. (2010). Optimum load matching by an array reconfiguration in photovoltaic generators. *Energy & Fuels*, *24*(11), 6126–6130. doi:10.1021/ef1006275

ADDITIONAL READING

Antchev, M. H. (2010a). Energy and energy efficiency. In *Technologies for Electrical Power Conversion, Efficiency, and Distribution: Methods and Processes* (pp. 1–9). Hershey, PA: IGI Global.

Antchev, M. H. (2010b). *Storage and usage of energy. Technologies for Electrical Power Conversion, Efficiency, and Distribution: Methods and Processes* (pp. 10–31). Hershey, PA: IGI Global.

Antchev, M. H. (2010c). DC/DC conversion. In *Technologies for Electrical Power Conversion, Efficiency, and Distribution: Methods and Processes* (pp. 134–166). Hershey, PA: IGI Global.

Antchev, M. H. (2010d). Conversion of electrical power from renewable energy sources. In *Technologies for Electrical Power Conversion, Efficiency, and Distribution: Methods and Processes* (pp. 231–247). Hershey, PA: IGI Global.

Aquila, J. O., Owolabi, S. W., & Olugbemiga, I. M. (2010). Emerging technology penetration: The case of solar electricity in Nigeria. In *Nanotechnology and Microelectronics: Global Diffusion, Economics and Policy* (pp. 430–449). Hershey, PA: IGI Global. doi:10.4018/978-1-61692-006-7.ch025

Cadman, T., & Hume, M. (2012). Developing sustainable governance systems for regional sustainability programmes and 'green' business practices: The case of 'green' timber. In *Sustainable Systems and Energy Management at the Regional Level: Comparative Approaches* (pp. 365–382). Hershey, PA: IGI Global.

Calò, C., Lay-Ekuakille, A., Vergallo, P., Chiffi, C., Trotta, A., Fasanella, A., & Fasanella, A. M. (2011). measurements and characterization of photovoltaic modules for tolerance verification. *International Journal of Measurement Technologies and Instrumentation Engineering*, *1*(2), 73–83. doi:10.4018/ijmtie.2011040106

Canlar, T. Y. (2010). *Sustainable urban and regional infrastructure development: Technologies, applications and management*. Hershey, PA: IGI Global.

Cebi, S., Kahraman, C., & Kaya, I. (2012). Soft computing and computational intelligent techniques in the evaluation of emerging energy technologies. In *Innovation in Power, Control, and Optimization: Emerging Energy Technologies* (pp. 164–197). Hershey, PA: IGI Global. doi:10.4018/978-1-61350-138-2.ch005

Doukas, H., Makarouni, I., Karakosta, C., Marinakis, V., & Psarras, J. (2012). EU—GCC clean energy cooperation: From concepts to action. In *Sustainable Systems and Energy Management at the Regional Level: Comparative Approaches* (pp. 288–308). Hershey, PA: IGI Global.

Fanetti, S. (2012). Renewable energy in Italy: Incentives, bureaucratic obstacles and Nimby syndrome. In *Sustainable Systems and Energy Management at the Regional Level: Comparative Approaches* (pp. 68–86). Hershey, PA: IGI Global.

Flamos, A., Karakosta, C., Doukas, H., & Psarras, J. (2010). Setting technology transfer priorities with CDM-SET: Development of sustainable energy technology transfer tool. In *Intelligent Information Systems and Knowledge Management for Energy: Applications for Decision Support, Usage, and Environmental Protection* (pp. 205–222). Hershey, PA: IGI Global.

Gelil, I. A. (2012). Globalization of the environmental issues: Response of the arab region. In *Sustainable Systems and Energy Management at the Regional Level: Comparative Approaches* (pp. 147–165). Hershey, PA: IGI Global.

Hilty, L. M. (2011). Information and communication technologies for a more sustainable world. In *Green Technologies: Concepts, Methodologies, Tools and Applications* (pp. 36–45). Hershey, PA: IGI Global. doi:10.4018/978-1-60960-472-1.ch104

Hin, L. T. W., & Subramaniam, R. (2012). Use of policy instruments to promote sustainable energy practices and implications for the environment: Experiences from Singapore. In *Sustainable Systems and Energy Management at the Regional Level: Comparative Approaches* (pp. 219–235). Hershey, PA: IGI Global.

Hocaoglu, F. P., Gerek, O. N., & Kurban, M. (2009). Solar radiation forecasting model. In *Encyclopedia of Artificial Intelligence* (pp. 1433–1438). Hershey, PA: IGI Global.

Kim, E.-H. (2012). Government policies to promote production and consumption of renewable electricity in the US business. In *Sustainable Systems and Energy Management at the Regional Level: Comparative Approaches* (pp. 1–18). Hershey, PA: IGI Global.

Miller, W., & Birkeland, J. (2011). Green energy: Sustainable energy sources and alternative technologies. In *Green Technologies: Concepts, Methodologies, Tools and Applications* (pp. 1–16). Hershey, PA: IGI Global. doi:10.4018/978-1-60960-472-1.ch101

Obara, S. (2010). Fuel reduction effect of the solar cell and diesel engine hybrid system with a prediction algorithm of solar power generation. In *Intelligent Information Systems and Knowledge Management for Energy: Applications for Decision Support, Usage, and Environmental Protection* (pp. 242–271). Hershey, PA: IGI Global. doi:10.1299/jpes.2.1166

Olla, P. (2012). *Global sustainable development and renewable energy systems*. Hershey, PA: IGI Global. doi:10.4018/978-1-4666-1625-7

Omer, A. M. (2012). Renewable energy and sustainable development. In *Innovation in Power, Control, and Optimization: Emerging Energy Technologies* (pp. 95–136). Hershey, PA: IGI Global.

Pérez-Plaza, M., & Linares, P. (2009). Strategic decisions for green electricity marketing: Learning from past experience. In *Web-Based Green Products Life Cycle Management Systems: Reverse Supply Chain Utilization* (pp. 2589–2605). Hershey, PA: IGI Global.

Sanseverino, E. R., Zizzo, G., & Scimemi, G. F. (2011). A new meta-heuristic multi-objective approach for optimal dispatch of dispersed and renewable generating units in power distribution systems. In *Knowledge-Based Intelligent System Advancements: Systemic and Cybernetic Approaches* (pp. 162–181). Hershey, PA: IGI Global. doi:10.4018/978-1-61692-811-7.ch009

Visconti, P., Romanello, D., Zizzari, G., Ventura, V., & Cavalera, G. (2011). Design, Measurements And Characterization Of Smart Electronic Board For PV Streetlight Based on LED and High Intensity Discharge Lamps. *International Journal of Measurement Technologies and Instrumentation Engineering*, *1*(3), 1–13. doi:10.4018/ijmtie.2011070101

Wang, H.-L., & Ghose, A. K. (2011). Green strategic alignment. In *Green Technologies: Concepts, Methodologies, Tools and Applications* (pp. 1435–1447). Hershey, PA: IGI Global. doi:10.4018/978-1-60960-472-1.ch606

KEY TERMS AND DEFINITIONS

Renewable Energy: Is energy which comes from natural resources which are renewable naturally.

Solar Cells Array: The modules are linked together to form an array.

Power's Management and Control Device: Is an equipment used to control the power between each component.

Photovoltaic Modules: A system which uses photovoltaic cells to convert sunlight into electricity.

Mean Square Error: EQM measures the average of the squares of the errors.

Fuel Cell: Is a device that converts the chemical energy into electricity.

Electrolysis: Is the process of decomposition of a substance into its chemical components by an electric current.

Proton Exchange Membrane (PEM): Fuel cells take hydrogen and air and produce power, heath and water.

Compilation of References

1E (2008). *Case study dell: Focus on energy efficiency.* Retrieved from http://www.1e.com/download/Whitepapers/Dell_Case%20Study_US.pdf.

Aanstoos, T., Torres, V., & Nichols, S. (1998). Energy model for end-of-life computer disposition. *IEEE Transactions on Components. Packaging and Manufacturing Technology*, *21*(4), 295–301. doi:10.1109/3476.739174

Abbasian, A., Hatami, S., Afzali-Kusha, A., & Pedram, M. (2008). Wavelet-based dynamic power management for nonstationary service requests. *ACM Transactions on Design Automation of Electronic Systems*, *13*(1), 1–41. doi:10.1145/1297666.1297679

Abowd, G. D., Rehg, J., Starner, T. E., & Arriaga, R. (2007). *Using information technologies to support detection of developmental delay*. Paper presented at the Assisted Cognition Workshop. Rochester, NY.

Abowd, G., & Mynatt, E. (2005). Designing for the human experience in smart environments. In Cook, D., & Das, S. (Eds.), *Smart Environments: Technology, Protocols, and Applications* (pp. 153–174). New York, NY: Wiley. doi:10.1002/047168659X.ch7

Abras, S., Ploix, S., Pesty, S., Jacomino, M., Cetto, J. A., & Ferrier, J. (2008). A multi-agent home automation system for power management. *Informatics in Control Automation and Robotics*, *15*, 59–68. doi:10.1007/978-3-540-79142-3_6

Accenture. (2008). *Green maturity assessment*. Retrieved from http://nstore.accenture.com/global/flash/gme/greenmaturity.htm.

ACPI. (2004). *Advanced configuration and power interface specification, rev. 3.0*. Retrieved from http://www.acpi.info/spec.htm.

ADEME. (2009). *Bâtiments et démarches HQE.* Retrieved from http://www.ecoresponsabilite.environnement.gouv.fr/IMG/BROCHURE_HQE.pdf.

ADEME. (2011). *ADEME, PDE, plan de déplacements d'entreprise.* Retrieved from http://www2.ademe.fr/servlet/KBaseShow?sort=-1&cid=96&m=3&catid=14263.

Aggarwal, J. K., & Park, S. (2004). Human motion: Modeling and recognition of actions and interactions. In *Proceedings of the 2nd International Symposium* on *3D Data Processing, Visualization and Transmission(3DPVT)*, (pp. 640-647). 3DPVT.

Agostinelli, M., Pae, S., Yang, W., Prasad, C., Kenche, D., Ramey, S., et al. (2005). Random charging effects for PMOS NBTI in ultra-small gate area devices. In *Proceedings of Annual International Reliability Physics Symposium.* Reliability Physics Symposium.

Agrawal, P., Chen, J. C., Kishore, S., Ramanathan, P., & Sivalingam, K. (1998). Battery power sensistive video processing in wireless networks. In *Proceedings of IEEE PIMRC 1998*, (vol. 1), (p. 116). Boston, MA: IEEE Press.

Akbari, J., Oyamada, K., & Saito, Y. (2001). LCA of machine tools with regards to their secondary effects on quality of machined parts. In *Proceedings EcoDesign 2001: Second International Symposium on Environmentally Conscious Design and Inverse Manufacturing*, (pp. 347-352). Tokyo, Japan: IEEE.

Alberta Infrastructure. (2007). *LEED project delivery process manual.* Retrieved from http://www.infrastructure.alberta.ca/Content/docType486/Production/LEED_PD_Manual.pdf.

Alekhnovich, M., & Arora, S. (2005). Towards strong nonapproximability results in the Lovasz-Schrijver hierarchy. In *Proceedings of the 37th Annual ACM Symposium on Theory of Computing*, (pp. 294–303). ACM Press.

Aleksic, S. (2008). Power consumption issues in future high-performance. In *Proceedings of the 10th Anniversary International Conference on Transparent Optical Networks*, (pp. 194-198). IEEE Press.

Aleksic, S. (2009). Analysis of power consumption in future high-capacity network nodes. *Journal of Optical Communications and Networking*, *1*(3), 245–258. doi:10.1364/JOCN.1.000245

Alliance-TIC. (2009). *L'eco-utilisateur, guide des bonnes pratiques de l'utilisateur informatique et telecom*. Retrieved from http://www.alliance-tics.org/dossiers/environnement/Guide%20éco-utilisateur/GUIDE%20Eco-24p%20BD.pdf.

Amin, M. (2004). Balancing market priorities with security issues: Interconnected system operations and control under the restructured electricity enterprise. *IEEE Power and Energy Magazine*, *2*(4), 30–38. doi:10.1109/MPAE.2004.1310871

Amin, M., & Stringer, J. (2008). The electric power grid: Today and tomorrow. *MRS Bulletin*, *33*, 399–407. doi:10.1557/mrs2008.80

Amin, M., & Wollenberg, B. (2005). Toward a smart grid. *IEEE Power and Energy Magazine*, *3*(5), 34–38. doi:10.1109/MPAE.2005.1507024

Andrae, A. S. G., & Anderson, O. (2010). Life cycle assessments of consumer electronics – Are they consistent? *International Journal of Life Cycle Assessment*, *15*, 827–836. doi:10.1007/s11367-010-0206-1

Android. (2012). *Android developers*. Retrieved March 6, 2012, from http://developer.android.com/.

Apache. (2011a). *Apache http server version 2.0 documentation*. Retrieved April 29, 20011, from http://httpd.apache.org/docs/2.0/en/.

Apache. (2011b). *Apache module mod_deflate*. Retrieved April 29, 2011, from http://httpd.apache.org/docs/2.0/en/mod/mod_deflate.html.

Appelbaum, J. (1989). The operation of loads powered by separate sources or by a common source of photovoltaic cells. *IEEE Transactions on Energy Conversion*, *4*(3), 351–357. doi:10.1109/60.43235

Apple Inc. (2010). iPhone 4 environmental report. *Apple Computer*. Retrieved May 7, 2011 from http://images.apple.com/environment/reports/docs/iphone_4_product_environmental_report.pdf.

Aravind, D., & Christmann, P. (2008). Institutional and resource-base determinants of substantive implementation of ISO 14001. In *Proceedings of the Academy of Management*. Anaheim, CA: Academy of Management.

Armbrust, M., Fox, A., Griffith, R., Joseph, A. D., Katz, R. H., & Kownwinski, A. ... Zaharia, M. (2009). *Above the clouds: A Berkeley view of cloud computing*. Technical Report, No. UCB/EECS-2009-28. Berkeley, CA: UC Berkeley.

Armengaud, E., Zoier, M., Baumgart, A., Biehl, M., Chen, D., Griessnig, G., et al. (2011). Model-based toolchain for the efficient development of safety-relevant automotive embedded systems. In *Proceedings of the SAE 2011 World Congress & Exhibition*. SAE.

Arora, S. (2002). Proving integrality gaps without knowing the linear program. In *Proceedings of the 43rd Symposium on Foundations of Computer Science*, (pp. 313–322). IEEE.

Arpinen, T., Salminen, E., Hämäläinen, T., & Hännikäinen, M. (2010). Extension to MARTE profile for modeling dynamic power management of embedded systems. In *Proceedings of the 1st Workshop on Model Based Engineering for Embedded Systems Design (M-BED), co-located with DATE 2010*. Dresden, Germany: IEEE.

Asanovic, K., Bodik, R., Catanzaro, B. C., Gebis, J. J., & Husbands, P. K. (2006). *The Landscape of parallel computing research: A view from Berkeley*. Berkeley, CA: University of California at Berkeley.

Assuncao, M. D. D., Costanzo, A. D., & Buyya, R. (2010). A cost-benefit analysis of using cloud computing to extend the capacity of cluster. *Journal of Cluster Computing*, *13*, 335–347. doi:10.1007/s10586-010-0131-x

Auer, G., Gódor, I., Hévizi, L., Imran, M., Malmodin, J., Fazekas, P., et al. (2010). Enablers for energy efficient wireless networks. In *Proceedings of the 72nd Vehicular Traffic Conference,* (pp. 1-5). IEEE Press.

Avram, O. (2010). *Machine tool use phase: Modeling and analysis with environmental considerations.* Ph.D. Dissertation. Lusanne, Switzerland: EPFL.

Avram, O., Stroud, I., & Xirouchakis, P. (2011). A multi-criteria decision method for sustainability assessment of the use phase of machine tool systems. *International Journal of Advanced Manufacturing Technology, 53,* 811–828. doi:10.1007/s00170-010-2873-2

Avram, O., & Xirouchakis, P. (2011). Evaluating the use phase energy requirements of a machine tool system. *Journal of Cleaner Production, 19,* 699–711. doi:10.1016/j.jclepro.2010.10.010

Aydinalp-Koksal, M., & Ugursal, V. I. (2008). Comparison of neural network, conditional demand analysis, and engineering approaches for modeling end-use energy consumption in the residential sector. *Applied Energy, 85*(4), 271–296. doi:10.1016/j.apenergy.2006.09.012

Aydinalp, M., Ugursal, V. I., & Fung, A. S. (2003). Modelling of residential energy consumption at the national level. *International Journal of Energy Research, 27*(4), 441–453. doi:10.1002/er.887

Bachour, N. (2008). *Green project management for green IT.* Unpublished.

Bachour, N. (2010a). Green IT project management: *How to optimize green IT projects for sustainable value creation within organizations?* Unpublished MBA Thesis. Paris, France: SKEMA Business School.

Bachour, N. (2010b). Optimizing the value of green IT projects within organizations. In *Proceedings of the Green Technologies Conference,* (pp. 1-10). Grapevine, TX: IEEE Press.

Balakrishnan, V., Boyd, S., & Balemi, S. (1991). Branch and bound algorithm for computing the minimum stability degree of parameter-dependent linear systems. *International Journal of Robust and Nonlinear Control, 1*(4), 295–317. doi:10.1002/rnc.4590010404

Baldi, M., & Ofek, Y. (2009). Time for a greener internet. In *Proceedings of Green Communications (IEEE ICC 2009).* Dresden, Germany: IEEE Press.

Baliga, J., Ayre, R., Hinton, K., & Tucker, R. (2007). Photonic switching and the energy bottleneck. In *Proceedings of International Conference on Photonics in Switching.* San Francisco, CA: IEEE.

Baluja, S., & Pomerlau, D. (1994). *Non intrusivegaze tracking using artificial neural networks.* Technical report CMU-CS-94-102. Pittsburgh, PA: Carnegie Mellon University.

Bambos, N. (1998). Toward power-sensitive network architectures in wireless communications: Concepts, issues, and design aspects. *IEEE Personal Communications, 5*(3), 50–59. doi:10.1109/98.683739

Barney, J. (1991). Company resources and sustained competitive advantage. *Journal of Management, 17*(1), 99–120. doi:10.1177/014920639101700108

Barnhart, C., & Johnson, E. (1998). Branch-and-price: Column generation for solving huge integer programs. *Operations Research, 46*(3), 316–329. doi:10.1287/opre.46.3.316

Barroso, L., & Holzle, U. (2007). The case for energy-proportional computing. *IEEE Computer, 40*(12), 33–37. doi:10.1109/MC.2007.443

Basal Action Network. (2010). *e-Stewards.* Retrieved from http://e-stewards.org/certification-overview/program-details/.

Baumgart, A. (2010). A common meta-model for the interoperation of tools with heterogeneous data models. In *Proceedings of the ECMFA 2010 - 3rd Workshop on Model-Driven Tool & Process Integration.* ECMFA.

Baumgart, A., Reinkemeier, P., Rettberg, A., Stierand, I., Thaden, E., & Weber, R. (2010). A model-based design methodology with contracts to enhance the development process of safety-critical systems. In *Proceedings of the Software Technologies for Embedded and Ubiquitous Systems,* (pp. 59-70). Berlin, Germany: Springer.

Bechini, A., Cimino, M. G. C. A., Marcelloni, F., & Tomasi, A. (2008). Patterns and technologies for enabling supply chain traceability through collaborative e-business. *Information and Software Technology Journal*, *50*, 342–359. doi:10.1016/j.infsof.2007.02.017

Belady, C., Rawson, A., Pfleuger, J., & Cader, T. (2008). *Green grid data center power efficiency metrics: PUE and DCIE.* Retrieved May 10, 2011, from http://www.thegreengrid.org/~/media/WhitePapers/White_Paper_6__PUE_and_DCiE_Eff_Metrics_30_December_2008.ashx?lang=en.

Bellaouar, A., Fridi, A., Elmasry, M., & Itoh, K. (1998). Supply voltage scaling for temperature-insensitive CMOS circuit operation. *IEEE Transaction on Circuits and System II*, *45*(3), 415–417. doi:10.1109/82.664253

Bellifemine, F., Poggi, A., & Rimassa, G. (2001). Developing multi-agent systems with a FIPA-compliant agent framework. *Software, Practice & Experience*, *31*(2), 103–128. doi:10.1002/1097-024X(200102)31:2<103::AID-SPE358>3.0.CO;2-O

Benini, L., Bogliolo, A., & De Micheli, G. (2000). A survey of design techniques for system-level dynamic power management. *IEEE Transactions on Very Large Scale Integration (VLSI). Systems*, *8*(3), 299–316.

Benmessaoud, M. T., Zerhouni, F. Z., Zegrar, M., Boudghene Stambouli, A., & Tioursi, M. (2010). New approach modeling and a maximum power point tracker method for solar cells. *Computers & Mathematics with Applications (Oxford, England)*, *60*(44), 1124–1134. doi:10.1016/j.camwa.2010.04.001

Berkhout, F., & Hertin, J. (2004). De-materialising and re-materialising: digital technologies and the environment. *Futures*, *36*, 903–920. doi:10.1016/j.futures.2004.01.003

Berrone, P., Gelabert, L., Fosfuri, A., & Gomez-Mejia, L. (2008). Can institutional forces create competitive advantage? An empirical investigation of environmental innovation. In *Proceedings of the Academy of Management.* Anaheim, CA: Academy of Management.

Bevilacqua, A. (1999). A dynamic load balancing metthod on a heterogeneous cluster of workstations. *Informatica*, *23*(1), 49–56.

Bevrani, H., Hiyama, T., & Mitani, Y. (2008). Power system dynamic stability and voltage regulation enhancement using an optimal gain vector. *Control Engineering Practice*, *16*(9), 1109–1119. doi:10.1016/j.conengprac.2008.01.001

Bhardwaj, S., Wang, W., Vattikonda, R., Cao, Y., & Vrudhula, S. (2006). Predictive modeling of the NBTI effect for reliable design. In *Proceedings of IEEE Custom Integrated Circuits Conference.* IEEE Press.

Bianchini, R., & Carrera, E. V. (2000). Analytical and experimental evaluation of cluster-based network servers. *World Wide Web (Bussum)*, *3*(4), 215–229. doi:10.1023/A:1018734706376

Bianchini, R., & Rajamony, R. (2004). Power and energy management for server systems. *Computer*, *37*(11), 68–76. doi:10.1109/MC.2004.217

Blaauw, D. T., Oh, C., Zolotov, V., & Dasgupta, A. (2003). Static electromigration analysis for on-chip signal interconnects. *IEEE Transactions on Computer-Aided Design of Integrated Circuits and Systems*, *22*(1), 39–48. doi:10.1109/TCAD.2002.805728

Blue Angel. (2009). *Computer monitors standards.* Retrieved from http://www.blauer-engel.de/en/products_brands/search_products/produkttyp.php?id=486.

Bokun, S. (2009). Vodafone kills two birds with one customer loyalty program. *Pyramid Research.* Retrieved May 6, 2011 from http://www.pyr.com/points/item/090820.htm.

Bomhof, F. W., Van Hoorik, P., & Donkers, M. (2009). Systematic analysis of rebound effects for "greening by ict" initiatives. *Communications & Strategies*, *76*, 77–96.

Borkar, S. (2005). Designing reliable systems from unreliable components: The challenges of transistor variability and degradation. *IEEE Micro*, *25*(6), 10–16. doi:10.1109/MM.2005.110

Boudghéne Stambouli, A. (2003). *Electricity from photovoltaics and fuel cells: The expectations for an enviromental-friendly and sustainable source of energy.* Paper presented at the Environment 2010 International Symposium. Porto, Portugal.

Boudghéne Stambouli, A. (2011a). Fuel cells: The expectations for an environmental-friendly and sustainable source of energy. *Renewable & Sustainable Energy Reviews*, *15*(9), 4507–4520. doi:10.1016/j.rser.2011.07.100

Boudghéne Stambouli, A. (2011b). Promotion of renewable energies in Algeria: Strategies and perspectives. *Renewable & Sustainable Energy Reviews*, *15*(2), 1169–1181. doi:10.1016/j.rser.2010.11.017

Boudghéne Stambouli, A. (2011c). Algerian renewable energy assessment: The challenge of sustainability. *Energy Policy*, *39*(8), 4507–4519. doi:10.1016/j.enpol.2010.10.005

Boudghéne Stambouli, A., & Traversa, E. (2002a). Solide oxide fuel cells (SOFCs): A review of an environmentally clean and efficient source of energy. *Renewable & Sustainable Energy Reviews*, *6*, 433–455. doi:10.1016/S1364-0321(02)00014-X

Boudghéne Stambouli, A., & Traversa, E. (2002b). Fuel cells, an alternative to standard souces of energy. *Renewable & Sustainable Energy Reviews*, *6*(3), 295–304. doi:10.1016/S1364-0321(01)00015-6

Boudreau, M., Chen, A., & Huber, M. (2008). *Green IS: Building sustainable business practices.* Retrieved from http://globaltext.terry.uga.edu/userfiles/pdf/Green.pdf.

Boughton, K. (2008). Intel's 45nm dual-core E8500: The best just got betters. *AnandTech.* Retrieved May 10, 2011 from http://www.anandtech.com/show/2468/6.

Boukraa, M., & Ando, S. (2002). A computer vision system for knowledge-based 3D scene analysis using radio-frequency tags. In *Proceedings of International Conference on Multimedia and Expo*, (vol 2), (pp. 245-248). Conference on Multimedia.

Bouquain, D., Blunier, B., & Miraoui, A. (2008). A hybrid fuel cell/battery wheelchair: Modeling, simulation and experimentation. In *Proceedings of the Vehicle Power and Propulsion Conference, 2008*. IEEE Press.

Boyd, S. B., Dornfeld, D., & Krishnan, N. (2006). Life cycle inventory of a CMOS chip. In *Proceedings of the 2006 International Symposium on Electronics and the Environment*. Scottsdale, AZ: IEEE.

Boyd, S. B., Horvath, A., & Dornfeld, D. (2009). Life-cycle energy demand and global warming potential of computational logic. *Environmental Science & Technology*, *43*, 7303–7309. doi:10.1021/es901514n

Boyd, S. B., Horvath, A., & Dornfeld, D. (2011). Life-cycle assessment of NAND flash memory. *IEEE Transactions on Semiconductor Manufacturing*, *24*(1), 117–124. doi:10.1109/TSM.2010.2087395

Brandic, I., Music, D., Leitner, P., & Dustdar, S. (2009). *VieSLAF framework: Enabling adaptive and versatile SLA-management*. Paper presented at the 6th International Workshop on Grid Economics and Business Models 2009 (Gecon09). Delft, The Netherlands.

Braspenning, R., de Haan, G., & Hentschel, C. (2002). Complexity scalable motion estimation. In *Proceedings of VCIP 2002*, (vol. 4671), (pp. 442–453). San Jose, CA: VCIP.

Bridge. (2010). Securing the electricity grid. *The Bridge*, *40*, 13-20.

Briscoe, G., & Marinos, A. (2009). Digital ecosystems in the clouds: Towards community cloud computing. In *Proceedings of the 3rd IEEE International Conference on Digital Ecosystems and Technologies (DEST 2009)*, (pp. 103-108). New York, NY: IEEE Press.

British Broadcasting Corporation. (2010, July 9). *News technology*. Retrieved May 4, 2011, from http://www.bbc.co.uk/news/10569081.

Brodkin, J. (2008). Economy driving green IT initiatives. *New World (New Orleans, La.)*, *25*(49), 16.

Brown, R., Webber, C., & Koomey, J. (2000). *Status and future directions of the ENERGY STAR program.* Washington, DC: ACEEE Summer Study.

Brown, D., Dillard, J., & Marshall, R. (2005). Strategically informed, environmentally conscious information requirements for accounting information systems. *Journal of Information Systems*, *19*(2), 79–103. doi:10.2308/jis.2005.19.2.79

Brueil, H., Burette, D., Flury-Hérard, B., Cueugniet, J., & Vignolles, D. (2008). *Rapport TIC et développement durable.* Conseil général de l'environnement et du développement durable, Conseil général des Technologies de l'Information.

Brundtland, G. H. (1987). *Our common future: Brundtland commission (WCED) report*. World Commission on Environment and Development.

Buford, F. J., Yu, H., & Lua, K. E. (2009). *P2P networking and applications*. New York, NY: Morgan Kaufmann.

Burd, T. D., & Broderson, R. W. (1995). Energy efficient CMOS microprocessor design. In *Proceedings of the 28th Annual Hawaii International Conference on System Sciences*. (pp. 288–297). Los Alamitos, CA: IEEE Press.

Business Wire. (2011). Research and markets: Smartphones: A worldwide view - More than half of US handset shipments will be smartphones by 2012 and will move to 1 billion by 2015 worldwide. *Business Wire: A Berkshire Hathaway Company*. Retrieved May 6, 2011 from http://www.businesswire.com/news/home/20110210005842/en/Research-Markets-Smartphones-Worldwide-View---Handset.

Busquet, S. (2003). *Etude d'un système autonome de production d'énergie couplant un champ photovoltaïque, un électrolyseur et une pile à combustible: Réalisation d'un banc d'essai et modélisation*. Ph. D Thesis. Paris, France: Ecole des Mines de Paris.

Busquet, S., Hubert, C. E., Labbé, J., Mayer, D., & Metkemeijer, R. (2004). A new approach to empirical electrical modelling of a fuel cell, an electrolyser or a regenerative fuel cell. *Journal of Power Sources*, *134*(1), 41–48. doi:10.1016/j.jpowsour.2004.02.018

Butler, T. (2011). Compliance with institutional imperatives on environmental sustainability: Building theory on the role of Green IS. *The Journal of Strategic Information Systems*, *20*, 6–26. doi:10.1016/j.jsis.2010.09.006

Buyya, R., Ranjan, R., & Calheiros, R. N. (2010). InterCloud: Utility-oriented federation of cloud computing environments for scaling of application services. *Lecture Notes in Computer Science*, *6081*, 13–31. doi:10.1007/978-3-642-13119-6_2

Buyya, R., Yeo, C. S., Venugopal, S., Broberg, J., & Brandic, I. (2009). Cloud computing and emerging IT platforms: Vision, hype, and reality for delivering computing as the 5th utility. *Journal of Future Generation Computer Systems*, *25*(6), 559–616. doi:10.1016/j.future.2008.12.001

California Department of Toxic Substances Control. (2004). *Electronic waste recycling act*. Retrieved from http://www.dtsc.ca.gov/HazardousWaste/EWaste/.

Cao, X., Cong, G., & Jensen, C. S. (2010). Mining significant semantic locations from GPS data. In *Proceedings of 36th International Conference on Very Large Data Bases*, (pp. 1009-1020). Singapore: IEEE.

Carrasco, J. M., Franquelo, L. G., Bialasiewicz, J. T., Galvan, E., Guisado, R. C. P., & Prats, M. A. M. (2006). Power-electronic systems for the grid integration of renewable energy sources: A survey. *IEEE Transactions on Industrial Electronics*, *53*(4), 1002–1016. doi:10.1109/TIE.2006.878356

Carrera, E., Pinheiro, E., & Bianchini, R. (2003). Conserving disk energy in network servers. In *Proceedings of 17th International Conference on Supercomputing*. San Francisco, CA: ACM.

Celik, A., Holliday, J., & Hurst, Z. (2006). *Data dissemination to a large mobile network: Simulation of broadcast clouds*. Paper presented at the 7th International Conference on Mobile Data Management (MDM 2006). Santa Clara, CA.

Cerovsky, Z., & Mindl, P. (2005). *Regenerative braking by electric hybrid vehicles using super capacitor and power splitting generator*. Paper presented at the Power Electronics and Applications. London, UK.

Ceuppens, L., Kharitonov, D., & Sardella, A. (2008). Power saving strategies and technologies in network equipment opportunities and challenges. In *Proceedings of IEEE Symposium on Applications and the Internet*. IEEE Press.

Chandrakasan, A. P., Sheng, S., & Brodersen, R. W. (1992). Low-power digital CMOS design. *IEEE Journal of Solid-state Circuits*, *27*(4), 473–484. doi:10.1109/4.126534

Chang, V. (2009). *Cloud storage framework – An integrated technical approach and prototype for breast cancer*. Paper presented at the All Hands Meeting 2009. Oxford, UK.

Chang, V., Bacigalupo, D., Wills, G., & De Roure, D. (2010a). *A categorisation of cloud computing business models*. Paper presented at the CCGrid 2010 IEEE Conference. Melbourne, Australia.

Chang, V., De Roure, D., Wills, G., Walters, R., Li, C. S., Barry, T., & Currie, W. (2011d). The development that leads to the cloud computing business framework. *Information and Management*.

Chang, V., Mills, H., & Newhouse, S. (2007). *From open source to long-term sustainability: Review of business models and case studies*. Paper presented at the UK e-Science All Hands Meeting. Nottingham, UK.

Chang, V., Wills, G., & De Roure, D. (2010b). *A review of cloud business models and sustainability*. Paper presented at the IEEE Cloud 2010, the Third International Conference on Cloud Computing. Miami, FL.

Chang, V., Wills, G., & De Roure, D. (2010c). *Cloud business models and sustainability: Impacts for businesses and e-research*. Paper presented at the e-Science All Hands Meeting 2010, Software Sustainability Workshop. Cardiff, UK.

Chang, V., Wills, G., & Walters, R. (2011e). *Towards business integration as a service 2.0 (BIaaS 2.0)*. Paper presented at the IEEE International Conference on e-Business Engineering, The 3rd International Workshop on Cloud Services - Platform Accelerating e-Business. Beijing, China.

Chang, C., Tran, V. H., Wang, J., Fuh, Y.-K., & Lin, L. (2010). Direct-write piezoelectric polymeric nanogenerator with high energy conversion efficiency. *Nano Letters*, *10*(2), 726–731. doi:10.1021/nl9040719

Chang, N., Choi, I., & Shim, H. (2004). DLS: Dynamic backlight luminance scaling of liquid crystal display. *IEEE Transactions on VLSI Systems*, *12*(8), 837–846. doi:10.1109/TVLSI.2004.831472

Chang, V., De Roure, D., Wills, G., & Walters, R. (2011c). Case studies and organisational sustainability modelling presented by cloud computing business framework. *International Journal of Web Services Research*, *8*(3), 26–53.

Chang, V., De Roure, D., Wills, G., Walters, R., & Barry, T. (2011b). Organisational sustainability modelling for return on investment: Case studies presented by a national health service (NHS) trust UK. *Journal of Computing and Information Technology*, *19*(3).

Chang, V., Li, C. S., De Roure, D., Wills, G., Walters, R., & Chee, C. (2011a). The financial clouds review. *International Journal of Cloud Applications and Computing*, *1*(2), 41–63. doi:10.4018/ijcac.2011040104

Chen, G., Chuah, K. Y., Li, M. F., Chan, D. S., Ang, C. H., Zheng, J. Z., et al. (2003). Dynamic NBTI of PMOS transistors and its impact on device lifetime. In *Proceedings of Annual International Reliability Physics Symposium*, (pp. 196–202). Reliability Physics Symposium.

Chen, X., Wills, G. B., Gilbert, L., & Bacigalupo, D. (2010). *Using cloud for research: A technical review*. Retrieved from http://eprints.ecs.soton.ac.uk/21273/.

Chen, Z., Shen, H. T., & Zhou, X. (2011). Discovering popular routes from trajectories. In *Proceedings of 2011 IEEE 27th International Conference on Data Engineering (ICDE)*. Hannover, Germany: IEEE Press.

Chen, Z., Zhou, P., & He, Y. (2002). Fast integer and fractional pel motion estimation for JVT. In *Proceedings of JVT of ISO/IEC MPEG and ITU-T VCEG 6th Meeting*. Awaji Island, Japan: JVT.

Cheng, W.-C., & Pedram, M. (2004). Power minimization in a backlight TFT-LCD display by concurrent brightness and contrast scaling. *Design Automation and Test in Europe*, *1*, 16–20.

Cheong, Y., Kim, Y., Yoo, S., Lee, H., Lee, S., Chae, S., & Choi, H. (2011). An ontology-based reasoning approach towards energy-aware smart homes. In *Proceedings: IEEE Conference on Consumer Communications and Networking Conference (CCNC)*, (pp. 850 – 854). IEEE Press.

Choi, I., Shim, H., & Chang, N. (2002). Low power color TFT LCD display for hand-held embedded systems. In *Proceedings of the International Symposium on Low-Power Electronics and Design*, (pp. 112-117). IEEE.

Chowhan, S. S., & Saxena, R. (2011). Customer relationship management from the business strategy perspective with the application of cloud computing. *Proceedings of DYNAA*, *2*(1).

Christensen, K., Gunaratne, C., Nordman, B., & George, A. (2005). Managing energy consumption costs in desktop PCs and LAN switches with proxying split TCP connections, and scaling of link speed. *International Journal of Network Management*, *15*(5), 297–310. doi:10.1002/nem.565

Church, K., Greenberg, A., & Hamilton, J. (2008). On delivering embarrasingly distributed cloud services. In *Proceedings of ACM HotNets*. Calgary, Canada: ACM.

CIOL. (2009). Smartphones ring in healthy growth in 2009. *CIOL: Information Technology News*. Retrieved May 15, 2011 from http://www.ciol.com/Biz-Watch/News-Reports/Smartphones-ring-in-healthy-growth-in-2009/53091168230/.

Cisco. (2010). *Cisco visual networking index*. Retrieved May 10, 2011, from http://www.cisco.com/en/US/solutions/collateral/ns341/ns525/ns537/ns705/ns827/white_paper_c11-481360_ns827_Networking_Solutions_White_Paper.html.

Clarkson University. (2011). *Clarkson sustainability initiatives*. Retrieved from http://www.clarkson.edu/green/.

Climate Group. (2008). *SMART 2020: Enabling the low carbon economy in the information age. Paper presented as Report on Behalf of the Global eSustainability Initiative (GeSI)*. London, UK: Climate Group.

CloudMade. (2012). *The CloudMade developer zone*. Retrieved March 21, 2012, from http://developers.cloudmade.com/.

Coburn, M. (2008, August 15). GE CEO declares: "Green is green". *The Dartmouth Staff*.

Cohen, H. (2011). *Mobile marketing: 56 must have facts*. Retrieved December 9, 2011, from http://heidicohen.com/mobile-marketing-must-have-facts/.

Colajanni, M., Cardellini, V., & Yu, P. S. (1998). Dynamic load balancing in geographically distributed heterogeneous web servers. In *Proceedings of the 18th IEEE International Conference on Distributed Computing Systems (ICDCS 1998)*, (pp. 295-303). Washington, DC: IEEE Computer Society.

Computer Industry Almanac Inc. (2011). *Computer industry almanac*. Retrieved May 4, 2011, from http://www.c-i-a.com/worldwideuseexec.htm.

Consumer Electronics Association. (2008). *Market research report: Trends in CE reuse, recycle and removal*. Washington, DC: Consumer Electronics Association.

Cook, D., & Das, S. (Eds.). (2004). *Smart environments: Technology, protocols and applications*. Hoboken, NJ: John Wiley & Sons. doi:10.1002/047168659X

Correia, L., Zeller, D., Blume, O., Ferling, D., Jading, Y., & Godor, I. (2010). Challenges and enabling technologies for energy aware mobile radio networks. *IEEE Communications Magazine*, *48*(11), 66–72. doi:10.1109/MCOM.2010.5621969

Creeger, M. (2009). CTO roundtable: Cloud computing. *Communications of the ACM*, *52*(8).

Croson, R., & Donohue, K. (2003). Impact of POS data sharing on supply chain management: An experimental study. *Production and Operations Management*, *12*(1), 1–11. doi:10.1111/j.1937-5956.2003.tb00194.x

Cuervo, E., Balasubramanian, A., Cho, D., Wolman, A., Saroiu, S., Chandra, R., & Bahl, P. (2010). MAUI: Making smartphones last longer with code offload. In *Proceedings of the 8th International Conference on Mobile Systems, Applications, and Services*, (pp. 49-62). New York, NY: ACM Press.

da Fonseca, T. A., & de Queiroz, R. L. (2009). Complexity-constrained H.264 HD video coding through mode ranking. In *Proceedings of PCS 2009*. Chicago, Illinois: PCS.

Dahmus, J., & Gutowski, T. (2004). An environmental analysis of machining. In *Proceedings of the 2004 ASME International Mechanical Engineering Congress*. Anaheim, CA: ASME.

Dai, X., & Raychandran, K. (2003). *Computer screen power management through detection of user presence*. US Patent 6650322. Washington, DC: US Patent Office.

Dalrymple, J. (2010). Apple: Over 300,000 iPads sold on first day. *CNET News*. Retrieved May 6, 2011 from http://news.cnet.com/8301-13579_3-10472602-37.html.

Delaware. (2010). *Go green*. Retrieved from http://dti.deleware.gov/greenit/.

Deng, L., Babbitt, C. W., & Williams, E. D. (2011). Economic-balance hybrid LCA extended with uncertainty analysis: Case study of a laptop computer. *Journal of Cleaner Production*, *19*(11), 1198–1206. doi:10.1016/j.jclepro.2011.03.004

Department of Climate Change. (2008). *Australia's national greenhouse accounts: National inventory by economic sector 2006.* Canberra, Australia: Commonwealth of Australia, Department of Climate Change.

Department of Energy. (2000). *Carbon dioxide emissions from the generation of electric power in United States.* Washington, DC: Environmental Protection Agency, Department of Energy.

Dietmann, A., Helbok, R., Lackner, P., Fischer, M., Reindl, M., & Lell, B. (2009). Pathophysiology and severe malaria – From basic knowledge to new therapeutic approaches. *European Journal of Tropical Medicine and International Health, 14*(2).

Dimaggio, P., & Powell, W. (1983). The iron cage revisited: Institutional isomorphism and collective rationality in organisational fields. *American Sociological Review, 48*(2), 147–160. doi:10.2307/2095101

Dinsmore, P., & Cooke-Davies, T. (2006). *The right projects done right.* San Francisco, CA: Jossey-Bass.

Dornkundwar, V., Krishnan, N., Bauer, D., & Sheng, P. (1998). A distributed environmental planning model for machined components. In *Proceedings 5th International Seminar on Life Cycle Engineering,* (pp. 227-239). Stockholm, Sweden: KTH. Eco-indicator-99. (2011). *Website.* Retrieved from http://www.pre.nl/content/eco-indicator-99.

Douxchamps, D., & Campbell, N. (2008). Robust real time face tracking for the analysis of human behavior. *Lecture Notes in Computer Science, 4892,* 1–10. doi:10.1007/978-3-540-78155-4_1

Dreslinski, R., Wieckowski, M., Blaauw, D., Sylvester, D., & Mudge, T. (2010). Near-threshold computing: Reclaiming Moore's law through energy efficient integrated circuits. *Proceedings of the IEEE, 98*(2), 253–266. doi:10.1109/JPROC.2009.2034764

Duchowski, A. T. (2002). A breadth-first survey of eye tracking applications. *Behavior Research Methods, Instruments, and Computers.* Retrieved from http://andrewd.ces.clemson.edu/research/vislab/docs/BET107cr.pdf.

Duckham, M., Winter, S., & Robinson, M. (2010). Including landmarks in routing instructions. *Journal of Location Based Services, 4*(1), 28–52. doi:10.1080/17489721003785602

Duda, R., Hart, P., & Stork, E. (2001). *Pattern classification* (2nd ed.). New York, NY: Wiley.

Dufaux, F., Gao, W., Tubaro, S., & Vetro, A. (2010). Distributed video coding: Trends and perspectives. *EURASIP Journal on Image and Video Processing.* Retrieved from http://www.merl.com/reports/docs/TR2010-026.pdf.

Dunn, T. (2010). *Identity management: Cloud and virtualisation.* Paper presented at the European Identity Conference, Munich Cloud 2010. Munich, Germany.

Durresi, A., Durresi, M., Paruchuri, V., & Barolli, L. (2011). Ad hoc communications for emergency conditions. In *Proceedings of the 25th IEEE International Conference on Advanced Information Networking and Applications (AINA-2011),* (pp. 787-794). Washington, DC: IEEE Computer Society.

EARTH. (2010). *Energy Aware Radio and Network Technologies.* Retrieved May 10, 2011, from https://www.ict-earth.eu/

Earth911. (2011). *Website.* Retrieved from http://earth911.com/.

Eclipse. (2012) *About the eclipse foundation.* Retrieved March 12, 2012, from http://www.eclipse.org/org/.

ECORYS. (2008). *Werkwijzer OEI bij MIT-planstudies, bijlage kengetallen.* Rotterdam, The Netherlands: ECORYS.

Elbatt, T., Goel, S. K., Holland, G., Krishnan, H., & Parikh, J. (2006). Cooperative collision warning using dedicated short range wireless communications. In *Proceedings of the 3rd International Workshop on Vehicular ad hoc Networks,* (pp. 1-9). New York, NY: ACM Press.

Electronics TakeBack Coalition. (2010). *Take it back, make it green, recycle responsibly.* Retrieved from http://www.electronicstakeback.com.

Elliot, S., & Binney, D. (2009). *Environmentally sustainable ICT: Developing corporate capabilities & an industry-relevant IS research agenda.* Retrieved from http://www.pacis-net.org/file/2008/PACIS2008_Camera-Ready_Paper_209.pdf.

Energy Consumption Rating Initiative. (2009). *About the ECR*. Retrieved May 10, 2011, from http://www.ecrinitiative.org/about/.

Energy Star. (2008). *Programme requirement for computers, version 5.0*. Retrieved from http://www.energystar.gov/index.cfm?c=about.ab_history.

Energy Star. (2009). *Program requirements for computers version 5.0*. Retrieved from http://www.energystar.gov.

Enokido, T., Aikebaier, A., & Takizawa, M. (2010). A model for reducing power consumption in peer-to-peer systems. *IEEE Systems Journal*, *4*(2), 221–229. doi:10.1109/JSYST.2010.2047296

Enokido, T., & Takizawa, M. (2010). A purpose-based synchronization protocol for secure information flow control. *Journal of Computer Systems Science and Engineering*, *25*(2), 25–32.

Enrique, M., Dura'n, E., Sidrach-de-Cardona, M., & Andu'jar, J. M. (2007). Theoretical assessment of the maximum power point tracking efficiency of photovoltaic facilities with different converter topologies. *Photovoltaic Energy*, *81*(3), 1–38.

Environmental Protection Agency. (2010). *ENERGY STAR overview of 2009 achievements.* Retrieved May 10, 2011, from http://www.energystar.gov/ia/partners/annualreports/2009_achievements.pdf.

Environmental Protection Agency. (2011). *Lean manufacturing & the environment*. Retrieved from http://www.epa.gov/lean/thinking/sixsigma.htm.

EPA & US Department of Energy. (2011). *Save energy at home*. Retrieved from http://www.energystar.gov/index.cfm?c=products.pr_save_energy_at_home.

EPA. (2007). *Do the "PC" thing: Donate computers.* Retrieved from http://www.epa.gov.wastes/partnerships/plugin/pdf/pcthing-con.pdf.

EPA. (2010). EPA highlights recycling opportunities during national cell phone recycling week. *US Environmental Protection Agency*. Retrieved May 7, 2011 from http://yosemite.epa.gov/opa/admpress.nsf/e77fdd4f5afd88a3852576b3005a604f/2f0711847e1f1c18852576f80053f5fd!OpenDocument.

EPA. (2010). *Responsible recycling practices*. Retrieved from http://www.epa.gov/wastes/conserve/materials/ecycling/r2practices.htm.

EPA. (2011). *EnergyStar*. Retrieved from http://www.energystar.gov/.

EPEAT. (2006). *EPEAT background.* Retrieved from http://www.epeat.net/AboutEPEAT.aspx.

Ericcson. (2007). *Sutainable energy use in mobile communications.* Retrieved May 8, 2011, from http://www.ericsson.com/campaign/sustainable_mobile_communications/downloads/sustainable_energy.pdf.

Ernst, D., Das, S., Lee, S., Blaauw, D., Austin, T., & Mudge, T. (2005). Razor: Circuit-level correction of timing errors for low-power operation. *IEEE Micro*, *24*(10), 10–20.

Esty, D., & Winston, A. (2006). *Green to gold.* New Haven, CT: Yale.

Etoh, M., Ohya, T., & Nakayama, Y. (2008). Energy consumption issues on mobile network systems. In *Proceedings of the IEEE International Symposium on Applications and the Internet*. IEEE Press.

Etro, F. (2009). The economic impact of cloud computing on business creation, employment and output in Europe. *Journal of Review of Business and Economics*, *54*(2), 1–40.

EU Information Society. (2011). *ICT for sustainable growth*. Retrieved from http://ec.europa.eu/information_society/activities/sustainable_growth/index_en.htm.

EU. (1994). *Directive 94/62/EC on P&PW.* Retrieved from http://www.gidos.net/_library/EC.94.62.EC.on.Packaging.and.Packaging.Waste.pdf.

EU. (2002a). *Directive 2002/96/EC on WEEE*. Retrieved from http://ec.europa.eu/environment/waste/weee/index_en.htm.

EU. (2002b). *Directive 2002/95/EC on ROHS*. Retrieved from http://www.rohs.eu/english/legislation/rohs-legislation.html.

EU. (2006). *Regulation EC 1907/2006 on REACH.* Retrieved from http://ec.europa.eu/environment/chemicals/reach/reach_intro.htm.

EU. (2009). *Code of conduct for data centres version 2*. Retrieved from http://re.jrc.ec.europa.eu/energyefficiency/html/standby_initiative_data_centers.htm.

EU. (2011). *Energy star programme*. Retrieved from http://www.eu-energystar.org/en/index.htm.

Faisal, A. M. (2008). *Microgrid modelling and online management*. Espoo, Finland: Helsinki University of Technology.

Faist Emmenegger, M., Frischknecht, R., Stutz, M., Guggisberg, M., Witschi, R., & Otto, T. (2006). Life cycle assessment of the mobile communication system UMTS. *International Journal of Life Cycle Assessment*, *11*, 265–276. doi:10.1065/lca2004.12.193

Falaki, H., Mahajan, R., Kandula, S., Dimitrios, L., Govindan, R., & Estrin, D. (2010). Diversity in smartphone usage. In *Proceedings of the 8th International Conference on Mobile Systems, Applications, and Services,* (pp. 179-194). New York, NY: ACM Press.

Farhangi, H. (2009). The path of the smart grid. *IEEE Power and Energy Magazine*, *8*(1), 18–28. doi:10.1109/MPE.2009.934876

Fei, Y., Zhong, L., & Jha, N. K. (2008). An energy-aware framework for dynamic software management in mobile computing systems. *ACM Transactions on Embedded Computer Systems*, *7*, 1–31. doi:10.1145/1347375.1347380

Ferling, D., Bohn, T., Zeller, D., Frenger, P., Gódor, I., & Jading, Y. (2010). Energy efficiency approaches for radio nodes. In [Florence, Italy: IEEE Press.]. *Proceedings of the Future Network & Mobile Summit*, *2010*, 1–9.

Ferraiolo, D. F., Kuhn, D., & Chandramouli, R. (2007). *Role-based access control*. Norwood, MA: Artech House.

Fierce Wireless. (2008). Global consumer survey reveals that majority of old mobile phones are lying in drawers at home and not being recycled. *FierceWireless.* Retrieved May 7, 2011 from http://www.fiercewireless.com/press-releases/global-consumer-survey-reveals-majority-old-mobile-phones-are-lying-drawers-home-an-0.

Fischer, M. J., Lynch, N. A., & Paterson, M. S. (1985). Impossibility of distributed consensus with one faulty process. *Journal of the ACM*, *32*(2), 374–382. doi:10.1145/3149.214121

Fleishman Hillard. (2012). *2012 digital influence study*. Retrieved from March 5, 2012, from http://www.factbrowser.com/facts/4671/.

Fleurey, F., & Solberg, A. (2009). A domain specific modeling language supporting specification, simulation and execution of dynamic adaptive systems. In *Proceedings of Model Driven Engineering and Systems* (pp. 606–621). Berlin, Germany: Springer. doi:10.1007/978-3-642-04425-0_47

Flinn, J., & Satyanarayanan, S. (1999). Energy-aware adaptation for mobile applications. In *Proceedings of the Symposium on Operating Systems Principles*, (pp. 48-63). Operating Systems Principles.

Flinn, J., Narayanan, D., & Satyanarayanan, M. (2001). Self-tuned remote execution for pervasive computing. In *Proceedings of the 8th Workshop on Hot Topics in Operating Systems (HotOS),* (pp 61-66). Schloss Elmau, Germany: HotOS.

Flinn, J., & Satyanarayanan, M. (2004). Managing battery lifetime with energy-aware adaptation. *ACM Transactions on Computer Systems*, *22*, 137–179. doi:10.1145/986533.986534

Forrester. (2009). *Market overview: The state of enterprise green IT adoption*. Forrester.

Fortier, P. J., & Michel, H. E. (2003). *Computer systems performance evaluation and prediction*. Burlington, MA: Digital Press.

Foundation for Intelligent Physical Agents. (2002a). *FIPA foundation for intelligent physical agents contract net interaction protocol specification*. FIPA.

Foundation for Intelligent Physical Agents. (2002b). *FIPA foundation for intelligent physical agents propose protocol specification*. FIPA.

Foundation for Intelligent Physical Agents. (2002c). *FIPA foundation for intelligent physical agents request protocol specification*. FIPA.

Frigo, M., & Johnson, S. G. (1998). FFTW: An adaptive software architecture for the FFT. In *Proceedings of the 1998 IEEE International Conference on Acoustics Speech and Signal Processing,* (pp. 1381-1384). IEEE Press.

Frischknecht, R., Jungbluth, N., Althaus, H.-J., Doka, G., & Dones, R. (2007). *Overview and methodology*. Dübendorf, Germany: Swiss Center for Life Cycle Inventories.

Fujitsu-Siemens. (2006). *Energy savings with personal computers*. Retrieved from http://www.fujitsu-siemens.nl/aboutus/sor/energy_saving/prof_desk_prod.html.

Gable, G. (2011, April 23). *The story of e-waste: What happens to tech once it's trash.* Retrieved from http://www.gizmodo.com.au/2011/04/the-story-of-e-waste-what-happens-to-tech-once-its-trash/.

Gale, S. (2009a). Green out. *PM Network, 23*(11).

Gale, S. (2009b). The real deal. *PM Network, 23*(12).

Gallant, M., Cote, G., & Kossentini, F. (1999). An efficient computation-constrained blockbased motion estimation algorithm for low bit rate video coding. *IEEE Transactions on Image Processing, 8*(12), 1816–1822. doi:10.1109/83.806627

Galluccio, L., Leonardi, A., Morabito, G., & Palazzo, S. (2008). Timely and energy-efficient communications in rural infostation systems. *IEEE Wireless Communications, 15*(3), 48–53. doi:10.1109/MWC.2008.4547522

Garg, S. K., & Buyya, R. (2010). *Green cloud computing and environmental sustainability*. Working Paper. Melbourne, Australia: University of Melbourne.

Gartner, Inc. (2007). *Gartner estimates ICT industry accounts for 2 percent of global CO2 emmisions*. Press Release. New York, NY: Gartner.

Gartner. (2009a). *Gartner says consumer location-based services market will more than double in 2009*. Retrieved June 12, 2010, from http://www.gartner.com/it/page.jsp?id=1059812.

Gartner. (2009b). *Gartner identifies the top 10 consumer mobile applications for 2012*. Retrieved June 09, 2011, from http://www.gartner.com/it/page.jsp?id=1230413.

Gatti, F., Acquaviva, A., Benini, L., & Ricco, B. (2002). Low-power control techniques for TFT LCD displays. In *Proceedings of the International Conference on Compilers, Architecture and Synthesis for Embedded Systems,* (pp. 218-224). IEEE.

Geihs, K., Khan, M. U., Reichle, R., Solberg, A., Hallenstein, S., & Merral, S. (2006). Modeling of component-based adaptive distributed applications. In *Proceedings of the 2006 ACM Symposium on Applied Computing*, (pp. 718-722). ACM Press.

GeoNames. (2012). *About GeoNames*. Retrieved February 17, 2012, from http://www.geonames.org/about.html.

Geyer, R., & Doctori-Blass, V. (2010). The economics of cell phone reuse and recycling. *International Journal of Advanced Manufacturing Technology, 47,* 515–525. doi:10.1007/s00170-009-2228-z

Ghanbari, M. (1990). The cross-search algorithm for motion estimation. *IEEE Transactions on Communications, 38*(1), 950–953. doi:10.1109/26.57512

Ghemawat, S., Gobioff, H., & Leung, S. (2003). The Google file system. In *Proceedings of the 19th ACM Symposium on Operating Systems Principles (SOSP 2003),* (pp. 29-43). New York, NY: ACM Press.

Ghiasi, S., & Grunwald, D. (2011). *Aide de camp: Asymmetric dual core design for power and energy reduction.* Retrieved May 10, 2011, from http://citeseerx.ist.psu.edu/viewdoc/summary?doi=10.1.1.86.6806.

Giannotti, F., Nanni, M., Pedreschi, D., & Pinelli, F. (2007). Trajectory pattern mining. In *Proceedings of the 13th ACM SIGKDD International Conference on Knowledge Discovery and Data Mining*, (pp. 330-339). San Jose, CA: ACM Press.

Gid'ofalvi, G., & Pedersen, T. B. (2006). Mining long, sharable patterns in trajectories of moving objects. In *Proceedings of the Third Workshop on Spatio-Temporal Database Management (STDBM)*. Seoul, Korea: STDBM.

Gilberto, C., & Margaret, M. (2005). *Power prediction for intel XScale processors using performance monitoring unit events*. Paper presented at the Proceedings of the 2005 International Symposium on Low Power Electronics and Design. New York, NY.

Global e-Sustainability Initiative. (2008). *SMART 2020: Enabling the low carbon economy in the information age.* Retrieved August 22, 2011, from http://www.smart2020.org/_assets/files/Smart2020UnitedStatesReportAddendum.pdf.

Global Reporting Initiative. (2011). *Sustainability reporting guidelines, version 3.1*. Retrieved from http://www.globalreporting.org.

Goodland, R., & Daly, H. (1996). Environmental sustainability: Universal and non-negotiable. *Ecological Applications*, *6*(4), 1002–1017. doi:10.2307/2269583

Governor's Energy Office. (2010). *Recharge Colorado*. Retrieved from http://rechargecolorado.com/.

Green Electronics Council. (2009). *Green electronics made easy*. Retrieved from http://www.greenelectronicscouncil.org/.

Green Grid. (2007). *The green grid power efficiency metrics: PUE and DCiE*. Retrieved from http://www.thegreengrid.org.

Green Grid. (2009). *The green grid library & tools*. Retrieved from http://www.thegreengrid.org/library-and-tools.aspx.

GRI. (2006a). *Sustainability reporting guidelines*. Retrieved from http://www.globalreporting.org/ReportingFramework/ReportingFrameworkDownloads/.

GRI. (2006b). *Indicator protocols set for the environment*. Retrieved from http://www.globalreporting.org/ReportingFramework/ReportingFrameworkDownloads/.

Grinnell College. (2007). *Environmental stewardship*. Retrieved from http://www.grinnell.edu/etal/green.

Grossman, R. L. (2009). The case for cloud computing. *IT Professional*, *11*(2), 23–27. doi:10.1109/MITP.2009.40

Gruber, M., Blume, O., Ferling, D., Zeller, D., Imran, M., & Strinati, E. (2009). EARTH — Energy aware radio and network technologies. In *Proceedings of the 20th International Symposium on Personal, Indoor and Mobile Radio Communications*, (pp. 1-5). Cannes, France: IEEE Press.

Guinée, J. B., Gorrée, M., Heijungs, R., Huppes, G., Kleijn, R., & Koning, A. … Huijbregts, M. A. J. (2002). *Handbook on life cycle assessment: Operational guide to the ISO standards*. Dordrecht, The Netherlands: Kluwer Academic Publishers.

Gupta, M., & Singh, S. (2003). Greening of the internet. In *Proceedings of ACM SIGCOMM*, (pp. 361-371). Karlsruhe, Germany: ACM Press.

Hagelüken, C., & Buchert, M. (2008). The mine above ground - Opportunities & challenges to recover scarce and valuable metals from EOL electronic devices. In *Proceedings of 7th International Electronics Recycling Congress*. Salzburg, Austria: IEEE.

Hamann, H. (2008). A measurement-based method for improving data center energy efficiency. In *Proceedings of the International Conference on Sensor Networks, Ubiquitous and Trustworthy Computing*, (pp. 312-313). IEEE Press.

Hamzaoglu, F., Ye, Y., Keshavarzi, A., & Zhang, K., S., N., Borkar, S., et al. (2002). Analysis of dual-VT SRAM cells with full-swing single-ended bit line sensing for on-chip cache. *IEEE Transactions on Very Large Scale Integration (VLSI). Systems*, *10*(2), 91–95.

Han, Z., Himsoon, T., Siriwongpairat, W., & Liu, K. (2005). Energy-efficient cooperative transmission over multiuser OFDM networks: Who helps whom and how to cooperate. In *Proceedings of the Wireless Communications and Networking Conference*, (pp. 1030-1035). IEEE Press.

Handheld Computing Research. (2012). *Worldwide: Computer and device sales*. Retrieved March 2, 2012, from http://people.cs.und.edu/~wenchen/handheldresearch/facts/all.html.

Hanna, A., Windebank, J., Adams, S., Sowerby, S., Rance, S., & Cartlidge, A. (2009). *ITIL V3 foundation handbook: Pocketbook from the official publisher of ITIL*. London, UK: Stationery Office.

Hanna, N. K. (2010). *E-transformation: Enabling new development strategies*. Berlin, Germany: Springer Science. doi:10.1007/978-1-4419-1185-8

Hansen, D. W., & Ji, Q. (2010). In the eye of the beholder: A survey of models for eyes and gaze. *IEEE Transactions on Pattern Analysis and Machine Intelligence*, *32*(3), 478–500. doi:10.1109/TPAMI.2009.30

Hanssmann, F., & Hess, S. W. (1960). A linear programming approach to production and employment scheduling. *Management Technology*, *1*, 46–52.

Harald, B. (2010). *Results of the expert group on e-invoicing*. Paper presented on Electronic Invoicing in Europe. Madrid, Spain.

Harel, D. (1987). Statecharts: A visual formalism for complex systems. *Science of Computer Programming*, *8*, 231–274. doi:10.1016/0167-6423(87)90035-9

Harmon, R., Demirkan, H., Auseklis, N., & Reinoso, M. (2010). From green computing to sustainable IT: Developing a sustainable service orientation. In *Proceedings of the 43rd Hawaii International Conference on System Sciences*, (pp. 1-10). IEEE.

Hart, S. (1995). A natural-resource-based view of the company. *Academy of Management Review*, *20*(4), 986–1014.

Harvard Office for Sustainability. (2011). *Green tip of the month*. Retrieved from http://green.harvard.edu/green-tip.

Heath, T., Diniz, B., Carrera, E. V., Meira, W. J., & Bianchini, R. (2005). Energy conservation in heterogeneous server clusters. In *Proceedings of the 10th ACM SIGPLAN Symposium on Principles and Practice of Parallel Programming (PPoPP 2005)*, (pp. 186-195). New York, NY: ACM Press.

Heddeghem, W., Vereecken, W., Pickavet, M., & Demeester, P. (2009). Energy in ICT - Trends and research directions. In *Proceedings of the 3nd International Symposium on Advanced Networks and Telecommunication Systems*. New Delhi, India: IEEE Press.

Hemmert, S. (2010). Green HPC: From nice to necessity. *Computing in Science & Engineering*, *12*(6), 8–10. doi:10.1109/MCSE.2010.134

Henkel, J., & Parameswaran, S. (2007). *Designing embedded processors: A lower persepctive*. Dordrecht, The Netherlands: Springer. doi:10.1007/978-1-4020-5869-1

Hennessy, J. L., & Patterson, D. A. (2003). *Computer architecture: A quantitiative approach* (3rd ed.). London, UK: Elsevier Science Ltd.

Herzberg, F. (1971). *Work and the nature of man*. New York, NY: World Publishing.

Hewlett-Packard Co. (2006). *Global citizenship report*. Retrieved from www.hp.com/hpinfo/globalcitizenship/gcreport/pdf/hp2006gcreport_lowres.pdf.

Hewlett-Packard, I. Microsoft, Phoenix Technologies, & Toshiba. (2010). *Advanced configuration and power interface specification, revision 4.0a*. Washington, DC: HP, Intel, Microsoft, Phoenix, Toshiba.

Higgs, T. (2007). Energy efficient computing. In *Proceedings of the International Symposium on Electronics & the Environment*, (pp. 210-215). Orlando, FL: IEEE Press.

Hillier, F. S., & Lieberman, G. J. (2001). *Introduction to operations research*. New York, NY: McGraw-Hill.

Hilty, L. M. (2006). The relevance of ICTs for environmental sustainability – A prospective simulation study. *Environmental Modelling & Software*, *21*, 1618–1629. doi:10.1016/j.envsoft.2006.05.007

Hilty, L. M. (2008). *Information technology and sustainability - Essays on the relationship between ICT and sustainable development*. Norderstedt, The Netherlands: Books on Demand GmbH.

Hinton, K., Baliga, J., Feng, M., Ayre, R., & Tucker, R. (2011). Power consumption and energy efficiency in the internet. *IEEE Network*, *25*(2), 6–12. doi:10.1109/MNET.2011.5730522

Hontani, H., Baba, K., Kugimiya, T., Sato, K., & Nakagawa, M. (2003). Visual tracking system using an ID-tag and the network. In *Proceedings of SICE 2003 Annual Conference*, (vol 3), (pp. 2375-2380). SICE.

Hu, W.-C., Wiggen, T., & Yang, H.-J. (2005). Mobile commerce systems and payment methods. In *Proceedings of the 2005 Information Resources Management Association International Conference (IRMA 2005)*, (pp. 769-771). San Diego, CA: IRMA.

Huang, A. (2008). *A sustainable systems development life cycle*. Retrieved from http://www.pacis-net.org/file/2008/PACIS2008_Camera-Ready_Paper_081.pdf.

Hull, J. C. (2009). *Options, futures, and other derivatives* (7th ed.). Upper Saddle River, NJ: Pearson Prentice Hall.

IBM. (2007). *How to optimise your carbon management through green sigma*. Retrieved from http://www-05.ibm.com/ie/emerging_business_centre/pdf/Green_Sigma_final_Strategy.pdf.

IBM. (2008, August 18). *A measured approach to going green: IBM "green sigma (TM)" consulting offering to help clients reduce energy and water usage*. IBM.

IBM. (2009a). *Green IT assessment and ROI calculator*. Retrieved from http://www-304.ibm.com/businesscenter/cpe/html0/181575.html.

IBM. (2011). *Energy, the environment & IBM*. Retrieved from http://www.ibm.com/ibm/green/index5.shtml.

Idaho Department of Environmental Quality. (2011). *Waste management: Recycling in Idaho*. Retrieved from http://www.deq.idaho.gov/waste/recycling/recycling.cfm.

IDC. (2011). IDC forecasts worldwide smartphone market to grow by nearly 50% in 2011. *IDC Home: The Premier Global Market Intelligence Firm*. Retrieved May 6, 2011 from http://www.idc.com/getdoc.jsp?containerId=prUS22762811.

IEA. (2009). *Gadgets and gigawatts -- Policies for energy efficient electronics*. IEA Press.

Ikeda, M., Kulla, E., Hiyama, M., Barolli, L., & Takizawa, M. (2011). Experimental results of a MANET testbet in indoor stairs environment. In *Proceedings of the 25th IEEE International Conference on Advanced Information Networking and Applications (AINA-2011)*, (pp. 779-786). Washington, DC: IEEE Computer Society.

Institute of Scrap Recycling Industries, Inc. (2010). *Recycling industry operation standards*. Retrieved from http://www.isri.org/iMIS15_PROD/ISRI/_Program_and_Services/.

International Energy Agency. (2010a). *CO2 emssions from fuel combustion highlights*. Paris, France: International Energy Agency.

International Energy Agency. (2010b). *World energy outlook 2010 factsheet*. Paris, France: OECD/IEA.

International Technology Roadmap for Semiconductors. (2010). *ITRS 2010 update*. Retrieved May 5, 2011 from http://www.itrs.net/links/2010itrs/home2010.htm.

International Telecommunication Union. (2008). *ICTs and climate change*. Geneva, Switzerland: ITU.

International Telecommunication Union. (2010). *ITU*. Retrieved April 5, 2011, from http://www.itu.int/newsroom/press_releases/2010/pdf/PR08_ExecSum.pdf.

International Trade Administration. (2011). *Website*. Retrieved from http://export.gov/europeanunion/weeerohs/.

International, P. E. (2011). *GaBi professional database. GaBi Software, version 4.131, Service Pack 18*. PE International.

Intille, S. S., Larson, K., Beaudin, J., Munguia Tapia, E., Kaushik, P., Nawyn, J., & McLeish, T. J. (2005). A live-in laboratory for pervasive computing research (video). In *Proceedings of Pervasive 2005 Video Program. Pervasive Video Program*. The PlaceLab.

Irani, S., Singh, G., Shukla, S. K., & Gupta, R. K. (2005). An overview of the competitive and adversarial approaches to designing dynamic power management strategies. *IEEE Transactions on Very Large Scale Integration (VLSI). Systems*, *13*(12), 1349–1361.

Iranli, A., Lee, W., & Pedram, M. (2006). HVS-aware dynamic backlight scaling in TFT-LCDs. *IEEE Transactions on Very Large Scale Integration Systems*, *14*(10), 1103–1110. doi:10.1109/TVLSI.2006.884151

ISACA. (2009). *Cloud computing: Business benefits with security, governance and assurance perspectives*. White Paper. ISACA.

ISO 14649. (2001). *TC184/SC1/WG7, ISO 14649/FDIS, data model for computerized numerical controllers*. New York, NY: International Standards Organization.

ISO. (2002). *ISO 14000/14001 environmental management standards*. Retrieved from http://www.iso14000-iso14001-environmental-management.com/index.htm.

Jang, Y.-C., & Kim, M. (2010). Management of used & end-of-life mobile phones in Korea: A review. *Resources, Conservation and Recycling*, *55*, 11–19. doi:10.1016/j.resconrec.2010.07.003

Jayal, A. D., & Balaji, A. K. (2007). On a process modeling framework for sustainable manufacturing: A machining perspective. In *Proceedings of the ASME International Mechanical Engineering Congress and Exposition*, (pp. 301-307). ASME.

JD Power and Associates Reports. (2010). *Average length of time wireless customers keep their mobile phones increases notably*. JD Power and Associates.

Jenkin, T., & McShane, L. (2009). *Green information technologies and systems in organisations: The state of practice*. Retrieved from http://www.albany.edu/business/csr_Jenkin_SpecialIssue_BlindPaper_GreenITS.pdf.

Jenkin, T., Webster, J., & McShane, L. (2011). An agenda for green information technology and system research. *Information and Organization, 21*(1), 17–40. doi:10.1016/j.infoandorg.2010.09.003

Jeon, B., & Lee, J. (2003). Fast mode decision for H.264. In *Proceedings of JVT of ISO/IEC MPEG and ITU-T VCEG 10th Meeting*. Waikoloa, HI: JVT.

Jian, W., Longyun, K., Binggang, C., & Liuchen, C. (2005). *Reliability study of a distributed generation system based on renewable energy.* Paper presented at the Electrical and Computer Engineering, 2005. Toronto, Canada.

Ji, Q., & Zhu, Z. (2004). Eye and gaze tracking for interactive graphic display. *Journal of Machine Vision and Applications, 15*(3), 139–148.

JISC. (2011). *Cloud computing increasingly attractive to universities, says JISC*. JISC.

JiWire. (2011). *JiWire mobile audience insights report Q2 2011*. Retrieved from February 18, 2012, from http://www.factbrowser.com/facts/3196/.

Kairer, R. (2005). Mobile phone sales to exceed one billion in 2009. *Palm Infocenter*. Retrieved May 6, 2011 from http://www.palminfocenter.com/view_story.asp?ID=7978.

Kang, K., Kim, K., Islam, A. E., Alam, M. A., & Roy, K. (2007). Characterization and estimation of circuit reliability degradation under NBTI using on-line IDDQ measurement. In *Proceedings of Design Automation Conference*, (pp. 358–363). Design Automation Conference.

Kang, K., Kufluoglu, H., Alam, M. A., & Roy, K. (2007). Efficient transistor-level sizing technique under temporal performance degradation due to NBTI. In *Proceedings of IEEE International Conference on Computer Design*, (pp. 934–939). IEEE Press.

Kannangara, C. S., Richardson, I. E. G., Bystrom, M., Solear, J. R., Zhao, Y., Maclennan, A., & Cooney, R. (2006). Low complexity skip prediction for H.264 through Lagrangian cost estimation. *IEEE Transactions on Circuits and Systems for Video Technology, 16*(2), 202–208. doi:10.1109/TCSVT.2005.859026

Kannangara, C. S., Richardson, I. E., & Miller, A. J. (2008). Computational complexity management of a real-time H.264/AVC encoder. *IEEE Transactions on Circuits and Systems for Video Technology, 18*(9), 1191–1200. doi:10.1109/TCSVT.2008.928881

Kansal, A., & Zhao, F. (2008). Fine-grained energy profiling for power-aware application design. In *Proceedings of the First Workshop on Hot Topics in Measurement and Modeling of Computer Systems (HotMetrics08) at ACM Sigmetrics.* ACM Press.

Kansal, A., Zhao, F., Liu, J., Kothari, N., & Bhattacharya, A. A. (2010). *Virtual machine power metering and provisioning*. Retrieved from http://research.microsoft.com/pubs/120435/JoulemeterVM.pdf.

Karl, E., Blaauw, D., Sylvester, D., & Mudge, T. (2006). Reliability modeling and management in dynamic microprocessor-based systems. In *Proceedings of Design Automation Conference*, (pp. 1057–1060). Design Automation Conference.

Karnik, T., Borkar, S., & De, V. (2002). Sub-90nm technologies: Challenges and opportunities for CAD. In *Proceedings of International Conference on Computer Aided Design*, (pp. 203 - 206). Conference on Computer Aided Design.

Kawato, S., & Ohya, J. (2000). Two-step approach for real-time eye-tracking with a new filtering technique. In *Proceedings of the IEEE International Conference on Systems, Man and Cybernetics,* (pp. 1366-1371). IEEE Press.

Kawato, S., Tetsutani, N., & Osaka, K. (2005). Scale-adaptive face detection and tracking in real time with SSR filters and support vector machine. *IEICE Transactions on Information & Systems. E (Norwalk, Conn.), 88-D*(12), 2857–2863.

Khajeh-Hosseini, A., Greenwood, D., & Sommerville, I. (2010). *Cloud migration: A case study of migrating an enterprise IT system to IaaS*. Paper presented at the 3rd IEEE International Conference on Cloud Computing, Cloud 2010. Miami, FL.

Kiczales, G., Lamping, J., Mendhekar, A., Maeda, C., Lopes, C., & Loingtier, J.-M. (1997). Aspect-oriented programming. In *Proceedings of the Object-Oriented Programming,* (pp. 220-242). Berlin, Germany: Springer.

Kihwan, C., Soma, R., & Pedram, M. (2005). Fine-grained dynamic voltage and frequency scaling for precise energy and performance tradeoff based on the ratio of off-chip access to on-chip computation times. *IEEE Transactions on Computer-Aided Design of Integrated Circuits and Systems*, *24*(1), 18–28. doi:10.1109/TCAD.2004.839485

Kim, K., Chalidabhongse, T., Harwood, D., & Davis, L. (2005). Real-time foreground-background segmentation using codebook model. In *Proceedings of the Sixth Annual IEEE Conference on Real-Time Imaging*. IEEE Press.

Kleissl, J., & Agarwal, Y. (2010). Cyber-physical energy systems: Focus on smart buildings. In *Proceedings of Design Automation Conference*, (pp. 749 - 754). Design Automation Conference.

Klems, M., Nimis, J., & Tai, S. (2008). *Do cloud compute? A framework for estimating the value of cloud computing*. Paper presented at Cloud Computing and its Application CCA 08. Chicago, IL.

Koga, T., Iinuma, K., Hirano, A., Iijima, Y., & Ishiguro, T. (1981). Motion compensated interframe coding for video conferencing. In *Proceedings of the National Telecommunication Conference 1981*. (pp. G5.3.1-G5.3.5). New Orleans, LA: NTC.

Kolodziej, K., & Hjelm, J. (2006). *Local positioning systems: LBS applications and services*. Boca Raton, FL: CRC Taylor & Francis.

Koomey, J. G. (2008). Worldwide electricity used in data centers. *Environmental Research Letters*. Retrieved from http://files.campus.edublogs.org/uowblogs.com/dist/d/75/files/2010/04/Koomey.pdf.

Krikke, J. (2008). Recycling e-waste: The sky is the limit. *IT Professional*, *10*(1), 51–55. doi:10.1109/MITP.2008.16

Kumar, S. V., Kim, C. H., & Sapatnekar, S. S. (2006). Impact of NBTI on SRAM read stability and design for reliability. In *Proceedings of International Symposium on Quality Electronic Design*. Quality Electronic Design.

Kupper, A. (2005). *Location-based services: Fundamentals and operation*. New York, NY: Wiley. doi:10.1002/0470092335

Kwon, J.-M., Nam, K. H., & Kwon, B. H. (2006). Photovoltaic power conditioning system with line connection. *Transactions on Industrial Electronic*, *53*(4), 1048–1054. doi:10.1109/TIE.2006.878329

Lachenmann, A., Marrón, P. J., Minder, D., & Rothermel, K. (2007). Meeting lifetime goals with energy levels. In *Proceedings of the 5th International Conference on Embedded Networked Sensor Systems*, (pp. 131-144). ACM Press.

Ladou, J., & Lovegrove, S. (2008). Export of electronics equipment waste. *International Journal of Hygiene and Environmental Health*, *14*(1), 1–10.

Lafond, S., & Lilius, J. (2006). An energy consumption model for an embedded Java virtual machine. In *Proceedings of Architecture of Computing Systems - ARCS 2006* (pp. 311–325). Berlin, Germany: Springer. doi:10.1007/11682127_22

Lagorse, J., Paire, D., & Miraoui, A. (2010). A multi-agent system for energy management of distributed power sources. *Renewable Energy*, *35*(1), 174–182. doi:10.1016/j.renene.2009.02.029

Lawler, E. L., & Wood, D. E. (1966). Branch-and-bound methods: A survey. *Operations Research*, *14*, 699–719. doi:10.1287/opre.14.4.699

Leem, H., Baek, S. Y., & Sung, D. K. (2010). The effects of cell size on energy saving, system capacity, and per-energy capacity. In *Proceedings of the Wireless Communications and Networking Conference*, (pp. 1-6). IEEE Press.

Lee, U., Rimac, I., Kilper, D., & Hilt, V. (2011). Toward energy-efficient content dissemination. *IEEE Network*, *25*(2), 14–19. doi:10.1109/MNET.2011.5730523

Lengwehasatit, K., & Ortega, A. (1998). A novel computationally scalable algorithm for motion estimation. In [San Jose, CA: VCIP.]. *Proceedings of VCIP*, *1998*, 68–79.

Li, J., Poulton, G., & James, G. (2007). Agent-based distributed energy management. *Advances in Artificial Intelligence*, *4830*, 569–578.

Lim, K. P., Sullivan, G., & Wiegand, T. (2004). Joint model reference encoding methods and decoding concealment methods. In *Proceedings of JVT of ISO/IEC MPEG and ITU-T VCEG*. Munich, Germany: JVT.

Lim, K. P., Wu, S., Wu, D. J., Rahardja, S., Limn, X., Pan, F., & Li, Z. G. (2003). Fast inter mode selection. In *Proceedings of JVT of ISO/IEC MPEG and ITU-T VCEG 9th Meeting*. San Diego, CA: JVT.

Lim, S. E., & Schoenung, J. M. (2010). Toxicity potentials from waste cellular phones, and a waste management policy integrating consumer, corporate, and government responsibilities. *Waste Management (New York, N.Y.)*, *30*, 1653–1660. doi:10.1016/j.wasman.2010.04.005

Lin, K., Kansal, A., Lymberopoulous, D., & Zhao, F. (2010). Energy-accuracy trade-off for continuous mobile device location. In *Proceedings of the 8th ACM/USENIX Annual International Conference on Mobile Systems, Applications and Services (MobiSys 2010)*, (pp. 285-298). New York, NY: ACM Press.

Liu, D. M. C. (2009). Recommend touring routes to travelers according to their sequential wandering behaviours. In *Proceedings of the 10th International Symposium on Pervasive Systems, Algorithms, and Networks (ISPAN 2009)*, (pp. 350-355). ISPAN.

Liu, X., Zhao, H., & Li, X. (2011). EPC: Energy-aware probability-based clustering algorithm for correlated data gathering in wireless sensor networks. In *Proceedings of the 25th IEEE International Conference on Advanced Information Networking and Applications (AINA-2011)*, (pp. 419-426). Washington, DC: IEEE Computer Society.

Liu, Z., & McGaughy, B. W. (2006). Design tools for reliability analysis. In *Proceedings of Design Automation Conference*, (pp. 182–187). Design Automation Conference.

Liu, X., & Karimi, H. A. (2006). Location awareness through trajectory prediction. *Computers, Environment and Urban Systems*, *30*(6), 741–756. doi:10.1016/j.compenvurbsys.2006.02.007

Liu, X., Tanaka, M., & Matsui, Y. (2006). Generation amount prediction and material flow analysis of electronic waste: A case study in Beijing, China. *Waste Management & Research*, *24*(5), 434–445. doi:10.1177/0734242X06067449

Lobo, L. O., & Arthur, J. D. (2005). *An objectives-driven process for selecting methods to support requirements engineering activities*. Paper presented at the 29th Annual IEEE / NASA Software Engineering Workshop (SEW-29 2005). Greenbelt, MD.

Logica Survey. (2007). *Turning concern into action: Energy efficiency and the European consumer.* Logica.

Lucia, U., & Gervino, G. (2006). Thermoeconomic analysis of an irreversible Stirling heat pump cycle. *The European Physical Journal B - Condensed Matter and Complex Systems, 50*(1), 367-369.

Lunix Virtual Server. (2011). *Job scheduling algorithms in linux virtual server*. Retrieved April 29, 2011, from http://www.linuxvirtualserver.org/docs/scheduling.html.

Lu, Z., Lach, J., Stan, M. R., & Skadron, K. (2005). Improved thermal management with reliability banking. *IEEE Micro*, *25*(6), 40–49. doi:10.1109/MM.2005.114

Macal, C. M., & North, M. J. (2010). Tutorial on agent-based modelling and simulation. *Journal of Simulation*, *4*(3), 151–162. doi:10.1057/jos.2010.3

Mahesri, A., & Vardhan, V. (2005). Power consumption breakdown on a modern laptop. *Lecture Notes in Computer Science*, *3471*, 165–180. doi:10.1007/11574859_12

Mai, H., & Meyer, T. (2010). *E-invoicing: Final step of an efficient invoicing process*. Berlin, Germany: Deutsche Bank Research.

Maine Department of Environmental Protection. (2005). *Guidelines for the environmentally sound management of e-waste*. Retrieved from http://www.maine.gov/dep/rwm/recycle/tvcomputerguidelines.htm.

Maine Department of Environmental Protection. (2008). *Recycling of electronic waste from households in Maine*. Retrieved from http://maine.gov/dep/rwm/ewaste/pdf/2008ewastereport.pdf.

Maine Department of Environmental Protection. (2011). *E-waste*. Retrieved from http://www.maine.gov/dep/rwm/ewaste/.

Malmodin, J. (2004). *Summary of the study, "life cycle assessment of a third generation (3G) system at Ericsson*. Ericsson.

Malvar, H. S., Hallapuro, A., & Karzewicz, M. (2003). Low-complexity transform and quantization in H.264/AVC. *IEEE Transactions on Circuits and Systems for Video Technology*, *13*(7), 598–603. doi:10.1109/TCSVT.2003.814964

Maryland Department of the Environment. (2007). *Maryland's statewide electronics recycling program*. Retrieved from http://mde.maryland.gov/programs/Land/RecyclingandOperationsprogram/.

Maryland Energy Administration. (2010). *EmPower Maryland: You owe it to yourself*. Retrieved from http://energy.maryland.gov/events.html.

Maryland Energy Administration. (2011). *Residential energy saving tips: Learn 15 things that will empower you to save money*. Retrieved from http://energy.maryland.gov/Residential/index.html.

Massey, J. G. (2004). NBTI: What we know and what we need to know - A tutorial addressing the current understanding and challenges for the future. In *Proceedings of the IEEE International Integrated Reliability Workshop*, (pp. 199–211). IEEE Press.

McArthur, S. D. J., Davidson, E. M., Catterson, V. M., Dimeas, A. L., Hatziargyriou, N. D., & Ponci, F. (2007). Multi-agent systems for power engineering applications part I: Concepts, approaches, and technical challenges. *IEEE Transactions on Power Systems*, *22*(4), 1743–1752. doi:10.1109/TPWRS.2007.908471

McDaniel, T., Kahol, K., Villanueva, D., et al. (2008). Integration of RFID and computer vision for remote object perception for individuals who are blind. In *Proceedings of the 1st International Conference on Ambient Media and Systems*, (pp. 1-10). Ambient Media and Systems.

McLeavey, D. W., & Narasimhan, S. L. (1985). *Production planning and inventory control*. Boston, MA: Prentice Hall.

Mellor, S. J., Scott, K., Uhl, A., & Weise, D. (2004). *MDA distilled: Principles of model-driven architecture*. Boston, MA: Addison Wesley.

Meskers, C., Hagelueken, C., & Van Damme, G. (2009). Green recycling of EEE: Special and precious metal recovery from EEE. In *Proceedings of Extraction & Processing Division of The Minerals, Metals & Materials Society* (pp. 1131–1136). San Francisco, CA: TMS.

Meulenhoff, P., Bomhof, F. W., & Van Hoorik, P. M. (2010). *Assessing the positive and negative impacts of ICT on people, planet and profit*. Paper presented on ICGreen. Athens, Greece.

Meyer, J., & Rowan, B. (1977). Institutional organisations: Formal structure as myth and ceremony. *American Journal of Sociology*, *83*(2), 340–363. doi:10.1086/226550

Microsoft. (2011). *Partners Show case*. Retrieved from http://www.microsoft.com/Environment/products-and-solutions/partner-showcase.aspx.

Microsoft. (2012). *Bing maps tile system*. Retrieved December 17, 2011, from http://msdn.microsoft.com/en-us/library/bb259689.aspx.

Miguelez, E. L., Rodriguez, L. R., Roman, T. G. S., Cerezo, F. M. E., Fernandez, M. I. N., & Lafarga, R. C. (2004). A practical approach to solve power system constraints with application to the Spanish electricity market. *IEEE Transactions on Power Systems*, *19*(4), 2029–2037. doi:10.1109/TPWRS.2004.831704

Mihailidis, A., Hoey, J., Boutilier, C., & Boger, J. (2007). *Towards the development of an intelligent home to support aging-in-place*. Paper presented at the Assisted Cognition Workshop. Rochester, NY.

Mishra, S., Pecht, M., & Goodman, D. L. (2002). In-situ sensors for product reliability monitoring. In *Proceedings of SPIE* (pp. 10–19). SPIE. doi:10.1117/12.462807

Mitra, S. (2007). Circuit failure prediction enables robust system design resilient to aging and wearout. In *Proceedings of the IEEE International On-Line Testing Symposium*, (p. 123). IEEE Press.

Moller, A., & Schaltegger, S. (2005). The sustainability balanced scorecard as a rramework for eco-efficiency analysis. *Journal of Industrial Ecology*. Retrieved from http://nbis.org/nbisresources/metrics/sustainability_balanced_scorecard.pdf.

Montana Department of Environmental Quality. (2010). *Why recycle electronics*. Retrieved from http://deq.mt.gov/Recycle/Electronics/whyelectronics.mcpx.

Moore, R. E. (1991). Global optimization to prescribed accuracy. *Computers & Mathematics with Applications (Oxford, England)*, *21*(6), 25–39. doi:10.1016/0898-1221(91)90158-Z

Morimoto, C., & Mimica, M. R. M. (2004). Eye gaze tracking techniques for interactive applications. *Computer Vision and Image Understanding*, *98*(1), 4–24. doi:10.1016/j.cviu.2004.07.010

Morin, B., Barais, O., Nain, G., & Jézéquel, J.-M. (2009). Taming dynamically adaptive systems using models and aspects. In *Proceedings of the 31st International Conference on Software Engineering (ICSE)*, (pp. 122-132). Washington, DC: IEEE Computer Society.

Moshnyaga, V. G., & Morikawa, E. (2005). LCD display energy reduction by user monitoring. In *Proceedings of the IEEE International Conference on Computer Design*, (pp. 94-97). IEEE Press.

Moshnyaga, V. G., Hashimoto, K., Suetsugu, T., & Higashi, S. (2009). A hardware implementation of the user-centric display energy management. *Lecture Notes in Computer Science*, *5953*, 56–65. doi:10.1007/978-3-642-11802-9_10

Mostafa, H., Anis, M., & Elmasry, M. (2011). Adaptive body bias for reducing the impacts of NBTI and process variations on 6T SRAM cells. *IEEE Transactions on Circuits and Systems. I, Regular Papers*, *58*(12), 2859–2871. doi:10.1109/TCSI.2011.2158708

Mozer, M. C. (2005). Lessons from an adaptive house. In Cook, D., & Das, R. (Eds.), *Smart Environments: Technologies, Protocols, and Applications* (pp. 273–294). Hoboken, NJ: John Wiley & Sons. doi:10.1002/047168659X.ch12

Msanjila, S. S., Tewoldeberhan, T., Bockstael-Blok, W., Janssen, M., & Verbraeck, A. (2005). *E–supply chain orchestration using web service technologies: A case using BPEL4WS*. Paper presented at IRMA 2005 Conference. San Diego, CA.

Mudd, G. (2007). Global trends in gold mining: Towards quantifying environmental and resource sustainability? *Resources Policy*, *32*(1-2), 42–56. doi:10.1016/j.resourpol.2007.05.002

Munoz, A. A., & Sheng, P. (1995). An analytical approach for determining the environmental impact of machining processes. *Journal of Materials Processing Technology*, *53*, 736–758. doi:10.1016/0924-0136(94)01764-R

Murph, D. (2010). Study: We're keeping our mobiles longer, despite sinking prices. *Engadget*. Retrieved May 6, 2011 from http://www.engadget.com/2010/09/27/study-were-keeping-our-mobiles-longer-despite-sinking-prices/.

Murphy, R., Sterling, T., & Dekate, C. (2010). Advanced architectures and execution models to support green computing. *Computing in Science & Engineering*, *12*(6), 38–47. doi:10.1109/MCSE.2010.124

Mutoh, N., & Inoue, T. (2007). A control method to charge series-connected ultraelectric double-layer capacitors suitable for photovoltaic generation systems combining MT control method. *Transactions on Industrial Electronics*, *54*(1), 374–383. doi:10.1109/TIE.2006.885149

Myslewski, R. (2009). iPhone conquers half the (smartphone) world. *The Register: Sci/Tech News for the World*. Retrieved May 6, 2011 from http://www.theregister.co.uk/2009/11/24/admob_mobile_metrics_report/.

Narita, H., Kawamura, H., Norms, T., Lian-Yi, C., Fujimoto, H., & Hasebe, T. (2006). Development of prediction system for environmental burden for machine tool operation. *JSME International Journal Series C*, *49*(4), 1188–1195. doi:10.1299/jsmec.49.1188

Navas, J., Mendez-Lojo, M., & Hermenegildo, M. (2008). Safe upper-bounds inference of energy consumption for Java bytecode applications. In *Proceedings of The Sixth NASA Langley Formal Methods Workshop*, (pp. 29-32). NASA.

Navas, J., Méndez-Lojo, M., & Hermenegildo, M. (2006). *Customizable resource usage analysis for Java bytecode - Deliverable 2.6 - Preliminary report on advanced resource policies*. Java.

Nemhauser, G. L., & Laurence, A. (1988). *Integer and combinatorial optimization*. New York, NY: John Wiley & Sons.

Nevada Division of Environmental Protection. (2011). *Nevada electronic waste reuse & recycling study*. Retrieved from http://nevadarecycles.gov/study2010/index.html.

New Jersey. (2011). *Living green: What you can do*. Retrieved from http://www.nj.gov/nj/green/living/.

Nick, J. M., Cohen, D., & Kaliski, B. S. (2010). Key enabling technologies for virtual private clouds. In *Handbook of Cloud Computing 2010, Part 1* (pp. 47–63). London, UK: Springer. doi:10.1007/978-1-4419-6524-0_3

Nordman, B., Piette, M.-A., & Kinney, K. (1996). *Measured energy savings and performance of power-managed personal computers and monitors*. LBL-38057. Retrieved from http://eetd.lbl.gov/EA/Reports/38057/.

Nordman, B., Piette, M.-A., Kinney, K., & Webber, C. (1997). *User guide to power management for PCs and monitors*. LBNL-39466. Retrieved from http://eetd.lbl.gov/EA/Reports/39466.

North Carolina Office of Information Technology Services. (2010). *Cyber security awareness*. Retrieved from http://www.esrmo.scio.nc.gov/CyberSecurity/.

Notter, D. A., Gauch, M., Widmer, R., Wagner, P., Stamp, A., Zah, R., & Althaus, H.-J. (2010). Contribution of li-ion batteries to the environmental impact of electric vehicles. *Environmental Science & Technology*, *44*, 6550–6556. doi:10.1021/es903729a

Object Management Group. (2011). *UML profile for MARTE: Modeling and analysis of real-time embedded systems, version 1.1*. Retrieved September 9, 2011, from http://www.omg.org/spec/MARTE/.

Office of Government Commerce. (2007). *Official ITIL lifecycle publication suite (5 books)*. TSO Publisher.

Ogasa, M. (2008). Energy saving and environmental measures in railway technologies: Example with hybrid electric railway vehicles. *IEEJ Transactions on Electrical and Electronic Engineering*, *3*(1), 15–20. doi:10.1002/tee.20227

OGC. (2005). *Managing successful projects with PRINCE2*. TSO.

OGC. (2007). *Managing successful programs*. TSO.

Ohio Environmental Protection Agency. (2010). *Electronic equipment from your home ... don't just throw it away*. Retrieved from http://www.epa.ohio.gov/ocapp/p2/mercury_pbt/ewaste/index.aspx.

Ohno, T., Mukawa, N., & Kawato, S. (2003). Just blink your eyes: A head-free gaze tracking system. In *Proceedings of the Computer-Human Interface*, (pp. 950-951). Computer-Human Interface.

Olfati-Saber, R., Fax, J. A., & Murray, R. M. (2007). Consensus and cooperation in networked multi-agent systems. *Proceedings of the IEEE*, *95*, 215–233. doi:10.1109/JPROC.2006.887293

Oliver, N., Horvitz, E., & Garg, A. (2002). Layered representations for human activity recognition. In *Proceedings of the IEEE International Conference on Multimodal Interfaces*, (pp. 3-8). IEEE Press.

Olson, A. R., & Langois, D. J. (2008). Solid state drives data reliability and lifetime. *Imitation*. Retrieved May 10, 2011 from http://www.imation.com/PageFiles/83/SSD-Reliability-Lifetime-White-Paper.pdf.

OpenStreetMap. (2012). *CloudMade developers zone*. Retrieved January 12, 2012, from http://developers.cloudmade.com/projects.

Oracle. (2009). *Platform-as-a-service private cloud with Oracle fusion middleware*. White Paper. Oracle.

Oracle. (2010). *BTrace user's guide*. Retrieved July 21, 2011, from http://kenai.com/projects/btrace/pages/UserGuide.

Oracle. (2010). *Oracle cloud computing*. White Paper. Oracle.

Oregon Department of Environmental Quality. (2010). *Oregon e-cycles*. Retrieved from http://www.deq.state.or.us/lq/ecycle/manufacturers/index.htm.

Oregon Department of Environmental Quality. (2011). *News release: Oregon e-cycles program show big gains in 2010*. Retrieved from http://www.oregon.gov/DEQ/.

Ortega, A., & Ramchandran, K. (1998). Rate distortion methods for image and video compression. *IEEE Signal Processing Magazine*, *15*(6), 23–50. doi:10.1109/79.733495

Ortiz, C. (2008). Worldwide and US mobile subscriber penetration. *About Mobility*. Retrieved May 15, 2011 from http://weblog.cenriqueortiz.com/mobility/2008/12/29/worldwide-and-us-mobile-subscriber-penetration-dec-2008/.

OSGi Alliance. (2011). *OSGi service platform release 4*. Retrieved July 29, 2011, from http://www.osgi.org/Release4/.

Osman, S., Subhraveti, D., Su, G., & Nieh, J. (2002). The design and implementation of zap: A system for migrating computing environments. In *Proceedings of the Fifth Symposium on Operating Systems Design and Implementation (OSDI)*, (pp. 361-376). New York, NY: ACM Press.

Owens, J. D., Luebke, D., Govindaraju, N., Harris, M., Krüger, J., Lefohn, A. E., & Purcell, T. J. (2007). A survey of general-purpose computation on graphics hardware. *Computer Graphics Forum*, *26*(1), 80–113. doi:10.1111/j.1467-8659.2007.01012.x

Pajorova, E., & Hluchy, L. (2010). *3D visualization the results of complicated grid and cloud-based applications*. Paper presented at the 14th International Conference on Intelligent Engineering Systems (INES). Las Palmas, Spain.

Papazoglou, M. P., & Georgakopoulos, D. (2003). Service oriented computing. *Communications of the ACM*, *46*(10).

Paradiso, J. A., & Starner, T. (2005). Energy scavenging for mobile and wireless electronics. *IEEE Pervasive Computing / IEEE Computer Society [and] IEEE Communications Society*, *4*(1), 18–27. doi:10.1109/MPRV.2005.9

Park, C., John, J., Klein, K., Teplik, J., Caravella, J., Whitfield, J., et al. (1995). Reversal of temperature dependence of integrated circuits operating at very low voltages. In *Proceedings of the International Electron Devices Meeting*, (pp. 71 - 74). Electron Devices Meeting.

Park, S., & Kautz, H. (2008a). Hierarchical recognition of activities in daily living using multi-scale, multi-perspective vision and RFID. In *Proceedings: IET International Conference on Intelligent Environments*. Seattle, WA: IET.

Park, S., & Kautz, H. (2008b). Privacy-preserving recognition of activities in daily living from multi-view silhouettes and RFID-based training. In *Proceedings of the AAAI Fall 2008 Symposium*. Washington, DC: AAAI.

Park, W. I. (1999). *Power saving in a portable computer*. EU Patent EP0949557. Geneva, Switzerland: European Union.

Park, W., Choi, C., Lee, H., & Park, K. (2008). Energy efficient home gateway based on user service traffic in always-on home network environment. In *Proceedings of International Conference Advances in Electronics and Micro-Electronics*, (pp. 121-125). Valencia, Spain: IEEE.

Park, R., & Kim, J. (2005). Real-time facial and eye gaze tracking system. *IEICE Transaction on Information & Systems. E (Norwalk, Conn.)*, *88-D*(6), 1231–1238.

Parliamentary Office of Science and Technology. (2006). *Carbon footprint of electricity generation*. London, UK: UK Parliament.

Pasricha, S., Luthra, M., Mohapatra, S., Dutt, N., & Venkatasubramanian, N. (2004). Dynamic backlight adaptation for low-power handheld devices. *IEEE Design and Test Magazine*, *21*(5), 398–405. doi:10.1109/MDT.2004.57

Pathak, J., Li, Y., Honavar, V., & McCalley, J. (2007). A service-oriented architecture for electric power transmission system asset management. In *Proceedings of Service-Oriented Computing ICSOC 2006* (*Vol. 4652*, pp. 26–37). Springer. doi:10.1007/978-3-540-75492-3_3

Pattanai, K. S. N., Tumblin, J. E., Yee, H., & Greenberg, D. P. (2000). Time dependent visual adaptation for realistic image display, in *Proceedings of SIGGRAPH*, (pp. 47-54). SIGGRAPH.

Patterson, D., Fox, D., Kautz, H., & Philipose, M. (2005). Fine-grained activity recognition by aggregating abstract object usage. In *Proceedings: IEEE International Symposium on Wearable Computers*. IEEE Press.

Paul, B. C., Kang, K., Kufluoglu, H., Alam, M. A., & Roy, K. (2006). Temporal performance degradation under NBTI: Estimation and design for improved reliability of nanoscale circuits. In *Proceedings of European Design and Test Conference*, (pp. 780–785). European Design and Test Conference.

Pedram, M. (2009). Green computing: Reducing energy cost and carbon footprint of information processing systems. In *Proceedings of the ACM Great Lakes symposium on VLSI.* ACM Press.

Perera, G. (2008). *New search paradigms and power management for peer-to-peer file sharing*. Saarbrücken, Germany: VDM Verlag.

Petri, C. A. (1962). *Kommunikation mit Automaten.* Bonn, Germany: Institut für instrumentelle Mathematik der Universität Bonn.

Pickavet, M. Vereecken., W., Demeyer, S., Audenart, P., Vermeulen, B., Develder, C., et al. (2008). Worldwide energy needs for ICT: The rise of power-aware networking. In *Proceedings of the 2nd International Symposium on Advanced Networks and Telecommunication Systems,* (pp. 1-3). Mumbai, India: IEEE Press.

Plug Load Solutions. (2011). *80 plus certified power supplies & manufacturers*. Retrieved August 21, 2011, from http://www.plugloadsolutions.com/80PlusPowerSupplies.aspx.

PMI, Inc. (2008). *A guide to the project management body of knowledge (PMBOK)* (4th ed.). Newtown Square, PA: PMI.

Pomona College Sustainability Integration Office. (2010). *The little green book: A guide to sustainable living at Pomona College* (2nd ed). Retrieved from http://www.pomona.edu/adminstration/sustainability/resources/publications/littlegreenbook.pdf.

Prechter, R. R. Jr, & Parker, W. D. (2007). The financial and economic dichotomy in social behavioral dynamics: The socionomic perspective. *Journal of Behavioral Finance*, *8*(2), 84–108. doi:10.1080/15427560701381028

Pretorius, M., Ghassemian, M., & Ierotheou, C. (2010). An investigation into energy efficiency of data centre virtualisation. In *Proceedings of the International Conference on P2P, Parallel, Grid, Cloud and Internet Computing,* (pp. 157-163). IEEE Press.

Protégé. (2011). *The protégé ontology editor and knowledge acquisition system*. Retrieved from http://protege.stanford.edu/.

PTM. (2011). *PTM technology*. Retrieved from http://ptm.asu.edu/.

Puckett, J., Westervelt, S., Gutierrez, R., & Takamiya, Y. (2005). *The digital dump: Exporting re-use and abuse to Africa*. Seattle, WA: Basel Action Network.

Qi, Z., & Stan, M. R. (2007). Accurate back-of-the-envelope transistor model for deep sub-micron MOS. In *Proceedings International Conference on Microelectronic Systems Education.* Microelectronic Systems Education.

Qi, Z., & Stan, M. R. (2008). NBTI resilient circuits using adaptive body biasing. In *Proceedings of the ACM Great Lakes symposium on VLSI*, (pp. 285-290). ACM Press.

Qi, Z., Meyer, B., Huang, W., Ribando, R., Skadron, K., & Stan, M. (2010). Temperature-to-power mapping. In *Proceedings of IEEE International Conference on Computer Design*, (pp. 384 - 389). IEEE Press.

Quadri, I., Gamatié, A., Boulet, P., & Dekeyser, J. (2010). *Modeling of configurations for embedded system implementations in MARTE*. Paper presented at the 1st Workshop on Model Based Engineering for Embedded Systems Design (M-BED), co-located with DATE 2010. Dresden, Germany.

Ra, M., Paek, J., Sharma, A. B., Govindan, R., Krieger, M. H., & Neely, M. J. (2010). Energy-delay tradeoffs in smartphone applications. In *Proceedings of the 8th International Conference on Mobile Systems, Applications, and Services,* (pp. 255-270). New York, NY: ACM Press.

Rabiner, L. (1989). A tutorial on hidden Markov models and selected applications in speech recognition. *Proceedings of the IEEE*, *77*(2), 257–286. doi:10.1109/5.18626

Rahimi, S. K., & Haug, F. S. (2010). *Distributed database management systems*. Hoboken, NJ: John Wiley & Sons. doi:10.1002/9780470602379

Rajamani, K., & Lefurgy, C. (2003). On evaluating request-distribution schemes for saving energy in server clusters. In *Proceedings of the 2003 IEEE International Symposium on Performance Analysis of Systems and Software (ISPASS 2003),* (pp. 111-122). Washington, DC: IEEE Computer Society.

Ranganathan, P., & Nygard, K. (2010). An optimal resource assignment problem in smart grid. In *Proceedings of the Second International Conference on Future Computational Technologies and Applications (FUTURE COMPUTING),* (pp. 28-34). Lisbon, Portugal: IARIA.

Ranganathan, P., Leech, P., Irwin, D., & Chase, J. (2006). Ensemble-level power management for dense blade servers. In *Proceedings of the 33rd International Symposium on Computer Architecture*, (pp. 66-77). IEEE Press.

Rao, K. R., & Yip, P. (1990). *Discrete cosine transform: Algorithms, advantages, applications*. San Diego, CA: Academic Press.

Rassweiler, A. (2010). *Google nexus one carries $174.15 materials cost, iSuppli teardown reveals*. Press Release. Retrieved May 15, 2011 from http://www.isuppli.com/Teardowns/News/Pages/Google-Nexus-One-Carries-$17415-Materials-Cost-iSuppli-Teardown-Reveals.aspx.

Ray, L. E., Lever, J. H., Streeter, A. D., & Price, A. D. (2007). Design and power management of a solar-powered "cool robot" for polar instrument networks. *Journal of Field Robotics*, *24*(7), 581–599. doi:10.1002/rob.20163

Rebitzer, G., Ekvall, T., Frischknecht, R., Hunkeler, D., Norris, G., & Rydberg, T. (2004). Life cycle assessment: Part 1: Framework, goal and scope definition, inventory analysis, and applications. *Environment International*, *30*(5), 701–720. doi:10.1016/j.envint.2003.11.005

Reed College. (2011). *Computing and information services: Green computing at Reed*. Retrieved from http://www.reed.edu/cis/facilities/green_computing.html.

Rekimoto, J., & Ayatsuka, Y. (2000). Designing augmented reality environments with visual tags. In *Proceedings of Designing Augmented Reality Environments (DARE). DARE*. CyberCode.

Ren, Z., Krogh, B. H., & Marculescu, R. (2005). Hierarchical adaptive dynamic power management. *IEEE Transactions on Computers*, *54*(4), 409–420. doi:10.1109/TC.2005.66

Report, O. E. C. D. (2009). *Smart sensor networks: Technologies and applications for green growth*. OECD.

Research, I. M. S. (2010). Smartphone shipments to quadruple in Asia by 2015. *IMS Research: Excellence in Market Intelligence*. Retrieved May 6, 2011 From http://imsresearch.com/news-events/press-template.php?pr_id=1350.

Richardson, I. E. G. (2010). *High efficiency video coding / HEVC / H.265: Beyond H.264*. Retrieved from http://www.vcodex.com/h265.html.

Richardson, I. E. G. (2003). *H.264 and MPEG-4 video compression: Video coding for next-generation multimedia*. New York, NY: John Wiley & Sons Ltd. doi:10.1002/0470869615

Richardson, I. E. G., & Zhao, Y. (2002). Adaptive management of video encoder complexity. *Real-Time Imaging*, *8*(4), 291–301. doi:10.1006/rtim.2001.0275

Rikhardsson, P. (1998). *Information systems for corporate environmental management accounting and performance measurement*. Retrieved from http://www.highbeam.com/doc/1G1-55308837.html.

Ritthoff, M., Rohn, H., & Liedtke, C. (2003). *Calculating MIPS – Resource productivity of products and services*. Wuppertal, Germany: Wuppertal Institute.

Roberts, L. (2009). A radical new router. *IEEE Spectrum*, *46*(7), 34–39. doi:10.1109/MSPEC.2009.5109450

Robertson, J., Homan, G. K., Mahajan, A., Nordman, B., Webber, C. A., & Brown, R. E. … Koomey, J. G. (2002). *Energy use and power levels in new monitors and personal computers*. LBNL-48581. Berkeley, CA: UC Berkeley.

Robins, F. (2006). The challenge of TBL: A responsibility to whom? *Business and Society Review*, *111*(1), 1–14. doi:10.1111/j.1467-8594.2006.00258.x

Robinson, B. H. (2009). E-waste: An assessment of global production and environmental impacts. *The Science of the Total Environment*, *408*, 183–191. doi:10.1016/j.scitotenv.2009.09.044

Rowsell-Jones, A. (2008). *Going green: The CIO's role in enterprisewide environmental sustainability*. Gartner.

Roy, N., Roy, A., & Das, S. (2006). Context-aware resource management in multi-inhabitant smart homes: A Nash h-learning based approach. In *Proceedings of the Fourth Annual IEEE International Conference on Pervasive Computing and Communications*. IEEE Press.

Rungtusanatham, M., Salvador, F., Forza, C., & Choi, T. Y. (2003). Supply-chain linkages and operational performance: A resource-based perspective. *International Journal of Operations & Production Management*, *23*, 1084–1099. doi:10.1108/01443570310491783

Saha, A. K., & Johnson, D. B. (2004). Modeling mobility for vehicular ad-hoc networks. In *Proceedings of the 1st ACM International Workshop on Vehicular ad hoc Networks,* (pp. 91-92). New York, NY: ACM Press.

Sakurai, T., & Newton, A. R. (1991). A simple MOSFET model for circuit analysis. *IEEE Journal of Solid-state Circuits, 38*(4), 887–893.

Salameh, Z. M., & Chaozi, L. (1990a). Optimum switching points for array reconfiguration controller. In *Proceedings of the 21st IEEE Photovoltaic Conference*, (vol 2), (pp. 971-976). IEEE Press.

Salameh, Z. M., Mulpur, A. K., & Dagher, F. (1990b). Two stage electrical array reconfiguration controller. *Photovoltaic Energy, 44*, 51–55.

Saluja, K., Vijayakumar, S., Sootkaneung, W., & Yang, X. (2008). NBTI degradation: A problem or a scare? In *Proceedings of the VLSI Design, 2008: VLSID 2008 21st International Conference*, (pp. 137 - 142). VLSI.

Sayood, K. (1996). *Introduction to data compression.* New York, NY: Morgan Kaufmann Publishers.

Scharnhorst, W. (2008). Life cycle assessment in the telecommunication industry: A review. *International Journal of Life Cycle Assessment, 13*, 75–86.

Schmidt, M., & Lipson, H. (2009). Distilling free-form natural laws from experimental data. *Science, 324*(5923), 81–85. doi:10.1126/science.1165893

Schmidt, W. (2006). Unfair trade - E-waste in Africa. *Environmental Health Perspectives, 114*(4), 232–235. doi:10.1289/ehp.114-a232

Schubert, H., Jeffery, K., & Neidecker-Lutz, B. (2010). *The future for cloud computing: Opportunities for European cloud computing beyond 2010*. Retrieved from http://cordis.europa.eu/fp7/ict/ssai/docs/cloud-report-final.pdf.

Scott, W. (1995). *Institutions and organisations*. Thousand Oaks, CA: Sage.

Segars, S. (2001). *Low power design techniques for microprocessors.* Retrieved May 10, 2011, from http://www.asicfpga.com/site_upgrade/asicfpga/pds/ip_pds_files/SS_ISSCC2001.pdf.

Seo, C., Malek, S., & Medvidovic, N. (2007). An energy consumption framework for distributed Java-based systems. In *Proceedings of the 22nd IEEE/ACM International Conference on Automated Software Engineering.* Atlanta, GA: IEEE Press.

Seo, C., Malek, S., & Medvidovic, N. (2008). Estimating the energy consumption in pervasive Java-based systems. In *Proceedings of the 2008 Sixth Annual IEEE International Conference on Pervasive Computing and Communications.* IEEE Press.

Shakeri, K., & Meindl, J. D. (2003). Temperature variable supply voltage for power reduction. In *Proceedings of IEEE Computer Society Annual Symposium on VLSI.* IEEE Press.

Shang, S., Ding, R., Yuan, B., Xie, K., Zheng, K., & Kalnis, P. (2012). Users oriented trajectory search for trip recommendation. In *Proceedings of the 15th International Conference on Extending Database Technology (EDBT 2012)*. Berlin, Germany: EDBT.

Sharp Microelectronics of the Americas. (2002). *Display modes: Transmissive/reflective/transflective*. Retrieved from http://www.sharpsma.com/sma/Products/displays/AppRefGuide/DisplayModes.htm.

Sharp, W. F. (1990). *Capital asset prices with and without negative holdings*. Nobel-Prize Economics Lecture. Retrieved from http://www.nobelprize.org/nobel_prizes/economics/laureates/1990/sharpe-lecture.pdf.

Sheriff, I., Acharya, P. A. K., Sampath, A., Zhao, B. Y., & Belding, E. M. (2007). Integrated data location in multihop wireless networks. In *Proceedings of the 2nd International Conference on Communication Systems Software and Middleware*, (pp. 1-10). IEEE Press.

Shim, H., Chang, N., & Pedram, M. (2004). A backlight power management framework for the battery-operated multi-media systems. *IEEE Design and Test Magazine, 21*(5), 388–396. doi:10.1109/MDT.2004.43

Shing-Wa, W., Valcarenghi, L., She-Hwa, Y., Campelo, D., Yamashita, S., & Kazovsky, L. (2009). Sleep mode for energy saving PONs: Advantages and drawbacks. In *Proceeding of 2nd Green Communications Workshop.* Honolulu, HI: IEEE Press.

Shivappa, S., Mohan, M., Trivedi, M. M., & Rao, B. D. (2010). Audio-visual information fusion in human computer interfaces and intelligent environments: A survey. *Proceedings of the IEEE*, *98*(10), 1692–1715. doi:10.1109/JPROC.2010.2057231

Shye, A., Scholbrock, B., & Memik, G. (2009). Into the wild: Studying real user activity patterns to guide power optimizations for mobile architectures. In *Proceedings of the 42nd Annual IEEE/ACM International Symposium on Microarchitecture,* (pp. 168-178). New York, NY: ACM Press.

Shye, A., Scholbrock, B., Memik, G., & Dinda, P. A. (2010). Characterizing and modeling user activity on smartphones: SUMMARY. In *Proceedings of the ACM SIGMETRICS International Conference on Measurement and Modeling of Computer Systems,* (pp. 375-376). New York, NY: ACM Press.

Silveira, G. T. R., & Chang, S.-Y. (2010). Cell phone recycling experiences in the United States and potential recycling options in Brazil. *Waste Management (New York, N.Y.)*, *30*, 2278–2291. doi:10.1016/j.wasman.2010.05.011

Silvius, A. J., & Schipper, R. (2010). *A maturity model for integrating sustainability in projects and project management.* Paper presented on 24th IPMA World Congress. Istanbul, Turkey.

Singhal, P. (2004). *Stage I final report: Life cycle environmental issues of mobile phones. Integrated Product Policy Pilot Project.* Espoo, Finland: Nokia Corporation.

Socolof, M. L., Overly, J. G., & Geibig, J. R. (2005). Desktop computer displays- A life-cycle assessment. *Journal of Cleaner Production*, *13*, 13–14. doi:10.1016/j.jclepro.2005.05.014

Soeleman, H., & Roy, K. (1999). Ultra-low power digital subthreshold logic circuits. In *Proceedings of the International Symposium on Low Power Electronics Design,* (pp. 94-96). IEEE/ACM.

Solak, S. B., & Labeau, F. (2010). Complexity scalable encoding for power-aware application. In *Proceedings of IEEE IGCC 2010*. Chicago, IL: IEEE Press.

Somavat, P., Jadhav, S., & Namboodiri, V. (2010). Energy consumption of personal computing including portable communication devices. In *Proceedings of 1st International Conference on Energy-Efficient Computing and Networking,* (pp. 141-149). New York, NY: ACM Press.

SPEEDS Consortium. (2009). *D.2.1.5 SPEEDS L-1 meta-model.* Retrieved from http://speeds.eu.com/downloads/SPEEDS Meta-Model.pdf.

SQLite. (2012a). *About SQLite*. Retrieved from March 5, 2012, from http://www.sqlite.org/about.html.

SQLite. (2012b). *The architecture of SQLite*. Retrieved January 12, 2012, from http://www.sqlite.org/arch.html.

Srikantaiah, S., Kansal, A., & Zhao, F. (2008). Energy aware consolidation for cloud computing. In *Proceedings of the 2008 Conference on Power Aware Computing and Systems.* IEEE.

Srinivasan, M., & Sheng, P. (1999a). Feature-based process planning for environmentally conscious machining - Part 1: Microplanning. *Robotics and Computer-integrated Manufacturing, 15*, 257–270. doi:10.1016/S0736-5845(99)00017-4

Srinivasan, M., & Sheng, P. (1999b). Feature based process planning in environmentally conscious machining- Part 2: Macroplanning. *Robotics and Computer-integrated Manufacturing, 15*, 271–281. doi:10.1016/S0736-5845(99)00018-6

Stan, M., Cabe, A., Ghosh, S., & Qi, Z. (2007). Teaching top-down ASIC/SoC design vs bottom-up custom VLSI. In *Proceedings of the International Conference on Microelectronic Systems Education (MSE).* MSE.

Stanford Office of Sustainability. (2010). *A student's guide to sustainable living at Stanford.* Retrieved from http://ssu.stanford.edu/sites/sustainable.stanford.edu/files/documents/Sustainable_at_Stanford_New.pdf.

Starner, T., & Paradiso, J. (2004). Human generated power for mobile electronics. In Piguet, C. (Ed.), *Low Power Electronics Design*. Boca Raton, FL: CRC Press.

Starzyk, J., & He, H. (2007). A novel low-power logic circuit design scheme. *IEEE Transactions on Circuits and Wystems. II, Express Briefs*, *54*(2), 176–180. doi:10.1109/TCSII.2006.883093

Steiniger, S., Neun, M., & Edwardes, A. (2006). *Foundations of location-based services*. Retrieved January 13, 2012, from http://www.geo.unizh.ch/publications/cartouche/lbs_lecturenotes_steinigeretal2006.pdf.

Steinmetz, R., & Nahrstedt, K. (2004). *Multimedia systems*. New York, NY: Springer-Verlag.

Stern, N. (2007). *The economics of climate change: The stern review*. Cambridge, UK: Cambridge University Press.

Suhring, K. (2011). *JM reference software*. Retrieved from http://iphome.hhi.de/suehring/tml/.

Suh, S.-H., Kang, S.-K., Chung, D.-H., & Stroud, I. A. (2008). *Theory and design of CNC systems*. Berlin, Germany: Springer.

Sullivan, D. (2006). Recycled cell phones - A treasure trove of valuable metals. *US Geological Survey*. Retrieved May 7, 2011 from http://pubs.usgs.gov/fs/2006/3097/.

Sullivan, J. L., & Gaines, L. (2010). A review of battery life-cycle analysis: State of knowledge and critical needs. *Argonne National Laboratory Technical Report ANL/ESD/10-7*. Retrieved from http://www.ipd.anl.gov/anlpubs/2010/11/68455.pdf.

Sullivan, G. J., & Wiegand, T. (1998). Rate distortion optimization for video compression. *IEEE Signal Processing Magazine*, *15*(6), 74–90. doi:10.1109/79.733497

Sun Microsystems. (2009). *Introduction to cloud computing architecture*. White Paper. Sun Microsystems.

Sun. (2008). *Solaris dynamic tracing guide*. Retrieved May 10, 2011, from http://download.oracle.com/docs/cd/E19253-01/817-6223/.

Sweet Briar College. (2011). *Green computing*. Retrieved from http://www.sbc.edu/ac/green-computing.

Swiss Centre for Life Cycle Inventories. (2006). *Ecoinvest centre*. Retrieved May 10, 2011, from http://www.ecoinvent.ch/.

SWRL. (2011). *Website*. Retrieved from http://www.w3.org/Submission/SWRL/.

Takemura, K., Ohara, K., Ohba, K., Chong, N. Y., Hirai, S., & Tanie, K. (2004). Knowledge distributed tag-based vision system. In *Proceedings of the 1st International Workshop on Networked Sensing Systems*. Networked Sensing Systems.

TAMUZ. (2007). Technical bulletin: General information about TFT-LCD displays. *TAMUZ Broadcast Monitors*. Retrieved May 10, 2011 from http://www.tamuz.tv/manuals/tamuz-tb-lcd-effects.pdf.

Tan, Y. H., Lee, W. S., Tham, J. Y., & Rahardja, S. (2009). Complexity rate distortion optimization for real-time H.264/AVC encoding. In *Proceedings of the IEEE ICCCN 2009*. San Francisco, CA: IEEE Press.

Tang, S., Narendra, S., & De, V. (2003). Temperature and process invariant MOS-based reference current generation circuits for sub-1V operation. In *Proceedings of of International Symposium on Low Power Electronics and Design*. Low Power Electronics and Design.

TCO. (2006). *Displays 5.0*. Retrieved from http://www.tcodevelopment.com/.

TechSoup. (2009). *Reduce your environmental impact and save money*. Retrieved from http://www.techsoup.org/greentech/.

Tenhunen, M., & Penttinen, E. (2010). Assessing the carbon footprint of paper vs. electronic invoicing. In *Proceedings of the 21st Australasian Conference on Information Systems*. Brisbane, Australia: IEEE.

TenStep, Inc. (2010). *Green project management articles*. Retrieved from http://www.green-pm.com/greenarticles.html.

Terrachoice. (2009). *A marketer's guide to avoid the sins of green washing*. Retrieved from http://sinsofgreenwashing.org/solutions/marketers-guide/.

Theocharides, T., Link, G., Vijakrishnan, N., Irwin, M. J., & Wolf, W. (2004). Embedded hardware face detection. In *Proceedings of the 17th IEEE International Conference on VLSI Design*, (pp. 33-138). IEEE Press.

TIA. (2011). *E-cycling begins with you*. Retrieved from http://ecyclingcentral.com.

Tolia, N., Wang, Z., Marwah, M., Bash, C., Ranganathan, P., & Zhu, X. (2008). Delivering energy proportionality with non energy-proportional systems: Optimizing the ensemble. In *Proceedings of Conference on Power Aware Computing and Systems*. ACM Press.

Tourapis, A. M., Cheong, H. Y., & Topiwala, P. (2005). Fast ME in the JM reference software. In *Proceedings of the JVT of ISO/IEC MPEG and ITU-T VCEG, 16th Meeting*. Poznan, Poland: JVT.

Tschanz, J. W., Kao, J. T., Narendra, S. G., & Nair, R. A. A., D., Chandrakasan, A. P., et al. (2002). Adaptive body bias for reducing impacts of die-to-die and within-die parameter variations on microprocessor frequency and leakage. *IEEE Journal of Solid-state Circuits, 37*(11), 1396–1402. doi:10.1109/JSSC.2002.803949

Tsirogiannis, D., Harizopoulos, S., & Shah, M. A. (2010). Analyzing the energy efficiency of a database server. In *Proceedings of the 2010 International Conference on Management of Data (SIGMOD 2010),* (pp. 231–242). New York, NY: ACM Press.

Tsuchiya, H. (2008). Innovative renewable energy solutions for hydrogen vehicles. *International Journal of Energy Research, 32*(5), 427–435. doi:10.1002/er.1374

Tumblin, J. E., Hodgins, J. K., & Guenter, B. K. (1999). Two methods for display of high contrast images. *ACM Transactions on Graphics, 18*(1), 56–94. doi:10.1145/300776.300783

United Nations Environment Programme. (2009). *Recycling - From e-waste to resources*. Berlin, Germany: Oktoberdruck AG.

US Census Bureau. (2011). *Statistical abstract of the United States: 2011*. Washington, DC: US Census Bureau.

US Department of Energy. (2011). *Earth day 2011*. Retrieved from http://www.energy.gov/earthday/.

US Department of Energy. (2011). *Federal energy management program*. Retrieved May 6, 2011, from http://www1.eere.energy.gov/femp/program/dc_energy_consumption.html.

US Environmental Protection Agency. (2007). *Report to congress on server and data center energy efficiency*. Washington, DC: EPA.

US Green Building Council. (2010). *What is LEED?* Retrieved from http://www.usgbc.org/DisplayPage.aspx?CMSPageID=1988.

van Dam, S. S., Bakker, C. A., & Van Hal, J. D. M. (2010). Home energy monitors: Impact over the medium-term. *Building Research and Information, 38*(5), 458–469. doi:10.1080/09613218.2010.494832

van Hoorik, P. M., & Bomhof, F. W. (2010). *Sustainability impact of ICT – E-invoicing as a case for a generic framework*. Paper presented at the CMI Conference. Copenhagen, Denmark.

van Hoorik, P. M., van Harmelen, T., Head, M., & Meulenhoff, P. (2010). *Energie- en CO_2-besparing door elektronisch factureren*. Paper presented as a TNO report for the Dutch ministry of Economic Affairs. Delft, The Netherlands.

Vattikonda, R., Wang, W., & Cao, Y. (2006). Modeling and minimization of PMOS NBTI effect for robust nanometer design. In *Proceedings of Design Automation Conference*, (pp. 1047–1052). Design Automation Conference.

VESA. (2011). *Display power management signaling specification (DPMS)*. Retrieved from http://vesa.org/vbe3.pdf.

Viola, P., & Jones, M. (2001). Rapid object detection using a boosted cascade of simple features. In *Proceedings of the IEEE Conference on Computer Vision and Pattern Recognition*. IEEE Press.

Wang, S., Min, J., & Yi, B. K. (2008). Location based services for mobiles: technologies and standards. In *Proceedings of the IEEE International Conference on Communication (ICC)*. Beijing, China: IEEE Press.

Wang, W., Yang, S., Bhardwaj, S., Vattikonda, R., Vrudula, S., Liu, F., et al. (2007). The impact of NBTI on the performance of combinational and sequential circuits. In *Proceedings of Design Automation Conference*, (pp. 364-369). Design Automation Conference.

Wang, Y., Pang, S., & Cheng, K. (2010). A GPU-accelerated face annotation system for smartphones. In *Proceedings of the International Conference on Multimedia,* (pp. 1667-1668). New York, NY: ACM Press.

Wang, X., Dong, J. S., Chin, C. Y., Hettiarachchi, S. R., & Zhang, D. (2004). Semantic space: An infrastructure for smart spaces. *IEEE Pervasive Computing / IEEE Computer Society [and] IEEE Communications Society*, *3*(3), 32–39. doi:10.1109/MPRV.2004.1321026

Washburn, D. (2009). *Q&A: The economics of green IT, what are the financial benefits of green IT*. Forrester.

Washington State Department of Ecology. (2010). *2010 annual report*. Retrieved from http://www.ecy.wa.gov/programs/swfa/eproductrecycle/.

Watts Up Meters. (2011). *Watts up? Products*. Retrieved April 29, 2011, from https://www.wattsupmeters.com/secure/products.php?pn=0&wai=0&spec=2.

Wattsup. (2011). *Website*. Retrieved from https://www.wattsupmeters.com.

Wedi, T. (2003). Motion compensation in H.264/AVC. *IEEE Transactions on Circuits and Systems for Video Technology*, *13*(7), 577–586. doi:10.1109/TCSVT.2003.815171

Weidong, X., Nathan, O., & William, D. G. (2007). Topology study of photovoltaic interface for maximum power point tracking. *Transactions on Industrial Electronics*, *54*(3), 1696–1704. doi:10.1109/TIE.2007.894732

Weinhardt, C., Anandasivam, A., & Blau, B., & StoBer, J. (2009). Cloud computing – A classification, business models, and research direction. *Journal of Business and Information Systems Engineering*, *1*(5), 391–395. doi:10.1007/s12599-009-0071-2

Wernerfelt, B. (1984). A resource based view of the company. *Strategic Management Journal*, *5*, 171–180. doi:10.1002/smj.4250050207

Whaley, R. C., & Dongarra, J. (1997). *Automatically tuned linear algebra software*. Knoxville, TN: University of Tennessee.

Whitney, L. (2009). Jobs: Over 1 million new iPhones sold. *CNET News*. Retrieved May 6, 2011 from http://news.cnet.com/8301-13579_3-10269991-37.html.

Wiegand, T., Sullivan, G. J., Bjontegaard, G., & Luthra, A. (2003). Overview of the H.264/AVC video coding standard. *IEEE Transactions on Circuits and Systems for Video Technology*, *13*(7), 560–575. doi:10.1109/TCSVT.2003.815165

Williams, E. (2004). Energy intensity of computer manufacturing: Hybrid assessment combining process and economic input-output methods. *Environmental Science & Technology*, *38*(22), 6166–6174. doi:10.1021/es035152j

Williams, E. D., Ayres, R. U., & Heller, M. (2002). The 1.7 kilogram microchip: Energy and material use in the production of semiconductor devices. *Environmental Science & Technology*, *36*, 5504–5510. doi:10.1021/es025643o

Williams, E., Ayres, R., & Heller, M. (2002). The 1.7 kilogram microchip: Energy and material use in the production of semiconductor devices. *Environmental Science & Technology*, *36*(2), 5504–5510. doi:10.1021/es025643o

Wisconsin Department of Natural Resources. (2011a). *E-cycle Wisconsin*. Retrieved from http://dnr.wi.gov/org/aw/wm/ecycle/.

Wisconsin Department of Natural Resources. (2011b). *E-cycle Wisconsin 2010 report*. Retrieved from http://dnr.wi.gov/org/aw/wm/publications/anewpub/wa1529/pdf.

Wolsey, L. A. (1998). *Integer programming*. New York, NY: John Wiley & Sons.

World in Balance. (1987). *Our common future*. Oxford, UK: Oxford University Press. Retrieved from http://worldinbalance.net/intagreements/1987-brundtland.php.

Wu, J., Osuntogun, A., Choudhury, T., Philipose, M., & Rehg, J. M. (2007). A scalable approach to activity recognition based on object use. In *Proceedings of International Conference on Computer Vision*. IEEE.

Wu, S., Lim, K. P., Wu, D. J., Rahardja, S., Limn, X., Pan, F., & Li, Z. G. (2003). Fast mode decision for intra prediction. In *Proceedings of the JVT of ISO/IEC MPEG and ITU-T VCEG 7th Meeting*. Pattaya II, Thailand: JVT.

X10. (2011). *Website*. Retrieved from http://www.x10.com.

Xiaoling, H., & Hodgson, J. W. (2002a). Modeling and simulation for hybrid electric vehicles. *IEEE Transactions on Intelligent Transportation Systems*, *3*(4), 235–243. doi:10.1109/TITS.2002.807781

Xiaoling, H., & Hodgson, J. W. (2002b). Modeling and simulation for hybrid electric vehicles. *IEEE Transactions on Intelligent Transportation Systems*, *3*(4), 244–251.

Xu, M., Hu, Z., Wu, J., & Zhou, Y. (2008). A hybrid society model for simulating residential electricity consumption. *International Journal of Electrical Power & Energy Systems, 30*(10), 569–574. doi:10.1016/j.ijepes.2008.08.005

Xun, L., Ortiz, P., Browne, J., Franklin, D., Oliver, J., Geyer, R., et al. (2010). Smartphone evolution and reuse: Establishing a more sustainable model. In *Proceedings of the 39th International Conference on Parallel Processing,* (pp. 476-484). San Diego, CA: IEEE Press.

Yale Office of Sustainability. (2010). *Student taskforce for environmental partnership: Recycling made easy.* Retrieved from http://www.yalestep.com.

Yamamoto, S., & Moshnyaga, V. G. (2009). Algorithm optimizations for low-complexity eye tracking. In *Proceedings of the IEEE International Conference on Systems, Man and Cybernetics,* (pp. 18-22). IEEE Press.

Yang, T., Mino, G., Spaho, E., Barolli, L., Durresi, A., & Xhafa, F. (2011). A simulation system for multi mobile events in wireless sensor networks. In *Proceedings of the 25th IEEE International Conference on Advanced Information Networking and Applications (AINA-2011),* (pp. 411-418). Washington, DC: IEEE Computer Society.

Yang, X., & Saluja, K. (2007). Combating NBTI degradation via gate sizing. In *Proceedings of International Symposium on Quality Electronic Design,* (pp. 47–52). Quality Electronic Design.

Yang, Y., Xiong, N., Aikebaier, A., Enokido, T., & Takizawa, M. (2009). Minimizing power consumption with performance efficiency constraint in web server clusters. In *Proceedings of the 12th International Conference on Network-Based Information Systems (NBiS2009),* (pp. 45-51). Washington, DC: IEEE Computer Society.

Yi, X., Zhang, J., Ling, N., & Shang, W. (2005). Improved and simplified motion estimation for JM. In *Proceedings of JVT of ISO/IEC MPEG and ITU-T VCEG, 16th Meeting.* Poznan, Poland: JVT.

Yoon, H., Zheng, Y., Xie, X., & Woo, W. (2010). Smart itinerary recommendation based on user-generated GPS trajectories. *Lecture Notes in Computer Science, 6406,* 19–34. doi:10.1007/978-3-642-16355-5_5

Zeng, J., Liu, J. F., Ngan, H. W., & Wu, J. (2009). *A multi-agent solution to energy management of distributed hybrid renewable energy generated system.* Paper presented at the Advances in Power System Control, Operation and Management (APSCOM 2009). New York, NY.

Zerhouni, F. Z. (1996). *Adaptation optimale d'une charge à un générateur photovoltaïque.* Master Thesis. Oran, Algeria: University of Sciences and Technology Mohamed Boudiaf.

Zerhouni, F. Z. (2005). *Contribution dans l'amélioration du fonctionnement d'un système photovoltaïque.* Paper presented at the International Conference, ICEL05. Oran, Algeria.

Zerhouni, F. Z. (2009). *Développement et optimisation d'un générateur énergétique hybride propre à base de PV-PAC.* Doctoral Thesis. Oran, Algeria: University of Sciences and Technology Mohamed Boudiaf Oran.

Zerhouni, F. Z., Benmessaoud, M. T., & Boudghene Stambouli, A. (2006b). Développement de modèle de système à énergie renouvelable pour son utilisation en électronique de puissance et optimization. In *Proceedings of JNEA 2006.* Oran, Algeria: JNEA.

Zerhouni, F. Z. (2006a). Etude des l'influence de différents agents sur les caractéristiques d'un générateur photovoltaïque. *Technological and Scientific Review, 4,* 39–44.

Zerhouni, F. Z., Zegrar, M., Benmessaoud, M. T., Boudghene Stambouli, A., & Midoun, A. (2008). A novel method to optimize photovoltaic generator operation. *International Journal of Energy Research, 32*(15), 1444–1453. doi:10.1002/er.1461

Zerhouni, F. Z., Zegrar, M., Benmessaoud, M. T., Boudghene Stambouli, A., & Midoun, A. (2010). Optimum load matching by an array reconfiguration in photovoltaic generators. *Energy & Fuels, 24*(11), 6126–6130. doi:10.1021/ef1006275

Zhang, L., & Zhou, Q. (2009). CCOA: Cloud computing open architecture. In *Proceedings of the IEEE International Conference on Web Services,* (pp. 607-616). Los Angeles, CA: IEEE Computer Society.

Zhang, Y., & Stan, M. (2007). Temperature-aware circuit design using adaptive body biasing. In *Proceedings of the ACM Great Lakes Symposium on VLSI*, (pp. 84–89). ACM Press.

Zhang, H., Liu, L., & Li, T. (2011). Designing IT systems according to environmental settings: A strategic analysis framework. *The Journal of Strategic Information Systems*, *20*, 80–95. doi:10.1016/j.jsis.2011.01.001

Zhang, Y., Chowdhury, P., Tornatore, M., & Mukerjee, B. (2010). Energy efficiency in telecom optical networks. *IEEE Communications Surveys & Tutorials*, *12*(4), 441–458. doi:10.1109/SURV.2011.062410.00034

Zhao, W., & Cao, Y. (2006). New generation of predictive technology model for sub-45nm early design exploration. *IEEE Transactions on Electron Devices*, *53*(11), 2816–2823. doi:10.1109/TED.2006.884077

Zheng, Y., & Xie, X. (2011). Learning travel recommendations from user-generated GPS traces. *ACM Transactions on Intelligent Systems and Technology*, *2*(1).

Zhi, G. J., Hua, Z., & Xiao, L. (2008). Study on the cutting technology based on green manufacturing. *Key Engineering Materials*, *375-376*, 158–162. doi:10.4028/www.scientific.net/KEM.375-376.158

Zhou, Q., & Mohanram, K. (2006). Gate sizing to radiation harden combinational logic. *IEEE Transactions on Computer-Aided Design of Integrated Circuits and Systems*, *25*(1), 155–166. doi:10.1109/TCAD.2005.853696

Zhu, Q., Shankar, A., & Zhou, Y. (2004). PB-LRU: A self-tuning power aware storage cache replacement algorithm for conserving disc energy. In *Proceedings of the 18th International Conference on Supercomputing*. Malo, CA: ACM.

Zhuang, Z., Kim, K., & Singh, J. P. (2010). Improving energy efficiency of location sensing on smartphones. In *Proceedings of the 8th International Conference on Mobile Systems, Applications, and Services*, (pp. 315-330). New York, NY: ACM Press.

Zirpins, C., & Emmerich, W. (2008). A reference model of virtual service production networks. *International Journal on Service Oriented Computing and Applications*, *2*(2-3), 145–166. doi:10.1007/s11761-008-0025-4

Zoeteman, B. C. J., Krikke, H. R., & Venselaar, J. (2009). Handling WEEE waste flows: On the effectiveness of producer responsibility in a globalizing world. *International Journal of Advanced Manufacturing Technology*, *47*, 415–436. doi:10.1007/s00170-009-2358-3

About the Contributors

Wen-Chen Hu received a BE, an ME, an MS, and a PhD, all in Computer Science, from Tamkang University, Taiwan, the National Central University, Taiwan, the University of Iowa, and the University of Florida, in 1984, 1986, 1993, and 1998, respectively. He is currently an Associate Professor in the Department of Computer Science of the University of North Dakota. He is the Editor-in-Chief of the *International Journal of Handheld Computing Research* (IJHCR), and has served as editor and editorial advisory/review board members for over 20 international journals/books and chaired more than 20 tracks/sessions and program committees for international conferences. Dr. Hu has been teaching more than 10 years at US universities and has advised more than 50 graduate students. He has published over 100 articles in refereed journals, conference proceedings, books, and encyclopedias, edited five books and proceedings, and solely authored a book. His current research interests include handheld computing, electronic and mobile commerce systems, Web technologies, and databases.

Naima Kaabouch received her B.S., M.S., and Ph.D., all in Electrical Engineering, from the University of Paris 11 and the University of Paris 6, Paris, France, in 1982, 1986, 1990, respectively. She is currently an Assistant Professor, in the Department of Electrical Engineering, University of North Dakota. Dr. Kaabouch's research interests include configurable computing, real-time systems, and signal/image processing. She has published over 80 articles in refereed journals, proceedings, and books. In the green computing field, she has co-edited two books. In teaching, she developed and taught several courses in the field of configurable computing and real-time systems involving microcontrollers and Field Programmable Gate Array based devices.

* * *

Ailixier Aikebaier received his BE degree in Computers and Systems Engineering from XinJiang University, China, in 2000, and ME Degree in Computers and Systems Engineering from Tokyo Denki University, Japan, in 2009. He received his DE Degree in Computer Science from Seikei University, Japan, in 2011. He won the best paper award at CISIS2008 and CISIS2010. His research interests include distributed systems, P2P networks, consensus problems, and fault-tolerant systems.

Uwe Assmann studied Computer Science at Karlsruhe Institute of Technology, Germany, where he got his diploma (M.Sc.) in 1989, and also received his Ph.D. in 1995, focusing on compiler optimization techniques, as well as his postdoctoral qualification (habilitation) in 2002. From 2001 until 2004, Uwe Assmann was Lecturer and, starting in 2003, full Professor of Software Composition at Linköpings

Universitet, Sweden. Since 2004, Uwe Assmann is full Professor and Head of the Software Engineering Chair at Technische Universität Dresden. His research covers a broad range of software engineering topics, including software composition, the application of graph transformation systems, self-adaptive systems, and the integration of ontologies with metamodeling techniques.

Oliver Avram obtained his M.Sc. in Integrated Engineering in 2005 from the University Polithenica of Timisoara, Romania, before moving to the CAD/CAM Laboratory (LICP) of the Ecole Polytechnique Fédérale de Lausanne (EPFL) in Switzerland, where he obtained his Ph.D. in 2010. For his Ph.D. work, Avram developed techniques for measuring consumption in milling processes. This work was done in the framework of a European project on the next generation of machine tools (NEXT). Avram's pioneering work in this field is part of the movement in industry towards cleaner production, enabling quantification of environmental effects during manufacturing. Avram's work involved extensive measurement of machine tools in operation as well as theoretical analyses and modelling of machine tools. Avram is currently responsible for research at Dixi Polytools of Le Locle, Switzerland.

Nathalie Bachour is a Senior Consultant at Devoteam Consulting, Paris, France. She has earned a BE in Computer and Communication from Notre Dame University (Lebanon) in 2001, an MS and MBA in Project and Program Management from SKEMA Business School (France) in 2010. Ms. Bachour has more than 10-years experience in IT consulting. She is currently assigned as a project manager for the IT department of a large industrial group in France. She is also an active member of the Green IT pole at Devoteam Consulting. She is member of the PMI and APM associations and is certified as PRINCE2, MSP practitioner, Six Sigma Yellow Belt, and ITIL Foundation.

Freek Bomhof completed his MSc Electrical Engineering in 1990. During his work on optimization of business processes through innovative ICT's, he became interested in the relation between technology and human behaviour in organizations, especially when considering the sustainability effects of that behaviour. With his background in pattern recognition and project management, he tries to quantify this behaviour and its effects in terms of indicators that can be managed from a sustainability viewpoint. He has been doing this for the Amsterdam Smart City programme that ran from 2009 to 2011, where the main goal was to reach a very substantial carbon reduction for the City of Amsterdam.

Sebastian Cech studied Computer Science at Technische Universität Dresden and got his diploma (M.Sc.) in 2006. Since then, he is working as a Research Assistant and Ph.D. student at the TU Dresden, Software Technology Group. Sebastian Cech focuses in his research on the topic of reuse in metamodelling and reuse during the development process of domain-specific languages. Besides his contribution to the Cool Component Model (CCM) and the reconfiguration planning, he is currently working on the CoolSoftware modeling workbench as well as on THEATRE. Next to the CoolSoftware project, he worked in the Modeling and Simulation Task of the MODELPLEX project, as well as in a project about predictive maintenance and conflict handling (PreCon).

Victor Chang was an IT Manager leading and managing several projects in one of NHS Trusts, UK. He has extensive experience and has successfully delivered many projects ranging from implementation to strategy focus. He is currently working as an Assistant Professor and IT Manager at Business School,

University of Greenwich, while doing his part-time research with the University of Southampton. CCBF is part of his research to help organisations in achieving good Cloud design, deployment, and services. He has 20 publications in 24 months, and always offers 100% effort for his full-time work.

Lei Chen received his PhD in Computer Science and Software Engineering from Auburn University in 2007. He is currently an Assistant Professor in the Department of Computer Science at the Sam Houston State University. His research interests include wireless routing, wireless communications, QoS, and Multimedia.

Fred Chong is the Director of Computer Engineering and a Professor of Computer Science at UCSB. He also directs the Greenscale effort in Energy-Efficient Computing, which involves over 20 multi-disciplinary faculty. Chong received his Ph.D. from MIT in 1996 and was a faculty member and Chancellor's fellow at UC Davis from 1997-2005. He is a recipient of the NSF CAREER award and his research interests include emerging technologies for computing, multicore and embedded architectures, computer security, and sustainable computing.

Mary V. Connolly is an Associate Professor of Mathematics and Computer Science at Saint Mary's College, Notre Dame, Indiana. She is a graduate of Manhattanville College, and holds an A.M. from Brown University in Mathematics and a M.S. from the University of Illinois at Chicago in Computer Science. Prior to coming to Saint Mary's, she taught at Fordham University, Villanova University, and Indiana University at South Bend. She is a past president of the Association of Small Computer Users in Education (ASCUE). She has long been interested in ethical issues in computer science and, in particular, in raising the awareness of these issues for undergraduate students.

Wendy Currie is a Professor in Information Systems at Business School, University of Greenwich. She is the Editor-in-Chief in Health Policy and Technology. Her research, consultancy, and publications focus on policy-making for large-scale ICT projects in health, financial services, and government. She has been a Full Professor since May 1997, with professorial positions at the universities of Greenwich, Warwick, Brunel, and Sheffield.

Tomoya Enokido received BE and ME Degrees in Computers and Systems Engineering from Tokyo Denki University, Japan, in 1997 and 1999, respectively. After that he worked for NTT Data Corporation, he joined Tokyo Denki University in 2002. He received his DE Degree in Computer Science from Tokyo Denki University in 2003. After that, he worked for Computers and Systems Engineering as a Research Associate; he joined Faculty of Business Administration of Rissho University in 2005. He is an Associate Professor in the Faculty of Business Administration, Rissho University. His research interests include distributed systems. He is a member of IEEE.

Diana Franklin is tenured teaching faculty and Director of the Center for Computing Education and Diversity at UCSB. Franklin received her Ph.D. from UC Davis in 2002 and is a recipient of the NSF Career award. She was an Assistant Professor (2002-2007) and Associate Professor (2007) of Computer Science at the California Polytechnic State University, during which she held the Forbes Chair (2002-2007). Her research interests include parallel programming and architecture, computing education, and ethnic and gender diversity in computing.

Juan Manuel García-Chamizo is a Ph D in Computer Science from the University of Alicante. He is a Professor of Architecture and Computer Technology and director of seventeen doctoral theses and a similar number of funded public research projects as well as many other private. He is the author of dozens of publications in scientific journals and a reviewer for several of them. He is the founder and director of the research group of Computer Engineering and Computer Networks.

Sebastian Götz studied Computer Science at Technische Universität Dresden and got his diploma (M.Sc.) in 2010, focusing on self-managed persistency of class-based, object-oriented applications. Since 2010, he is working as a Research Assistant and Ph.D. student at the TU Dresden, Software Technology Group. His research focuses on self-adaptive systems, with special focus on optimization of non-functional properties. Besides his contribution to contract checking and negotiation as well as to theatre, he is currently working on dynamic runtime updates for class-based, object-oriented applications, the support for object roles on the level of implementation, their applicability to specify collaborative behavior of distributed systems and the applicability of predicate dispatch to implement autonomous, self-adaptive systems.

Maarten Hoeve After finishing his PhD in Mathematics (Algebraic Geometry) in 2010, Maarten Hoeve joined TNO at the department strategic business analysis as a Research Scientist. There he started working on a variety of quantitative models, ranging from a model on the costs of telecommunications networks, a model on the market shares in the market for business phone lines, to a model on the effect of emotions on consumer behaviour, and a model on the participation of the unemployed. The thread connecting these diverse subjects is trying to quantify effects that are hard to measure so they can be taken into account in decisions.

Paula van Hoorik has completed her MSc Psychology in 1999. In her work, she focuses on how innovations could really work in practice. Starting with user research, later working on change management and implementation of new business processes, she gained experience in the telecommunications sector. Later she started working on sustainability effects of ICT and processes; for example, she conducted an analyses for the Dutch Ministry of Economic Affairs on the sustainability effects of e-invoicing. Currently, she works in the energy market focusing on new business models and implications concerning the energy transition and smart grids.

Marko Junkkari got Master's Degree in Computer Science from the University of Tampere in 1997. He achieved Licentiate Degree in 2002 and PhD in 2007 in Computer Science. His research activities cover conceptual structures, graph-based methods, development and integration of data models, query languages, and XML information retrieval. He has been a reviewer in several international journals and a member of several programme committees. Since 2006, he has had a position as an Associate Professor of Data Management in the School of Information Sciences in the University of Tampere. His teaching activities include database programming, XML, special topics on data management, and fuzzy logic.

Henry Kautz received the A.B. degree in Mathematics from Cornell University, Ithaca, NY, the M.A. degree in Creative Writing from the The Johns Hopkins University, Baltimore, MD, the M.Sc. in Computer Science from the University of Toronto, Toronto, ON, Canada, and the Ph.D. degree in

Computer Science from the University of Rochester, Rochester, NY. He had 13 years of industrial experience at AT&T Bell Laboratories, where he was head of the AI Principles Research Department. In 2000, he became a faculty member at the University of Washington, Seattle. From 2006 to 2007, he founded and directed the Intelligent Systems Center at Kodak Research Laboratories, Rochester, NY. He is currently Chair of the Department of Computer Science at the University of Rochester, where he directs the Laboratory for Assisted Cognition Environments. His research encompasses artificial intelligence, pervasive computing, and assistive technology. His contributions include efficient algorithms for logical and probabilistic reasoning, the planning as satisfiability framework, and methods for behavior recognition from sensor data. He is the author of more than 70 refereed publications. Dr. Kautz is a Fellow of the American Association for the Advancement of Science, a Fellow of the Association for Advancement of Artificial Intelligence, and winner of the Computers and Thought Award from the International Joint Conference on Artificial Intelligence.

Brandon Kuczenski is a Postdoctoral Scholar at the University of California, Santa Barbara. He received the B.S. degree in Mechanical Engineering from the University of California, Berkeley, CA, where he studied as a Regents' and Chancellor's Scholar. In 2008, he received the Ph.D. degree in Mechanical Engineering from Carnegie Mellon University, Pittsburgh, PA, where he studied the dynamics of biological cellular signaling. Dr. Kuczenski now studies Industrial Ecology at UCSB's Bren School of Environmental Science and Management. His research focuses on the use and recycling of high-volume products, from plastic bottles to cellular phones. He also studies recycling policy, green chemistry, and life cycle assessment methodology.

Fabrice Labeau is an Associate Professor in the Electrical and Computer Engineering Department at McGill University, Montreal, Quebec, Canada. He received the Electrical Engineer degree in 1995 from Université Catholique de Louvain, Belgium, and the Diplôme D'études Spécialisées en Sciences Appliquées, Orientation Télécommunications also from UCL in 1996. From 1996 to 2000, he was with the Communications and Remote Sensing Laboratory of UCL. From January to March 1999, he was a visiting scientist to the Signal and Image Department (TSI) of ENST Paris. He received a Ph.D. degree in September 2000 from UCL. His research interests include signal-processing applications to e-health and energy management, multirate processing, joint source channel coding, data compression, and error-control coding. He was part of the organizing committee of ICASSP 2004 in Montreal and is/was Technical Program Committee co-chair for the IEEE Vehicular Technology Conference in the Fall of 2006 and 2012, and the IEEE International Conference on Image Processing 2015. He is a Senior Member of IEEE.

Xun Li is a PhD candidate from the Department of Computer Science, University of California, Santa Barbara. His research interests cover a wide range of areas in computer architecture, including computer security, memory hierarchy, and sustainable computing. His research is under the advice of Professors Fred Chong and Tim Sherwood.

Vasily G. Moshnyaga received the Computer Engineering Degree with Honors from Technical State University, Sevastopol, USSR, in 1980, and Ph.D. in Computer Engineering from Moscow Aviation Institute in 1986. From 1986 to 1992 he was a Senior Lecturer and an Associate Professor of Technical University of Moldova, Chisinau, Moldova. From 1991 to 1992, he was a Visiting Scholar at the Department of Electronics and Communication, Kyoto University, Japan. From 1992 to 1998, he was a Lecturer at the Department of Electronics and Communication, Kyoto University, Japan. Since 1998, he has been with the Department of Electronics Engineering and Computer Science, Fukuoka University, Japan, where he is currently a Professor. In 2005-2006, he was a Visiting Professor with the Computer Science Department, UCLA. His current research interests are in the areas of video processing, embedded systems, application-specific processors, VLSI design, and design methodologies with a particular emphasis on energy-efficient design techniques. He has authored or co-authored over 170 referred journal and conference publications and holds five patents. Prof. Moshnyaga served as Vice-Chairman of the IEEE Circuits and Systems Society, Fukuoka Chapter from 2008 to 2010, and an Associate Editor of the *IEICE Transactions on Fundamentals of Electronics, Communication and Computer Sciences* from 2005 to 2008. He is now a member of IEEE CAS Technical Committee on VLSI and a member of Technical Program Committees of several conferences and symposia including IEEE International Symposium on Circuits and Systems, ACM/IEE International Symposium on Low-Power Electronics and Design, IEEE International Conference on Systems, Man, and Cybernetics, etc. He received the Nikkei LSI IP Design Award in 2001, Best Paper Award from SCI 2000/ISAS 2000 in 2000 and Encouragement Award from the Embedded Systems Symposium in 2011. Prof. Moshnyaga is a member of the Institute of Electronics, Information, and Communication Engineers of Japan, and a member of the Information Processing Society of Japan and senior member of IEEE.

Vinod Namboodiri is currently an Assistant Professor at the Department of Electrical Engineering and Computer Science at Wichita State University, USA. He graduated with a Ph.D. in Electrical and Computer Engineering from the University of Massachusetts, Amherst, in 2008. He has served or is currently serving on the technical program committees of IEEE GLOBECOM, IEEE ICC, IEEE IPCCC, and IEEE GREENCOM, and is an active reviewer for numerous journals and conferences in the wireless networking and mobile computing areas. His research interests include designing algorithms and protocols for energy-intelligent and sustainable computing, and designing an effective communication architecture for smart electric grids. In the past, he has worked on designed MAC and routing protocols, and developing energy-efficient protocols and algorithms for different wireless technologies like Wireless LANs, RFID Systems, Wireless Sensor Networks, and Wireless Mesh Networks.

Kendall E. Nygard is Professor of Computer Science and Operations Research at North Dakota State University where he has served on the faculty since 1977. From 1996 to 2005, he served as Department Chair. In 1986 and 1987, he was on the adjunct faculty in operations research at the Naval Postgraduate School in Monterey, California. In 1994 and 1995, he was Director of the Scientific Computing Center at the University of North Dakota in Grand Forks. In 1984, he was a visiting scientist at the Air Force Logistics Command at Wright-Patterson AFB in Ohio. In 2000, he was a research fellow at the Air Vehicle Directorate of the Air Force Research Lab. His research areas involve combinatorial optimization methods, with applications to management of networks and sensor networks, cooperative mission control for Unmanned Air Vehicles (UAVs), computer-based transportation analysis, and bioinformatics.

Pablo Ortiz studied Computer Science at UC Santa Barbara and graduated with highest honors. Currently, he is working on his doctorate at MIT. His research interests include green computing, mobile systems, and parallel architectures.

Sangho Park received the B.S. degree in Electronics and Computer Engineering at Yonsei University, Seoul, Korea. He received the M.A. degree in Psychology from the University of Texas at Austin, specializing in human vision, and the Ph.D. degree in Electrical and Computer Engineering from the University of Texas at Austin, specializing in computer vision. He worked as a Postdoctoral Researcher at the Electrical and Computer Engineering Department, University of California at San Diego and as a Research Associate at the Computer Science Department, University of Rochester, NY. He is a Research Professor at the Electrical Engineering Department, University of North Dakota, Grand Forks. His interests include computer vision, pattern recognition, image/signal processing, and sensor-based intelligent systems. He is working on projects involving video surveillance, visual navigation for UAV, activity analysis in intelligent systems, and sensor-based intelligence. He is a member of the IEEE Computer Society.

Zhenyu Qi received the B.S. degree from Fudan University, Shanghai, in Jun. 2003, the M.S. degree from University of California, Riverside, in Aug. 2004, and Ph.D. degree from the University of Virginia, Charlottesville, in 2010, all in Electrical Engineering. He has been with Broadcom Corp. since 2010. His research interests include high performance low power custom and ASIC digital circuit design and SoC system integration. He has also worked on CAD algorithm and tool development, including power/thermal analysis, decoupling capacitance optimization and model order reduction. He is the author of over 30 peer-refereed papers and a book chapter.

Prakash Ranganathan is Assistant Professor of Electrical Engineering at University of North Dakota, Grand Forks, ND, since 2007. He has over 8 years of teaching experience in the Electrical Engineering areas. He currently teaches Electric Circuits, Computer Aided Measurements and Controls and Digital Electronics. His research interests are in Simulation and Modeling, Smart Grid Technologies, Optimization Methods, and Software Engineering. He is an active IEEE Senior Member.

Antti Sirkka got Master's Degree in Computer Science from the University of Tampere, in 2001, and he is finalizing his dissertation on supply chain performance management using traceability. His main expertises are OLAP, RFID, and environmental management systems. He works in Tieto building data management systems, and is currently working on the Tivit Cloud Software program, which pioneers in building new cloud business models, lean software enterprise models, and open cloud software infrastructure.

Serdar Burak Solak received the B.Eng. degree in Electrical Engineering in 2008 and the M.Eng. degree in Telecommunication and Signal Processing in 2010, both from McGill University, Montreal, QC, Canada. From 2007 to 2010, he was a Research Assistant with the Multimedia Signal Processing Group at the Telecommunications and Signal Processing Laboratories of McGill University where he worked on the development and implementation of error resilient, low complexity, and complexity scalable encoding algorithms for H.264 video coding architecture. He is currently a Technology Associate at Morgan Stanley, Montreal, QC, Canada. His research interests include image and video compression, digital signal processing, and developing applications for financial markets.

Pavel Somavat is currently working as an Assistant Professor in Electrical Engineering at College of Agricultural Engineering and Technology, CCS Haryana Agricultural University, India. He obtained M. S. in Information and Communications Engineering from University of Applied Sciences Giessen Friedberg, Friedberg, Germany. He has worked as a researcher in fiber optics industry developing dynamic attenuation testing mechanism for high attenuation fibers. His research interests' concern wired and wireless networking, energy efficiency issues in ICT, smart grids, and optical fiber based sensors. He is a keen volleyball player and takes interest in global environmental issues.

Amine Boudghene Stambouli's research interests include: Photovoltaics, Fuel cells, hybrid systems, and environment impacts. He is a Full Professor of Optoelectronics and Material Science for Environment and Energy Applications at the Department of Electronics, USTO, Algeria. He is United Nations consultant (Index 382958), member of many scientific and industrial organizations and director of several doctoral courses. His studies are documented by 1 book and 3 polycopies. He is a Reviewer and an Editorial Board Member of several international journals. He was co-responsible for the research team "Photovoltaics and fuel cells" between the University of Roma Tor Vergata and the University of Sciences and Technology of Oran. He was awarded the prize of the best publication of the year 2009, delivered by CDER and the ministry of higher education and scientific research of Algeria. He is co-responsible (Algerian side) of the Sahara Solar Breeder (SSB) project.

Mircea R. Stan received the Diploma from the Politechnica University, Bucharest, Romania, and the M.S. and Ph.D. degrees in Electrical and Computer Engineering from the University of Massachusetts at Amherst. Since 1996, he has been with the Charles L. Brown Department of Electrical and Computer Engineering, University of Virginia, where he is now a Professor. He is teaching and doing research in the areas of high-performance and low-power VLSI, temperature-aware and reliability-aware circuits and architecture, storage, embedded systems, and nanoelectronics. He has more than eight years of industrial experience, was a visiting scholar at UC Berkeley, in 2004–2005, a visiting faculty member at IBM in 2000, and at Intel in 2002 and 1999. Dr. Stan received the NSF Career Award in 1997 and was coauthor on Best Paper Awards at the 2008 ISQED, the 2006 GLSVLSI, the 2003 ISCA, and Workshop on Self-Healing, Adaptive, and self-MANaged Systems (SHAMAN) 2002. He is a member of the IEEE, ACM, Eta Kappa Nu, Phi Kappa Phi, and Sigma Xi.

Ian Stroud is an expert on boundary representation modelling, the currently popular technique used for CAD/CAM. He has been working in this area since 1977, as well as on several applications based on solid models. From 1977 until 1980, he worked at Cambridge University, UK, developing many basic modelling operations. From 1980 until 1985, he worked at the Royal Institute of Technology (KTH) in Stockholm Sweden, developing methods for product and hybrid modelling. From 1986 until 1989, he worked at Cranfield Institute of Technology before moving to the Computer and Automation Research Institute of the Hungarian Academy of Sciences, where he gained his Ph.D. in 1993. In 1997, he moved to the Ecole Polytechnique Fédérale de Lausanne where he is now Adjoint Scientifique in the CAD/CAM laboratory (LICP).

Makoto Takizawa is a Professor in the Department of Computer and Information Science, Seikei University. He was a Professor and the Dean of the Graduate School of Science and Engineering, Tokyo

Denki University. He was a Visiting Professor at GMD-IPSI, Keele University, and Xidian University. He was on the Board of Governors and is a Golden Core member of IEEE CS and is a fellow of IPSJ. He received his DE in Computer Science from Tohoku University. He chaired many international conferences like IEEE ICDCS, ICPADS, and DEXA. He founded IEEE AINA. His research interests include distributed systems and computer networks.

Rafael J. Valdivieso-Sarabia was born in Alicante, Spain in 1983. He received his Computer Engineering degree from University of Alicante in 2006. He received a M.Eng. in Home Automation from the same university in 2007. Currently, he is integrated in the research group of Computer Engineering and Computer Networks and he is a Ph. D student on Technologies for Information Society program in the Computer Technology Department at University of Alicante, Spain, thanks to a predoctoral grant, which has been partially supported by the same university Caja de Ahorros de Mediterraneo and Unión Fenosa. His fields of interest are related with multi-agent systems, power management, and control technologies.

Robert John Walters is an Assistant Professor in Computer Science at the University of Southampton. His research interests include middleware, distributed computing, hypermedia, and graphical formal modeling language.

Claas Wilke studied Media Computer science at the TU Dresden and got his diploma (M.Sc.) in 2009, focusing on contract-based runtime verification for OSGI components. Since 2010, Claas Wilke is working as a Research Assistant and Ph.D. student at the TU Dresden, Software Technology Group. His research focuses on model-based testing, especially for optimizing software, i.e. energy consumption. Besides his contribution to the Energy Contract Language (ECL), he is currently working on an energy testing extension for JUnit called JouleUnit. Furthermore, he is a contributor of the Dresden OCL toolkit for parsing and evaluation of OCL (Object Constraint Language) constraints.

Gary Wills is an Associate Professor in Computer Science at the University of Southampton. He is a Visiting Professor at the Cape Peninsular University of Technology. Gary's research projects revolve around System Engineering and applications for industry, medicine, and education. These systems are underpinned by SOA, adaptive systems, advanced knowledge technologies, and Cloud Computing.

Paul Xirouchakis is directing the Computer-Aided Design and Computer-Aided Manufacturing (CAD/CAM) laboratory, Institute of Production and Robotics at the Swiss Federal Institute of Technology in Lausanne, Switzerland. His research interests are in the areas of (1) product modelling and reasoning for manufacture/assembly/remanufacture, (2) manufacturing information systems, and (3) informatics for planning and scheduling for manufacture/assembly/remanufacture. Current project activities include: knowledge-based product development methodologies using ontologies, zero defect manufacturing adaptive process planning, and planning and scheduling of remanufacturing systems. Professor Xirouchakis has industrial experience with CAD industries in the USA where he directed computer-aided simulation-based design projects. He has been a professor at the National Technical University of Athens and the Massachusetts Institute of Technology in the area of Structural Mechanics. He obtained his Ph.D. in Structural Mechanics in 1978 from Massachusetts Institute of Technology.

Ming Yang is currently an Associate Professor in the School of Computing and Software Engineering, Southern Polytechnic State University, GA. He received his Ph.D. degree in Computer Science from Wright State University in 2006, and his Master's and Bachelor's degrees in Electrical Engineering from Tianjin University (China) in 2000 and 1997, respectively. He is interested in research on digital video coding, H.264/AVC, multimedia communication and networking, and steganography/cryptography. He has published over thirty peer-reviewed journal/conference papers and book chapters. He has also obtained various external/internal research grants to support his research.

Mansour Zegar received his M.Sc. degree in Electrical Engineering from the University of Sciences and Technology USTO Mohamed Boudiaf of Oran, where he is a professor and researcher. He is currently pursuing his PhD in the field of multilevel inverter, and solar energy. Recent activities in teaching involve courses in switched-mode converters, power electronic components, switched-mode converter dynamics and EMC as well as photovoltaic interfacing issues. The main focus of his recent research is on the dynamic issues in the renewable energy systems and their grid integration including photovoltaic and fuel-cell systems. His areas of interest are renewable energies and power electronics.

Fatima Zohra Zerhouni obtained her PhD degree from University of Sciences and Technology Mohamed Boudiaf, Oran, in Algeria where she is a professor and researcher. Her studies started in the field of power electronics and changed to environmental friendly production of energy. She was awarded the prise of the best publication of the year 2010, delivered by CDER and the ministry of higher education and scientific research of Algeria. Her studies are documented by several papers mainly published on International Journals and on Proceedings of International and National Conferences. Her interest research fields are: renewable energies, fuel cells, power electronics, electrical storage systems and soft computing control.

M'hamed Houari Zerhouni received his M.Sc. degree in electrical engineering from the University of Sciences and Technology, USTO, Mohamed Boudiaf of Oran, where he is a professor and researcher. Since 2005, he has been a Researcher with the Department of Electronic Engineering. His fields of interest are in power electronics applied to photovoltaic energy systems. He has coauthored papers on International Journals, on Proceedings of International, National Conferences and in technical journals. He is currently pursuing his PhD in the field of PEMFC and solar energy. His areas of interest are renewable energies, photovoltaic cells and proton exchange membrane fuel cells PEMFC.

Yan Zhang received the BS and MS degrees in electrical engineering from Tsinghua University, Beijing, in 1997 and 2000, respectively, and the MS and PhD degrees in electrical and computer engineering from the University of Virginia in 2003 and 2006, respectively. His research interests include power-aware and temperature-aware computing, low-power VLSI design, and architecture-level power modeling. He is currently working as a circuit design engineer at Qualcomm, San Diego.

Index

A

B

C

D

E

T

U

V

W

www.ingramcontent.com/pod-product-compliance
Lightning Source LLC
LaVergne TN
LVHW062137200926
848484LV00007B/1825